Lecture Notes in Electrical Engineering

Volume 750

The book series *Lecture Notes in Electrical Engineering* (LNEE) publishes the latest developments in Electrical Engineering - quickly, informally and in high quality. While original research reported in proceedings and monographs has traditionally formed the core of LNEE, we also encourage authors to submit books devoted to supporting student education and professional training in the various fields and applications areas of electrical engineering. The series cover classical and emerging topics concerning:

- Communication Engineering, Information Theory and Networks
- Electronics Engineering and Microelectronics
- Signal, Image and Speech Processing
- Wireless and Mobile Communication
- Circuits and Systems
- Energy Systems, Power Electronics and Electrical Machines
- Electro-optical Engineering
- Instrumentation Engineering
- Avionics Engineering
- Control Systems
- Internet-of-Things and Cybersecurity
- Biomedical Devices, MEMS and NEMS

For general information about this book series, comments or suggestions, please contact leontina.dicecco@springer.com.

To submit a proposal or request further information, please contact the Publishing Editor in your country:

China

Jasmine Dou, Editor (jasmine.dou@springer.com)

India, Japan, Rest of Asia

Swati Meherishi, Editorial Director (Swati.Meherishi@springer.com)

Southeast Asia, Australia, New Zealand

Ramesh Nath Premnath, Editor (ramesh.premnath@springernature.com)

USA, Canada:

Michael Luby, Senior Editor (michael.luby@springer.com)

All other Countries:

Leontina Di Cecco, Senior Editor (leontina.dicecco@springer.com)

**** This series is indexed by EI Compendex and Scopus databases. ****

More information about this series at http://www.springer.com/series/7818

Santhosh K V · K. Guruprasad Rao
Editors

Smart Sensors Measurements and Instrumentation

Select Proceedings of CISCON 2020

Springer

Editors
Santhosh K V
Department of Instrumentation and Control Engineering
Manipal Academy of Higher Education
Manipal, India

K. Guruprasad Rao
Department of Mechanical Engineering
National Institute of Technology Karnata
Surathkal, Karnataka, India

ISSN 1876-1100 ISSN 1876-1119 (electronic)
Lecture Notes in Electrical Engineering
ISBN 978-981-16-0335-8 ISBN 978-981-16-0336-5 (eBook)
https://doi.org/10.1007/978-981-16-0336-5

This Springer imprint is published by the registered company Springer Nature Singapore Pte Ltd.
The registered company address is: 152 Beach Road, #21-01/04 Gateway East, Singapore 189721, Singapore

Contents

Editors and Contributors

About the Editors

Dr. Santhosh K V is working as Associate Professor in the Department of Instrumentation and Control at Manipal Institute of Technology, Manipal Academy of Higher Education, Manipal, and Dy. Director Research (Technical) at Manipal Academy of Higher Education. He received Bachelors and Masters from VTU, Belgaum, and Ph.D. from National Institute of Technology, Silchar, in 2013. His research interest is in the area of sensors, wireless sensor network, multi-sensor data fusion and smart sensors. He has published around 80 publications and delivered several technical talks. He is serving as reviewer and editorial board member for several peer-reviewed journals. He is also BOE and BOS Member at various universities and active member of several professional societies.

Dr. K. Guruprasad Rao is presently serving as a teaching faculty since October 2000 at the Department of Mechanical Engineering, National Institute of Technology Karnataka, Surathkal, India. Dr Guruprasad obtained his Ph.D. in the year 2008 from IISc, Bangalore, and postdoctoral fellowship from the University of Nebraska, USA. His research interests are robotics—dynamics, control and path planning; multi-agent (robot) systems—deployment, coverage, search, cooperative and distributed control; location optimization; Voronoi partition—application to multi-agent systems and sensor networks, distributed algorithms for Voronoi cell computation and generalization of Voronoi partition; physics of musical instruments. He has published over 30 publications and is serving as reviewer in peer-reviewed journals. He is the recipient of several awards and fellowship like the INAE fellowship during June–July 2010.

Contributors

Shivani G. Aithal Department of ICT, Manipal Institute of Technology, Manipal Academy of Higher Education, Manipal, India

S. Akshay Department of Electrical and Electronics Engineering, Muthoot Institute of Technology and Science Kochi, Kochi, India

K. B. Akshaya Department of Instrumentation and Control Engineering, NSS College of Engineering (Affliated to APJ Abdul Kalam Technological University Kerala), Palakkad, India

Alaka Ananth Department of Information Science and Engineering, NMAM Institute of Technology, Nitte, Karnataka, India

Md. Sayeed Ansari Ramaiah Institute of Technology, Bengaluru, India

T. P. Ashwini Mangalore Institute of Technology and Engineering, Mangalore, India

S. H. Bharathi Reva University, Bengaluru, India

Sachin Bhat Department of Electronics and Communication Engineering, Shri Madhwa Vadiraja Institute of Technology and Management, Udupi, Karnataka, India

C. S. Asha Mechatronics Engineering, Manipal Institute of Technology, Manipal Academy of Higher Education, Manipal, India

Jiso Chacko Department of Electrical and Electronics Engineering, Muthoot Institute of Technology and Science Kochi, Kochi, India

P. K. Chandrashekhar VTU Research Centre, CPRI, Bengaluru, India

Vidit Vijay Chaturvedi Ramaiah Institute of Technology, Bengaluru, India

Ron Chaudhuri Electronics and Communication Department, Manipal Institute of Technology, MAHE, Manipal, India

Shreesha Chokkadi Instrumentation and Control Engineering, Manipal Institute of Technology, Manipal Academy of Higher Education, Manipal, India

Nivedita Daimiwal Cummins College of Engineering for Women, Pune, India

Pawan Kumar Dhakal Department Electrical and Computer Engineering, University of Oviedo, Ovideo, Spain

Abraham Dolly Mary Department of Electrical Engineering, Rajiv Gandhi Institute of Technology, Kottayam, Kerala, India

A. V. Gayathri Department of Instrumentation and Control Engineering, NSS College of Engineering (Affliated to APJ Abdul Kalam Technological University Kerala), Palakkad, India

Jim George Department of Electrical and Electronics Engineering, Muthoot Institute of Technology and Science Kochi, Kochi, India

Vadakkekkara Itty George Instrumentation & Control Engineering, Manipal Institute of Technology, Manipal, Karnataka, India

S. P. Ghanegaonkar College of Engineering, Pune, India

G. Aayushii Department of Electronics and Instrumentation, R.V. College of Engineering, Bangalore, Karnataka, India;
Springer-Verlag, Computer Science Editorial, Heidelberg, Germany

Anand Gudnavar Research Scholar, Department of ECE, UBDTCE, Davanagere, India

Rahul J. Gujaran Robonics Control Solutions, Bangalore, India

Veerayya Hiremath SDM College of Engineering and Technology, Dharwad, India;
Belagavi, India

Shamli S. Jadhav College of Engineering, Pune, India

B. Jayalakshmi Department of Instrumentation and Control Engineering, NSS College of Engineering, Palakkad, Kerala, India

Aditya Y. Jeppu RV College of Engineering, Bengaluru, India

Yogananda Jeppu Manipal Institute of Technology, Manipal, India

Edwin Jerald Department of Electrical and Electronics Engineering, Muthoot Institute of Technology and Science Kochi, Kochi, India

Dolbin Jojo Department of Instrumentation and Control Engineering, NSS College of Engineering (Affliated to APJ Abdul Kalam Technological University Kerala), Palakkad, India

Adit V. Kaneria Department of ICT, Manipal Institute of Technology, Manipal Academy of Higher Education, Manipal, India

Mathew K. Kanthi Department of Electrical Engineering, Rajiv Gandhi Institute of Technology, Kottayam, Kerala, India

J. Shwetha Katari Department of E&EE, Siddaganga Institute of Technology, Tumakuru, Karnataka, India

Sai Kate MKSSS's Cummins College of Engineering for Women, Pune, India

Pauroosh Kaushal Department of Instrumentation and Control, College of Engineering Pune, Pune, India

Kavyahsree Manipal School of Architecture and Planning, Manipal Academy of Higher Education, Manipal, India

S. S. Kerur SDM College of Engineering and Technology, Dharwad, India; Belagavi, India

Ravi Kiran VTU Extension Centre, Honeywell Technology Solutions, Bangalore, India

Narasimha Kollaparti KPIT Technologies Ltd, Bengaluru, India

Rakesh Kumar JNU, Delhi, India

VTNS Pavan Kumar Electronics and Communication Department, Manipal Institute of Technology, MAHE, Manipal, India

Shailaja Kurode Department of Electrical Engineering, College of Engineering, Pune, India

I. S. Kushagra Mechatronics Engineering, Manipal Institute of Technology, Manipal Academy of Higher Education, Manipal, India

B. N. Madhukar Reva University, Bengaluru, India

Vaishnavi Malkapure MKSSS's Cummins College of Engineering for Women, Pune, India

N. Manjanaik Professor, Department of ECE, UBDTCE, Davanagere, India

Doddabasappa Marebal Automation & Robotics, KLE Technological University, Hubballi, India

Akshaya Mhatre College of Engineering Pune, Pune, India

K. Mohith JSS Science and Technology University, Mysuru, India

Rohini P. Mudhalwadkar Department of Instrumentation and Control, College of Engineering Pune, Pune, India

B. Murugan Department of Electrical and Electronics Engineering, RAAK College of Engineering & Technology, Puducherry, India

M. G. Nagarjun Department of Electronics and Instrumentation, R.V. College of Engineering, Bangalore, Karnataka, India;
Springer-Verlag, Computer Science Editorial, Heidelberg, Germany

Rushali Naik Department of Electronics and Communication Engineering, Shri Madhwa Vadiraja Institute of Technology and Management, Udupi, Karnataka, India

A. Nandhan Department of Instrumentation and Control Engineering, NSS College of Engineering, Palakkad, Kerala, India

Bhagyashree Narkhede MKSSS's Cummins College of Engineering for Women, Pune, India

Tushar S. Narsimpur Ramaiah Institute of Technology, Bengaluru, India

Gurunayk Nayak Department of Electrical and Electronics, M. S. Ramaiah Institute of Technology, Bangalore, India

C. Nikhil Binoy Department of Instrumentation and Control Engineering, NSS College of Engineering, Palakkad, Kerala, India

Ashwini Pachare College of Engineering Pune, SPP University, Pune, India

Smitha N. Pai Department of ICT, Manipal Institute of Technology, Manipal Academy of Higher Education, Manipal, India

Vijay N. Pande College of Engineering, Pune, India

Akash Panhale Department of Electrical Engineering, College of Engineering, Pune, India

Manjusha R. Parodkar Instrumentation and Control Engineering, Manipal Institute of Technology, Manipal Academy of Higher Education, Manipal, India

C. G. Parvathi Department of E&EE, Siddaganga Institute of Technology, Tumakuru, Karnataka, India

Rahul B. Patil Electronics and Communication Department, Manipal Institute of Technology, MAHE, Manipal, India

Shantharam Patil Manipal School of Architecture and Planning, Manipal Academy of Higher Education, Manipal, India

Sudarshan Patilkulkarni JSS Science and Technology University, Mysuru, India

U. Pooja Nair Department of Electronics and Communication Engineering, Shri Madhwa Vadiraja Institute of Technology and Management, Udupi, Karnataka, India

S. Akshay Prakash Department of Instrumentation and Control Engineering, NSS College of Engineering (Affliated to APJ Abdul Kalam Technological University Kerala), Palakkad, India

K. Praveen Shenoy Mangalore Institute of Technology and Engineering, Mangalore, India

N. Komal Raj Department of E&EE, Siddaganga Institute of Technology, Tumakuru, Karnataka, India

R. Rakshith Mechatronics Engineering, Manipal Institute of Technology, Manipal Academy of Higher Education, Manipal, India

S. Ramya Electronics and Communication Department, Manipal Institute of Technology, MAHE, Manipal, India

Abishek B. Rao Department of ICT, Manipal Institute of Technology, Manipal Academy of Higher Education, Manipal, India

Vidya S. Rao Manipal Institute of Technology, Manipal Academy of Higher Education, Manipal, India

Abhilash Salian Mangalore Institute of Technology and Engineering, Mangalore, India

C. Sandhra Department of Instrumentation and Control Engineering, NSS College of Engineering (Affliated to APJ Abdul Kalam Technological University Kerala), Palakkad, India

Abhijeet Sarkar Ramaiah Institute of Technology, Bengaluru, India

Rajashree V. Sarwadnya Instrumentation Engineering Department, S.G.G.S. Institute of Engineering and Technology, Nanded, Maharashtra, India

Divya Sasidharan Department of Instrumentation and Control Engineering, NSS College of Engineering (Affliated to APJ Abdul Kalam Technological University Kerala), Palakkad, India

Surajkumar Sawai College of Engineering, Pune, India

P. Shamli Fathima Department of Instrumentation and Control Engineering, NSS College of Engineering (Affliated to APJ Abdul Kalam Technological University Kerala), Palakkad, India

Milankumar Sheta Solutions Architect, KPIT Technologies Pvt. Ltd, Bangalore, India

Prakash K. Shetty A. J. Institute of Engineering & Technology, Mangalore, India

Shrikanth K. Shirakol SDM College of Engineering and Technology, Dharwad, India;
Belagavi, India

B. Shoba Department of Electronics and Communication Engineering, Rajiv Gandhi College of Engineering & Technology, Puducherry, India

Revati Shriram MKSSS's Cummins College of Engineering for Women, Pune, India

N. Sidharth Department of Electronics and Communication Engineering, NSS College of Engineering, Palakkad, India

Vivek Kumar Singh Ramaiah Institute of Technology, Bengaluru, India

Mrinalini P. Sirsat Instrumentation Engineering Department, S.G.G.S. Institute of Engineering and Technology, Nanded, Maharashtra, India

Srilakshmi U. Sirurmath Department of Electronics and Instrumentation, R.V. College of Engineering, Bangalore, Karnataka, India;
Springer-Verlag, Computer Science Editorial, Heidelberg, Germany

Meera Sivadas Department of Electrical and Electronics Engineering, Muthoot Institute of Technology and Science Kochi, Kochi, India

Mohamed Hannan Sohail Department of Electrical Engineering, National Institute of Technology, Patna, India

Prashant Kumar Soori Heriot Watt University, Dubai, UAE

M. Sowmya Kini Mangalore Institute of Technology and Engineering, Mangalore, India

S. Sreehari Department of Instrumentation and Control Engineering, NSS College of Engineering (Affliated to APJ Abdul Kalam Technological University Kerala), Palakkad, India

S. G. Srivani Department of E & EE, RVCE, Bengaluru, India

Suresh Kumar Suraj Instrumentation & Control Engineering, Manipal Institute of Technology, Manipal, Karnataka, India

Suraksha Department of Electronics and Communication Engineering, Shri Madhwa Vadiraja Institute of Technology and Management, Udupi, Karnataka, India

Kendaganna Swamy Department of Electronics and Instrumentation, R.V. College of Engineering, Bangalore, Karnataka, India;
Springer-Verlag, Computer Science Editorial, Heidelberg, Germany

Amit Talli Automation & Robotics, KLE Technological University, Hubballi, India

Arti. V. Tare College of Engineering, Pune, India

S. Thasna Department of Instrumentation and Control Engineering, NSS College of Engineering, Palakkad, Kerala, India

I. Thirunavukkarasu Department of Instrumentation and Control Engineering, MAHE (Institute of Eminence), Manipal Institute of Technology, Manipal, India

Archana Thosar College of Engineering Pune, Pune, India

Shubham Toraskar Mechatronics Engineering, Manipal Institute of Technology, Manipal Academy of Higher Education, Manipal, India

Tinu Valsa Paul Instrumentation & Control Engineering, Manipal Institute of Technology, Manipal, Karnataka, India

Aiswarya Varghese Department of Instrumentation and Control Engineering, NSS College of Engineering (Affliated to APJ Abdul Kalam Technological University Kerala), Palakkad, India

Ajay Varghese Department of Electrical and Electronics Engineering, Muthoot Institute of Technology and Science Kochi, Kochi, India

G. Venugopal Department of Instrumentation and Control Engineering, NSS College of Engineering (Affliated to APJ Abdul Kalam Technological University Kerala), Palakkad, India

C. Vighneshwara Mangalore Institute of Technology and Engineering, Mangalore, India

K. U. Vinayaka Department of E&EE, Siddaganga Institute of Technology, Tumakuru, Karnataka, India

R. Vinodha Department of Electronics and Instrumentation Engineering, Annamalai University, Annamalai Nagar, Tamil Nadu, India

Vinutha Department of Electronics and Communication Engineering, Shri Madhwa Vadiraja Institute of Technology and Management, Udupi, Karnataka, India

P. Vishnu Department of Instrumentation and Control Engineering, NSS College of Engineering, Palakkad, Kerala, India

Azem Iqbal Vora Heriot Watt University, Dubai, UAE

Chandra Shekhar Yadav STQC, MeitY, New Delhi, India

Kesari Singh Yadav Sanskriti University, Mathura, India

Meenu Yadav SLIET, Longowal, Punjab, India

Prem Shankar Singh Yadav JNU, Delhi, India

Sandeep Yadav SLIET, Longowal, Punjab, India

Toward Sustainable Cooling Strategy for Developing Nations: Opportunities for an Off-Grid Solar DC Refrigerator

Azem Iqbal Vora, Prashant Kumar Soori, and Prakash K. Shetty

1 Introduction

The ever-increasing concern of climate change due to increased use of fossil fuels is promoting interest in researchers to study and improvise the ways of harnessing clean energy. There are various ways to produce clean energy such as the use of wind turbines to produce energy from wind or geothermal which uses ground's heat, but due to abundance of solar insolation, solar photovoltaic (PV) panels provide a better alternative to fossil fuels.

Solar PV panels have no moving parts and require minimal maintenance, hence making it an easy energy source to use. Continuous research and development in the PV technology have enabled companies to reduce the price of solar panels, meanwhile increasing their efficiency. The graph shown in Fig. 1 indicates the decrease in cost per Mega Watt hour (MWh) as the technology matures [1]. The graph also indicates a further decrease in cost in the following years.

Energy consumed by commercial and residential sector was around 39% of the overall energy produced in the United States of America (USA) and from that almost 35% of the energy was used for cooling and heating purposes. This is depicted in Fig. 2. Further breaking down the numbers obtained, refrigeration utilizes 51% of the energy used by residential buildings [2].

The rapid progress in developing countries leads to more energy usage in the buildings, the majority of which is used in refrigeration. The increase in energy usage increases carbon emissions and contributes to climate change. The refrigeration systems can utilize renewable energy such as solar energy to decrease the use of

A. I. Vora · P. K. Soori
Heriot Watt University, Dubai, UAE
e-mail: p.k.soori@hw.ac.uk

P. K. Shetty (✉)
A. J. Institute of Engineering & Technology, Mangalore, India
e-mail: shettyprakashk@ajiet.edu.in

S. K V and K. Guruprasad Rao (eds.), *Smart Sensors Measurements and Instrumentation*, Lecture Notes in Electrical Engineering 750,
https://doi.org/10.1007/978-981-16-0336-5_1

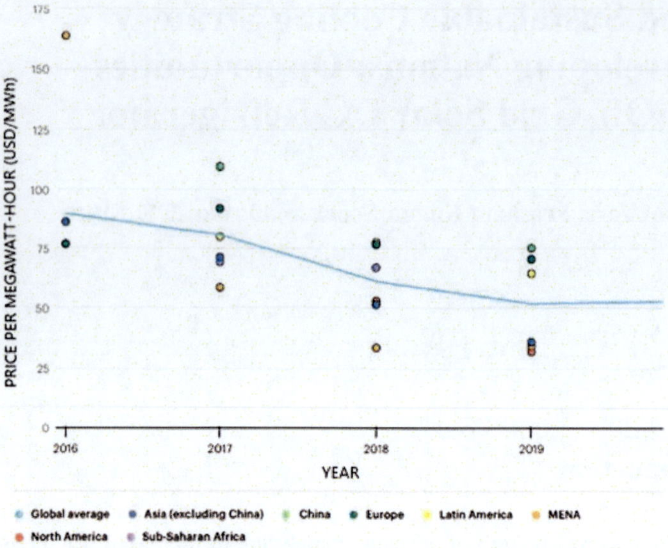

Fig. 1 Price per MWh of solar generated energy

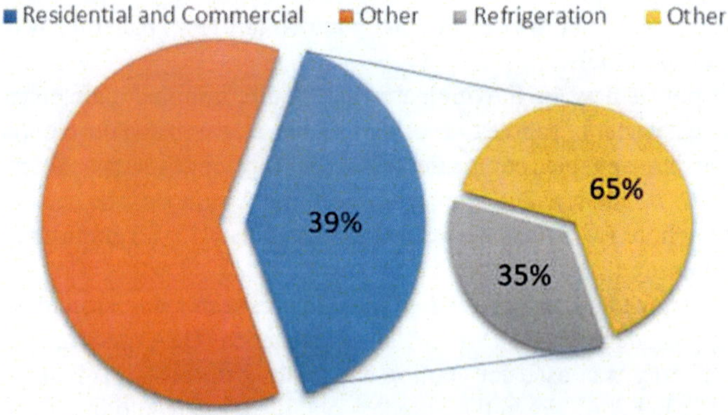

Fig. 2 Energy consumption by refrigerators in residential and commercial buildings

energy generated by fossil fuels and thus reducing the carbon emissions. Furthermore, solar run refrigeration systems can be utilized off-grid in places such as Sub-Saharan countries where the electricity is scarce, relying on diesel generators for powering the various electrical appliances [3]. The overall quality of life shall improve as refrigeration is essential to store perishable goods, particularly in hospitals, where constant refrigeration is required to store life-saving medicines, blood and vaccines.

The refrigeration system utilizing solar energy can be implemented either using a vapor compression refrigeration (VCR) system or vapor absorption refrigeration (VAR) system. VCR system consists of a compressor, condenser and a control unit

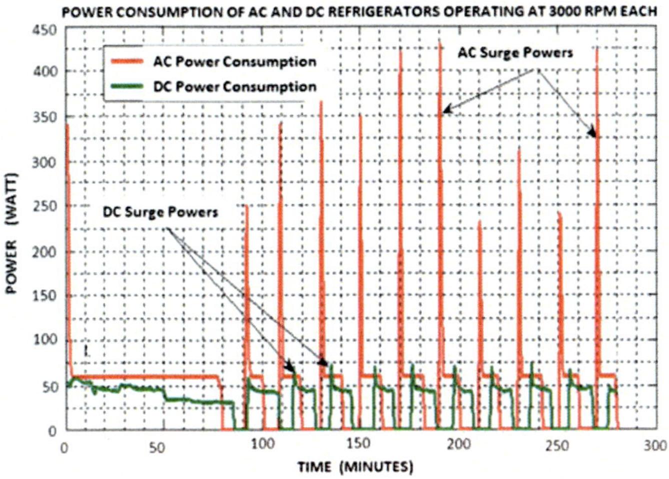

Fig. 3 AC versus DC power surge during operation

running on alternating current (AC) (using DC-AC converter); meanwhile, VAR system consists of an absorber, pump, generator, evaporator and condenser. The mature technology, along with ease of use, compact size and minimal maintenance, makes the VCR systems ideal for refrigerators utilizing solar power. The solar power produced by solar PV panels is direct current (DC) power; therefore, the DC VCR system can be utilized in the prototype as no inverters are required. An added benefit of utilizing DC VCR is decreased power spikes compared to AC VCR system as depicted in Fig. 3 [4].

The novelty of the presented work is to design and implement a refrigerator capable of running directly on solar power by utilizing 24 V DC VCR with variable speed control and microcontroller with sensors and display for smart control and operation.

2 Literature Review

The compressor in the VCR system can be operated using direct current (DC) or alternative current (AC). Researchers analyzed the overall performance of the refrigerator using AC and DC power [4]. The two systems were tested in Sub-Saharan countries to determine their efficiency and cost-effectiveness. The methodology used by researchers was to initially test the power consumption of two identical refrigerators with AC compressor and then determine the battery, inverter and PV panel rating. It was followed by replacing AC compressor to 12 V DC compressor. Voltage and current were monitored for a day for both the refrigerators in the same environment, and the data were logged on a data logger. The data from the experiment showed higher power consumption (over 20%) in the AC compressor in comparison with DC compressor. Furthermore, during start-up of the compressor, the power surges in the AC compressor were in the range of 250–400 W. In contrast, in DC, it stayed at around 75 W.

Another study reported an efficient 12 V DC compressor-based refrigerator with PV panels, solar charge controller and deep cycle lead-acid batteries [5]. The prototype was tested for a day to determine its performance. It was noted that after the initial start-up, the power consumption reduced to a minimum value throughout the testing phase.

Authors used a DC refrigerator and conducted tests with and without phase change material (PCM) in hot arid conditions to study the difference the PCM makes in off-grid applications [6].

Researchers substituted AC compressor in FV100 refrigerator to utilize DC compressor. They found that daily energy consumption reduced from 4.4 to 2.2 kWh/day. Additionally, insulation was added inside the refrigerator to reduce the heat loss resulting in a 30% decrease in volume. There were two tests performed, first test to determine the voltage to speed ratio and starting compressor current, and the second test was done with the load to determine the maximum power utilized by the system. The maximum power consumed was 102 W initially, dropping to 84 W at steady state at 3000 rpm. It was observed that power consumption was highly dependent on rpm and the set temperature [7].

In another study, a conventional AC refrigerator was used and paired it with an inverter to run from solar panels along with a solar charge controller and two 12 V batteries [8]. The test was conducted to determine the feasibility of using a refrigerator on solar energy. It was determined that 140 Wp was enough to run 165 L refrigerators even in a high ambient temperature of about 40 °C. Although the system performed well, it was reported that the payback time was high, and subsidies are required to bring down the cost.

An experiment with AC refrigerator coupled with inverter was proposed by scholars [9]. The researchers used 50 L refrigerator powered by two solar PV panels of 70 W.

The idea of using STM32 microcontroller running on Android as an operating system to read the data from the sensors and transmit them to a mobile device to display the data was presented by authors [10]. The proposed project is meant for domestic consumers, but this could be particularly helpful in the medical application where medicines, vaccines or blood are stored.

Authors used Arduino UNO as the microcontroller to collect data from a DHT11 temperature and humidity sensor along with MQ3 gas sensor. The data are analyzed, and the user is alerted via SMS when measured data are different from set parameters [11].

3 Proposed Methodology

The block schematics of the proposed solar-powered DC refrigerator are shown in Fig. 4. The prototype consists of 55 L refrigerator, 24 V DC compressor and brushless direct current (BLDC) motor controller. DHT11 temperature and humidity sensor, three load sensors, ESP8266 WiFi module, Arduino Mega and a TFT display were added for IoT application. All these features were added to make smart VCR system capable of running directly on DC power without the use of any other converters.

Initially, the existing AC compressor in the 55L refrigerator is replaced with a 24 V DC compressor. Furthermore, evaporating coil, condenser coil and two 24 V DC fans are also added to the refrigerator. This is followed by electrical and IoT part of the project, which is divided into two sections:

- Compressor control using Arduino MEGA and BLDC motor controller.
- Controlling refrigerator using TFT display and mobile application.

3.1 Compressor Control Using Arduino MEGA and BLDC Motor Controller

The compressor control using a microcontroller (Arduino Mega) was done with the help of temperature and humidity sensor (DHT11) placed in the refrigerator. The microcontroller monitors the cabinet temperature continuously and cross-checks with the set temperature. When temperature difference was noted, it determines the

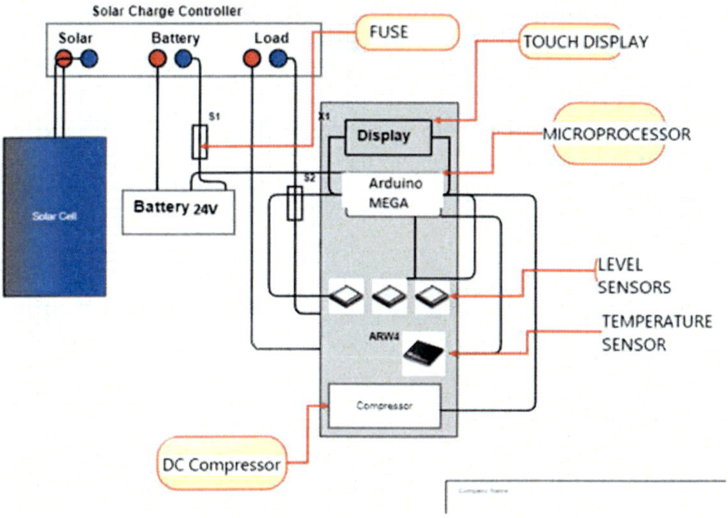

Fig. 4 DC refrigerator schematics

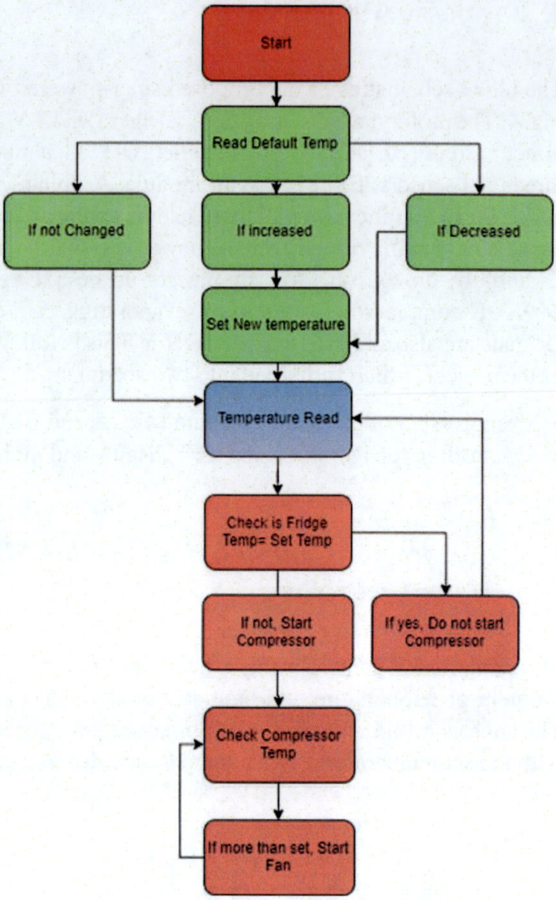

Fig. 5 Flow chart of compressor control

difference between the temperature and accordingly varies the duty cycle of the pulse width modulated (PWM) signal to the BLDC motor controller. The sequence is depicted in Fig. 5.

3.2 *Controlling Refrigerator Using TFT Display and Mobile Application*

For the successful control of DC compressor using a microcontroller, TFT display, load sensors and WiFi module were connected to Arduino. The TFT display is a UART display which allows serial connection with Arduino Mega; hence, the display works as a slave to the microcontroller. The display was programmed to show temperature and humidity from the DHT11 sensors, vary the set temperature

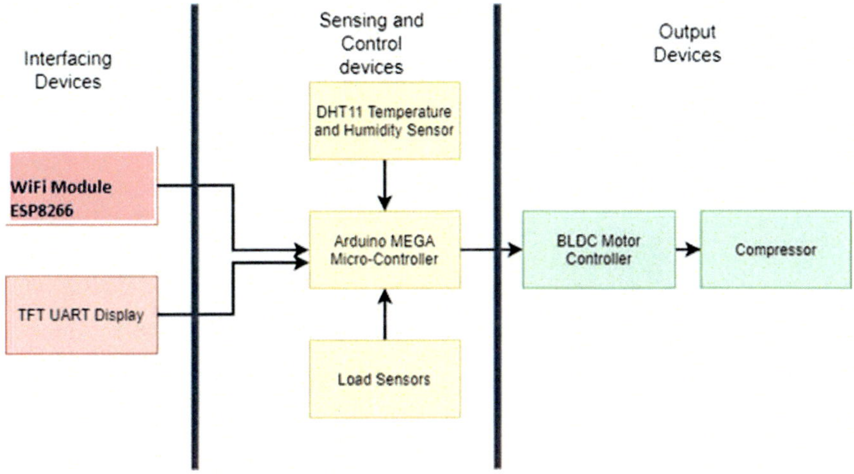

Fig. 6 System block diagram

and load sensors values. The WiFi module was configured in the Arduino code, and the interface was developed using RemoteXY open-source software [12]. The WiFi module enabled control of refrigerator remotely from a connected mobile device. Figure 6 shows the system block diagram.

3.3 Testing of the Prototype

The power rating of the entire system was calculated to compare the effectiveness of the developed system with an existing AC refrigerator. Figure 7 shows a schematic of the test rig-up used. The testing of the system was done using a high precision digital multimeter (DMM 4020), a Windows PC with putty and LabVIEW installed, solar PV panel and 24 V battery. Digital multimeter (DMM4020) is a high precision multimeter which can measure current and voltage and can be operated via a computer using RS-232 interface. For the purpose of this experiment, only current was measured as the voltage was constant. The accuracy of the DMM is ±0.18 A at 10 A and 23 °C. LabVIEW software was used to read the data from DMM and store it in a database. Along with that, putty was used to store serial interface data from the microcontroller. The microcontroller sends temperature and humidity data which were essential for data analysis.

Two tests were conducted on the developed prototype. The purpose of first test was to determine the current consumption of the prototype when the temperature inside the cabinet was dropped from 22 to 4 °C. The second test was conducted to determine the time taken to drop cabinet temperature from 22 to 4 °C. The first test was essential to determine the average power consumption by the prototype.

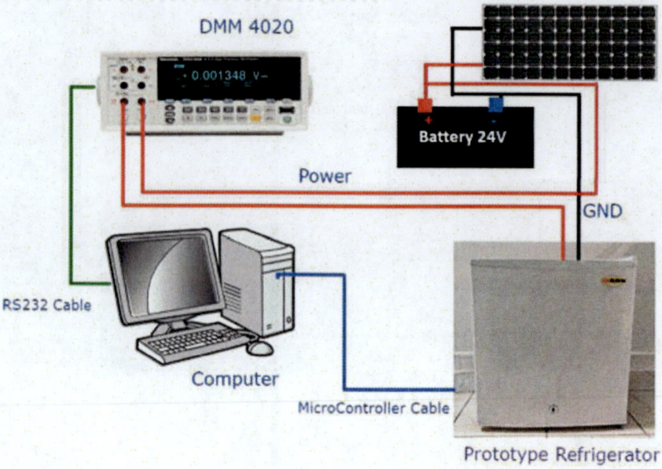

Fig. 7 Test rig set-up

Average power consumption helps to determine ratings of solar panels, batteries and solar charge controller. Both the tests were conducted at no load at temperature of 23 °C.

4 Results and Discussions

The developed smart DC refrigerator is shown in Fig. 8. Fabricated acrylic parts were added to house the BLDC motor controller, Arduino MEGA microcontroller and TFT display. The swapped compressor was bolted on to the refrigerator body to

Fig. 8 Completed prototype

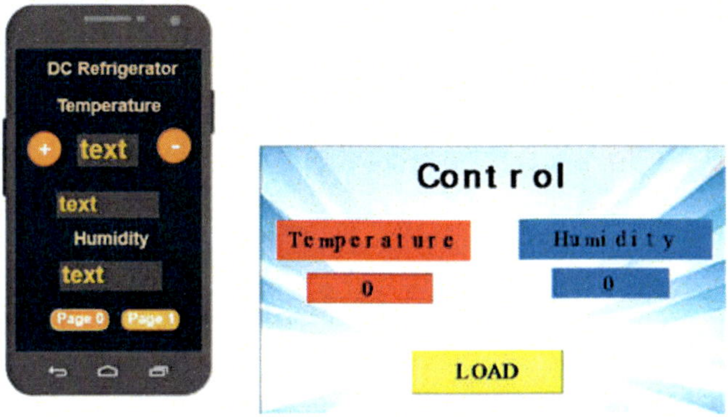

Fig. 9 Mobile and display GUI

reduce vibrations during operation. Figure 9 shows the display and mobile application interface.

4.1 *Testing Results*

The testing results from DMM 4020 were stored by LabVIEW software on the computer and sorted in Microsoft Excel to produce the graph shown in Fig. 10. It can be noted that the voltage remained constant at 24 V as the unit was powered by a DC source. Meanwhile, the current varied depending on the compressor operation.

Initially, the current was low as the microcontroller initializes all the variables and starts the compressor after noting the difference in temperature from the set temperature. The compressor operated for 11.8 min until the cabinet temperature reached 4 °C, and it is shown in Fig. 11. After the initial pull-down phase, the system maintained the internal cabinet temperature. High peaks can be seen at the time of

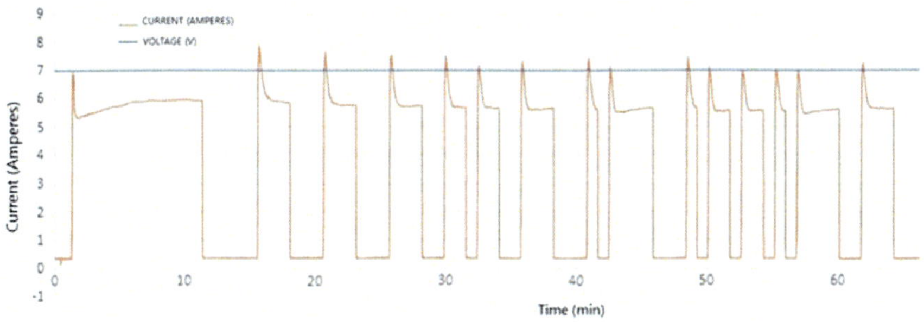

Fig. 10 Current and voltage graph

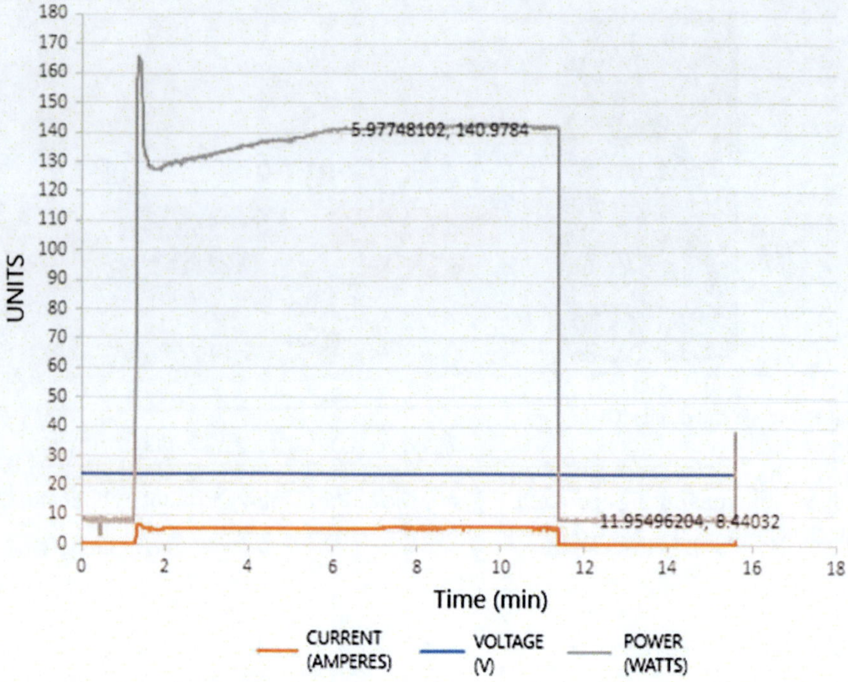

Fig. 11 Power graph during start-up

the start of the compressor unit. The average current is found to be 5.6 A, but at the starting, the current reading was up to 8 A.

The second test was conducted to determine the temperature variation over time, and it is presented in Fig. 12. The data were taken from the microprocessor using putty and plotted using Microsoft Excel. The initial temperature was 22 °C, and the

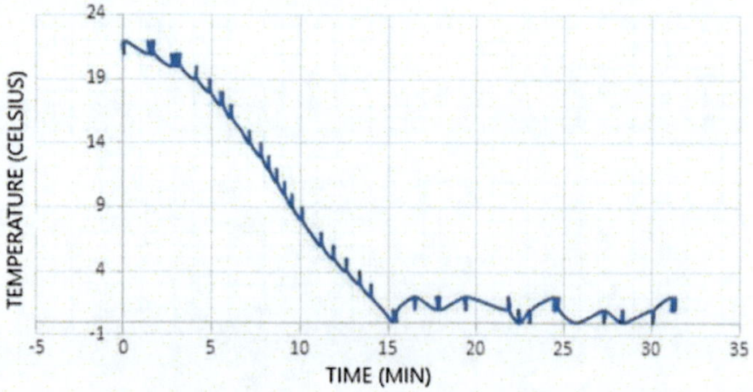

Fig. 12 Temperature variation over time

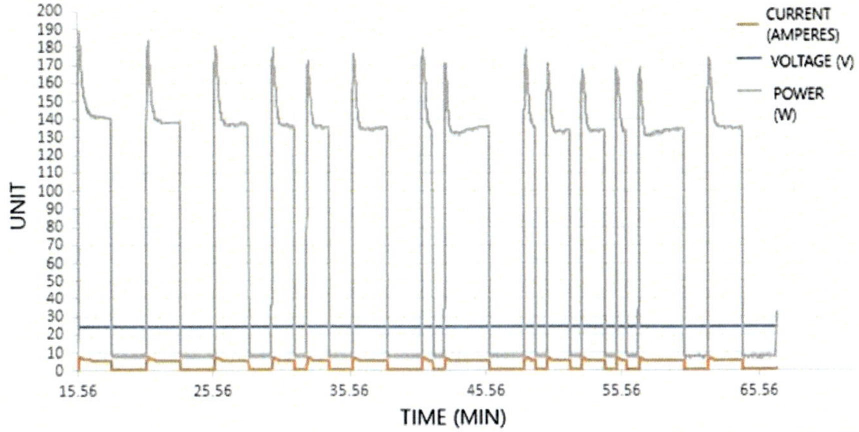

Fig. 13 Power consumption graph

temperature gradually reduced to 0 °C in 14 min at no load. It was observed that the variation of one-degree Celsius temperature resulted in compressor operation again.

The power consumption graph is depicted in Fig. 13. The average power consumption of the fabricated prototype was found to be 139.12 W. The duty cycle, which depicts the ratio of compressor operation time over total run time, was calculated to be 0.555. The daily energy consumption was found to be 1.85 kWh. Hence, the annual energy consumption of the fabricated prototype was expected to be 676.4 kWh under no load condition.

5 Conclusions

The average power consumption of the developed prototype is in line with conventional AC refrigerator available in the market. The main advantages of using a DC compressor are the reduced current spikes at the starting of the compressor. A conventional refrigerator could have current spikes 100–150% higher than normal current usage, which would elicit a higher rated inverter [4]. The duty cycle of the developed refrigerator was in line with the AC refrigerators. Additionally, added display, WiFi module and load sensors allow constant control and monitoring of the refrigerator using a paired mobile device. Further advancements such as improved insulation or addition of PCM can be done to improve the efficiency of the developed refrigerator. To conclude, a smart DC refrigerator prototype was successfully built and tested, which proved to be better suited to work in conjunction with a solar PV system.

References

1. International Energy Agency, https://www.iea.org/reports/renewables-2019
2. Energy Information Administration, https://www.eia.gov/energyexplained/use-of-energy/commercial-buildings.php
3. SciDev, https://www.scidev.net/sub-saharan-africa/pollution/feature/africa-grapples-with-diesel-generators-health-problems.html
4. Opoku, R., Anane, S., Edwin, I., Adaramola, M., Seidu, R.: Comparative techno-economic assessment of a converted DC refrigerator and a conventional AC refrigerator both powered by solar PV. Int. J. Refrig. **72**, 1–11 (2016)
5. Tsado, J., Mahmood, M., Raji, A., Usman, A. and Jiya, I.: Solar powered DC refrigerator with a monitoring and control system. In: 2018 IEEE PES/IAS PowerAfrica (2018)
6. El-Bahloul, A., Ali, A., Ookawara, S.: Performance and sizing of solar driven dc motor vapor compression refrigerator with thermal storage in hot arid remote areas. Energy Procedia **70**, 634–643 (2015)
7. Kaplanis, S., Papanastasiou, N.: The study and performance of a modified conventional refrigerator to serve as a PV powered one. Renewable Energy **31**, 771–780 (2006)
8. Modi, A., Chaudhuri, A., Vijay, B., Mathur, J.: Performance analysis of a solar photovoltaic operated domestic refrigerator. Appl. Energy **86**, 2583–2591 (2009)
9. Alshqirate, A., Tarawneh, M., Hammad, M.: Performance study of a domestic refrigerator powered by a photovoltaic generator. Appl. Solar Energy **51**, 1–5 (2015)
10. Zhongmin, W., Yanan, Y.: Design of an interactive smart refrigerator based on embedded system. In: 2018 International Conference on Sensing, Diagnostics, Prognostics, and Control (SDPC) (2018)
11. Nasir, H., Aziz, W., Ali, F., Kadir, K. Khan, S.: The implementation of IoT based smart refrigerator system. In: 2018 2nd International Conference on Smart Sensors and Application (ICSSA) (2018)
12. Remotexy, https://remotexy.com/en/editor/30212ea4e8ca57b4c795f2e8d0a33912/

Performance Improvement of Conical Tank Process Applying Fractional-Order Control

Edwin Jerald, Ajay Varghese, Jiso Chacko, S. Akshay, Jim George, Meera Sivadas, and I. Thirunavukkarasu

1 Introduction

Conical tank systems find uses in a variety of industrial applications such as process control as well as chemical manufacturing industries. This is because of its ability to provide excellent drainage of the liquid present inside. Due to the fact that at different heights, the diameter of the tank is varied, conical tank is considered as nonlinear systems. Due to the difficulties in control on nonlinear systems, fractional-order controller may be employed in the conical tank system, which can improve response of the system. Fractional-order proportional integral derivative (FOPID) controllers have two extra tunable parameters λ and μ, which results in improved system stability, reduced peak overshoot and integral square error (ISE).

2 Introduction to Fractional-Order Calculus

Fractional calculus is an extension to integral and derivative calculus where the integer order is replaced by any real or complex order. In this paper, we shall discuss

E. Jerald · A. Varghese · J. Chacko · S. Akshay · J. George (✉) · M. Sivadas
Department of Electrical and Electronics Engineering, Muthoot Institute of Technology and Science Kochi, Kochi 682308, India
e-mail: jimgeorge@mgits.ac.in

M. Sivadas
e-mail: meerasivadas@mgits.ac.in

I. Thirunavukkarasu
Department of Instrumentation and Control Engineering, MAHE (Institute of Eminence), Manipal Institute of Technology, Manipal 576104, India
e-mail: it.arasu@manipal.edu

S. K V and K. Guruprasad Rao (eds.), *Smart Sensors Measurements and Instrumentation*, Lecture Notes in Electrical Engineering 750,
https://doi.org/10.1007/978-981-16-0336-5_2

only the aspects of the real, non-integral order on fractional calculus. The following equations are obtained from [1].

Consider Cauchy's formula for repeated integration, where we repeatedly take the nth order anti-derivative of a function.

$$I^n f(t) = \frac{1}{(n-1)!} \int_0^t (t-T)^{n-1} f(\tau) d\tau \tag{1}$$

Taking $\gamma(n) = (n-1)!$, we can generalize Cauchy's formula to include real order α, which is strictly greater than zero. This is called the left Riemann–Liouville integral.

$$I^n f(t) = \frac{1}{\Gamma(\alpha)} \int_0^t (t-T)^{n-1} f(\tau) d\tau \tag{2}$$

R–L integration obeys the following relations:

$$(I_\beta f) = I_{\alpha+\beta} f \tag{3}$$

$$\frac{\mathrm{d}}{\mathrm{d}x} I^{\alpha+1} f = I^\alpha f \tag{4}$$

We know, taking the derivative n times after integrating n times is equivalent to the identity operation.

$$\frac{\mathrm{d}^n}{\mathrm{d}t^n}\left(I^n f(t)\right) = f(t) \tag{5}$$

Hence, on assuming the fractional derivative to order to have the similar property,

$$D^\alpha (I^\alpha f(t)) = f(t) \tag{6}$$

$$D^\alpha f = \frac{\mathrm{d}^{[\alpha]}}{\mathrm{d}t^{[\alpha]}}\left(I^{[\alpha]-\alpha} * f\right) \tag{7}$$

where $[\alpha]$ is called the ceiling function of α, which is the result of rounding up α to the next integer.

Substituting above equation, we get

$$D^\alpha f(t) = \frac{1}{\Gamma([\alpha]-\alpha)} * \frac{\mathrm{d}^{[\alpha]}}{\mathrm{d}t^{[\alpha]}} \int_0^t (t-T)^{[\alpha]-\alpha-1} f(\tau) d\tau \tag{8}$$

This is the left Riemann–Liouville fractional derivative. The special case of $\alpha = \frac{1}{2}$ is called the semi-derivative.

3 Modeling of Conical Tank

The process considered here is the conical tank system as shown in Fig. 1, in which liquid level is maintained at a constant rate by controlling the inflow of the tank F-in. Liquid passes through the tank from the top and leaves through the bottom. The modeling equations are derived from [2].

The physical parameters of the conical tank are as follows:

- H is the height of the tank
- R is the radius at the top of the tank
- F-in is the inlet flow rate
- F-out is the outlet flow rate
- h is the height of liquid level at any time t
- r is the radius of liquid level at a given height h

Here,

$$\tan\theta = \frac{r}{h} = \frac{R}{H} \tag{9}$$

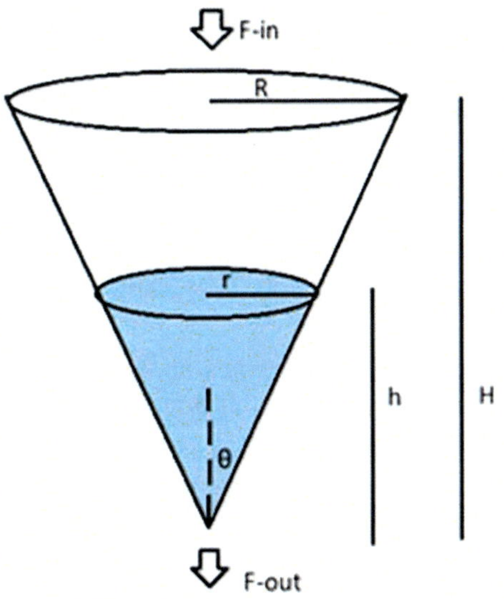

Fig. 1 Modeling of conical tank

Hence,

$$r = R * \frac{h}{H} r = R * \frac{h}{H} \tag{10}$$

The area of a cone is given by the equation,

$$\text{Area} = \pi r^2 \tag{11}$$

Now, substituting the equation for r in area, and differentiating area with respect to time,

$$\frac{\mathrm{d}a}{\mathrm{d}t} = \pi \left[\frac{R^2}{H^2} \right] * 2h \frac{\mathrm{d}h}{\mathrm{d}t} \frac{\mathrm{d}a}{\mathrm{d}t} = \pi \left[\frac{R^2}{H^2} \right] * 2h \frac{\mathrm{d}h}{\mathrm{d}t} \tag{12}$$

The volume of a cone is given by the equation,

$$V = \frac{1}{3} \pi r^2 h \tag{13}$$

Differentiating volume with respect to time,

$$\frac{\mathrm{d}v}{\mathrm{d}t} = \frac{1}{3} \left[A \frac{\mathrm{d}h}{\mathrm{d}t} + h \frac{\mathrm{d}a}{\mathrm{d}t} \right] \tag{14}$$

Now, substituting $\frac{\mathrm{d}a}{\mathrm{d}t}$ in above equation,

$$\frac{\mathrm{d}v}{\mathrm{d}t} = \frac{1}{3} \frac{\mathrm{d}h}{\mathrm{d}t} \left[A + 2h^2 \pi \frac{R^2}{H^2} \right] \tag{15}$$

On applying Newton's law,

$$F_{\text{in}} - F_{\text{out}} = \frac{1}{3} \frac{\mathrm{d}h}{\mathrm{d}t} \left[A + 2h^2 \pi \frac{R^2}{H^2} \right] \tag{16}$$

where $F_{\text{out}} = k\sqrt{h}$.
Hence, we get the following equation,

$$F_{\text{in}} - k\sqrt{h} = \frac{1}{3} \frac{\mathrm{d}h}{\mathrm{d}t} \left[A + 2h^2 \pi \frac{R^2}{H^2} \right] \tag{17}$$

Rearranging this equation in terms of rate of change of height of liquid level with respect to time,

$$\frac{dh}{dt} = \frac{3\left[F_{in} - k\sqrt{h}\right]}{A + 2h^2\pi\frac{R^2}{H^2}} \tag{18}$$

After substituting for the value of area in the above equation,

$$\frac{dh}{dt} = \frac{3\left[F_{in} - k\sqrt{h}\right]}{\pi\left[\frac{R^2}{H^2}\right] + 2h^2\pi\frac{R^2}{H^2}} \tag{19}$$

Now, we separate the numerator terms, such that the equation becomes,

$$\frac{dh}{dt} = \frac{F_{in}}{\pi h^2\left[\frac{R^2}{H^2}\right]} - \frac{kh^{\frac{1}{2}}}{\pi h^2\left[\frac{R^2}{H^2}\right]} \tag{20}$$

On integrating the above equation,

$$h = \int \left(F_{in} - k\sqrt{h}\right) * \frac{H^2}{h^2 r^2 \pi} dt \tag{21}$$

Here, we can see that this is a nonlinear equation and is a function of F_{in}, k, h and A. Hence, it can be expressed as,

$$f(h, F) = \frac{F_{\text{in}} - k\sqrt{h}}{A} \tag{22}$$

This model has two nonlinear functions: $F_{\text{in}} * h^{-2}$, and $h - \frac{1}{3}$. These two functions have to be linearized. The most popular technique for obtaining the linear approximation is based on Taylor series expansion of the nonlinear variables in the process model. We know, the area of cross section of the liquid surface is dependent on the height of liquid level.

By Taylor series expansion around $h = h_s$,

$$f(h, F) = \frac{F_{\text{in}} - k\sqrt{h}}{A} + \frac{1}{A}\left(F - F_{\text{in},s}\right) - \frac{k(h - h_s)}{2A\sqrt{h_s}} \tag{23}$$

Here, the first term on RHS becomes 0, as linearization is done about a steady state. Hence, the equation becomes,

$$f(h, F) = \frac{1}{A}F' - \frac{kh'_s}{2A\sqrt{h_s}} \tag{24}$$

This equation resembles the first-order differential equation,

$$\tau \frac{dy}{dt} + y = ku \tag{25}$$

Substituting the values in the above equations, the transfer function of Eq. (24) is expressed as,

$$\frac{H(s)}{F_{in}(s)} = \frac{k}{\tau s + 1} \tag{26}$$

where

$$\tau = \frac{2A\sqrt{h}}{k} \quad k = \frac{2\sqrt{h}}{k} \tag{27}$$

Now, on assuming the following parameters for a conical tank, and introducing a delay time, we obtain the transfer function for the model. $K = 0.925$, $T = 25.025$, $L = 1.095$

Hence,

$$G(s) = \frac{0.925\,e^{-1.095\,s}}{25.025\,s + 1} \tag{28}$$

where $e^{-1.095s}$ is the internal time delay present in the system. Figure 2 shows step response of conical tank system given by Eq. (28). Here, the system shows long settling time.

4 Modeling of PID Controller

PID controller is a closed-loop system which has feedback control. It compares the process variable with set point and generates an error signal. According to the error signal, it adjusts the output of system.

4.1 Ziegler–Nichols Tuning Using Stability Limit

The controller parameters can be tuned using Ziegler–Nichols tuning formula derived from [3], shown in Table 1.

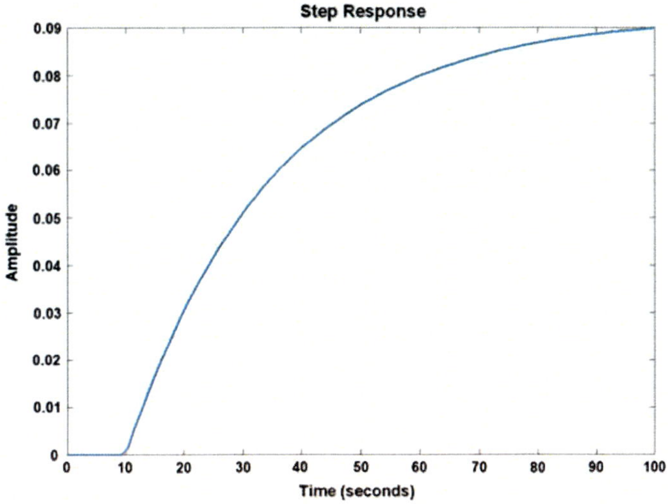

Fig. 2 Step response of conical tank

Table 1 Ziegler–Nichols tuning table

Tuning Table			
Controller	Kp (Kc)	Ti	Td
P	0.5		
PI	0.4	0.8Pc	
PID	0.6	0.5Pc	0.125Pc

Using the above formulae, the gain values can be obtained. The transfer function of the PID controller is obtained as,

$$\text{PID} = K_p + \frac{K_i}{s} + K_d * s \tag{29}$$

where

$$K_p = K_c \tag{30}$$

$$K_i = \frac{K_c}{T_i} \tag{31}$$

$$K_d = K_c T_d \tag{32}$$

In this paper, using MATLAB–SIMULINK, a model is designed and simulated, where the gain K is varied until the system shows oscillatory behavior as shown in Fig. 3. From this behavior, the critical period of oscillation Pc and the corresponding

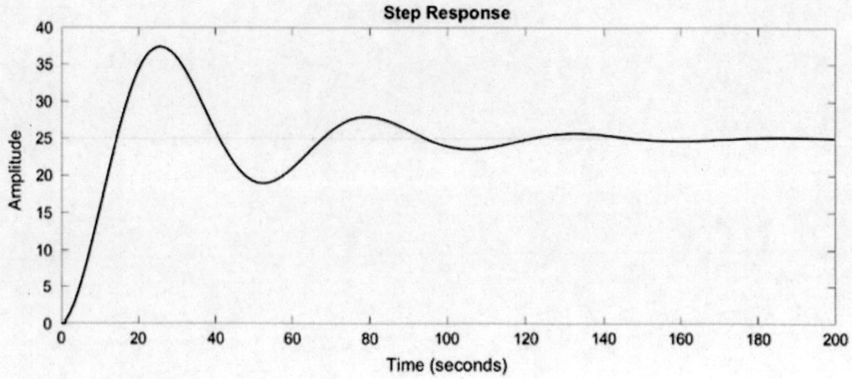

Fig. 3 Step response of system with PID controller

critical gain Kc is determined. We get the required values as, Kc = 1.3 and Pc = 4. Substituting these values in Ziegler—Nichols tuning formula, we get Kp = 0.78, Ki = 0.39 Kd = 0.37

Hence,

$$C(s) = 0.78 * s + \frac{0.39}{s} + 0.37 * s \tag{33}$$

4.2 *Step Response with PID*

Simulating the closed-loop response for the system using MATLAB at a set point of 25, we obtain the step response for the system as shown in Fig. 3.

Here, we observe that the system has considerable overshoot.

5 Modeling of Fractional-Order PID Controller

The fractional-order PID controller is manifested in the form of I^{λ} and D^{μ} and has transfer function as follows:

$$\text{FOPID} = K_p + \frac{K_i}{s^{\lambda}} + K_d * s^{\mu} \tag{34}$$

where

λ- fractional integral order

μ-fractional differential order.

5.1 *Tuning of FOPID Controller*

In order to obtain Kp, Ki, and Kd, we take values of [λ] and [μ] as 1. [4] Using Ziegler–Nichols tuning table, identical values are obtained as of integral order controller. Here, Kp = 0.78, Ki = 0.39, Kd-0.37. For a system with unity feedback, $G_c(s) = \frac{G}{1+GC} G_c(s) = \frac{G}{1+GC}$ to be stable, poles of the system should lie on left half ofs-plane (i.e., when the denominator, $1 + G(s)C(s) = 0$, the real part of s should be negative). This can be simulated in MATLAB. For values of λ and μ between 0.7 and 1.7, the step response of the system is plotted, as shown in Table 2:

On substituting values of λ and μ for which system is stable, we compare the different step responses obtained and select and corresponding to step response which satisfies the necessary conditions. From Fig. 4, it is observed that system is stable for the following set of values for λ and μ: (λ = 0.7, u = 0.7), (λ = 0.7, μ = 1) and (λ = 0.7, μ = 1.7).

On further considering the range of values for which system is stable and simulating in MATLAB, Fig. 5 shows that system has lower overshoot for (λ = 0.8, μ = 0.1), whereas system has slightly lower steady-state error for (λ = 0.9, μ = 0.8).

Table 2 Table for stability analysis

Stability analysis								
λ	0.7	0.7	0.7	1	1	1.7	1.7	1.7
μ	0.7	1	1.7	0.7	1.7	0.7	1	1.7
Stable	y	y	y	y	y	n	n	n

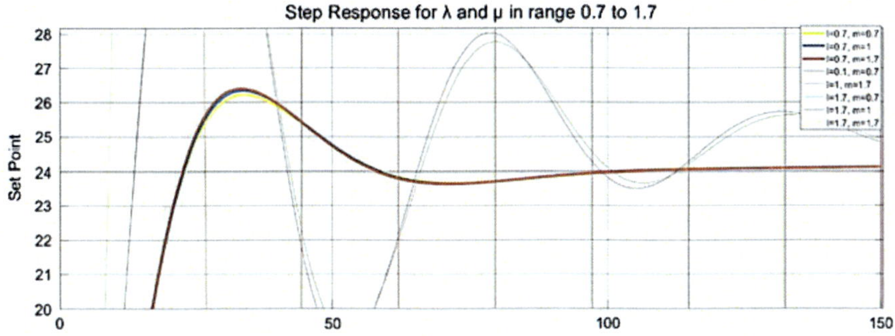

Fig. 4 Step response of fractional-order PID controller where 0.4 < λ < 0.9 and 0.8 < μ < 1.3

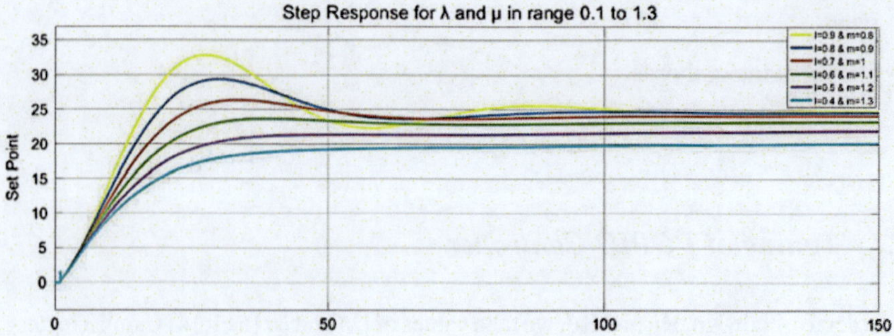

Fig. 5 Step response of fractional-order PID controller where $0.7 < \lambda < 1.7$ and $0.7 < \mu < 1.7$

For this experiment, preference is given to reduce overshoot. Hence, integral and differential order for the fractional-order controller is taken as, $\lambda = 0.8$ $\mu = 0.1$

$$C(s) = 0.78 + \left(\frac{0.39}{s^{0.8}} + 0.37s^{0.1}\right) \tag{35}$$

5.2 *Step Response with FOPID Controller*

Simulating the closed-loop response for the system using MATLAB for a set point of 1, we obtain the step response for the system as shown in Fig. 6.

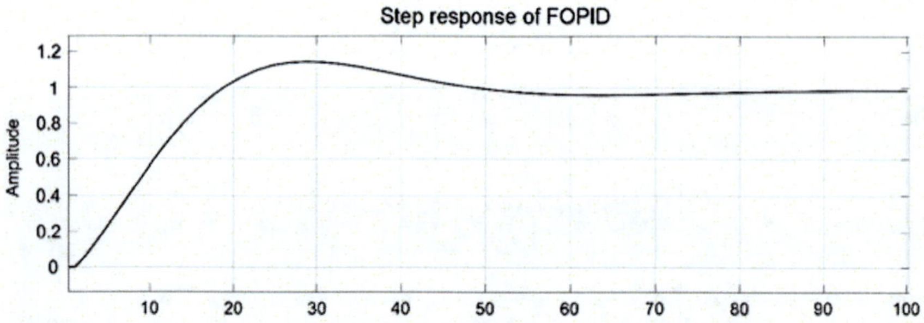

Fig. 6 Step Response of tank with fractional-order PID

6 Results and Discussions

In this section, set point tracking for the conical tank system is observed. Time domain indices like peak overshoot (M_p), steady-state error (e_{ss}), integral square error (ISE) and integral absolute error (IAE) are compared in Table 3. Set point tracking of nonlinear conical tank system employing PID controller and FOPID controller is shown in Fig. 7 and 8, respectively.

From Figs. 7and 8 and the corresponding Table 3, it is inferred that for the system with conical tank, when applying FOPID controller where ISE is set as performance index, peak overshoot is significantly lower, whereas rise time and steady-state error is slightly higher.

Table 3 Time domain indices

Stability Analysis					
	IAE	ISE	Mp (%)	Tr (s)	Ess
PID	24.55	10.53	50.75	9.75	0
ISE-FOPID	24.41	6.98	17.059	12.67	0.018

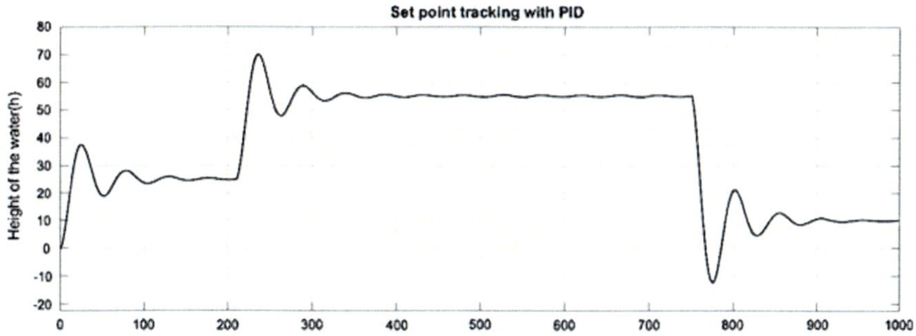

Fig. 7 Set point tracking of conical tank using PID

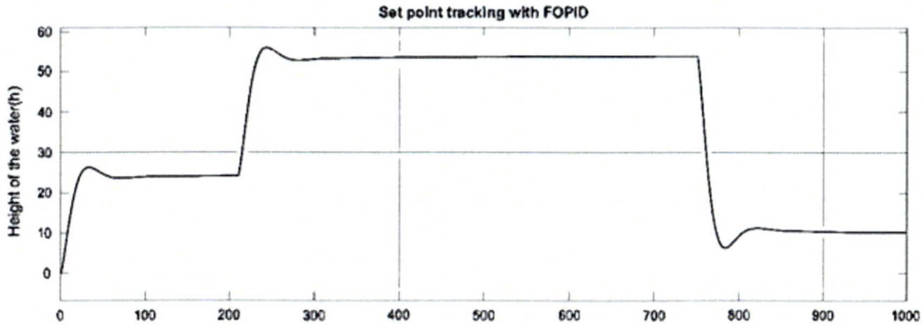

Fig. 8 Set point tracking of conical tank using ISE tuned FOPID

7 Conclusions

In this paper, simulated results for level control of a conical tank have been obtained with PID and FOPID controllers. The parameters of PID and FOPID controller were tuned using Ziegler–Nichols stability limit method. λ and μ for FOPID controller were optimized taking ISE as the performance index.

From this study, it is possible to conclude that FOPID controller is able to provide a response having lower ISE and overshoot, at the cost of a slightly higher steady-state error (E_{ss}). PID controller provides response with lower IAE and rise time, at the expense of higher overshoot.

Considering the above results, a comparison between PID and FOPID controllers may seem unfair. For future research, an optimization technique which involves both ISE and IAE as performance index may be considered.

Acknowledgements This project was performed at the Process Control Laboratory, at the Department of Instrumentation and Control Engineering, Manipal Institute of Technology, Manipal Academy of Higher Education, Manipal-576104.

References

1. Saravanakumar, G., Dinesh, S., Preteep, S., Sridhar, S., Suresh, S.: Controller tuning method for non-linear conical tank system. Asian J. Appl. Sci. Technol. (AJAST) 1(2):224–228 (2017)
2. Vavilala, S.K., Vinopraba, T.: Design of fractional order controllers for conical tank process. In: Proceedings of 2018 International Conference on Emerging Trends and Innovation in Engineering and Technological Research (2018)
3. Jauregui, C., Duarte-mermoud, M.A., Orostica, R., Travieso-Torres, J.C. Beyt, O.: Conical tank level control using fractional order PID controllers:a simulated and experimental study. Control Theory Tech. 14(4):369–384 (2016)
4. Ashutha, K., Yadav, E.S., Indiran, T., and Shreesha, C.: Implementation of fuzzy control for a nonlinear system—conical level process. In: 2017 9th International Conference on Information Technology and Electrical Engineering (ICITEE), Phuket, Thailand

End-Effector Position Analysis of SCARA Robot by Using MATLAB

Amit Talli and Doddabasappa Marebal

1 Introduction

Kinematics is the science of motion which deals with the way things move without consideration of forces/torque. Under the kinematics, one studies the position, displacement, rotation speed, velocity and acceleration (concerning time and other variables) [1, 2].

A manipulator is a device which consists of links and is connected with the help of joints, which produces relative displacement between the rigid links [3]. Industrial robots are also called as serial manipulator due to serially connected joints between the rigid links [4]. The two types of widely used joints in the industrial manipulators are, the revolute and sliding joint. The revolute joint is also known as a pin joint, and it allows only pure rotary motion. The sliding joint is also called a prismatic joint, and it allows only straight-line motion [5]. The one end of the robot manipulator is connected to a supporting base or fixed frame while the other end is open and attached to the tool (end-effector). The joints are driven by the actuators such as electric, hydraulic, or pneumatic[6].

A robotic manipulator with multiple degrees of freedom is designed to perform a wide variety of tasks in the spatial workspace or 3D workspace[7]. The trajectory of the tool or end-effector is planned to achieve a smooth and steady transition from one location to another location with respect to time. The control design performs an essential role in the positioning of end-effector of the robot manipulator. A mathematical representation of the robot arm is required to describe the geometrical and trajectory of the motion[8]. The kinematic model represents the geometrical description of rigid links and joint with respect to the reference frame.

A. Talli (✉) · D. Marebal
Automation & Robotics, KLE Technological University, Hubballi 580031, India
e-mail: amit@kletech.ac.in

D. Marebal
e-mail: doddabasappa@kletech.ac.in

S. K V and K. Guruprasad Rao (eds.), *Smart Sensors Measurements and Instrumentation*, Lecture Notes in Electrical Engineering 750,
https://doi.org/10.1007/978-981-16-0336-5_3

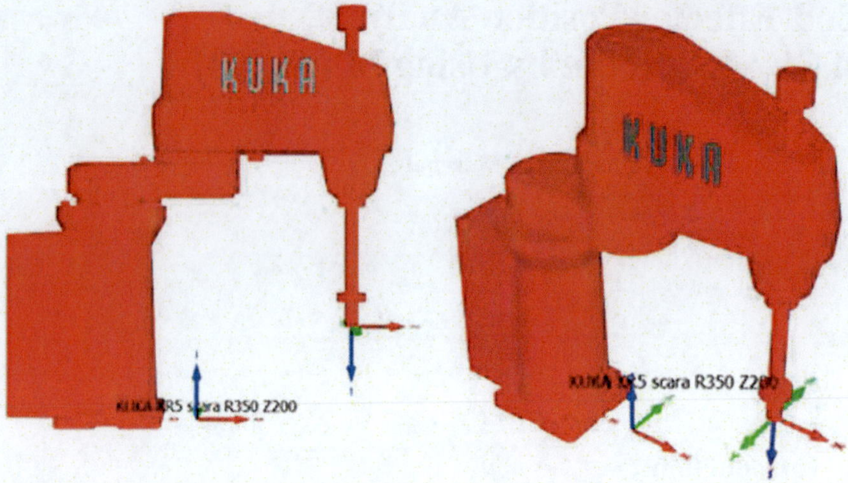

Fig. 1 KUKA KR5 R350 SCARA Robot

2 Kinematic Model of KUKA KR5 R350 SCARA Robot

The KUKA KR5 R350 SCARA robot is a pick and place robot, with four joints consisting of three revolute joint and one prismatic joint. The SCARA robot is widely used in electronic PCB industries for placing electronic parts. Figures 1 and 2 show the structure and workspace of the SCARA robot [9].

The full form of SCARA is selective compliance articulated robot arm and comes with different specifications. The SCARA robot is widely designed and manufactured by various companies such as FANUC, ABB, Stäubli and KUKA. The prismatic joint helps to create a cylindrical workspace, as shown in Fig. 2. The SCARA robot is a lightweight four degree of freedom robot as compared to the six degrees of freedom industrial robots. The fourth joint of the SCARA robot is a revolute joint which helps the end-effector to rotate $\pm 360°$ approximately. Figure 2 shows the wireframe structure of the workspace for the SCARA robot. The workspace resembles the cylindrical shape due to the prismatic joint of the SCARA robot. The workspace is also called as the work envelope for the SCARA robot.

3 Denavit - Hartenberg (D-H) Parameters

Denavit–Hartenberg convention is one of the most popular standards used for the systematic establishment of coordinate frames for each rigid link [10]. The D–H parameters allow the allocation of the coordinate frame with the help of standard parameters. The Denavit–Hartenberg convention is widely used for kinematic analysis of industrial robotic manipulators. Figure 3 shows a pair of adjacent links labelled

Fig. 2 Workspace of KUKA KR5 R350

Fig. 3 Denavit–Hartenberg (D-H) description of a rigid link

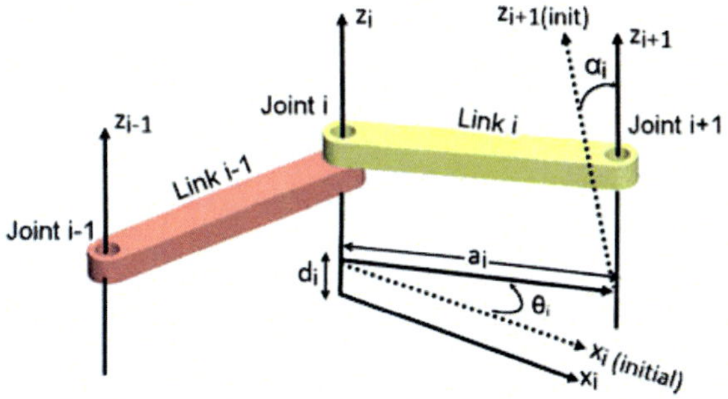

as link i and link $i-1$; their associated joints are labelled as joint $(i-1)$, joint i and joint $(i+1)$. A-frame $\{i\}$ is assigned to link i as follows:

- The z-axis lies along the axis of motion of the ith joint.
- The x-axis is normal to the z-axis and pointing away from it.
- The y-axis is always normal or perpendicular to the screen.

The Denavit–Hartenberg (D–H) description of a rigid link depends on four standard parameters associated with each link [11]. These four parameters fully define the structure of links, and the orientation of joints is given below:

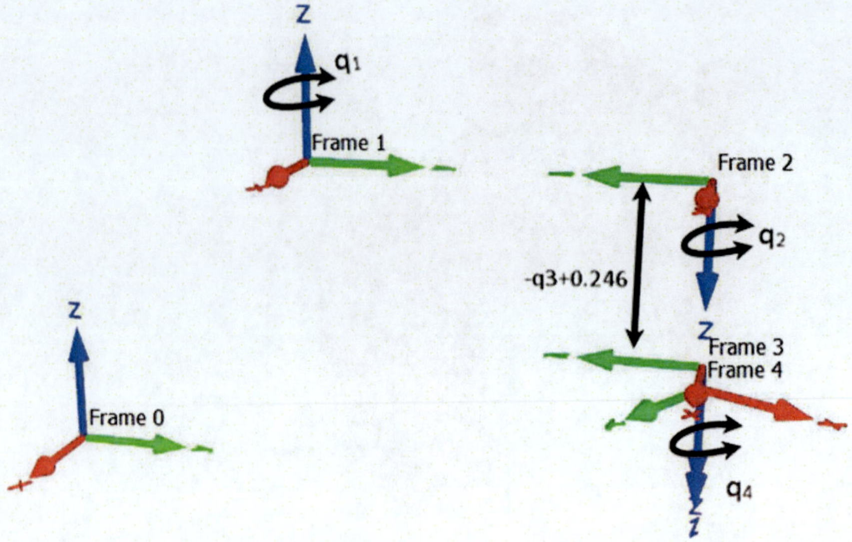

Fig. 4 Frame allocation for KUKA KR5 R350

Table 1 D-H parameter for SCARA robot

Joint	θ	d	a	α	Operating range
1	q_1	0.246	0.125	0	−155° to 155°
2	q_2	0	0.225	π	−145° to 145°
3	0	$-q(3)$ +0.246	0	π	0.046–0.200 m
4	q_4	0	0	0	−358° to 358°

1. θ_i is the joint angle.
3 d_i is the joint offset.
4 a_i is the link length.
5 α_i is the twist angle.

The frame allocation for KUKA KR5 R350 as per the Denavit–Hartenberg (D–H) notations is shown in Fig. 4. The (D-H) parameters, according to this model, are given in Table 1.

4 Kinematic Relationship Between Adjacent Links

Once the coordinate frames are allocated for each link according to (D–H) parameters, a 4 × 4 homogeneous transformation matrix can easily be derived considering the data available from the (D-H) parameter table [12]. This transformation consists of four matrices having joint and link information, as shown in Fig. 3.

1. Rotation about the z-axis by an angle θ_i
2. Translation along the z-axis by distance d_i
3. Translation by distance a_i along the x-axis and
4. Rotation by angle α_i about the x-axis

$$^{i-1}T_i = T_{\text{rot}}(\theta_i, z) * T_{\text{trans}}(d_i, z) * T_{\text{trans}}(a_i, x) * T_{\text{rot}}(\alpha_i, x) \tag{1}$$

Which can be expanded to homogeneous matrix form:

$$^{i-1}T_i = \begin{bmatrix} \cos(\theta_i) & -\sin(\theta_i)\cos(\alpha_i) & \sin(\theta_i)\cos(\alpha_i) & a_i\cos(\theta_i) \\ \sin(\theta_i) & \cos(\theta_i)\cos(\alpha_i) & -\cos(\theta_i)\cos(\alpha_i) & a_i\sin(\theta_i) \\ 0 & \sin(\alpha_i) & \cos(\alpha_i) & d_i \\ 0 & 0 & 0 & 1 \end{bmatrix} \tag{2}$$

The elements of the last column of the matrix in Eq. 2 describe the position of the end-effector, and the remaining elements describe the orientation[13]. The homogeneous transformation equation is obtained by performing matrix multiplication from base to end-effector as per the D-H parameter table is given by:

$$^{b}T_{ee} =^{0} T_1 *^{1} T_2 *^{2} T_3 *^{3} T_4 \tag{3}$$

$$^{0}T_1 = \begin{bmatrix} C_1 & -S_1 & 0 & 0.125 \\ S_1 & C_1 & 0 & 0 \\ 0 & 0 & 1 & 0.246 \\ 0 & 0 & 0 & 1 \end{bmatrix} \tag{4}$$

$$^{0}T_1 = \begin{bmatrix} C_2 & S_2 & 0 & 0.225 * C_2 \\ S_2 & -C_2 & 0 & 0.225 * S_2 \\ 0 & 0 & -1 & 0 \\ 0 & 0 & 0 & 1 \end{bmatrix} \tag{5}$$

$$^{2}T_3 = \begin{bmatrix} 1 & 0 & 0 & 0 \\ 0 & -1 & 0 & 0 \\ 0 & 0 & -1 & -q_3 + 0.246 \\ 0 & 0 & 0 & 1 \end{bmatrix} \tag{6}$$

$$^{b}T_{ee} = \begin{bmatrix} r_{11} & r_{12} & r_{13} & x \\ r_{21} & r_{22} & r_{23} & y \\ r_{31} & r_{32} & r_{33} & z \\ 0 & 0 & 0 & 1 \end{bmatrix} \tag{7}$$

where the elements of the matrix are:

$$
\begin{aligned}
r_{11} &= C_4 * (C_{12}) - S_4 * (S_{12}) \\
r_{12} &= -C_4 * (S_{12}) - S_4 * (C_{12}) \\
r_{13} &= 0 \\
r_{21} &= C_4 * (S_{12}) + S_4 * (C_{12}) \\
r_{22} &= C_4 * (C_{12}) - S_4 * (S_{12}) \\
r_{23} &= 0 \\
r_{31} &= 0 \\
r_{32} &= 0 \\
r_{33} &= 1 \\
x &= 0.225 * (C_{12}) + 0.125 * C_1 \\
y &= 0.225 * (S_{12}) + 0.125 * S_1 \\
z &= -q_3
\end{aligned}
$$

For simplicity, we use short notations for trigonometric identities, $S_i = \text{Sin}\ (\theta_i)$, $C_i = \text{Cos}\ (\theta_i)$, $S_{ij} = \text{Sin}\ (\theta_i + \theta_j)$, $C_{ij} = \text{Cos}\ (\theta_i + \theta_j)$. The last column of the homogeneous transformation matrix represents the position in terms of x, y and z-axis, and the remaining elements in the matrix represents the orientation [14].

5 Simulation Results and Discussions

The kinematic model is developed using the above methodology in MATLAB for KUKA KR5 SCARA robot. Frame assignment for KUKA KR5 R350 robot in MATLAB GUI is shown in Fig. 5. The developed model is used to determine the position and orientation of the end-effector. The joint values of $q_1 = \text{pi}/4$ or (45°), $q_2 = -\text{pi}/4$ or (−45°), $q_3 = 0.1$ m or 100 mm, and $q_4 = \text{pi}$ or (180°) are selected within the operating range as per the D-H parameters as shown in Table 1. The position of the end-effector of the KUKA KR5 R350 is obtained by substituting the joint values in the following (Eqs. 8–10).

$$x = 0.225 * (C_{12}) + 0.125 * C_1 \tag{8}$$

$$y = 0.225 * (S_{12}) + 0.125 * S_1 \tag{9}$$

$$z = -q_3 \tag{10}$$

Substituting the joint values in the above equation yields, $x = 0.3133$ m, $y = 0.0883$ m, $z = 0.1000$ m. In order to validate the derived equation of the forward kinematic, the position of end-effector of the robot is calculated by using ARTE toolbox[15] in MATLAB as shown in Fig. 5.

The analytical and MATLAB results are compared to validate the kinematic model of the SCARA robot. Table 2 shows the comparison between the MATLAB and

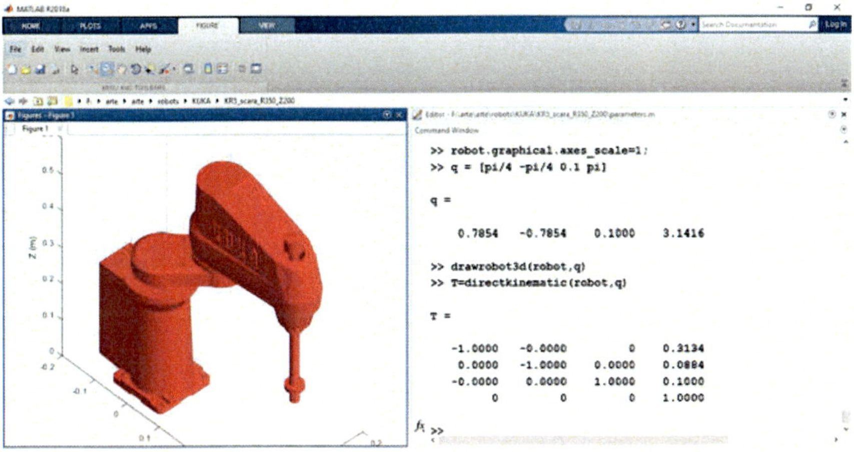

Fig. 5 Frame allocation for KUKA KR5 R350

analytical results for different joint angles within the limited operating range. The first column represents the figure of actual pose for the specified joint values. The result column is divided into two columns for showing the position matrix of MATLAB and analytical results. The last column represents the error between MATLAB and analytical results.

6 Conclusion

The kinematic model of a 4-DOF robotic arm is presented in this paper. Forward kinematic equations are established for the KUKA KR5 R350 robotic arm. The ARTE toolbox is implemented to simulate the forward kinematic in MATLAB. The analysed position of the end-effector is determined analytically and compared with simulation results. It has been shown that the simulation results correlate well with the analytical results. Consequently, the results presented herein provide useful information about the kinematic analysis of a robotic arm, and the information can be used for trajectory planning and control system design of the robot's operations for further study.

Table 2 MATLAB and analytical results

Pose and joint values	Results		Error
	MATLAB	Analytical	
 $q = [0\ 0\ 0\ 0]^T$	T = $\begin{matrix} 1.0000 & 0 & 0 & 0.3500 \\ 0 & 1.0000 & 0.0000 & -0.0000 \\ 0 & -0.0000 & 1.0000 & 0 \\ 0 & 0 & 0 & 1.0000 \end{matrix}$	$\begin{bmatrix} 0.3500 \\ 0 \\ 0 \end{bmatrix}$	0
 $q = [\text{pi}/4\ -\text{pi}/4\ 0.5\ \text{pi}]^T$	T = $\begin{matrix} -1.0000 & -0.0000 & 0 & 0.3134 \\ 0.0000 & -1.0000 & 0.0000 & 0.0884 \\ -0.0000 & 0.0000 & 1.0000 & 0.5000 \\ 0 & 0 & 0 & 1.0000 \end{matrix}$	$\begin{bmatrix} 0.3134 \\ 0.0844 \\ 0.5000 \end{bmatrix}$	0
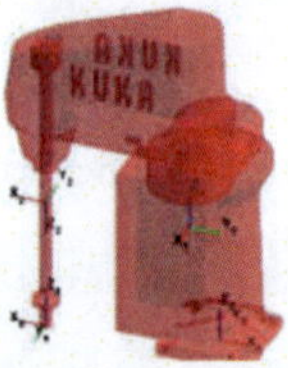 $q = [-\text{pi}/2\ -\text{pi}/2\ 0.046\ 0]^T$	T = $\begin{matrix} -1.0000 & 0.0000 & 0.0000 & -0.2250 \\ -0.0000 & -1.0000 & -0.0000 & -0.1250 \\ 0 & -0.0000 & 1.0000 & 0.0460 \\ 0 & 0 & 0 & 1.0000 \end{matrix}$	$\begin{bmatrix} -0.2250 \\ -0.1250 \\ 0.0460 \end{bmatrix}$	0
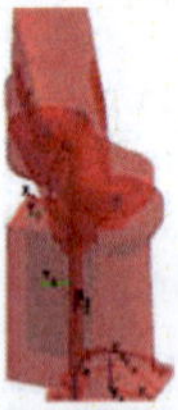 $q = [2.70\ 2.53\ 0.046\ 6.24]^T$	T = $\begin{matrix} 0.4568 & 0.8896 & 0.0000 & -0.0017 \\ -0.8896 & 0.4568 & 0.0000 & -0.1421 \\ 0.0000 & -0.0000 & 1.0000 & 0.0460 \\ 0 & 0 & 0 & 1.0000 \end{matrix}$	$\begin{bmatrix} -0.0017 \\ -0.1421 \\ 0.0460 \end{bmatrix}$	0
 $q = [-2.70\ -2.53\ 0.2\ -6.24]^T$	T = $\begin{matrix} 0.4568 & -0.8896 & -0.0000 & -0.0017 \\ 0.8896 & 0.4568 & 0.0000 & 0.1421 \\ -0.0000 & -0.0000 & 1.0000 & 0.2000 \\ 0 & 0 & 0 & 1.0000 \end{matrix}$	$\begin{bmatrix} -0.0017 \\ 0.1421 \\ 0.2000 \end{bmatrix}$	0

References

1. Sadanand, R.O.M., Chittawadigi, R.G., Saha, S.K.: Virtual robot simulation in roboanalyzer. In: 1st International and 16th National Conference on Machines and Mechanisms, iNaCoMM 2013. pp. 686–693 (2013).
2. Verma, A., Deshpande, V.: End-effector position analysis of SCORBOT-ER Vplus robot. Int. J. Smart Home. **5** (2011).
3. Liu, J., Jian, F., Liu, J.: Four degrees of freedom SCARA robot modeling and simulation. In: Proceedings of 2014 International Symposium Computer Consumer and Control, IS3C 2014, 410–412 (2014). https://doi.org/10.1109/IS3C.2014.113
4. Özer, A., Arai, H.: Robotic metal spinning-experimental implementation using an industrial robot arm. In: Proceedings of IEEE International Conference on Robototic Automation, 140–145 (2009). https://doi.org/10.1109/ROBOT.2009.5152265
5. Lokesh, R., Chittawadigi, R.G., Saha, S.K.: MechAnalyzer: 3D simulation software to teach kinematics of machines. 2nd International 17th National Conference Machanism and Mechine 1–9 (2015).
6. Zudilova, T.V., Ivanov, S.E.: Mathematical modeling of the robot manipulator with four degrees of freedom. Glob. J. Pure Appl. Math. **12**, 4419–4429 (2016)
7. Nguyen, M.T., Yuan, C., Huang, J.H.: Kinematic Analysis of A 6-DOF Robotic Arm. Presented at the (2019). https://doi.org/10.1007/978-3-030-20131-9_292
8. Othayoth, R.S., Chittawadigi, R.G., Joshi, R.P., Saha, S.K.: Robot kinematics made easy using RoboAnalyzer software. Comput. Appl. Eng. Educ. **25**, 669–680 (2017). https://doi.org/10.1002/cae.21828
9. KUKA: The compact robot for low payloads KR 5 scara.
10. Denavit, J.: A kinematic notation for lower-pair mechanisms based on matrices. Trans. ASME. J. Appl. Mech. **22**, 215–221 (1955).
11. Sadanand, R., Joshi, R.P., Chittawadigi, R.G., Saha, S.K.: Virtual experiments for integrated teaching and learning of robot mechanics using roboanalyzer. Lect. Notes Mech. Eng. 59–68 (2016). https://doi.org/10.1007/978-81-322-2740-3_7
12. Talli, A., V Meti, V.K.: Design, simulation, and analysis of a 6-axis robot using robot visualization software. IOP Conf. Ser. Mater. Sci. Eng. **872**, 012040 (2020). https://doi.org/10.1088/1757-899X/872/1/012040
13. Constantin, D., Lupoae, M., Baciu, C., Buliga, D.-I.: Forward Kinematic Analysis of an Industrial Robot. New Dev. Mech. Mech. Eng. 90–95 (2015)
14. Anciferov, S.I., Gaponenko, E.V., Kulakov, L.S.: Kinematic analysis of robotic complex within the solidworks motion system. **157**, 27–31 (2018). https://doi.org/10.2991/aime-18.2018.6
15. ARTE (A Robotic Toolbox for Education), https://arvc.umh.es/arte/index_en.html. Last Accessed 16 Feb 2020

Prediction of Song Popularity Using Machine Learning Concepts

Adit V. Kaneria, Abishek B. Rao, Shivani G. Aithal, and Smitha N. Pai

1 Introduction

The music companies invest a huge amount in search of song which can make an impact. An ideation on what makes a given song to be influential can bring huge profit to the music industry. This idea can help in predicting the success of the given song before the launch of the album. Good amount of work is carried out on the lyrics of the song. Automatic predicting a song to be hit is a challenge. With digitization, large amount of information about the song can be captured, reproduced, reworked and remodeled. Musical tastes vary from person to person and in today's world, it is highly individualistic. Streaming services like Spotify have realized this and have employed algorithms to understand an individuals' music taste and recommend music based on their listening history. This helps the user in getting a personalized playlist. However, as things are getting easier for consumers, artists have a tough time churning out good music every now and then. A hit song can do wonders for their careers and launch them straight into the spotlight. There is a lot of uncertainty when they launch a new song as to whether it will be accepted by consumers worldwide. It is a heavy investment for producers too. With the advent of data science and machine learning, industries have changed their approach to business. The availability of data has given them an opportunity to predict outcomes more accurately and with greater certainty. The music industry can also apply machine learning algorithms to see if a song will be a hit or no. This way they can make minor tweaks, keeping the artists music vision in mind, and appeal to a greater audience. The music industry around the globe consists of companies and individuals that make huge money by creating songs and selling them to the people by promoting their songs in live concert, shows, etc. There is no guarantee that a signal in the song might become popular resulting

A. V. Kaneria · A. B. Rao · S. G. Aithal · S. N. Pai (✉)
Department of ICT, Manipal Institute of Technology, Manipal Academy of Higher Education, Manipal, India
e-mail: smitha.pai@manipal.edu

S. K V and K. Guruprasad Rao (eds.), *Smart Sensors Measurements and Instrumentation*, Lecture Notes in Electrical Engineering 750,
https://doi.org/10.1007/978-981-16-0336-5_4

in high profit while some signal may turn out be not so popular which may lead to losses. In the project, analytics is applied to predict if a song can be classified as belonging to the top 100 of the official chart so that it can be classified as a hit song. Popularity of a track can be affected by other aspect like the social, economic, etc.

The prediction of song popularity can be used in various applications. Some of them are as follows: As a gesture of recognition, playing the popular song of the famous personalities as they arrive randomly at a gathering on recognizing their face. Recognizing the emotional expression of the person and play music as per their mood. Guessing the age range of the person and play music according to their generation. Popular music based on their native region can also be played on recognizing the person's place of origin. Music can also be played as per the situation, sick, patriotic, gaming, etc. to enhance their mood.

The work currently focuses on the features of the audio of the music track.

2 Related Work

Machine learning uses various algorithms for the purpose of predictions. Good amount of work is carried out to predict the popularity of a song. A study by Stanford students on the problem "Hit Song Science" [1] used dataset of 4000 songs and extracted features of the song from Spotify API. Among the different algorithms used, the best results were obtained using neural network with one hidden layer and logistic regression. It provided an accuracy up to 75%. Another study [2] in the same topic used a database of songs collected between the years 1985 and 2013. Music features of the songs were extracted from the Echo Nest [3], the world's leading music analytics organization with over a trillion data points on millions of popular songs. Five classifier models are used. All the models are tested with and without feature selection. Validation process of tenfold is used for verification on the models. The paper addressed that hit songs can be classified as hit or non-hit using audio features without any metadata. The authors had built an application, where users can fill in audio features of a new song and classify it as a hit or not a hit. Another study [4–6] tried to determine the change in popularity of songs over time to predict how popular songs in the future will turn out to be. The paper takes attributes of popular songs from the 1900s to today and analyzes how the attributes contribute to the song's genre and year of release. Study makes a comparison of hit songs from UK and China. Features for the track are obtained from the work [3]. The work [7] uses lyrics and acoustic-based features. Dataset contained songs by prominent artists from different generation. During the study, the range of number of tracks in the chart in the selection of hit and non-hit is taken differently by each of the author. Author [8] uses wide and deep neural network for and uses the features of audio that are low and high levels. This process is used is used to analyze the hit songs. The work [9] emphasizes the various moods revolving each of the songs. Songs could be of dancebability, rock, relaxed songs and the various moods party-like, electronic, etc. associated with it is addressed. In the work [10], two models support vector machine

(SVM) and long short-term memory (LSTM) are used to find the popularity of the artist. Many features are used as the input, and if the features are too fine grained, then the results could be counterproductive. Results showed that performance of LSTM is better compared to the LSTM. Deep learning technique using the Mahalanobis distance vector scores is used to find the next song which need to be played in the playlist which is recommended to the user. The work [11] uses the feature extraction from the lyrics, deep neural network autoencodes to carry out the decision for popular songs. In the work [12], the authors have found that there is difference in readability of the lyrics and its repetitiveness between the genre. The top artists lyrics are repetitive but may not be simple when it comes to readability. The author [13] discusses the tools and techniques used in developing context-aware services. Context-awareness involves cognitive psychology, perception of music by human, social computing and recognition of emotion in music. The author [14] uses the wide and deep neural network which exploits the high- and low-level audio features to predict hit songs. The study [15] revealed that it was easy to predict Chinese hit song more easily than the UK. This suggested that the characteristic for hit song varied with region of residence. Linear regression and support vector machine (SVM) were used for comparison. Hit songs were taken from upper half of the top chart and non-hit the lower half from the non-hit list. Using SVM and classifiers, they distinguished the top 1% of the hit song from others with promising result. Authors have also used other matrix to measure the popularity like the kurtosis, debut, standard length, mean, maximum skewness and sum [16]. A relation between artist and song is brought about in the work [17]. LSTM and SVM are used for comparison, and SVM is proved to be better.

Emotional well-being of a person is vital for the cognitive process. It has its influence on the learning, memory and other factors. Emotions can be captured using various methods like electroencephalography, electrocardiography, electromyography and bio-signals. Music can influence for the well-being of the individual. Happiness and sadness are the easily recognized emotions using speech recognition system [18]. There are ways of communicating emotions like usage of works, voice tone, body language, facial emotions, variations in heart rate and conductance in the skin [19]. The author addresses how emotions can be recognized using the highly discriminative features and combining the orientation and texture of the images captured. The process is effective even in the case of occlusion and illumination along with noise.

The current papers implement certain machine learning algorithms on a dataset consisting of features extracted from the Spotify database. Data preprocessing techniques were used to handle missing values and normalization. A total of four models were used–random forest, logistic regression, naïve Bayes and decision tree [20, 21]. The main work is focused on supervised learning, as the dataset has labeled output. The dataset contains hit and non-hit songs. Models should be able to recognize the pattern which can categories them along with their label.

3 Background Theory

3.1 Logistic Regression [22]

Logistic regression as the name implies computes the log of the features from the music track. The real values obtained are mapped to values between 0 and 1 as per (1) to obtain the output y.

$$y = 1/(1 + e^{-x}) \quad (1)$$

x forms the sum of the combination of the weights and feature vector values. Alternatively, output y is represented as in (2) as

$$y = e^{(w0 + w1*x1)} / \left(1 + e^{(w0+w1*x1)}\right) \quad (2)$$

w_0 is the intercept term, and w_1 is the weight associated with the feature vector $\times 1$. Similarly, a single weight value is associated with each of the column feature vector. The weight value is obtained after training the model.

3.2 Decision Tree [23]

Decision trees take bulk of data and are split in groups recursively as in a tree, until the data is classified into the required class at the leaves. It is also a supervised learning technique.

Mainly, two types of trees whose decisions are categorical and regression (continuous values).

Shannon entropy $E(S)$ measures how much of uncertainty or randomness exists in a data for a finite set S.

$$E(S) = \sum_{x \in X} p(x) \log_2 \frac{1}{p(x)} \quad (3)$$

It intimates about the predictability of a certain event. Higher uncertainty is due to higher values and lower with lesser value.

Kullback–Leibler divergence KL(S, A) (Information gain) with set S and choosing the attribute A shows the change in the entropy. It is a measure of changes in the entropy with regard to the independent variables.

$$\mathrm{KL}(S, A) = E(S) - E(S, A) \quad (4)$$

Alternatively,

$$\mathrm{KL}(S, A) = E(S) - \sum_{i=0}^{n} P(x) * E(x) \tag{5}$$

Entropy for the full set is $E(S)$ and $E(S, A)$ on applying the feature A. Probability of the event x is $E(x)$.

3.3 Naïve Bayes [24]

Naive Bayes classifier is based on the Bayes' theorem, which makes a presumption that input variables or features are not dependent on each other. Bayes' theorem, which is centered on likelihood or conditional probability, which says if an event (C) has occured, with what likelihood the event (B) would occur is given by (6).

$$P(B|C) = (P(C|B)P(B))/P(C) \tag{6}$$

P is the probability.

P (B|C) implies given with what probability will the event B occur, given that event C has occurred already.

P (C|B) implies given that B has occurred already, with what probability C is going to occur.

$P(B)$ signifies the probability of the event B occurring.

$P(C)$ signifies the probability of event C occurring.

The training time in the case of naïve Bayes is low. If the conditional dependence between the attributes holds, it will converge faster than the discriminative models like that of logistic regression. The size of the model is small.

3.4 Random Forest [25]

Random forest consists of large collection of decision tree which work like an ensemble. Each tree provides the class to which the features belong. The class with majority of vote signifies the class to which it belongs. The work suggests that a collective decision of large number of unrelated models of tree will provide a better result than a single tree. The error if introduced by individuals will get suppressed unless all of the models are providing erroneous results. The results obtain are better if models are created around specific information of the features rather than randomly selection. Results obtained could also improve if the error introduced by tree does not correlate with others. The model is large. It has to rebuild the model with quick changes in the data.

3.5 *K-Nearest Neighbor (kNN) [26]*

The k neighbors are chosen to form cluster. The choice of *k* has a large image on the results obtained from the algorithm. With large vale of *k*, the effect of variance because of random error is reduced but may lose out on small information which might be importance. The correct value of *k* brings a balance between under and over fitting.

3.6 *XGBoost [27]*

This algorithm used gradient boosted decision tree for increasing the speed and performance. New models are added to existing model in order to reduce the errors introduced by them. It uses the gradient descent algorithm to reduce the errors.

4 Methodology

The project uses dataset for the audio features from Spotify Web API. The following features are available [28]:

a. **Danceability**—It provides a value which signifies how much it is meant for dancing. This value is based on the elements of music like beat strength, tempo, overall regularity and rhythm stability. Zero value signifies low hit, and value of one indicates high hit for the dance.
b. **Energy Level**—Measures how intense the activity is. This can be related to loud, noisy and fast track. The overall value is based on features like timbre, dynamic range, onset rate, loudness and general entropy. The values can be one or zero based on the liking or not.
c. **Key**—Specifies the track the key is in. Corresponding integer values are suitably chosen.
d. **Loudness**—Average loudness of the track is measured in decibels (dB). Values are between the range −60 and 0 dB.
e. **Mode**—The scale from which the melodic content is obtained. Values are one and zero for major and minor values.
f. **Speechiness**—It identifies if there exist any spoken words in the track. A value below 0.33 represents only music without the speech content. Between 0.33 and 0.66 signifies both music and speech are present in sections. A value of 0.66 to 1 indicates that the track is entirely made of spoken words.
g. **Acoustisness**—It gives a measure between 0 and 1 about the behavior of sound waves (acoustic) of the track.
h. **Liveness**—Detects for vocal content. Value close to 1 depicts that the track has no vocal content. Value above 0.5 is for tracks with instrumental content.

i. **Valence**—Cheerfulness and positivity are indicated by this value. High value of 1 represents positivity and 0 with sad and other negative aspects.
j. **Tempo**—It is the estimated average beats in the track for every minute. It also suggests the pace or speed of the song.
k. **Duration**—The length of the track in milliseconds.
l. **Time Signature**—Specifies how many beats are present in each measure.
m. **Instrumentalness**—This includes some vocal words. Closer is this value to the 1, implies it does not have the vocal words.

Some other features which are present in the dataset like genre, artisit_name, track_name and track_id are not used for predicting popularity. There is a feature called "popularity" which is used as the true output. This results obtained by using various models are compared with this feature which is available. The machine learning work flow is separated in five parts: extraction of data, preprocessing the data, building the model, evaluation with the test set and prediction the results [29]. Following is the steps carried out in the current work.

4.1 Extract Data

Spotipy extracts music features from the Spotify database and stores them in a csv file. The API takes the artists names as input and returns the following attributes—acousticness, danceability, loudness, tempo, speechiness, liveness and so on. The API will require a developer account to be created with Spotify to access their database. On logging the execution of the content at the given URL is proceeded. The attributes are fetched and stored into the destination csv file. The dependent variable values are obtained by scraping the data from an official chart website. For a song to be a hit, it must have reached the top 100 in the official chart.

4.2 Preprocess Data

During data preprocessing, null values have to be removed. This is especially in the case of random forest. Datasets with only metadata are chosen. Audio features within the specified range are selected. Duplicates in the dataset were avoided. Dataset is split into 80 percent training and 20 percent testing set with both the set containing equal proportion of hit and non-hit songs.

4.3 Model Building

Machine learning model, code is written in Python (v3.6) in sublime text editor. It is an open-source text editor that can implement data visualization, statistical modeling, machine learning, etc. The machine learning part of the code is in Python, and the library ScikitLearn is used.

Four different algorithms decision trees, logistic regression, support vector machine and random forest are used. Both linear and nonlinear models are used. The seed is fixed to have the same split in the data with each run. This enables uniformity and accuracy while the results are compared across different models. Models are built to classify the song into hit and non-hit using the rules that are obtained on training with the various algorithms.

4.4 Evaluation of the models

The models obtained on training have to be cross-checked for expected results. To incorporate this, the dataset is divided into train and test set. There are various validation techniques, of which, K-fold cross validation is one of them. The dataset is partitioned into K subset each of same size. Each of these is referred to as fold. Algorithms are trained on the K-1 folds and tested on the remaining one fold. This process is done K times each time with a different partition as test and train. Average of this is used for future prediction. Currently, a tenfold validation is used for estimating the accuracy. Accuracy refers to the ratio of correctly identified class to the total number of instances in the correct class. All the four models in the current project use the tenfold cross-validation process.

Confusion matrices are useful in finding the accuracy. The information provided in the matrix is recall, precision, f1-score and the support for each model. The four region of the matrix stores information about the predicted outcome and actual outcome. Figure 1 shows the confusion matrix with the false negative (FN), true positive (TP), true negative (TN) and false positive (FP) values. The actual explanation with respect to hit and non-hit is also provided.

The metrics of evaluation of the model like f1-score, accuracy, recall and support can be obtained from the confusion matrix.

		Predicted class	
		P	N
Actual Class	P	**True Positive(TP)** Songs that are actually hit and rightly predicted by the model.	**False Negative (FN)** Songs predicted as non-hit but are actually a hit.
	N	**False Positive (FP)** Songs that are actually not a hit but predicted as hit by the model.	**True Negative(TN)** Songs that are not a hit actually and are rightly predicted by the model.

Fig. 1 Confusion matrix

The **precision** of the model specifies how correctly it is able to not label a non-hit song as a hit song. It is the ratio of TP to (TP + FP) from values of confusion matrix.

Recall of the model specifies how correctly it is able to find all the hit songs. It is the ration of TP to (TP + FN).

Accuracy suggests what percent of the prediction carried out by the model were correct. It is given by the ration of (TP + TN) to (TP + TN + FP + FN). Accuracy is the most preferred metric to evaluate a model.

F1-score provides a balance between the precision and recall. F1 - score or measure = 2 × ((precision × recall)/(precision + recall)). It is the harmonic mean of precision and recall.

Area under curve is used to check how capable the model is, to distinguish between the classes.

4.5 Extension of Project for Related Application

Recognizing of face process involves creation of individual folder with their names. Within each person's folder, various subfolders are created one each for each facial emotion. The emotions that are currently recognized are surprise, fear, anger, happiness, sadness, disgust and neutral. For illustration purpose, Japanese Female Facial Expression (JAFFE) [30, 31] database is used in the implementation. During implementation, live feed of faces are carried out. But results in the paper are confined to only JAFFE database due to the restriction to reveal any body part. The result shows 96% accuracy on the still images. The actual implementation has people from other country originals as well. The accuracy of the result could reduce if the subject is on the move and not directly facing the camera. The face region of interest is obtained using the multi-task cascaded neural network (MTCNN) [32, 33]. Figure 2 shows the CNN model used in the implementation of the work.

The face recognition is carried out using the classifier and is carried out using Fisher discriminant analyzer [34]. Song pertaining to a particular region can be played on recognizing the face contour. Figure 3 shows the high level diagram representing the flow of events in the project. Classifying the song based on emotion is a part which

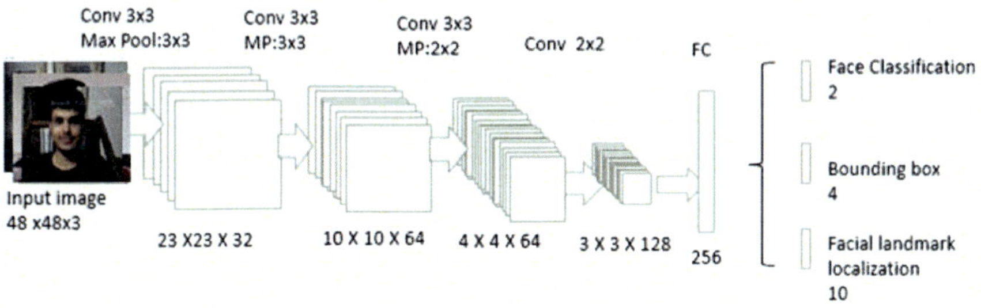

Fig. 2 Facial recognition using MTCNN[28]

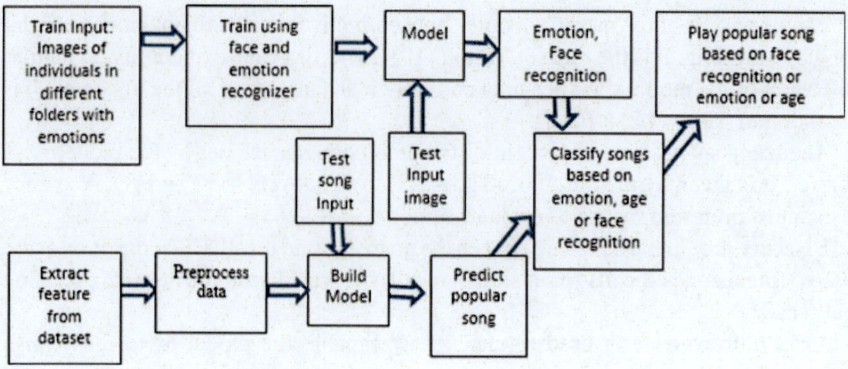

Fig. 3 Overall work flow of the project

is proposed and not yet completed. The emotion recognition is carried out using the trained model mini-Xception. [35]. Songs for different age group can be played as soon as face is recognized and age is computed. Age and gender can be computed using the CNN model [36]. It has three layers with 96 nodes in the first layer with kernel size 7, second layer with 256 nodes with kernel size 5 and third layer of 384 nodes with kernel size 3. The photos available in the dataset are 26580 with number of subjects as 2,284 annotated with age and gender. The age is grouped into groups of (0–2, 4–6, 8–13, 15–20, 25–32, 38–43, 48–53, 60-). The younger group has their facial features changing faster, so smaller size variation in age is used. The problem is changed from regression to classification by using this grouping.

5 Results and Discussion

A correlation between every feature with every other feature is analyzed in Fig. 4. Energy and loudness have a strong correlation. From the values that are visible, popularity is better correlated with loudness, followed by danceability, energy, valence, tempo and duration. Popularity is negatively correlated with to acoustics, speechiness, liveness and instrumentalnes.

The experiments were carried out for all the four models using the tenfold validation. The results of the prediction are shown in Table 1 in the form of a confusion matrix. The result shows that random forest with highest value of true positive and true negative among all other models is the best algorithm. The size of the dataset is 37236 songs.

Table 2 depicts the precision, recall, $f1$-score, accuracy and AUC statistic of the various models. Most promising among this is random forest with an accuracy of 89%.

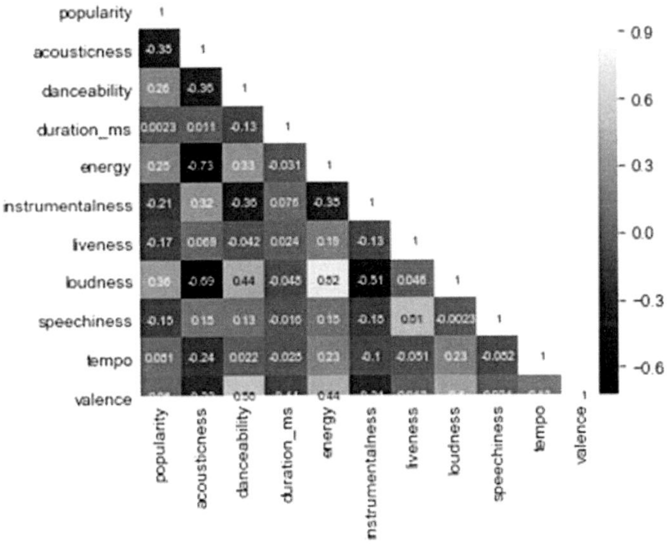

Fig. 4 Correlation between various features

Table 1 Confusion matrix using various models

Predicted class				
	Model		Hit	Non-hit
Actual class	Decision tree	Hit	23,869	4097
		Non-hit	2361	6909
	Logistic regression	Hit	27,349	617
		Non-hit	8495	775
	Naïve Bayes	Hit	14,363	13,603
		Non-hit	1997	7273
	Random forest	Hit	26,739	1227
		Non-hit	3015	6255
	K-nearest neighbor	Hit	23,842	4124
		Non-hit	4909	4361
	XGBoost	Hit	27,344	622
		Non-hit	8366	904

The data distribution is such that random forest provides highest accuracy. Decision tree also works on a similar concept, hence provides good accuracy and is comparable with random forest. Naïve Bayes works with the assumption that features are independent of each other. The popularity of music is dependent on many features in the song as seen from the correlation matrix. So, naïve Bayes cannot be used for this classification problem. It has the lowest accuracy. KNN presumes that the

Table 2 Classification report of various models

Model		Precision	Recall	$F1$-score	Accuracy	AUC
Decision tree	Hit	0.93	0.88	0.90	0.85	0.81
	Non-hit	0.62	0.74	0.68		
Logistic regression	Hit	0.79	0.99	0.88	0.79	0.51
	Non-hit	0.55	0.03	0.05		
Naïve Bayes	Hit	0.90	0.52	0.66	0.85	0.64
	Non-hit	0.30	0.78	0.43		
Random forest	Hit	0.91	0.97	0.94	0.91	0.81
	Non-hit	0.87	0.66	0.75		
KNN	Hit	0.85	0.88	0.87	0.79	0.65
	Non-hit	0.49	0.41	0.45		
XGBoost	Hit	0.79	0.99	0.88	0.79	0.51
	Non-hit	0.56	0.03	0.05		

neighboring data within the same cluster/class have similar characteristic. There are some features which are not correlated with each other, and hence, nearest neighbor do not provide as good result as the decision tree algorithm. Logistic regression is a simple classifier which presumes there is a linearity between dependent and independent variable which is not true in this example, and hence, it does not provide the required accuracy. Though XGBoost has provided the good accuracy, it has low false negatives.

6 Conclusion

The study showed that it is possible to predict hit songs with reasonable accuracy. Inclusion of metadata could enhance the accuracy of the results. Inclusion of additional features could provide improvement on the results. Some of them include the popularity of the artist, the lyrics, etc. The inclusion of age factor helps in predicting the song desired by that age group. Identification of face using the face recognition technique helps in playing the song of the popular artist. The songs can be played based on the emotional state of the individual. This can help the sick or emotionally depressed people to recover at a faster rate. A relative comparison of various protocols to compare the popularity of the result is carried out using decision tree, random forest, naïve Bayes, logistic regression, KNN and XGBoost. Random forest provides good results as compared to other algorithm in terms of accuracy. The distribution of data is better suited for decision rule and hence works well both with random forest and decision tree.

7 Future Scope

The study presumed some pre-assumed features for carrying out the project. Since a number of features are present in a song, the use of principal component analysis or enhanced feature extractor or selector could help in obtaining better results. Work can be substantiated with other database which includes several other features. The artist and lyrics also contribute to the popularity of the song. These features can be incorporated to make better predictions.

References

1. Francois, P.: In: Li, T Tzanetakis, G., Ogihara, M. (eds.) Music Data Mining, pp. 305–326. CRC Press (2011)
2. Braheny, J.: Craft and Business of Songwriting, 3rd edn. F & W Publications, New York (2007)
3. The EchoNest, Available: https://the.echonest.com.
4. Leikin, M.A.: How to Write a Hit Song, 5th edn. Hal Leonard (2008)
5. Perricone J.: Melody in Songwriting: Tools and Techniques for Writing Hit Songs. Berkley Press (2000)
6. Borg, N., Hokkanen, G.: What Makes For a Hit Pop Song? What Makes For a Pop Song? CS229 Report (2011)
7. Interiano, M.: Muscial trends and predictability of success in contemporary songs in and out of the top charts. J. Royal Soc **5**(5) (2018)
8. Schedl, M.: Deep learning in music recommendation system. Frontiers Frontiers Appl. Mathe. Statist. **5**(44), 1–9 (2019)
9. Yu, H., et al.: Popularity Prediction for Artists Based On User Songs Dataset, pp. 17–24. ICCAI, Indonesia, ACM (2019)
10. Tran, T., Seeney, R., Lee, K.: Adversarial Mahalanobis Distance-based Attentive Song Recommender for Automatic Playlist Continuation", SIGIR' 19, pp. 245–254. ACM, France (2019)
11. Martín-Gutiérrez, D., Hernández, P.G., Belmonte-Herández, A., Álvarez, G F.: A multimodal end-to-end deep learning architecture for music popularity prediction. IEEE Access **8**, 39361–39374 (2020)
12. Schedl, M.: Genre Differences of Song Lyrics and artist Wikis: An Analysis of Popularity, Length, Repetitiveness and Readability, pp. 3201–3207. IW3C2, ACM (2019)
13. Kaminskas, M., Ricci, F.: Contextual Music Information Retrieval and Recommendation: State of The Art and Challenges. Comput. Sci. Rev. 1–31 (2012)
14. Zangerle, E., et al.: Hit song prediction: leveraging low-high level audio features. IN: 20th ISMIR Conference, Netherlands (2019)
15. Fan, J., Casey, M.A.: Study of Chinese and UK hit songs prediction. In: Proceedings of International Symposium on Computer Music Multi-Disciplinary Research, pp. 640–652 (2013)
16. Lee, J., Lee, J.: Music popularity: metrics, characteristics, and audio-based prediction. IEEE Trans. Multimedia **20**(11), 3173–3182 (2018)
17. Yu, H., Li, Y., Zhang, S., Liang, C.: Popularity Prediction for Artists Based on User Songs Dataset, ICCAI '19, April 19–22, 2019, p. 1734. Indonesia ACM, Bali (2019)
18. Maria, E., Matthias, L., Sten, H.: Emotion recognition from physiological signal analysis: a review. Electron. Notes Theoret. Comput. Sci. **343**, 35–55 (2019)
19. Mahmood, A., Hussain, S., Iqbal, K., Elkilan, W.S. Recognition of facial expressions under varying conditions using dual-feature fusion. J. Comput. Intell. Image Process (2020)

20. Daumé, H.: III. Machine learning, A course in machine learning. CIML (2012)
21. Han, J., Kamber, M., Pei, J.: Data mining: concepts and techniques. Morgan Kaufmann, Elsevier (2016)
22. Peng, C., Kee, K.L., Ingersoll, G.M.: An introduction to logistic regression analysis and reporting. J. Educ. Res. **96**(01), 3–14 (2002)
23. Navada, A., Ansari, A.N., Patil, S., Sonkamble.: Overview of use of decision tree algorithms in machine learning. In: 2011 IEEE Control and System Graduate Research Colloquium, pp. 37–42. Shah Alam (2011). https://doi.org/10.1109/ICSGRC.2011.5991826
24. Rish, I.: An Empirical Study of the Naïve Bayes, pp. 41–46, Research Article (2001)
25. Biau, G.: Analysis of random forests model. J. Mach. Learn. Res. **13**, 1063–1095 (2012)
26. Zhang, Z.: Introduction to machine learning:k-nearest neighbor. Ann. Transl. Med. **4**(11), 1–7 (2016)
27. Chen, T., Guestrin, C.: XGBoost: A scalable tree boosting system. In: International Conference on Knowledge Discovery and Data Mining, pp. 785–794 (2016)
28. Spotify for Developers, https://developer.spotify.com/community/news/2016/03/29/audio-features-recommendations-user-taste
29. Herremans, D., David, M., Sörensen, K.: Dance hit song prediction. J. New Music Res. **43**(3), 291–302 (2014)
30. Lyons, M.J., Akamatsu, S., Kamachi, M., Gyoba, J.L Coding facial expressions with gabor wavelets. In: 3rd IEEE International Conference on Automatic Face and Gesture Recognition, pp. 200-205. The Japanese Female Facial expression (JAFFE) Dataset (1998). https://doi.org/10.5281/zenodo.3451524
31. Michael, L., Miyuki, K., Jiro, G.: Japanese Female Facial Expression (JAFFE) Database, figshare. J. Contrib https://doi.org/10.6084/m9.figshare.5245003.v2
32. Zhang, K., Zhang, Z., Li, Z., Qiao, Y.: Joint Face Detection and alignment using multi-task cascaded convolution networks. IEEE Signal Process. Lett. **23**(10), 1499–1503 (2016)
33. Aneesh, K.U., Hasan, A.Z., Pai, S.N.: Facial emotion recognition-a gift for the visionless. Int. J. Sci. Technol. Res. **9**(6), 228–233 (2020)
34. Wang, Z., Sun, X.: Manifold adaptive kernel local fisher discriminant analysis for face recognition. J. Multim. **7** (2012)
35. Arriaga, O., Paul, G.P, Matias, V.: Real-time Convolutional Neural Networks for Emotion and Gender Classification, pp 1–5 (2017)
36. Eidinger, E., Enbar, R., Hassner, T.: Age and Gender Estimation of Unfiltered Faces in IEEE. Trans. Inform. Forensi. Sec. **9**(12), 2170–2179 (2014)

Robot Motion Control Using OpenPose

C. Vighneshwara, M. Sowmya Kini, and K. Praveen Shenoy

1 Introduction

Human activity recognition model is continuously developing day by day. It helps to monitor human activities in various fields such as health care, surveillance, thief detection, anomalous behavior detections, yoga pose detection, and robot motion control [1]. But these fields usually require object interfaces like wearable sensors that should be in contact with the human body or video surveillance like image classification. Human pose data can be extracted by installing accelerometer and gyroscope sensors throughout the body. The detection of position is done by looking at the pattern over time [2].

For example, sleeping position can be detected by zero/negligible motion from the data that is collected by an accelerometer whereas in video surveillance, initially, it identifies the position and classifies the position under various groups and then identifies the position [3]. But the development of these systems is challenging. If wearable sensors are considered, then its installation and damage of sensors reduce the accuracy of a system [4]. Whereas in video surveillance, background noise, low light error, and subject occlusion cause difficulties in tracking a human [5]. Solving the above-mentioned problem leads to a new way of the human activity recognition approach [6]. The OpenPose algorithm performs real-time multi-people pose detection based on the data gathered by multiple cameras. This makes it easy by eliminating the usage of wearable sensors and also makes it convenient in any condition [7]. This open-source tool extracts the key-points in the form of a skeleton and it can handle conditions like low light distortion or background noise that makes the model robust in nature [8]. The work described in the following sections focuses on real-time human pose detection from a camera using OpenPose. Emphasis has been also given to elimination of wearable sensors to increase the scalability and

C. Vighneshwara · M. Sowmya Kini · K. Praveen Shenoy (✉)
Mangalore Institute of Technology and Engineering, Mangalore, India
e-mail: praveemks@mite.ac.in

S. K V and K. Guruprasad Rao (eds.), *Smart Sensors Measurements and Instrumentation*, Lecture Notes in Electrical Engineering 750,
https://doi.org/10.1007/978-981-16-0336-5_5

convenience in any conditions [9]. A robust approach by using computer vision and deep learning techniques has been proposed and tested for its efficacy.

2 Methodology

The methodology involves two main steps one of which comes under image processing and computer vision and the other comes under motion control. It involves human activity recognition and position transfer. Human activity recognition presents steps that help to estimate the activities. This section has been divided into subsections wherein Sect. 1 deals with the extraction of key-points, 2 deals with the continuous motion features, 3 deals with the gathering the datasets, 4 deals with classification, and 5 deals with activity recognition. Figure 1 shows the workflow of the proposed work.

2.1 Human Activity Recognition

2.1.1 Key—Extraction

The video inputs can be read with the help of the open-source library OpenCV [10]. Ten equal-spaced frames are generated per second. Each pixel value of an image ranges from 0 to 255 after which the pixel intensities from those images are converted from [0, 255] to [−1, 1] as shown in (1).

$$\text{Images} = \frac{\text{Images}}{255} - 1 \quad (1)$$

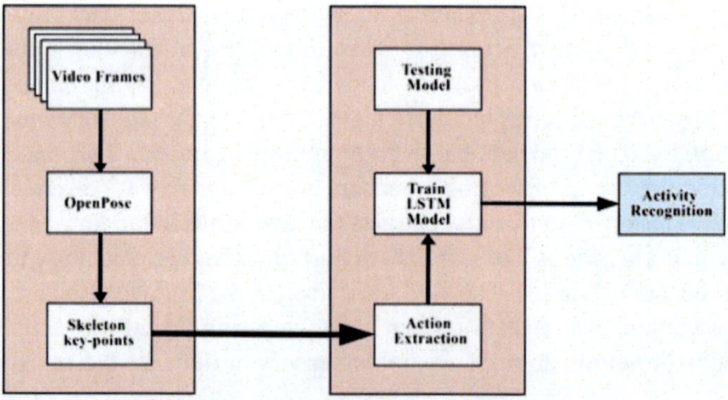

Fig. 1 Workflow

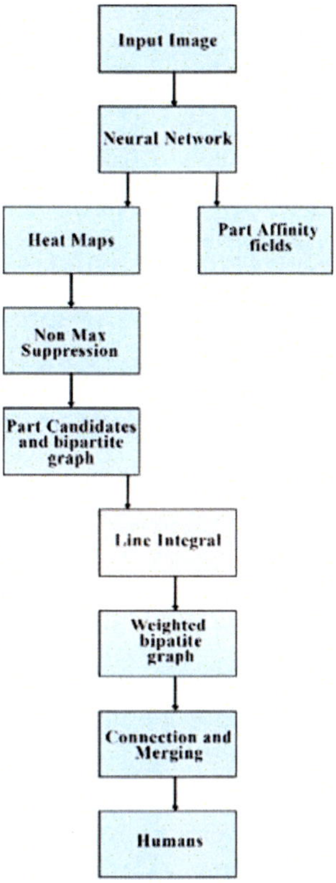

Fig. 2 OpenPose architecture

Extraction of anatomical key-points can be done with the help of OpenPose. OpenPose is an algorithm that performs real-time multi-people pose detection based on the data gathered by multiple cameras. The process flow of the same has been shown in Fig. 2.

It detects key-points, position, facial expressions, foot, and hand key-points. The pre-trained model gives 15 (without facial and foot key-points), 18 (without foot key-points), and 25 key descriptions to a human body which is depicted in Fig. 3. The OpenPose model is created with the help of COCO datasets to achieve 18 body key-point coordinates. The input images or video frames are read from a webcam or pre-recorded video and with the help of a neural network which returns 57 matrix tensors. The results of tensor mainly focus on two factors that is heat map and the part affinity fields.

Heat maps store the confidence of a pixel in the form of a matrix. The 18 heat maps are shown in the skeleton associated with each body part [11]. The position of each part is estimated with the help of these 18 matrices. Part affinity fields (PAF) consist of position and orientation of the body in the form of matrices. For all key-points, there are PAF, one being X-direction, and another one being Y-direction. To form a skeleton, it requires $38(18 + 18)$ such matrices.

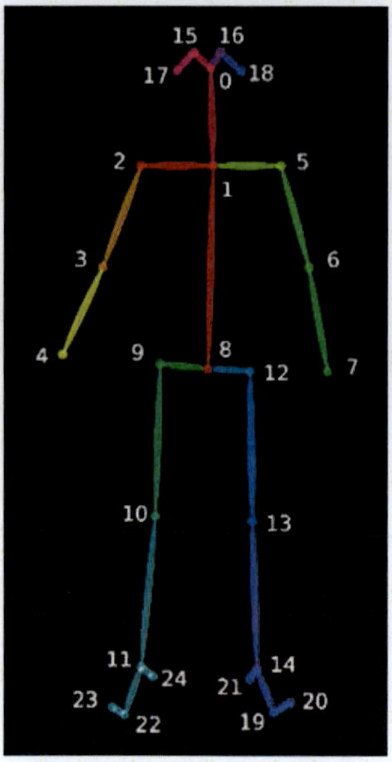

Fig. 3 OpenPose skeleton joints with 25 key-points

2.1.2 Motion Features

To determine the motion feature, we extract body joint data sets from several frames while performing as depicted in Fig. 4.

The formula is retrieved at joint numbers for magnitude M, at time t as given in Eq. (2) and the angle of the joint N for tth time is as given in Eq. (3).

$$M_t^N = \sqrt{\left(L_{x(t+1)} - L_{x(t)}\right)^2 + \left(L_{y(t+1)} - L_{y(t)}\right)^2} \tag{2}$$

$$A_{N(x,y)t} = \tan^{-1}\left(\frac{L_{y(t+1)} - L_{y(t)}}{L_{x(t+1)} - L_{x(t)}}\right) \tag{3}$$

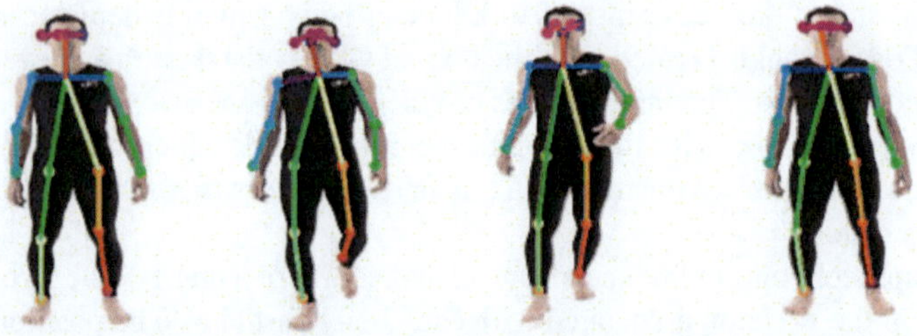

Fig. 4 Extracted body joints from several frames

In which L_x and L_y are distances in between two consecutive frames in the x-axis and y-axis [12]. For instance, here, several frames were extracted while the specimen is walking, where dotted colored lines indicate joint-to-joint motion feature connection as shown in Fig. 4. Hence, the motion feature MF at time t can be calculated as shown in Eq. (4).

$$T_t = \left[M_t^N, A_{N(x,y)t}\right] \tag{4}$$

2.1.3 Data Set Description

Datasets can be created with the help of a popular Python image processing tool called OpenCV which is an open-source library. With the help of OpenCV, the classification of human body position takes place and as a result, it forms a **.txt** file of all the frames that are captured through the webcam or a given video file [13]. The latter **.txt** file is converted into the required csvformat. The datasets are created from the above methodology for bending, stand-up, stand-down, jumping, sleeping, waving, punching, walking, push-up, and clapping positions. The dataset which was created has been shown in Fig. 5.

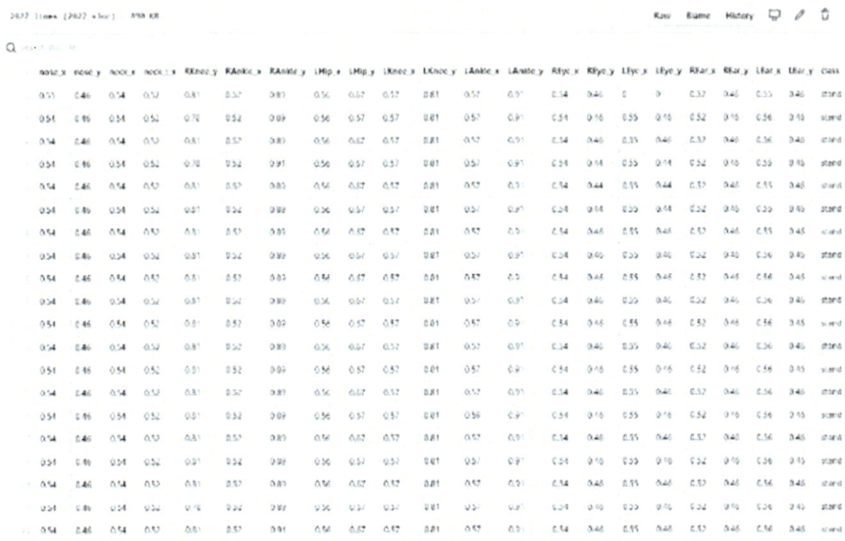

Fig. 5 Dataset in CSV format

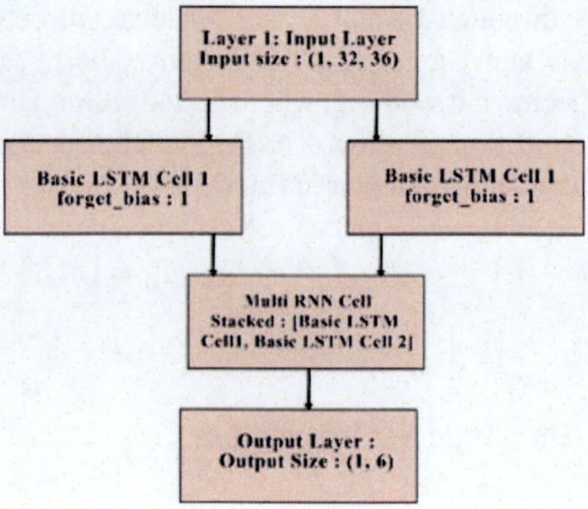

Fig. 6 LSTM architecture

2.1.4 Classification

Recurrent neural network (RNN) with long short-term memory cells (LSTM) is being utilized for the activity recognition because of its capability to sustain memory in several stepped iterations. Hence, this architecture is suitable for the human activity recognition (HAR) model [14]. This method also undergoes some limitations due to the longer time series. The exploding of gradients takes place during BPTT, also called back-propagation through a time, when the length of the sequence is too long. LSTMs solve the problem by adding multiple gates of learnable parameters as a weight to enable more control. This will help retain its memory of what it forgets in each iteration [15]. The LSTM architecture used to incorporate the classification has been provided in Fig. 6.

To classify, there are several other methods such as long short-term memory cells (LSTM). They are decision trees, support vector machine (SVM), and random forests.

2.1.5 Activity Recognition

After training the output of OpenPose, the human activity recognition model is done for the above ten activities. Eighteen anatomical key-points are extracted and are detected with the help of OpenPose. The output with 32 sequential frames is considered as an activity so while training a model, it should consist of 32 sequential frames and a label of activities associated with it. As a result, activities are recognized by LSTM recurrent neural network (Fig. 7).

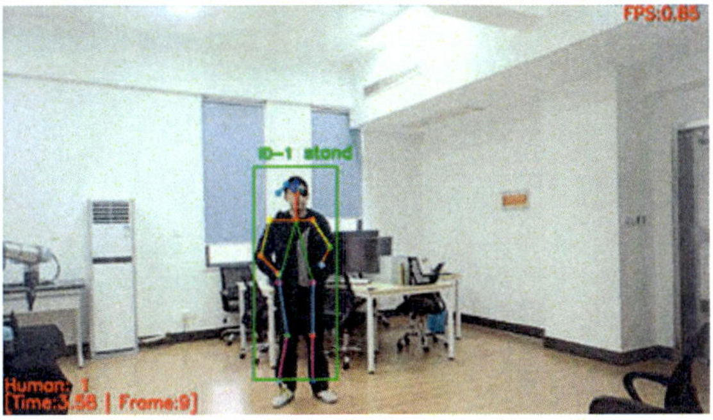

Fig. 7 Activity recognition

2.2 *Position Transfer*

The representation of OpenPose key-points are given below:

{0, "Nose"}, {1, "Neck"}, {2, "RShoulder"}, {3, "RElbow"}, {4, "RWrist"}, {5, "LShoulder"}, {6, "LElbow"}, {7, "LWrist"}, {8, "MidHip"}, {9, "RHip"}, {10, "RKnee"}, {11, "RAnkle"}, {12, "LHip"}, {13, "LKnee"}, {14, "LAnkle"}, {15, "REye"}, {16, "LEye"}, {17, "REar"}, {18, "LEar"},

{19, "LBigToe"}, {20, "LSmallToe"}, {21, "LHeel"},

{22, "RBigToe"}, {23, "RSmallToe"}, {24, "RHeel"} and {25, "Background"}.

In order to transfer the position into a robot, initially, the extraction of position data must be done with the help of OpenPose. To accomplish this, following step was followed.

The position of a person was saved in the Write_Iconflag using a custom JSON writer. So, it consists of people array of object, in which each object has:

- A Pose_keypoints_2d has the body part location and detects the confidence in the form $x_1, y_1, c_1, x_2, y_2, c_2$, and so on. The *x*- and *y*-coordinates can be normalized to [−1,1], [0,1], [0, Sourse_size], [0, output_size], etc.
- arrayshand_left_keypoints_2d, face_keypoints_2d, pose_keypoints_2d and hand_right_keypoints_2d.
- The analogous 3D arrays hand_right_keypoints_2d, hand_left_keypoints_2d, face_keypoints_3d, and body_keypoints_3d. In 3D, $x_1, y_1, c_1, x_2, y_2, c_2$ and so on are formatted instead of $x_1, y_1, c_1, x_2, y_2, c_2$, and so on.
- The parts identification before being assembled into people as part candidates.

Position from OpenPose output in a coordinate system can be transferred directly into the robot motion control directory. The real-time implementation of position transfer has been shown in Fig. 8.

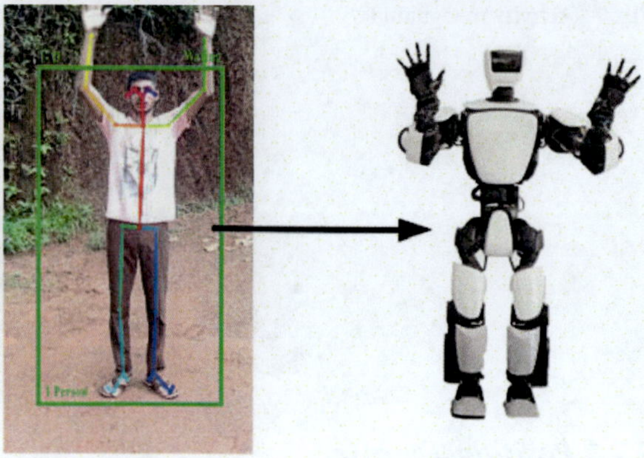

Fig. 8 Position transfer

3 Testing and Results

To train the LSTM model of above human activities, it requires sequential data inputs. It detects most of the activities that are trained with appreciable accuracy. Not only the above activities, it even identifies no activity when there is zero value in the array. A dataset is split into two. The 80% of the datasets are named as training datasets. Ten activity datasets were trained with a weight of $45((10 + 5) \times 3)$ filters with an accuracy of 93.25% and an average loss of 0.92%. The 20% of the datasets are named as testing datasets. While testing, the LSTM model achieved an accuracy of 84.25%. The working video of the testing can be seen here.

4 Conclusion

Image-based applications grow day by day, so it is a very important blend with the technology thereby reducing the usage of wearable sensors and the cost of its installation. The model not only can be used in humanoid robots it can also be used in industrial robots or any sort of robot by gesture control. In this work, the approach of human activity recognition in the existing system has been elaborated and explained. The work can be extended to suite any specific commercial oriented application which can potentially benefit the industrial sector.

Acknowledgements The authors would like to acknowledge Mangalore Institute of Technology and Engineering for the support extended toward the execution of this work.

References

1. Cao, Z., Hidalgo, G., Simon, T., Wei, S.E., Sheikh, Y.: OpenPose: realtime multi-person 2D pose estimation using Part Affinity Fields. arXiv (2018)
2. Qiao, S., Wang, Y., Li, J.: Real-time human gesture grading based on OpenPose. In: 2017 10th International Congress on Image and Signal Processing, BioMedical Engineering and Informatics (CISP-BMEI), pp. 1–6. IEEE (2017)
3. Fourie M., van der Haar D.: A Feature Importance Study in Ballet Pose Recognition with OpenPose. In: Degen H., Reinerman-Jones L. (eds.) Artificial Intelligence in HCI. HCII 2020. Lecture Notes in Computer Science, vol. 12217. Springer, Cham (2020)
4. Chen, W., Jiang, Z., Guo, H., Ni, X.: Fall Detection based on key points of human-skeleton using openpose. Symmetry 2020, **12**, 744. (2020)
5. Zhe, C., Hidalgo, G., Simon, T., Wei, S.-E., Sheikh, Y.: CMU-perceptual-computing-lab/openpose (2019)
6. Perera, A.G.; Law, Y.W.; Chahl, J. Drone-Action.: An outdoor recorded drone video dataset for action recognition. Drones, **3**, 82. (2019)
7. Rathod V.: Smart Surveillance and Real-Time Human Action Recognition Using OpenPose. In: Kumar A., Paprzycki M., Gunjan V. (eds.) ICDSMLA 2019. Lecture Notes in Electrical Engineering, vol. 601. Springer, Singapore (2020) .
8. Gupta V.: Multi-person pose estimation in OpenCV using OpenPose https://www.learnopencv.com/multi-person-pose-estimation-in-opencv-using-openpose/
9. Groos, D., Ramampiaro, H., Ihlen, E.A.: EfficientPose: Scalable single-person pose estimation. Appl Intell (2020)
10. CMU-Perceptual-Computing-Lab. https://github.com/CMU-Perceptual-Computing-Lab/openpose
11. Dutta, T.: Evaluation of the Kinect sensor for 3-D kinematic measurement in the workplace. Appl. Ergon. **43**(4), 645–649 (2012)
12. Cao, Z., Simon, T., Wei, S., Sheikh, Y.: Realtime multi-person 2D pose estimation using part affinity fields. In: CVPR 2017 (2017)
13. Wang, X.: An optimization method of basketball teaching and training system design based on motion capture technology. Revista De La Facultad De Ingenieria **32**, 498–504 (2017)
14. Noori F.M., Wallace B., Uddin M.Z., Torresen J.: A robust human activity recognition approach using OpenPose, motion features, and deep recurrent neural network. In: Felsberg M., Forssén PE., Sintorn IM., Unger J. (eds) Image Analysis. SCIA 2019. Lecture Notes in Computer Science, vo.l 11482. Springer, Cham (2019)
15. Chen, Y., Wang, Z., Peng, Y., Zhang, Z., Yu, G., Sun, J.: Cascaded pyramid network for multi-person pose estimation. In: CVPR, (2018)

Kinematic Analysis and Simulation of Robotic Manipulator Based on RoboAnalyzer

Amit Talli and Doddabasappa Marebal

1 Introduction

In the fast-growing world, demand for robotics is growing high in industries and is creating a vital need in various sectors. Robots are used in several industries such as assembly, welding and handling tasks, or even in medical operating rooms doing surgery [1]. The robots began working almost 60 years ago and were very successful in various operations,[2]. Learning the concepts about robotics is gaining momentum in the present era. The Denavit-Hartenberg parameters method is used to describe the architecture of an industrial robot.

There are two types of problem which comes under the robot kinematics: the forward or direct kinematics problem and inverse or reverse kinematics problem. Forward kinematics problem helps to determine the pose of the end-effector from the specified joint values. Inverse kinematics helps to determine the joint values from the specified pose of the end-effector.

The main focus of this paper is to derive the forward kinematic equations for a robot manipulator and to build a kinematic model by using RoboAnalyzer to validate the equations with the data provided by the software [3]. The information can be used for prototyping and optimizing the design parameters to improve the performance of the robot manipulator. An overview of the Denavit–Hartenberg (DH) parameter and conventions followed by forward kinematics is discussed in Sect. 2 and 3. Section 4 gives a brief overview of RoboAnalyzer version 7.5 interface, and other options are discussed. The mathematical model and verification of two-link planar robot, and Selective Compliance Assembly Robot Arm or Selective Compliance Articulated Robot Arm (SCARA) robot by using RoboAnalyzer are discussed in Sect. 4.1 and 4.2.

A. Talli (✉) · D. Marebal
Automation & Robotics, KLE Technological University, Hubballi 580031, India
e-mail: amit@kletech.ac.in

D. Marebal
e-mail: doddabasappa@kletech.ac.in

S. K V and K. Guruprasad Rao (eds.), *Smart Sensors Measurements and Instrumentation*, Lecture Notes in Electrical Engineering 750,
https://doi.org/10.1007/978-981-16-0336-5_6

Section 4.3 details, how inverse kinematics can be performed using RoboAnalyzer and finally followed by the conclusion in Sect. 5.

2 Denavit–Hartenberg Parameter

A serial robot arm is open-chain kinematics consisting of several rigid links connected by the joints. The joints allow relative motion between rigid links, and essentially, they are known as lower-order joints due to the area contact. The joints could be a revolute (R), which allows rotary motion and prismatic (P), which allows straight-line motion. The relation between the coordinate frame determines the location of the end-effector in three-dimensional space. For this, a coordinate frame, according to Denavit–Hartenberg parameter, is attached to the link of robot manipulator[2] 4 5. The systematic way of describing the Denavit–Hartenberg parameter has been presented in many research articles[6]. The four standard parameters are required to communicate the structure and orientation of links and joints, as illustrated in Fig. 1, and parameters are defined in Table 1.

The Denavit–Hartenberg consists of four standard parameters for defining the joints and rigid links. The joint angle and offset parameters are dedicated to defining the joint specification, such as joint angle rotation and joint offset. The link length and twist angle parameters are dedicated to defining the length and twist angle of the rigid link. The joint parameters are variable, and the link parameters remain constant. The axis of motion of joint rotation and joint offset occurs about and along the z-axis, whereas the axis direction of link length and twist angle occurs about and along the x-axis.

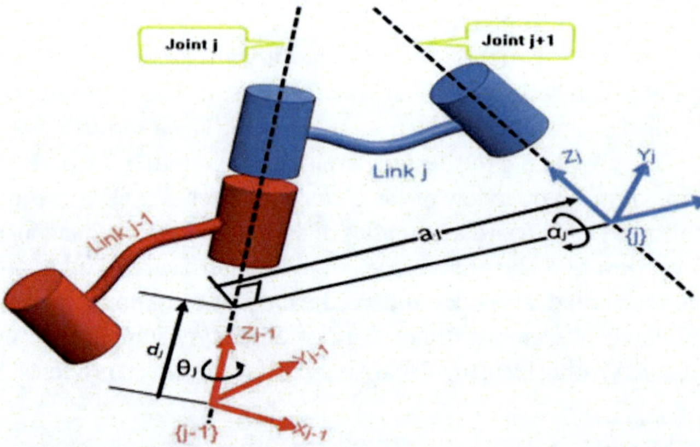

Fig. 1 Denavit–Hartenberg model [2]

Table 1 Denavit–Hartenberg link parameters [2]

Parameters	Symbol	description	Variable/constant
Joint angle	θ_j	The angle between the x_{j-1} and x_j axes about the z_{j-1} axis	Variable
Joint offset	d_j	The distance from the origin of the frame $\{j-1\}$ to the x_j axis along the z_{j-1} axis	Variable
Link length	a_j	The distance between the z_{j-1} and z_j axes along the x_j axis	Constant
Twist angle	α_j	The angle between the z_{j-1} and z_j axes about the x_j axis	Constant

3 Forward Kinematics

In forward or direct kinematics, the position and orientation of end-effector are determined by prescribing the joint variables. The same can be achieved by deriving homogeneous transformation matrix[7, 8]. The matrix is obtained by following the set of rules:

1 Assignment of coordinate frame.
2 Define Denavit–Hartenberg parameters.
3 Establish the chain of homogeneous transformation matrices to determine the position and orientation of end-effector as per the Denavit–Hartenberg parameters. The transformation from link coordinate frame $\{j-1\}$ to frame $\{j\}$ is defined in terms of elementary rotations and translations as:

$$^{j-1}T_j = T_{\text{rot}}(\theta_j, z) * T_{\text{trans}}(d_j, z) * T_{\text{trans}}(a_j, x) * T_{\text{rot}}(\alpha_j, x) \tag{1}$$

Which can be expressed in terms of homogeneous transformation matrix form:

$$^{j-1}T_j = \begin{bmatrix} \cos(\theta_j) & -\sin(\theta_j)\cos(\alpha_j) & \sin(\theta_j)\cos(\alpha_j) & a_j\cos(\theta_j) \\ \sin(\theta_j) & \cos(\theta_j)\cos(\alpha_j) & -\cos(\theta_j)\cos(\alpha_j) & a_j\sin(\theta_j) \\ 0 & \sin(\alpha_j) & \cos(\alpha_j) & d_j \\ 0 & 0 & 0 & 1 \end{bmatrix} \tag{2}$$

The position of the end-effector is represented by the fourth column of the matrix in (Eq. 2).

4 Overview of the RoboAnalyzer

RoboAnalyzer software tool is used for teaching and learning the concepts associated with robot kinematics and dynamics analysis[9]. This section presents a summary of the software tool.

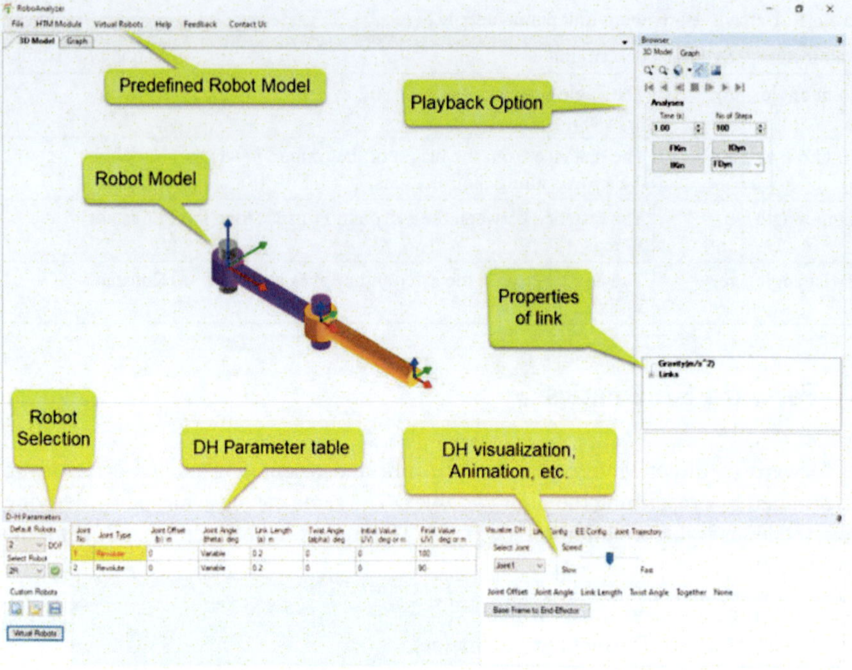

Fig. 2 RoboAnalyzer interface

As illustrated in Fig. 2, a user can select a robot manipulator available in the DOF list, or by selecting the type of joint, at the same time, the 3D model of the robot manipulator for the default values in the interface is displayed. The robot parameters can be altered in the Denavit–Hartenberg parameter section, such as link length, twist angle, and offset distance. The changes made in the user inputs are reflected as an updated 3D model of the robot manipulator. The simulation can be viewed after performing the forward or inverse kinematics operation. The path traced by the robot can be viewed by using the playback options provided in the GUI. The viewing options, such as zoom in, zoom out, isometric, and background colour, are provided. The tool can also generate graphical results for various analysis, and the data can be exported in excel format for further analysis [10].

4.1 Forward Kinematics of 2-Link Planar Robot

In this section, forward kinematic equations are derived and verified with RoboAnalyzer[11]. Figure 3 illustrates the two-link planar robot constructed in RoboAnalyzer by using Denavit–Hartenberg parameters, as shown in Table 2. The robot has two-link and two-revolute joints with one degree of freedom, which allows rotary motion

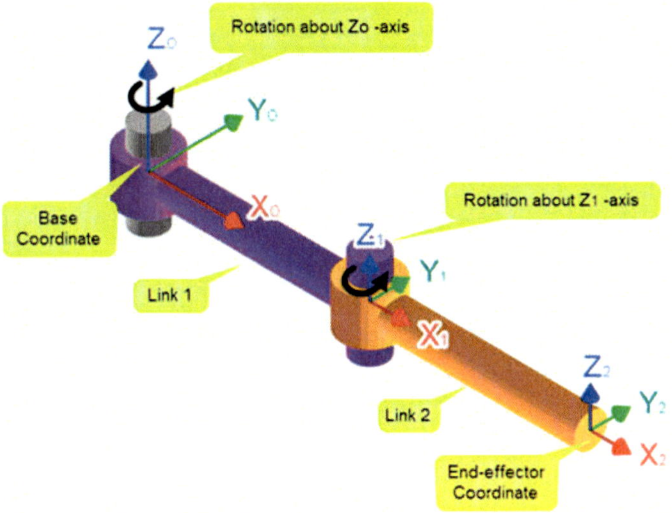

Fig. 3 Denavit–Hartenberg visualization

Table 2 Link parameters for two-link planar manipulator

Link, i	θ_i	d_i (m)	a_i (m)	α_i
1	θ_1, Variable	0	0.5	0
2	θ_2, Variable	0	0.3	0

about local z-axis[12]. The two-link planar robot has two degrees of freedom because the robot has two revolute joints and requires two motors to precisely position the rigid links with respect to the base frame. The motion of the two-link planar robot occurs in the two-dimensional plane, as shown in Fig. 3.

Table 2 shows the Denavit–Hartenberg parameters of two-link planar manipulator. According to (Eqs. 1–2), the homogeneous transformation for each link is derived:

$$^{b}T_{ee} = T_{\text{rot}}(\theta_1, z) * T_{\text{trans}}(0.5, x) * T_{\text{rot}}(\theta_2, z) * T_{\text{trans}}(0.3, x) \tag{3}$$

We will use the notation such as $s_i = \sin(\theta_i)$, $c_i = \cos\theta_i$, and $s_{ij} = \sin(\theta_i + \theta_j)$, for simplicity. After simplification of (Eq. 3), we get the homogeneous transformation matrix as shown below in (Eq. 4):

$$^{b}T_{ee} = \begin{bmatrix} c_{12} & -s_{12} & 0 & 0.3c_{12} + 0.5c_1 \\ s_{12} & c_{12} & 0 & 0.3s_{12} + 0.5s_1 \\ 0 & 0 & 1 & 0 \\ 0 & 0 & 0 & 1 \end{bmatrix} \tag{4}$$

Table 3 Two-Link planar robot position results

Analytical results in 'm'	RoboAnalyzer results in 'm'
$x = 0.48$ and $y = 0.54$	$x = 0.48511$ and $y = 0.54544$

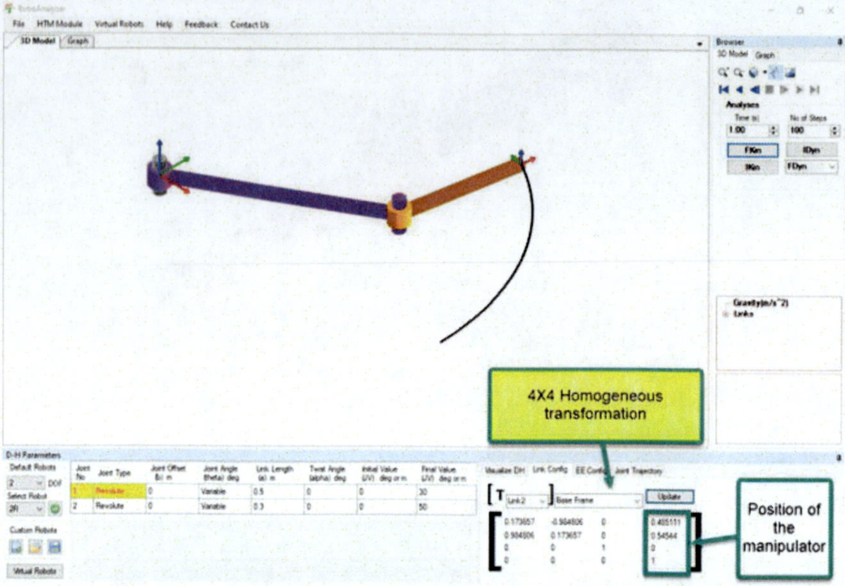

Fig. 4 Forward kinematics (FKin) module of RoboAnalyzer

The elements present in the last column represents the position of the end-effector in two-dimensional space. The results are analytically calculated and compared with the results displayed by the RoboAnalyzer. The joint angle θ_1 and θ_2 are varied by 30° and 50°. Table 3 shows the comparison between analytical results and RoboAnalyzer results.

The user can enter the values under the Denavit–Hartenberg parameter section, and forward kinematics can be performed by selecting the FKin option. Besides, the animation can be viewed by pressing the playback option. Figure 4 shows the position in the interface as well as the values in matrix form.

4.2 Forward Kinematics of SCARA Robot

In this section, the SCARA robot is used for the forward kinematics analysis. The SCARA robot is a four-degree of freedom industrial robot having three revolute joint and one prismatic joint. The SCARA robot finds its application in pick and place task. Figure 5 illustrates the SCARA robot constructed in RoboAnalyzer by

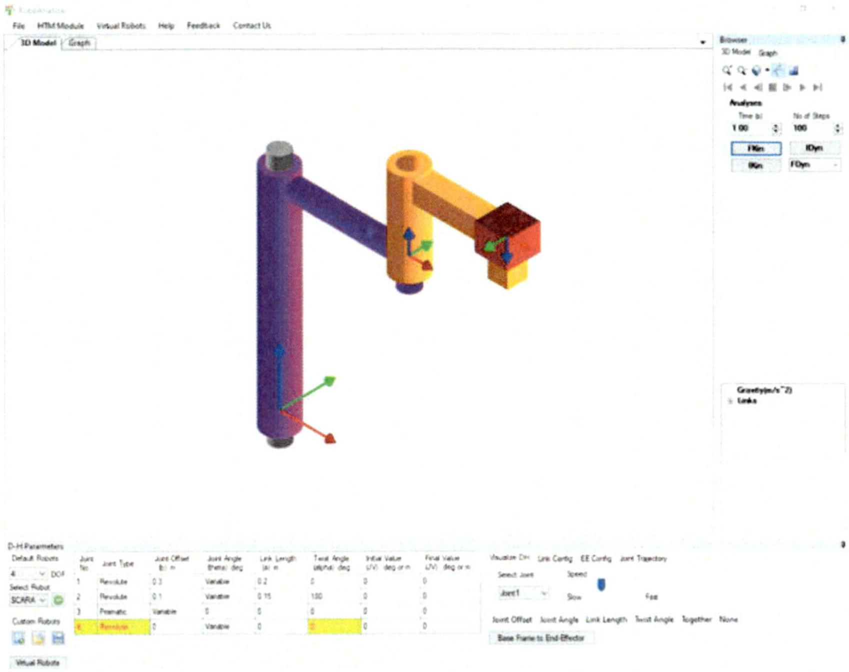

Fig. 5 SCARA robot in RoboAnalyzer

using Denavit–Hartenberg parameters[13] as shown in Table 4. The homogeneous transformation equations are derived in the following steps:

Figure 5 shows the isometric view of the SCARA robot in RoboAnalyzer.

Deriving the expressions for the components of the 4 × 4 homogeneous transformation matrix:

$$
\begin{aligned}
{}^{b}T_{ee} = {} & T_{\text{rot}}(\theta_1, z) * T_{\text{trans}}(0.3, z) * T_{\text{trans}}(0.2, x) * T_{\text{rot}}(\theta_2, z) * T_{\text{trans}}(0.1, z) \\
& * T_{\text{trans}}(0.15, x) * T_{\text{rot}}(pi, x) * T_{\text{trans}}(d_3, z) * T_{\text{rot}}(\theta_3, z)
\end{aligned}
\tag{5}
$$

Table 4 Denavit–Hartenberg parameters of SCARA robot

Link, i	θ_i	d_i (m)	a_i (m)	α_i
1	θ_1, Variable	0.3	0.2	0
2	θ_2, Variable	0.1	0.15	180°
3	0	d_3, Variable	0	0
4	θ_4, Variable	0	0	0

Table 5 SCARA robot position results

Analytical results in 'm'	RoboAnalyzer results in 'm'
$x = 0.199$, $y = 0.246$, and $z = 0.2$	$x = 0.199252$, $y = 0.247721$, and $z = 0.2$

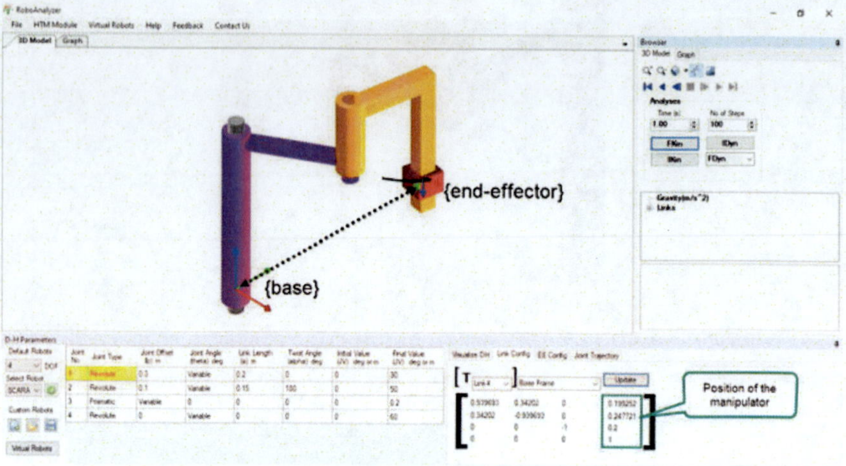

Fig. 6 Forward kinematics (FKin) of SCARA robot

$$
{}^{b}T_{ee} = \begin{bmatrix} c_{12}c_4 + s_{12}s_4 & s_{12}c_4 - c_{12}s_4 & 0 & 0.15c_{12} + 0.2c_1 \\ s_{12}c_4 - c_{12}s_4 & -c_{12}c_4 - s_{12}s_4 & 0 & 0.15s_{12} + 0.2s_1 \\ 0 & 0 & -1 & 0.4 - d_3 \\ 0 & 0 & 0 & 1 \end{bmatrix} \tag{6}
$$

The elements present in the last column represents the position of the end-effector in space. The results are analytically calculated and compared with the results displayed by the RoboAnalyzer. The joint angle θ_1, θ_2, and d_3 are varied by 30°, 50°, and 0.2 m. Table 5 shows a comparison between analytical results and RoboAnalyzer results.

Figure 6 shows the position of SCARA robot in the interface as well as the values in matrix form.

4.3 Inverse Kinematics

In inverse kinematics, the joint variables are determined by prescribing the position and orientation of end-effector. As compared to forward kinematics, the inverse kinematics is tedious and does not have a unique solution like forward kinematics [9]. Inverse kinematics requires rigorous mathematics and derivation. As the number

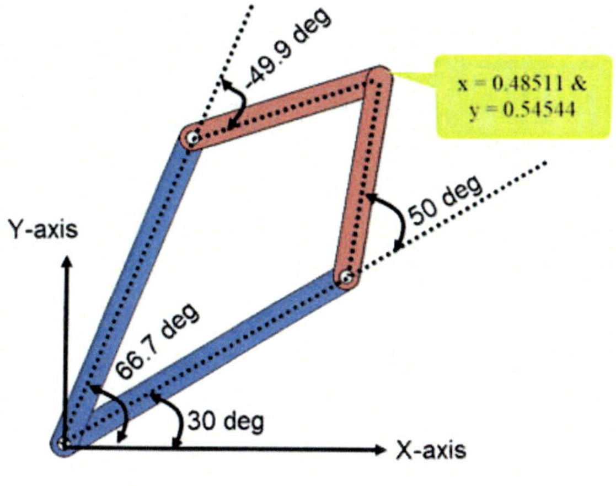

Fig. 7 Inverse kinematics of two-link planar robot

of degrees of freedom increases, the number of the possible solution to reach a particular location also increases. A two-degree of freedom planar robot arm, as shown in Fig. 7, is used to derive the inverse kinematics equations as an example and verified the results with the IKin module of the RoboAnalyzer. Assume that the end-effector needs to be at a particular position, say $[x,y]^T$. Inverse or reverse kinematics is used to determine the values of the joint variable for θ_1 and θ_2 such that the end-effector achieves the desired position.

Let the position of the robot be, $x = 0.48511$ m and $y = 0.54544$ m, as determined in the forward kinematics section. Now, using the (Eq. 4) or elements from the last column from matrix (Eq. 4):

$$x = 0.3c_{12} + 0.5c_1 \quad (7)$$

$$y = 0.3s_{12} + 0.5s_1 \quad (8)$$

Now, squaring and adding (Eqs. 7 and 8) yield :

$$\cos(\theta_2) = \left(x^2 + y^2 - 0.34\right)/0.3 \quad (9)$$

If the value of θ_2 lies between $-1 \leq \cos\theta_2 = (x^2 + y^2 - 0.34)/\,0.3 \leq 1$, then it has two possible solutions to the inverse kinematics, and the two possible solutions are:

$$\theta_2 = \cos^{-1}\left[\left(x^2 + y^2 - 0.34\right)/0.3\right] \quad (10)$$

$$\theta_2 = 2 * \text{pi} - \cos^{-1}\left[\left(x^2 + y^2 - 0.34\right)/0.3\right] \quad (11)$$

Table 6 Inverse kinematics results

Target position in 'm'	Possible solutions
$x = 0.48511$ and $y = 0.54544$	$\theta_1 = 30°$ and $\theta_2 = 50°$
	$\theta_1 = 66°$ and $\theta_2 =$ -50°

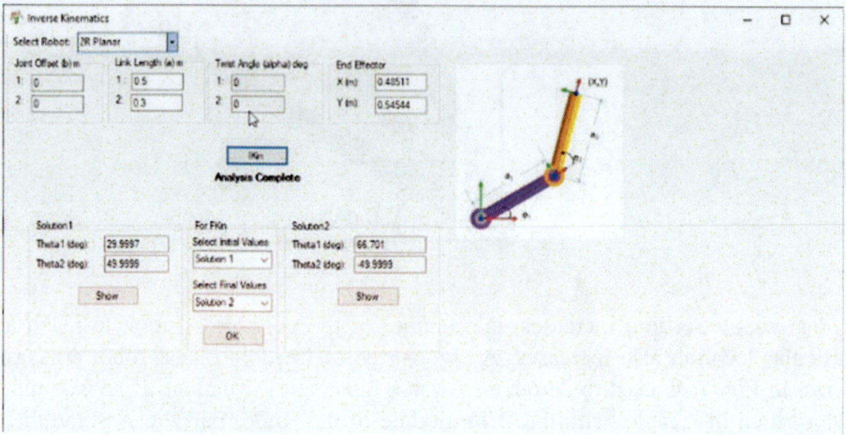

Fig. 8 Inverse kinematics interface

The results are analytically calculated and compared with the results given by the RoboAnalyzer. The joint angle θ_1 and θ_2 are calculated by keeping the position as, $x = 0.48511$ m and $y = 0.54544$ m. Table 6 shows a comparison between analytical results and RoboAnalyzer results.

In RoboAnalyzer, the inverse kinematics option can be accessed in the GUI by IKin option. The parameters such as link dimension and end-effector position should be entered. Based on the robot configuration, the RoboAnalyzer displays the possible solutions. The complexity increases as the number of links or degree of freedom increases. Figure 8 shows the solution for the two-DOF planar link robot with two solutions.

5 Conclusion

The kinematic analysis of planar and spatial robot manipulator is presented in this paper. The standard Denavit–Hartenberg parameters method was used to describe the geometry of the manipulator. The kinematics of the robotic manipulator is analysed and simulated in RoboAnalyzer. Results show that the developed kinematic model is capable of determining the position of end-effector for given input values of the joint variable. Also, this paper shows the process of RoboAnalyzer tool and its usage to study the kinematics aspects of the manipulator. The information provided by the software can be used for control system design and also for hardware-in-the-loop (HIL) for real-time simulation.

References

1. Özer, A., Arai, H.: Robotic metal spinning-experimental implementation using an industrial robot arm In:. Proceedings of IEEE International Conference on Robotics Automation, 140–145 (2009). https://doi.org/10.1109/ROBOT.2009.5152265
2. Corke, P.: Robot Arm Kinematics. In: Robotics, Vision and Control: Fundamental Algorithms In MATLAB® Second, Completely Revised, Extended And Updated Edition. pp. 193–228. Springer International Publishing, Cham (2017). https://doi.org/10.1007/978-3-319-54413-7_7
3. Gupta, V., Chittawadigi, R.G., Saha, S.K.: Robo analyzer: Robot visualization software for robot technicians. ACM Int. Conf. Proceeding Ser. Part **F1320**, 1–5 (2017). https://doi.org/10.1145/3132446.3134890
4. Flanders, M., Kavanagh, R.C.: Build-A-Robot: using virtual reality to visualize the Denavit-Hartenberg parameters. Comput. Appl. Eng. Educ. **23**, 846–853 (2015). https://doi.org/10.1002/cae.21656
5. Guo, S., Li, H.W., Ji, J.C., Ming, Z.F.: Kinematic analysis and simulation of a new-type robot with special structure. Adv. Manuf. **2**, 295–302 (2014). https://doi.org/10.1007/s40436-014-0094-x
6. Constantin, D., Lupoae, M., Baciu, C., Buliga, D.-I.: Forward kinematic analysis of an ındustrial robot. New Dev. Mech. Mech. Eng. 90–95 (2015)
7. Patwardhan, A., Prakash, A., Chittawadigi, R.G.: Kinematic analysis and development of simulation software for nex dexter robotic manipulator. Procedia Comput. Sci. **133**, 660–667 (2018). https://doi.org/10.1016/j.procs.2018.07.101
8. Ambuja Singh, Ratna Priya Kanchan, Saakshi Singh: A review paper on analysis and simulation of kinematics of 3r robot with the help of roboanalyzer. Int. J. Eng. Res. V5, 272–275 (2016). https://doi.org/10.17577/ijertv5is040535
9. Othayoth, R.S., Chittawadigi, R.G., Joshi, R.P., Saha, S.K.: Robot kinematics made easy using RoboAnalyzer software. Comput. Appl. Eng. Educ. **25**, 669–680 (2017). https://doi.org/10.1002/cae.21828
10. Sadanand, R.O.M., Chittawadigi, R.G., Saha, S.K.: Virtual robot simulation in RoboAnalyzer. In: 1st International and 16th National Conference on Machines and Mechanisms, iNaCoMM 2013. pp. 686–693 (2013)
11. Talli, A., V Meti, V.K.: Design, simulation, and analysis of a 6-axis robot using robot visualization software. IOP Conf. Ser. Mater. Sci. Eng. **872**, 012040 (2020). https://doi.org/10.1088/1757-899X/872/1/012040
12. Olwan, O., Matan, A., Abdullah, M., Abu-Khalaf, J.: The design and analysis of a six-degree of freedom robotic arm. In: ISMA 2015—10th International Symposium on Mechatronics and its Applications (2016). https://doi.org/10.1109/ISMA.2015.7373465
13. Rajeevlochana, C.G., Saha, S.K.: RoboAnalyzer : 3D Model Based Robotic. Learning. 3–13 (2011).

Internet of Things Enabled Electronic Tongue for Remote Monitoring of Water Quality

Pauroosh Kaushal and Rohini P. Mudhalwadkar

1 Introduction

Electronic tongue system is a sensor system which complements the gustatory sensing of humans. It consists of an array of cross-selective sensors to provide chemical information of sample under analysis. This information is used by pattern recognition algorithms to characterize and classify various foods and beverages. Voltammetry is an electrochemical method widely used to implement electronic tongue system in which an electrical potential is applied between several working electrodes and a reference electrode and the resulting current is measured. A plot of resulting current as a function of applied potential is called a voltammogram which indicates the total chemical content in the liquid sample. The measurement data can be seen as a fingerprint which is processed to provide qualitative and quantitative information of the sample like coffee [1], honey [2], fruit juices [3], and rice [4]. In the context of evaluating quality of water sample, many works have been reported. In [5], thick film electrodes based electronic tongue was used with fuzzy ARTMAP algorithm to classify drinking water samples. Pulsed voltammetric electronic tongue was employed for authentication of water sample. The acquired data is processed using slantlet-transform followed by different neural networks for classification purpose [6]. A voltammetric electronic tongue was used to detect the quality of water. The electronic tongue system acquires data which is processed using principal component analysis (PCA). Based on the PCA plot, different kinds of water samples could be discriminated [7]. An important aspect in designing a sensor system is the ease of interpretation of the system output. It means that a sensor system should provide the result of analysis of a given sample in the form of human perception which can be easily understood by an average user. In this direction, electronic tongues were reported which detect the changes in the quality of food samples and communicate to a human user in terms of linguistic symbols or warning signals [8, 9].

P. Kaushal (✉) · R. P. Mudhalwadkar
Department of Instrumentation and Control, College of Engineering Pune, Pune, India

S. K V and K. Guruprasad Rao (eds.), *Smart Sensors Measurements and Instrumentation*, Lecture Notes in Electrical Engineering 750,
https://doi.org/10.1007/978-981-16-0336-5_7

A remotely and continuously monitoring of electronic tongue are desirable which enable accessing the real-time analysis of sample from anywhere. The qualitative information can be monitored and it could lead to management of quality control of sample more efficiently. Furthermore, remote monitoring of electronic tongue brings automation to the analysis process and reduces effort for sample handling. The proposed work is focused on implementing Internet of things (IoT) platform on voltammetric electronic tongue for remote monitoring of water quality. It is studied that IoT platform has been implemented for various systems for remote and constant monitoring of processes. In [10], a mobile application is employed for real-time identification of plant deficiencies in oil palm plantations. The mobile data is send to Microsoft Azure for long-term storage and visualization of historic predictions. A complete IoT platform for structural health monitoring (SHM) system is developed to monitor structures and to determine the damage's location and size [11]. A smart energy monitoring system for home appliances is developed using IoT to monitor real-time energy consumption and to forcast electricity bill of the user [12]. It is studied that development of IoT platform for systems can lead to better management and improved efficiency. A potentiometric electronic tongue system with smartphone and cloud platform was developed for the classification of beverages and mineral water [13]. However, at present, commercial electronic tongue is not available which can be remotely accessed. This motivated use to carry out research for developing Internet of things platform for remote monitoring the output of electronic tongue and demonstrate it for quality assessment of water samples.

The present work is focused on implementing Internet of things platform on electronic tongue for remote monitoring the quality of water samples. The electronic tongue acquires data which characterizes the water sample and store it on a cloud platform. Principal component analysis and k-nearest classifiers are deployed on the cloud platform for feature extraction and quality assessment of water, respectively. Finally, the obtained qualitative information is sent to the user in terms of light signals using an android application.

2 System Design

2.1 Internet of Things Platform for Electronic Tongue

The Internet of things platform remotely acquires and analyzes the electronic tongue data to provide the information on water quality to the user. The electronic tongue is based on cyclic voltammetry in which a triangular electrical potential is applied between a noble working electrodes and a reference electrode and the current response between counter electrode and working electrode is converted to voltage and acquired. The block diagram of the system is shown in Fig. 1. The list of components required in the hardware design of electronic tongue and their specifications

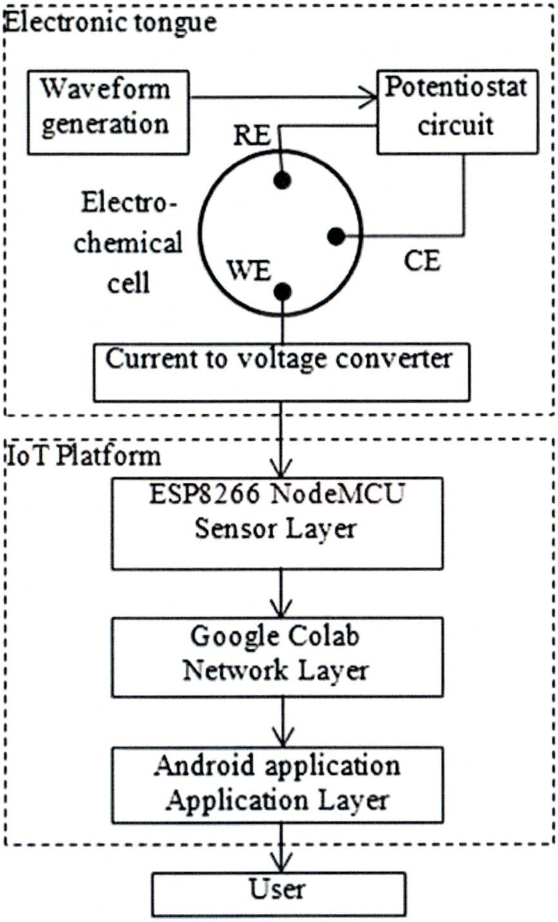

Fig. 1 Block diagram of IoT-enabled electronic tongue

is shown in Table 1. The measurements are performed by immersing working electrodes, counter electrode, and reference electrode in an electrochemical cell. The hardware setup of electronic tongue system is shown in Fig. 2. The system generates triangular waveform using ATmega328 microcontroller and digital-to-analog converter IC DAC0808 and applies it between working electrode and reference electrode. The potentiostat circuit uses a control amplifier to maintain the potential of working electrode constant relative to reference electrode by adjusting the current at counter electrode. This resulting current between counter electrode and working electrodes characterizes the chemical content of the sample which represents its quality. The obtained current is converted to voltage and fed to the IoT platform for further processing.

The architecture of IoT platform consists of three layers: Sensor layer, network layer, and application layer. In the sensor layer, the WiFi module of the development board ESP8266 NodeMCU acquires the output of electronic tongue at the

Table 1 Hardware components of electronic tongue with their specifications

Hardware component	Specifications
Metal wire-type working electrodes (iridium, rhodium, palladium, and platinum)	Diameter 0.5 mm and length 10 cm
Platinum counter electrode	Square shape electrode with area of 6.25 mm^2
Reference electrode	Length 12.5 cm
Electrochemical cell	Round shape with volume of 200 ml
LM324-based potentiostat	Quad operational amplifier, CMRR of 80 dB
ESP8266 NodeMCU development board	32-bit Xtensa LX106 microcontroller, ESP8266 WiFi SoC

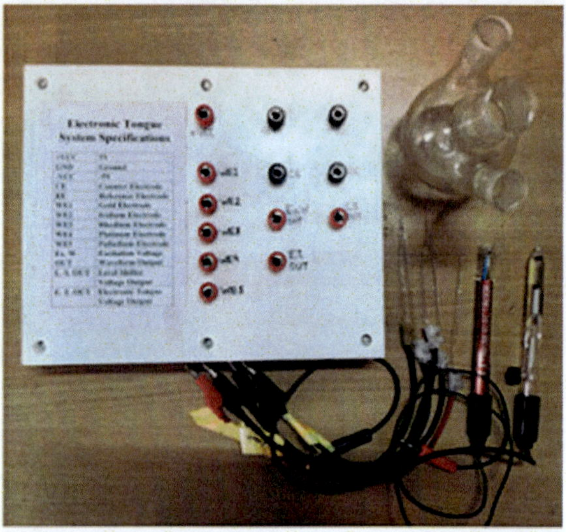

Fig. 2 Hardware setup of electronic tongue system

sampling rate of 10 Hz and stores it in Google Sheets which can be accessed by anyone anywhere. The ESP8266 NodeMCU development board is programmed using Arduino IDE (version 1.8.10). The network layer accesses the data on Google Sheet using Google Colab where the feature extraction and classification algorithm are deployed and returns the classification output to Google Sheet. In the Google Colab environment, features are extracted from data of electronic tongue system using principal component analysis (PCA). PCA decomposes data into principal components to account for maximum variance and minimum correlation in the data. The extracted features are fed to k-nearest neighbor (K-NN) classifier to predict the water quality in terms of numeric class label which is sent back to the Google Sheet. K-NN is a supervised classifier which stores the training features of a sample and classifies the features of unknown sample by computing the majority of class of k closest neighbor in the training features. The value of k is optimally selected by varying it from 4 to 10 and calculating the classification accuracy. The value of k at which the classification accuracy is highest is selected as the optimal value of k. Euclidean distance is used as distance metric and the performance is validated using ten-fold cross-validation method for the K-NN model. In the application layer, an android application on a smartphone accesses the numeric class label provided by the k-NN classifier to the Google Sheet and converts it into light signal. The light signal is an intuitive way to indicate the quality of water sample. The programing of android application is done on Android Studio (version 3.5).

2.2 *Water Samples*

Internet of things-enabled electronic tongue system is used for remote monitoring of quality of water and represents the result in terms of light signal. The water samples taken for measurement are of three different quality based on its consumption: Good, Bad, and Uncertain. Table 2 shows the source of water samples for three categories of water quality based on its consumption and light signal to indicate the quality. The color of light signal is selected in such a way that an average user can easily interpret the output of the system. As seen from the table, the green light indicates good quality of water and safe to consume, red light indicates bad quality of water and unsafe to consume, whereas yellow light indicates uncertain quality of water and may not be consumed. The yellow light signifies that further analysis of water sample must be conducted and informed decision regarding consumption of the sample should be taken.

3 Experimentation

The pH value of good, uncertain, and bad quality of water is measured at 30 °C and found to be at 6.8 ± 2.1, 7.25 ± 1.6, and 7.4 ± 2, respectively. A total of 90 measurements is recorded for water sample of each quality using four working electrodes. All the measurements is performed by generating triangular waveform with voltage amplitude ranging from -1.5 to $+ 1.5$ V at the scan rate of 80 mV/s

Table 2 Source of water samples for different quality and its indicating light signal

Source of water sample	Water quality	Light signal indication
Drinkable water from mineral bottle and boiled water	Good	Green
Mixture of drinkable and non-drinkable water	Uncertain	Yellow
Non-drinkable water from pond and lake	Bad	Red

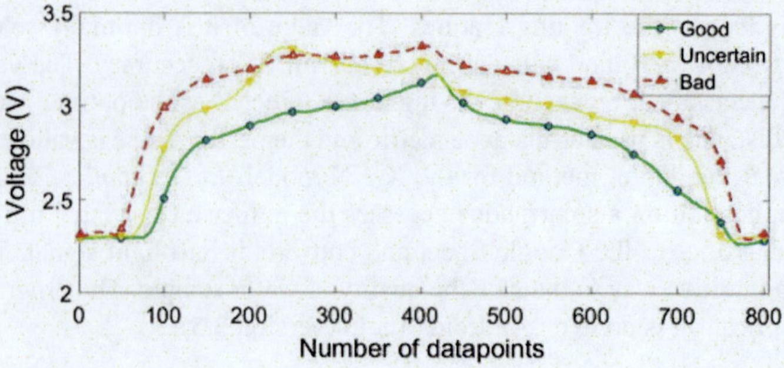

Fig. 3 Electronic tongue responses obtained on Google sheets for different quality of water

which is applied between working electrode and reference electrode. The resultant current is converted to voltage and stored to Google Sheets using ESP8266 NodeMCU development board. The acquired voltage contains 806 datapoints for a single measurement. The data collected for three types of water quality using palladium electrode is shown in Fig. 3. It can be seen from the figure that the acquired responses are different for different quality of water. The response around 420 datapoint has shown increase in amplitude as the water quality decreases for consumption. The transient response during increasing and decreasing triangular waveform is different for different water quality. Similar responses are obtained using other working electrodes. This data stored in Google Sheets is accessed by Google Colab which performs feature extraction and classifies the data in terms of water quality.

4 Result and Discussion

4.1 Data Analysis

Feature extraction and classification of water samples are implemented on Google Colab using principal component analysis (PCA) and k-nearest neighbor classifier, respectively. PCA is implemented using scikit-learn library to process data into two principal components which accounted for 96.52% of the total variance of the data. PCA plot for different types of water quality is shown in Fig. 4. The figure indicates that the water samples of good quality can be clearly discriminated from bad and uncertain quality. However, the clusters of bad and uncertain are close to each other and may not be discriminated easily.

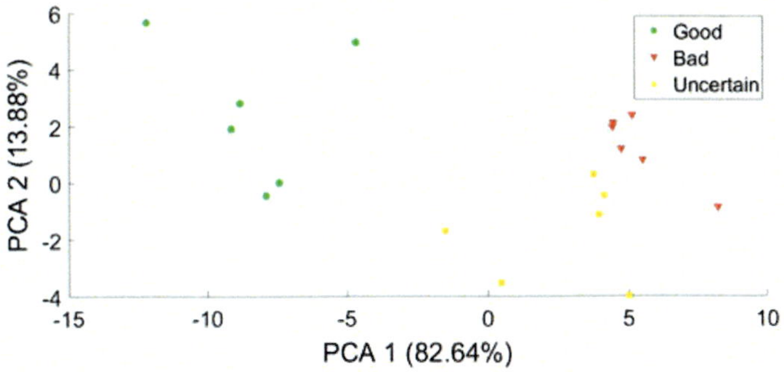

Fig. 4 PCA plot indicating discrimination between water samples of different quality

4.2 Assessment of Water Quality

The extracted two principal components are processed using k-NN classifier which outputs class label indicating water quality. The class 1, 2, and 3 are labeled to good, uncertain and bad quality of water. An optimal value of k is selected before deploying the model for remote monitoring of water quality. The value of k is varied from 4 to 10 and classification accuracy is computed. The highest classification accuracy is obtained at k equals 7 which is selected as optimal value. The model is used to classify unknown water sample and returns class label value to Google Sheet.

An android application reads the value of class label and indicates the water quality in terms on light signal on smartphone. The output of the application on screen is shown in Fig. 5. It can be seen that a user can easily interpret the result of the system. The accuracy of the system is evaluated by constructing confusion matrix as shown in Table 3. From the table, the classification accuracy is found to be 91.11%. It can be observed from the table that water samples of good quality are correctly classified whereas misclassification is seen between water samples of uncertain and bad quality.

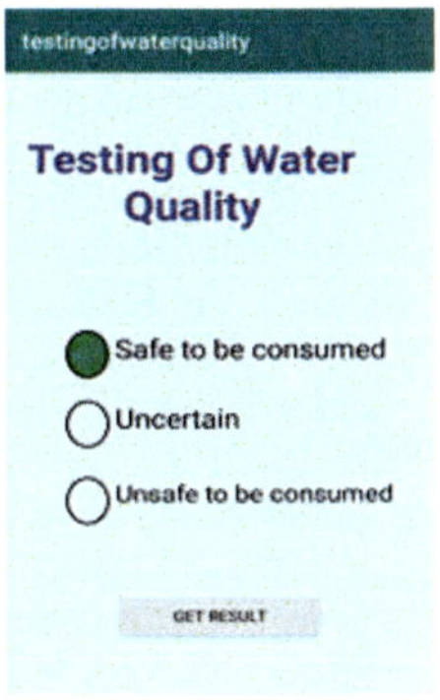

Fig. 5 Output of the system displayed on smartphone screen

Table 3 Confusion matrix to classify water samples based on its quality

	Green	Yellow	Red
Good quality of water	30	0	0
Uncertain quality of water	0	24	6
Bad quality of water	0	2	28

5 Conclusion

In this work, an Internet of things-enabled electronic tongue is developed for remote monitoring of water quality. The developed system classifies the quality of water sample as safe, uncertain, or unsafe to consume and indicate the result in terms of light signal on smartphone screen. The electronic tongue performs cyclic voltammetry using four working electrodes and acquires voltage data to characterize water sample. The data is remotely stored on Google Sheets and processed using principal component analysis and k-nearest neighbor classifier on Google Colab to classify water sample. The classification output of Google Colab is communicated to user using an android application. The system is able to classify water samples based on its quality with classification accuracy of 91.11%. Thus, Internet of things-enabled electronic tongue is implemented for remote analysis of water quality and can be used further for remote analysis of a variety of samples.

References

1. Morais, T.C.B., Rodrigues, D.R., Souto, U.T., et al.: A simple voltammetric electronic tongue for the analysis of coffee adulterations. Food Chem. **273**(1), 31–38 (2019)
2. Hassani, N.A., Tahri, K., Llobet, E., et al.: Emerging approach for analytical characterization and geographical classification of Moroccan and French honeys by means of a voltammetric electronic tongue. Food Chem. **243**, 36–42 (2018)
3. Yu, H., Zhang, Y., Zhao, J., et al.: Taste characteristics of Chinese bayberry juice characterized by sensory evaluation, chromatography analysis, and an electronic tongue. Int. J. Food Sci. Tech. **55**(5), 1624–1631 (2018)
4. Lu, L., Hu, X., Tian, S., et al.: Visualized attribute analysis approach for characterization and quantification of rice taste flavor using electronic tongue. Anal. Chim. Acta **919**, 11–19 (2016)
5. Garcia-Breijoa, E., Atkinsonb, J., Garriguesa, J., Gila, L., Ibañeza, J., Glancb, M., Olguina, C.: An electronic tongue for monitoring drinking waters using a fuzzy ARTMAP neural network implemented on a microcontroller. In: IEEE International Symposium on Industrial Electronics, Gdansk, pp. 1270–1275 (2011)
6. Kundu, P.K., Chatterjee, A., Panchariya, P.C.: Electronic tongue system for water sample authentication: a Slantlet-transform-based approach. IEEE Trans. Instrum. Meas. **60**(6), 1959–1966 (2011)
7. Lindquist, M., Wide, P.: New sensor system for drinking water quality. In: ISA/IEEE Sensors for Industry Conference, pp. 30–34 (2004)
8. Wide, P.: Human based sensing-sensor systems to complement human perception. Int. J. Smart Sens. Intell. Syst. **1**(1), 57–69 (2008)

9. Loutfi, A., Coradeschi, S., Wide, P.: An introductory concept to 'Human like' odor descriptions by grounding linguistic symbols. In: 19th IEEE Instrumentation and Measurement Technology Conference, pp. 55–60 (2002)
10. Culman, M. A., Gomez, J. A., Talavera, J., et al.: A novel application for identification of nutrient deficiencies in oil palm using the internet of things. In: 5th IEEE International Conference on Mobile Cloud Computing, Services, and Engineering, pp. 169–172 (2017)
11. Mahmud, M.A., Bates, K., Wood, T., Abdelgawad, A., Yelamarthi, K.: A complete internet of things (IoT) platform for structural health monitoring (SHM). In: IEEE 4th World Forum on Internet of Things, pp. 275–279 (2018)
12. Rashid, R.A., Chin, L., Sarijari, M.A., Sudirman, R., Ide, T.: Machine learning for smart energy monitoring of home appliances using IoT. In: 11th International Conference on Ubiquitous and Future Networks, pp. 66–71 (2019)
13. Iliev, B., Lindquist, M., Robertsson, L., Wide, P.: A fuzzy technique for food- and water quality assessment with an electronic tongue. Fuzzy Set Syst. **157**, 1155–1168 (2006)

Robust Control Using Modified Continuous Twisting Algorithm for Motion Control System

Akash Panhale and Shailaja Kurode

1 Introduction

Robust control is necessary for robust performance against parametric uncertainties, unmodeled dynamics, and perturbations[1]. Sliding mode control (SMC) is the most robust control technique. The SMC was evolved due to pioneering work of Emelyanov and Barbashin in variable structure control [2]. In SMC trajectories are steered to sliding surface. Once reached to this surface, the trajectory constrains to surface. This leads to key features of SMC, i.e., invariance to matched disturbance and order reduction.

In first-order sliding mode control (FOSMC), discontinuity term lies in first derivative of sliding variable. During sliding, the system dynamics are governed by sliding surface and finite time convergence is achieved due to discontinuity in control. However, the discontinuity leads to infinite switching frequency [3]. This results in undesirable chattering. Chattering limits practical applicability of SMC.

Chattering can be attenuated by approximation of discontinuous control by continuous term like sigmoid function, saturation function, etc. [4]. With this kind of control, chattering is alleviated but ideal sliding no longer takes place. Paradigm of higher-order sliding mode control (HOSMC) is evolved to deal with chattering [4].

HOSMC not only removes limitations of traditional approach but also improves tracking accuracy. In HOSMC, discontinuity term is shifted in higher derivative of control. Levant has proposed twisting algorithm (TA) that yields second-order SMC [5]. The developed control is applicable for any second-order system. For relative degree, one surface TA gives smooth control, but requires information of

A. Panhale (✉) · S. Kurode
Department of Electrical Engineering, College of Engineering, Pune, India
e-mail: panhaleaa18.elec@coep.ac.in

S. Kurode
e-mail: srk.elec@coep.ac.in

S. K V and K. Guruprasad Rao (eds.), *Smart Sensors Measurements and Instrumentation*, Lecture Notes in Electrical Engineering 750,
https://doi.org/10.1007/978-981-16-0336-5_8

sliding variable and its derivative. Derivative term in control increases noise [6], so to avoid this super-twisting algorithm (STA) was proposed [5]. STA has restriction that sliding surface should be of relative degree one. TA has advantage over STA that it can be used for developing the control with relative degree one as well as relative degree two surface. However, the TA-based control with relative degree two surface is discontinuous though it is continuous with relative degree one surface.

This issue in TA-based controller was researched by Moreno et al. He proposed continuous twisting algorithm (CTA) [7]. This algorithm gives continuous control for relative degree two surface but it requires four gains to be tuned. Driving motivation behind this research work is to investigate CTA which needs less no of gains with the enhanced degree of smoothness in control. The goal of this paper is to modify algorithm to yield more degree of smoothness in control with more accuracy and finite time convergence.

1.1 Outline of the Paper

This paper is organized as follows. Section 2 discusses convergence analysis of discontinuous differential equation. Modified continuous twisting algorithm is proposed in Sect. 3. Analysis of proposed algorithm in presence of perturbation is done in Sect. 4. Dynamics of typical motion control system are described in Sect. 5. Control development using MCTA is presented in Sect. 6. Subsequently simulation results are presented in Sect. 7. Finally conclusions are made in Sect. 8.

2 Convergence Analysis of Discontinuous Differential Equation

Generally controller design is done using mathematical model of system. Differential equations are used to represent system dynamics. Solution of these equations describes system behavior. It is found that several mechanical and electrical systems are represented by differential equations having discontinuous structure, e.g., mechanical contact, on–off control of relay, power converter, etc. To analyze asymptotic or finite time stability of any system, Lyapunov function is used. For any valid Lyapunov function V, if $\dot{V} < 0$, then the solution of the differential equation describing the system is asymptotically stable. If $\dot{V} \leq -\eta V^{\gamma}$, $\eta > 0$, $\gamma \in (0, 1)$, then solution of the differential equation is finite time stable. Bhat and Bernstein proposed another approach for analyzing finite time stability using homogeneity approach [8]. This approach states that a dynamic system with homogeneity of negative order and asymptotic stability is finite time stable.

Consider the following equation with discontinuous right hand side,

$$\dot{x} = -ksign(x), \qquad k > 0 \tag{1}$$

Vector field $f(x)$ is said to be homogeneous of degree r with respect to a dilation $d_l : x \rightarrow l^{m_1}x$, if and only if for $f(d_l x) = l^{r+m_1}f(x)$.

Consider function $f(x) = -ksign(x)$. To evaluate the order of homogeneity, consider dilation

$$d_l : x \rightarrow l^{m_1}x$$

Using $m_1 = 1, f(d_l x)$ becomes

$$\begin{aligned} f(d_l x) &= ksign(l^1 x), \\ &= l^0 f(x) = l^{-1+m1} f(x). \end{aligned} \tag{2}$$

Hence Eq. (1) is homogeneous of degree -1.

In order to analyze asymptotic convergence, consider valid lyapunov function

$$V(x) = \frac{x^2}{2}. \tag{3}$$

Differentiating (3) gives,

$$\dot{V}(x) = x\dot{x}. \tag{4}$$

Substituting $\dot{x}$ from (1) gives,

$$\begin{aligned} \dot{V}(x) &= x(-ksign(x)), \\ &= -k|x|. \end{aligned}$$

Hence, $\dot{V} < 0$ and ensures asymptotic stability. Negative homogeneity and asymptotic convergence guarantees finite time stability of (1).

3 Proposal of Modified Continuous Twisting Algorithm (MCTA)

MCTA is proposed as

$$\ddot{x} = -k_1|x|^{\frac{1}{p}} sign(x) - k_2|\dot{x}|^{\frac{1}{q}} sign(\dot{x}). \tag{5}$$

Defining $x = x_1, \dot{x} = x_2$, above Eq. (5) becomes

$$\begin{aligned} \dot{x}_1 &= x_2, \\ \dot{x}_2 &= -k_1|x_1|^{\frac{1}{p}} sign(x_1) - k_2|x_2|^{\frac{1}{q}} sign(x_2). \end{aligned} \tag{6}$$

where x_1, x_2 are states and constant $p > 0, q > 0, k_1 > 0, k_2 > 0$. Finite time convergence of this differential equation can be analyzed by using homogeneity and asymptotic convergence analysis.

3.1 Homogeneity Analysis

The homogeneity of a function is defined as the invariance of the differential inclusion with respect to time coordinate transformation.

Defining $f_1(\mathbf{x}) = x_2$ and $f_2(\mathbf{x}) = -k_1|x_1|^{\frac{1}{p}} sign(x_1) - k_2|x_2|^{\frac{1}{q}} sign(x_2)$ with $\mathbf{x} = [x_1\ x_2]^T$ Eq. (6) becomes

$$\dot{\mathbf{x}} = \begin{bmatrix} f_1(\mathbf{x}) \\ f_2(\mathbf{x}) \end{bmatrix}$$

Vector field $f_i(x)(i = 1, 2)$ is said to be homogeneous of degree r with respect to a dilation $d_l : (x_1, x_2) \rightarrow (l^{m_1}x_1, l^{m_2}x_2)$ if and only if for each i, the ith component $f_i(.)$ is homogeneous with degree $r + m_i$.

Using dilation $d_l : (x_1, x_2) \rightarrow (l^{m_1}x_1, l^{m_2}x_2)$ with weights $(m_1, m_2) = (p, q)$

$$\begin{aligned} f_1(d_l\mathbf{x}) &= l^q x_2, \\ f_2(d_l\mathbf{x}) &= -k_1|l^p x_1|^{\frac{1}{p}} sign(l^p x_1) - k_2|l^q x_2|^{\frac{1}{q}} sign(l^q x_2). \end{aligned}$$

simplifying further,

$$\left.\begin{aligned} f_1(d_l\mathbf{x}) &= l^{(q-p)+p} f_1(\mathbf{x}) = l^{(q-p)+m_1} f_1(\mathbf{x}) \\ f_2(d_l\mathbf{x}) &= l^1 f_2(\mathbf{x}) = l^{(1-q)+q} f_2(\mathbf{x}) = l^{(1-q)+m_2} f_2(\mathbf{x}) \end{aligned}\right\} \tag{7}$$

$f_i(d_l\mathbf{x})$ have negative homogeneity

$$if \qquad p > q. \tag{8}$$

Further if $(q - p) = (1 - q)$, both $f_i(d_l\mathbf{x})$ will have same negative homogeneity i.e.

$$\begin{aligned} 2q - p &= 1 \\ q &= \frac{p + 1}{2} \end{aligned} \tag{9}$$

Then there are many possible combination of p and q which satisfy the above, e.g., (p, q) chosen as $(3, 2)$, $(\frac{3}{2}, \frac{5}{4})$, $(2, \frac{3}{2})$, etc.

3.2 Asymptotic Convergence

For convergence analysis, consider a Lyapunov function

$$v(\mathbf{x}) = \frac{k_1}{\frac{p+1}{p}} |x_1|^{\frac{p+1}{p}} + \frac{x_2^2}{2}. \tag{10}$$

The derivative is obtained as,

$$\dot{v}(\mathbf{x}) = \frac{\frac{p+1}{p} k_1}{\frac{p+1}{p}} |x_1|^{\frac{1}{p}} \frac{\partial |x_1|}{\partial t} + x_2 \dot{x}_2,$$
$$\dot{v}(\mathbf{x}) = k_1 |x_1|^{\frac{1}{p}} \dot{x}_1 sign(x_1) + x_2 \dot{x}_2.$$

Substituting $\dot{x}_1$ and $\dot{x}_2$ from (6) gives,

$$\begin{aligned}\dot{v}(\mathbf{x}) =& k_1 |x_1|^{\frac{1}{p}} x_2 sign(x_1) \\ &+ x_2(-k_1 |x_1|^{\frac{1}{p}} sign(x_1) - k_2 |x_2|^{\frac{1}{q}} sign(x_2)), \\ \dot{v}(\mathbf{x}) =& - k_2 x_2 |x_2|^{\frac{1}{q}} sign(x_2),\end{aligned}$$

$$\dot{v}(\mathbf{x}) = -k_2 |x_2|^{\frac{1+q}{q}}. \tag{11}$$

It implies that $\dot{v}(\mathbf{x}) < 0$. Hence, (11) ensures asymptotic stability in $\mathbf{x}$. Negative homogeneity and asymptotic convergence of (5) imply finite time convergence of x and $\dot{x}$.

4 Modified Continuous Twisting Algorithm with Perturbations

MCTA in presence of disturbance becomes,

$$\ddot{x} = -k_1 |x|^{\frac{1}{p}} sign(x) - k_2 |\dot{x}|^{\frac{1}{q}} sign(\dot{x}) + d. \tag{12}$$

d is a uncertain, unknown but bounded function. It is assumed that $|\dot{d}| \leq L$.

Defining $x = x_1, \dot{x} = x_2,\ d = x_3$, above Eq. (12) becomes,

$$\begin{aligned}\dot{x}_1 &= x_2 \\ \dot{x}_2 &= -k_1 |x_1|^{\frac{1}{p}} sign(x_1) - k_2 |x_2|^{\frac{1}{q}} sign(x_2) + x_3 \\ \dot{x}_3 &= \dot{d}\end{aligned} \tag{13}$$

4.1 Homogeneity Analysis

Defining $f_1(\mathbf{x}) = x_2$, $f_2(\mathbf{x}) = -k_1|x_1|^{\frac{1}{p}} sign(x_1) - k_2|x_2|^{\frac{1}{q}} sign(x_2) + x_3$ and $f_3(\mathbf{x}) = \dot{d}$ with $\mathbf{x} = [x_1\, x_2\, x_3]^T$, above equation becomes

$$\dot{\mathbf{x}} = \begin{bmatrix} f_1(\mathbf{x}) \\ f_2(\mathbf{x}) \\ f_3(\mathbf{x}) \end{bmatrix}$$

Using dilation $d_l : (x_1, x_2, x_3) \to (l^{m_1}x_1, l^{m_2}x_2, l^{m_3}x_3)$ with weights $(m_1, m_2, m_3) = (p, q, 1)$

$$\begin{aligned} f_1(d_l\mathbf{x}) &= l^q x_2 \\ f_2(d_l\mathbf{x}) &= -k_1|l^p x_1|^{\frac{1}{p}} sign(l^p x_1) - k_2|l^q x_2|^{\frac{1}{q}} sign(l^q x_2) + l x_3 \\ f_3(d_l\mathbf{x}) &= l^0 \dot{d} \end{aligned}$$

simplifying further,

$$\left.\begin{aligned} f_1(d_l\mathbf{x}) &= l^{(q-p)+p} f_1(\mathbf{x}) = l^{(q-p)+m_1} f_1(\mathbf{x}) \\ f_2(d_l\mathbf{x}) &= l^1 f_2(\mathbf{x}) = l^{(1-q)+q} f_2(\mathbf{x}) = l^{(1-q)+m_2} f_2(\mathbf{x}) \\ f_3(d_l\mathbf{x}) &= l^{-1+m_3} f_3(\mathbf{x}) \end{aligned}\right\} \tag{14}$$

If conditions (8) and (9) satisfied, then from (14) $f(d_l\mathbf{x}) = l^{-1} f(\mathbf{x})$, hence $f(\mathbf{x})$ is homogeneous with degree -1.

4.2 Asymptotic Convergence

For convergence analysis, consider a Lyapunov function

$$v(\mathbf{x}) = \frac{k_1}{\frac{p+1}{p}} |x_1|^{\frac{p+1}{p}} + \frac{x_2^2}{2} + |x_3|. \tag{15}$$

The derivative is obtained as,

$$\begin{aligned} \dot{v}(\mathbf{x}) &= \frac{\frac{p+1}{p} k_1}{\frac{p+1}{p}} |x_1|^{\frac{1}{p}} \frac{\partial |x_1|}{\partial t} + x_2\dot{x}_2 + \dot{x}_3 sign(x_3), \\ \dot{v}(\mathbf{x}) &= k_1|x_1|^{\frac{1}{p}} \dot{x}_1 sign(x_1) + x_2\dot{x}_2 + \dot{x}_3 sign(x_3), \end{aligned}$$

substituting $\dot{x}_1$, $\dot{x}_2$ and $\dot{x}_3$ from (13) gives,

$$\begin{aligned}\dot{v}(\mathbf{x}) =& k_1|x_1|^{\frac{1}{p}}x_2 sign(x_1)\\ &+x_2(-k_1|x_1|^{\frac{1}{p}}sign(x_1)-k_2|x_2|^{\frac{1}{q}}sign(x_2)+x_3)\\ &+\dot{x}_3 sign(x_3),\\ \dot{v}(\mathbf{x}) =& -k_2x_2|x_2|^{\frac{1}{q}}sign(x_2)+x_2x_3+\dot{x}_3 sign(x_3).\end{aligned}$$

$$\dot{v}(\mathbf{x}) = -(k_2|x_2|^{\frac{1+q}{q}} - dx_2 - \dot{d}sign(x_3)). \tag{16}$$

If $k_2 > |d| + |\dot{d}|$ then it implies that $\dot{v}(\mathbf{x}) < 0$. Hence, (16) ensures asymptotic stability in $\mathbf{x}$. Negative homogeneity and asymptotic convergence of (12) imply finite time convergence of x and $\dot{x}$ in the presence of matched disturbance. Proposed MCTA is used to develop control in the next Section.

5 Motion Control Using Proposed MCTA

Typical motion control system consists of a load driven by motor. The dynamic equation of a typical current controlled motion control system is as follows:

$$\begin{aligned}\dot{x}_1 &= x_2\\ \dot{x}_2 &= 201.1u + d\end{aligned} \tag{17}$$

where x_1 is angular position of a motor, x_2 is angular velocity, and d is lumped disturbance that includes neglected dynamics such as friction and gear, parametric uncertainties, and external disturbances. It is required to develop control using MCTA.

6 Control Development

The proposed MCTA is used for control synthesis. The system (17) can be represented as

$$\left.\begin{aligned}\dot{\mathbf{x}} &= A\mathbf{x} + \mathbf{b}u + \mathbf{b}d\\ y &= \mathbf{c}\mathbf{x}\end{aligned}\right\} \tag{18}$$

where

$$\mathbf{x} = \begin{bmatrix} x_1 \\ x_2 \end{bmatrix}, A = \begin{bmatrix} 0 & 1 \\ 0 & 0 \end{bmatrix}$$
$$\mathbf{b} = \begin{bmatrix} 0 & 201.1 \end{bmatrix}^T, \mathbf{c} = \begin{bmatrix} 1 & 0 \end{bmatrix}$$

1. **Consideration of relative degree one surface**
   Consider linear sliding surface

$$\sigma = \mathbf{c}^T \mathbf{e}_{\mathbf{x}} = c_1 \mathbf{e}_{x1} + c_2 \mathbf{e}_{x2} \tag{19}$$

   where, $\mathbf{e}_{\mathbf{x}} = \mathbf{x}_d - \mathbf{x}$, $\mathbf{x}_d$ is desired vector.
   $\mathbf{c}^T$ is a surface matrix with $c_1, c_2 > 0$. This is relative degree one surface.
   Differentiating above equation (19) gives,

$$\begin{aligned}\dot{\sigma} &= \mathbf{c}^T \dot{\mathbf{e}}_{\mathbf{x}},\\ &= \mathbf{c}^T \frac{d}{dt}[\mathbf{x}_d - \mathbf{x}].\end{aligned}$$

   For constant desired value, its derivative is zero. Using nominal part of (18), $\dot{\sigma}$ becomes

$$\dot{\sigma} = -\mathbf{c}^T [A\mathbf{x} + \mathbf{b}u]. \tag{20}$$

   Differentiating above to get $\ddot{\sigma}$,

$$\ddot{\sigma} = -\mathbf{c}^T [A\dot{\mathbf{x}} + \mathbf{b}\dot{u}]. \tag{21}$$

   The MCTA in σ is

$$\ddot{\sigma} = -k_1 |\sigma|^{\frac{1}{p}} sign(\sigma) - k_2 |\dot{\sigma}|^{\frac{1}{q}} sign(\dot{\sigma}). \tag{22}$$

   From (21) and (22) synthesized control is

$$\begin{aligned}\dot{u} = (\mathbf{c}^T b)^{-1}[-\mathbf{c}^T A\dot{\mathbf{x}} + k_1 |\sigma|^{\frac{1}{3}} sign(\sigma)\\ + k_2 |\dot{\sigma}|^{\frac{1}{2}} sign(\dot{\sigma})]\end{aligned} \tag{23}$$

   Integrating equation gives,

$$\begin{aligned}u = (\mathbf{c}^T b)^{-1}[-\mathbf{c}^T A\mathbf{x} + \int k_1 |\sigma|^{\frac{1}{3}} sign(\sigma)\\ + \int k_2 |\dot{\sigma}|^{\frac{1}{2}} sign(\dot{\sigma})] \equiv u_1.\end{aligned} \tag{24}$$

   Thus, the control is smooth and continuous.
2. **Consideration of relative degree two surface**
   The following surface is of relative degree two

$$\sigma = c_1 e_{x1} \tag{25}$$

where $c_1 > 0$, $e_{x1} = x_{1d} - x_1$, x_{1d} is desired input.

Differentiating above (25) gives,

$$\dot{\sigma} = -c_1(\dot{x}_{1d} - \dot{x}_1),$$

for constant desired value $\dot{x}_{1d} = 0$, therefore

$$\begin{aligned}\dot{\sigma} &= -c_1\dot{x}_1,\\ &= -c_1 x_2.\end{aligned}$$

Differentiating above to get $\ddot{\sigma}$

$$\ddot{\sigma} = -c_1\dot{x}_2,$$

substituting $\dot{x}_2$ from (18) without consideration of disturbance gives,

$$\ddot{\sigma} = -c_1 201.1u. \tag{26}$$

From (26) and (22) synthesized control is

$$u = (c_1 201.1)^{-1}[k_1|\sigma|^{\frac{1}{3}}sign(\sigma) + k_2|\dot{\sigma}|^{\frac{1}{2}}sign(\dot{\sigma})] \equiv u_2. \tag{27}$$

Thus, the control is smooth and continuous for both relative degree one and relative degree two surface.

For comparative analysis, control is devised using well known control strategy in SOSM, i.e., conventional twisting algorithm (TA). It is defined as

$$\ddot{\sigma} = -k_1 sign(\sigma) - k_2 sign(\dot{\sigma}). \tag{28}$$

From (26) and (28) control input becomes,

$$u = (c_1 201.1)^{-1}[k_1 sign(\sigma) + k_2 sign(\dot{\sigma})] \equiv u_3. \tag{29}$$

Thus, TA gives discontinuous control for relative degree two surface.

7 Simulation Results

To validate the proposed algorithm, system dynamics in (18) were simulated in MATLAB. Controller was commanded to track square signal of amplitude 10 (rad) with frequency 0.1 Hz. Sliding surface gain designed as $c_1 = 1$ to have closed loop pole at -1 during sliding. To test robustness, disturbance of 1 $\sin(\omega t)$ with 1 Hz was added in input channel.

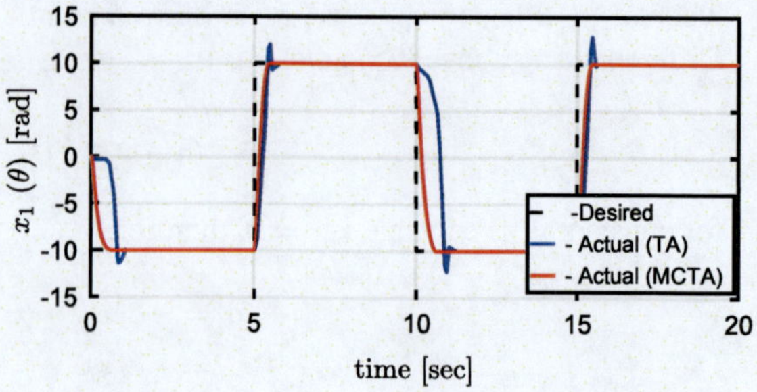

Fig. 1 Evolution of state x_1

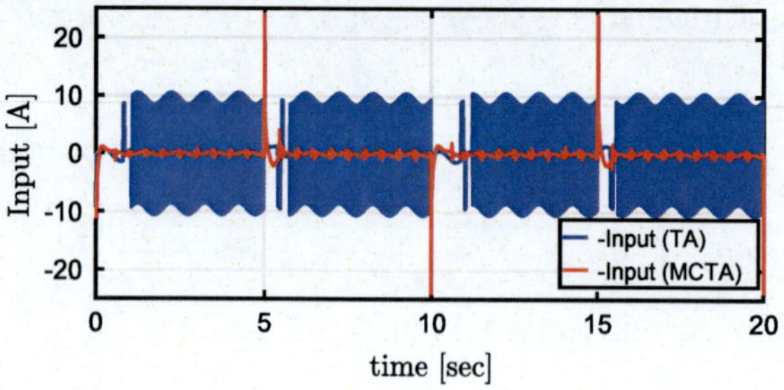

Fig. 2 Control action

Table 1 Comparison of control performance

Control algorithm	$\|\|e_{x1}\|\|_1$	$\|\|u\|\|_2$
TA	6268.9	747.73
MCTA	747.95	69.32

For comparative analysis, controller using two algorithms (TA (u_3) and MCTA (u_2)) for relative degree two surface was applied to system. The parameters were tuned such that the output response of the motion control system is same with both u_2 and u_3. The settling time for both control input was considered to be 0.5 s. Controller gains required for TA were $k_1 = 1000$, $k_2 = 920$ and for MCTA gains were $k_1 = 1050$, $k_2 = 300$. Figures 1 and 2 show evolution of state x_1 and associated control input, respectively.

The control effort required for MCTA compared to TA is smooth and clearly visible in Fig. 2. For quantitative analysis, 'norm 2' of control and output error 'norm 1' are tabulated in Table 1. $||u||_2$ gives measure of control energy required for control input. The controller with modified algorithm requires less energy compared to that with TA. $||e_{x1}||_1$ of output error gives measure of accuracy. It is very clear that MCTA-based controller provides more accuracy with less control energy.

8 Conclusion

Modified continuous twisting algorithm has been proposed in this paper. It gives continuous control for both relative degree one and relative degree two sliding surface. It is compared with the conventional TA-based controller. It has been found that the MCTA-based controller needs 90.73% less control energy than TA-based controller. The output accuracy with the proposed MCTA-based controller is 88.07% more than the conventional TA-based controller.

Relation of p and q parameters of the algorithm to degree of smoothness is being future scope of this research work.

References

1. Edwards, C., Spurgeon, S.: Sliding mode control: theory and applications. CRC Press (1998)
2. Utkin, Vadim: Variable structure systems with sliding modes. IEEE Trans. Autom. control **22**(2), 212–222 (1977)
3. Utkin, V., Guldner, J., Shijun, M.: Sliding mode control in electro-mechanical systems, vol. 34. CRC Press (1999)
4. Fridman, L., Levant, A.: Higher order sliding modes as a natural phenomenon in control theory. In: Robust Control Via Variable Structure and Lyapunov Techniques, pp. 107–133. Springer, Berlin, Heidelberg (1996)
5. Levant, A.: Sliding order and sliding accuracy in sliding mode control. Int. J. Control **58**(6), 1247–1263 (1993)
6. Boiko, I., Fridman, L., Castellanos, M.I.: Analysis of second-order sliding-mode algorithms in the frequency domain. IEEE Trans. Autom. Control **49**(6), 946–950 (2004)
7. Torres-González, V., Fridman, L.M., Moreno, J.A.: Continuous twisting algorithm. In; 2015 54th IEEE Conference on Decision and Control (CDC), pp. 5397–5401. IEEE (2015)
8. Bhat, S.P., Bernstein, D.S.: Finite-time stability of continuous autonomous systems. SIAM J. Control Opt. **38**(3), 751–766 (2000)

Comparative Analysis of Different Control Techniques for Six-Phase PMSM as an Application to HEV

K. Mohith, Sudarshan Patilkulkarni, and Narasimha Kollaparti

1 Introduction

Nowadays, due to the steep increase in the use of vehicles on road, there is an enormous rise in air pollution and fast depletion of fossil fuels. Therefore, both policy makers and autoindustry are showing the keen interest in development of electric transportation systems. Electric vehicles (EV) provide an advantage over other means of transport such as zero emission contributing in the reduction of air pollution and also providing an alternative for fossil fuels thus reducing the dependency on non-renewable energy. The propulsion power for electric vehicles is provided by the electric motor, and it acts as the heart of these vehicles. Various types of motors are used depending on their characteristics and the application it is being used. Permanent magnet synchronous motors are now being used because of the advantages they offer like high-power density, high-output torque, and good efficiency, and therefore, it saves energy when compare to other electric motors. Due to the advancement in power electronics, multiphase motor drive is used due to their advantages over three-phase drives like lower current ripple per phase or inverter leg, and it is able to operate even if there is a fault in one of the phases. Hence, it provides fault-tolerant capability, provides good efficiency, and extends range of the vehicles [1]. Various control techniques have been used with the motor to achieve the desired performance, out of which scalar control deals with only the magnitude control, i.e., by varying the voltage and frequency in proportion to achieve the desired speed. Field-oriented control deals with the decoupling of rotor flux and stator current

K. Mohith (✉) · S. Patilkulkarni
JSS Science and Technology University, Mysuru, India

S. Patilkulkarni
e-mail: sudarshan_pk@sjce.ac.in

N. Kollaparti
KPIT Technologies Ltd, Bengaluru, India
e-mail: Narasimha.Kollaparti@kpit.com

S. K V and K. Guruprasad Rao (eds.), *Smart Sensors Measurements and Instrumentation*, Lecture Notes in Electrical Engineering 750,
https://doi.org/10.1007/978-981-16-0336-5_9

through the transformation of stator phase quantities from the stationary frame to synchronously rotating frame, which is similar to DC motor control with benefits of faster response and less torque ripple [2]. Direct torque control provides direct torque and flux control, i.e., bang-bang control which provides with an advantage of faster dynamic response and eliminates the need of coordinate transformation [3]. It is crucial to find the best control technique for PMSM motor in terms of greater efficiency output, easier to implement, and cost effective, so that will have a positive impact on the control performance of the electric vehicles.

This paper begins by presenting the mathematical description of six-phase PMSM and then presents the working principle of three different control algorithms. The Simulink model diagram is developed for the three control algorithms along with six-phase PMSM, and then, the simulations are carried out under different working conditions to compare and analyze the output characteristics of the three control techniques in terms of speed and torque response output and finally highlighting the benefits and downsides of the utilization of these control techniques in the applications of hybrid electric vehicles.

2 Mathematical Modeling of Six-Phase PMSM

To derive the mathematical model of six-phase permanent magnet synchronous motor, following assumptions are made:

- Symmetrical arrangement of two three-phase windings.
- Parameters changes and magnetic saturation are neglected.
- Stator windings have the perfect sinusoidal distribution along the air gap.
- Presence of damper windings is neglected.

In this paper, six-phase PMSM with two three-phase windings *abc* and *xyz* are spatially shifted by an angle $\gamma = 60$ is considered, i.e., symmetrical configuration.

The equations that describe the electrical part of the six-phase machine in natural reference frame are as shown below:

Stator phase voltage equations

$$V_{abc} = (R_s * I_{abc}) + \mathrm{d}(\emptyset_{abc})/\mathrm{d}t \tag{1}$$

$$V_{xyz} = \left(R_s * I_{xyz}\right) + d\left(\emptyset_{xyz}\right)/dt \tag{2}$$

Flux linkage equations

$$\emptyset_{abc} = (L_{11} * I_{abc}) + (L_{12} * I_{xyz}) + \emptyset'_{mabc} \tag{3}$$

$$\emptyset_{xyz} = (L_{22} * I_{xyz}) + (L_{21} * I_{abc}) + \emptyset'_{mxyz} \tag{4}$$

where $R_s = \text{diag}\,[R_s, R_s, R_s]$ is stator resistance vector, $V_{abc} = [V_a V_b V_c]^T$ is stator phase voltage vector of windings abc, $V_{xyz} = [V_x V_y V_z]^T$ is stator phase voltage vector of windings *xyz*, $I_{abc} = [I_a I_b I_c]^T$ and $I_{xyz} = [I_x I_y I_z]^T$ are stator current vectors of *abc* and *xyz* windings, $\emptyset_{abc} = [\emptyset_a\ \emptyset_b\ \emptyset_c]^T$ and $\emptyset_{xyz} = [\emptyset_x\ \emptyset_y\ \emptyset_z]^T$ are flux linkage vectors of *abc* and *xyz* windings, L_{11} and L_{22} are the stator inductance vectors of *abc* and *xyz* windings, L_{12} and L_{21} are the mutual inductance vectors, $\emptyset'_{mabc}$ and $\emptyset'_{mxyz}$ are the permanent magnet flux linkage of *abc* and *xyz* windings.

These equations that describe the machine in natural reference frame are transformed into a new synchronously rotating reference frame using vector space decomposition transformation approach, and as a result, equations gets simplified and time-dependent coefficients in the equations are eliminated [4–6]. The transformation matrices used are Clark transformation matrix (CTM):

$$\begin{bmatrix} 1 & \cos(\alpha) & \cos(2\alpha) & \cos(\gamma) & \cos(\alpha+\gamma) & \cos(2\alpha+\gamma) \\ 0 & \sin(\alpha) & \sin(2\alpha) & \sin(\gamma) & \sin(\alpha+\gamma) & \sin(2\alpha+\gamma) \\ 1 & \cos(2\alpha) & \cos(\alpha) & -\cos(\gamma) & -\cos(\alpha+\gamma) & -\cos(2\alpha+\gamma) \\ 0 & \sin(2\alpha) & \sin(\alpha) & \sin(\gamma) & \sin(\alpha+\gamma) & \sin(2\alpha+\gamma) \\ 1 & 1 & 1 & 0 & 0 & 0 \\ 0 & 0 & 0 & 1 & 1 & 1 \end{bmatrix}$$

and Park transformation matrix (PTM):

$$\begin{bmatrix} \cos(\theta) & \cos(\theta-\alpha) & \cos(\theta-2\alpha) & \cos(\theta-\gamma) & \cos(\theta-\alpha-\gamma) & \cos(\theta-2\alpha-\gamma) \\ -\sin(\theta) & -\sin(\theta-\alpha) & -\sin(\theta-\alpha) & -\sin(\theta-\gamma) & -\sin(\theta-\alpha-\gamma) & -\sin(\theta-\alpha-\gamma) \\ 1 & \cos(2\alpha) & \cos(\alpha) & -\cos(\gamma) & -\cos(\alpha+\gamma) & -\cos(2\alpha+\gamma) \\ 0 & \sin(2\alpha) & \sin(\alpha) & \sin(\gamma) & \sin(\alpha+\gamma) & \sin(2\alpha+\gamma) \\ 1 & 1 & 1 & 0 & 0 & 0 \\ 0 & 0 & 0 & 1 & 1 & 1 \end{bmatrix}$$

After applying VSD approach on Eqs. 1–4, the transformed equation variables rare mapped into three orthogonal subspaces, namely α-β subspace that contains components that are involved in the electromechanical energy conversion of the electric motor, and the other two subspaces contains components relating to the losses, which includes harmonic components [7].

The Clark and Park transformation matrices transform the phase quantities from stationary frame into new rotating frame with an arbitrary rotor speed W_r, and the resulting stator voltage equations are as shown below:

$$V_d = (R_s I_d) + \frac{d(\emptyset_d)}{dt} - (W_e \emptyset_q) \tag{5}$$

$$V_q = \left(R_s I_q\right) + \frac{d(\emptyset_q)}{dt} + (W_e \emptyset_d) \tag{6}$$

$$V_{d1} = (R_s I_{d1}) + \frac{d(\emptyset_{d1})}{dt} \quad (7)$$

$$V_{q1} = \left(R_s I_{q1}\right) + \frac{d(\emptyset_{q1})}{dt} \quad (8)$$

where V_d, V_q are *d-q* axis voltages of windings *abc* and V_{d1}, V_{q1} are *d-q* axis voltages of windings *xyz*. I_d, I_q are *d-q* axis currents of windings *abc* and I_{d1}, I_{q1} are *d-q* axis currents of windings *xyz*.

The flux linkage equations are:

$$\emptyset_d = (L_d I_d) + \emptyset_m \quad (9)$$

$$\emptyset_q = \left(L_q I_q\right) + \emptyset_m \quad (10)$$

$$\emptyset_{d1} = (L_{d1} I_{d1}) + \emptyset_m \quad (11)$$

$$\emptyset_{q1} = \left(L_{q1} I_{q1}\right) + \emptyset_m \quad (12)$$

$$W_e = (p/2) W_r \quad (13)$$

where L_d, L_q are *d-q* axis inductance's of windings *abc* and L_{d1}, L_{q1} are *d-q* axis inductance's of windings *xyz*, W_r is the rotor angular velocity, W_e is the electrical angular velocity, $\emptyset_m$ is PM flux linkage, and p is the number of pole pairs of the motor.

The electromagnetic torque of six-phase PMSM is calculated using:

$$T_e = \ (3p/4) * (\emptyset_d I_q - \emptyset_q I_d \quad (14)$$

And the mechanical speed equation of six-phase PMSM is shown as:

$$W_m = \int (T_e - T_l - (B * W_{rm})/J).dt \quad (15)$$

where

J is the moment of inertia, B is damping coefficient, T_1 is the load torque,
W_{rm} is rotor mechanical speed [4–6].

3 Different Control Techniques for PMSM

Control techniques used for permanent magnet synchronous motor (PMSM) are classified as shown in Fig. 1. They are classified into two types: basic scalar control and vector control based on the quantities they control. Scalar control controls only the magnitude whereas vector control deals in controlling both magnitude and angle. The further classification of both these types leads to three other control techniques based on the control method that they incorporate [8].

3.1 Scalar Control

Scalar control is the most basic way of controlling PMSM, where it deals with controlling of only magnitudes. In this, motor speed is controlled by the variation of voltage and frequency in constant. It is used in applications where good dynamic performance is not of much importance, and it offers an advantage of a sensor less control and also making it easier to implement [9]. The block diagram of scalar control applied to a six-phase PMSM through a six-phase sine pwm inverter with supply voltage and frequency provided to it based on the required speed or torque is shown in Fig. 2.

3.2 Field-Oriented Control (FOC)

To achieve good dynamic performance from the motor, another control technique is used to control the PMSM. This control approach is made similar to DC motor control by using mathematical transformations, hence making it easier to control the AC motor, and its working principle is providing independent control of both torque

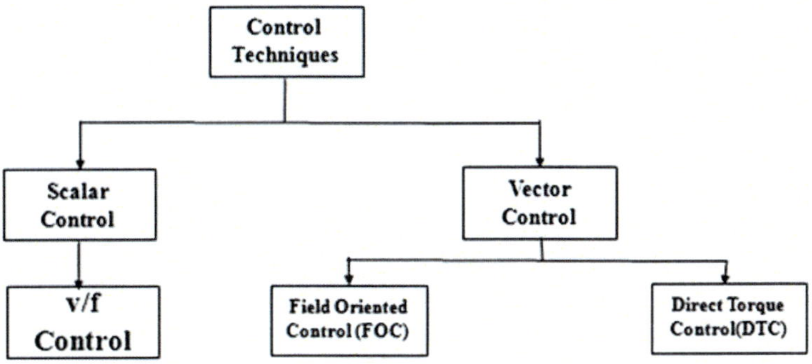

Fig. 1 Classification of control algorithms of PMSM

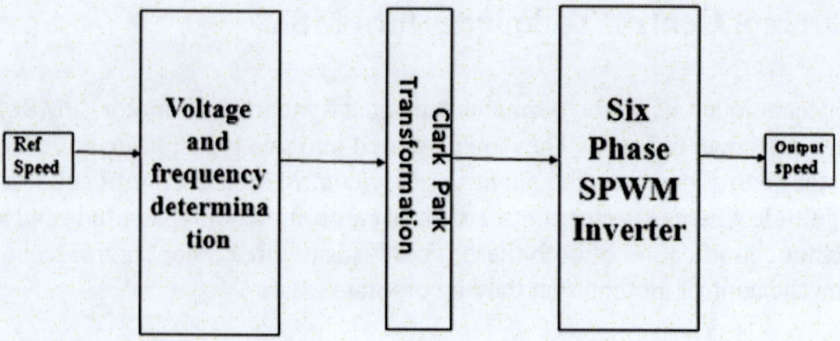

Fig. 2 Block diagram of scalar control

and flux producing currents. This control technique is known as field-oriented control (FOC). The aim of FOC used in AC machines is to control the motor similarly to separately excited DC motor control. The block diagram of FOC is shown in Fig. 3. The feedback currents (I_q and I_d) are compared with reference currents, i.e., based on the reference speed or reference torque, and the error signal is given to the two PI controllers. The controller outputs voltage in two axis rotor reference frame, which are transformed back into voltages in three-axis stationary frame using inverse Clark and Park transformation. These voltages are modulated onto the six-phase PMSM through the inverter [10].

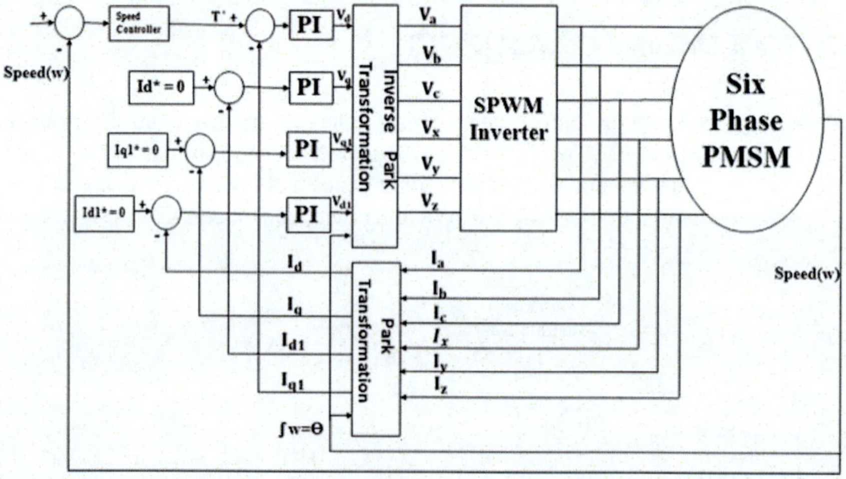

Fig. 3 Block diagram of FOC

3.3 Direct Torque Control

Direct torque control does not require the measurement of the rotors mechanical position, PI controllers, and also does not require coordinate transformation to transform from stationary to rotating frame making it simpler in implementation compared with the FOC technique. The disadvantages are higher amplitude of torque and current ripple, which produces higher noise and low performance at lower speed control. The working principle of DTC involves direct control of stator flux and electromagnetic torque by applying suitable voltage vectors onto the motor terminals. The block diagram of DTC applied to a six-phase PMSM through six-phase inverter with space vector modulation is shown in Fig. 4.

The stator flux is calculated using:

$$\varnothing_s = \int (V_s - R_s I_s).dt \tag{16}$$

The electromagnetic torque is calculated using:

$$T_e = (3p/4) * (\varnothing_{\alpha s} I_{\beta s} - \varnothing_{\beta s} I_{\alpha s}) \tag{17}$$

And the stator flux angle is given by:

$$\theta = \arctan[\varnothing_{\beta s} / \varnothing_{\alpha s}] \tag{18}$$

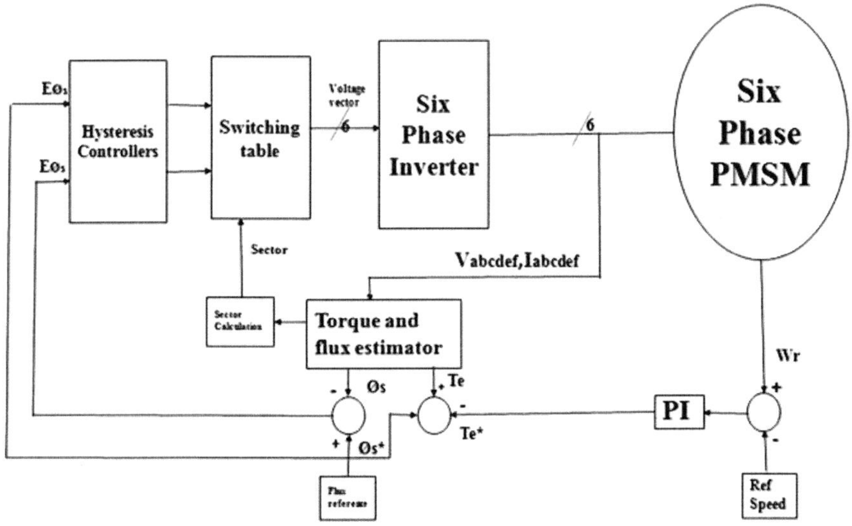

Fig. 4 Block diagram of DTC

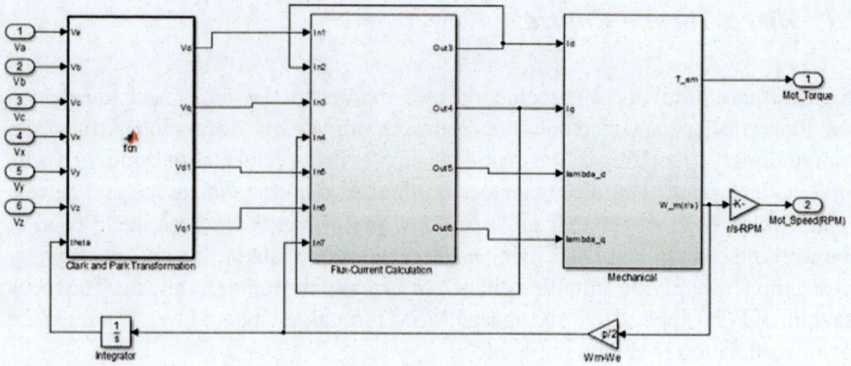

Fig. 5 Simulink model of scalar control of six-phase PMSM

where $\emptyset_{\beta s}$, $\emptyset_{\alpha s}$ are β-flux component and α-flux component of stator flux $\emptyset_s$ and $I_{\beta s}$, $I_{\alpha s}$ are β-current component and α-current component of stator current Is. Depending on this, angle sector is chosen and the controllers output suitable voltage vectors onto the six-phase inverter driving the six-phase PMSM [3, 11].

4 Simulation Implementation of Control Techniques of Six-Phase PMSM Using MATLAB/Simulink Software

4.1 Simulation Model Diagram of Six-Phase PMSM with Scalar Control

The Simulink model diagram developed of scalar control is applied to a six-phase PMSM through a six-phase sine pwm inverter with supply voltage provided to it based on the required speed, and the developed model is shown in Fig. 5.

4.2 Simulation Model Diagram of Six-Phase PMSM with Field-Oriented Control (FOC)

The mathematical model of six-phase PMSM expressed by equations along with six-phase sine pwm inverter has been used in implementing field-oriented control algorithm of a six-phase PMSM drive in matlab. The developed Simulink model diagram is shown in Fig. 6. The simulation model consists of subsystem which includes PI controller blocks, Clark and Park transformation blocks, six-phase inverter, and six-phase PMSM block.

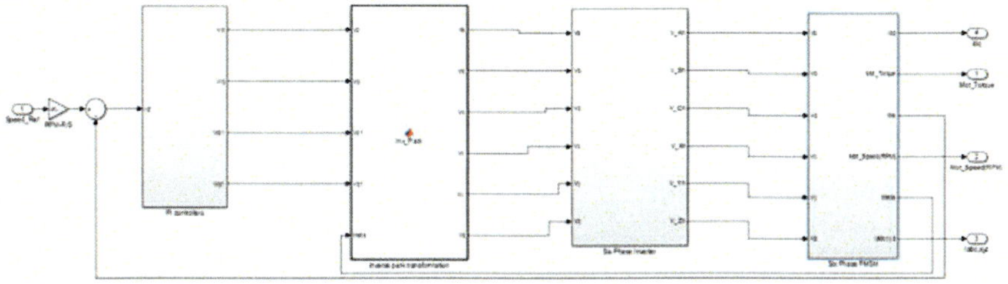

Fig. 6 Simulink model of FOC of 6-phase PMSM

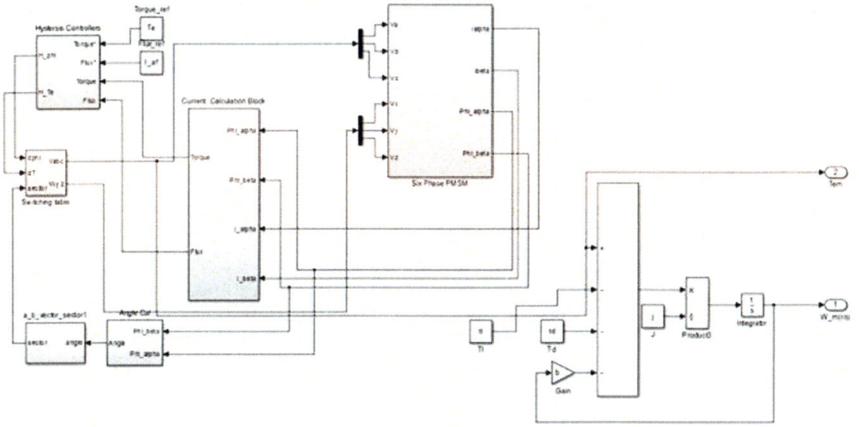

Fig. 7 Simulink model of DTC of six-phase PMSM

4.3 Simulation Model Diagram of Six-Phase PMSM with Direct Torque Control (DTC)

The Simulink model diagram developed of direct torque control is applied to a six-phase PMSM through a six-phase VSI with sinusoidal pulse width modulation technique is shown in Fig. 5. Here, the stator flux reference is set to 0.1461 Wb, and the hysteresis bandwidth of flux is 0.001 Wb and for torque is 0.01 Nm, which can be adjusted depending on the requirement of the accuracy of the control (Fig. 7).

5 Simulation Results

Simulations to compare scalar control, FOC, and DTC of six-phase PMSM have been carried out. The comparison includes the motor speed and motor torque response comparisons of the three control schemes, and each one of them was tested at no load, loaded, and dynamic working conditions. The reference speed was set at 1000 rpm for all working conditions. The motor parameters used for all the control algorithms simulation are: $V_{dc} = 300$ V, stator resistance $R_s = 1{:}4$, d-axis inductance $L_d = 6.774 \times 10e{-}3$ mH, q-axis inductance $L_q = 9.165 \times 10e{-}3$ mH, moment of inertia $J = 0.77 \times 10e{-}3$ Kgm^2, PM flux linkage = 0.1461 Wb, number of poles = 6.

Case 1: At no load condition

In the first set of simulations, the six-phase motor was operated with zero load (0 Nm). The speed and torque response of the six-phase motor for all the three control schemes under this condition are shown in Figs. 8 and 9. From Tables 1 and 2, it

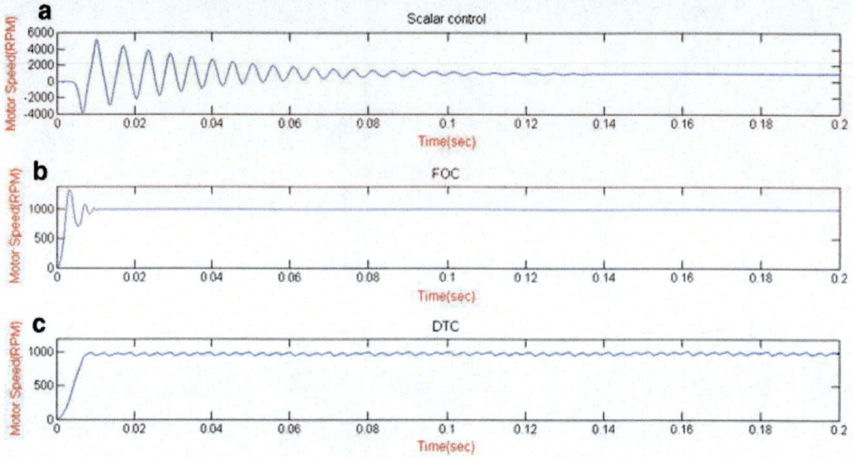

Fig. 8 Speed response of **a** scalar, **b** FOC and **c** DTC

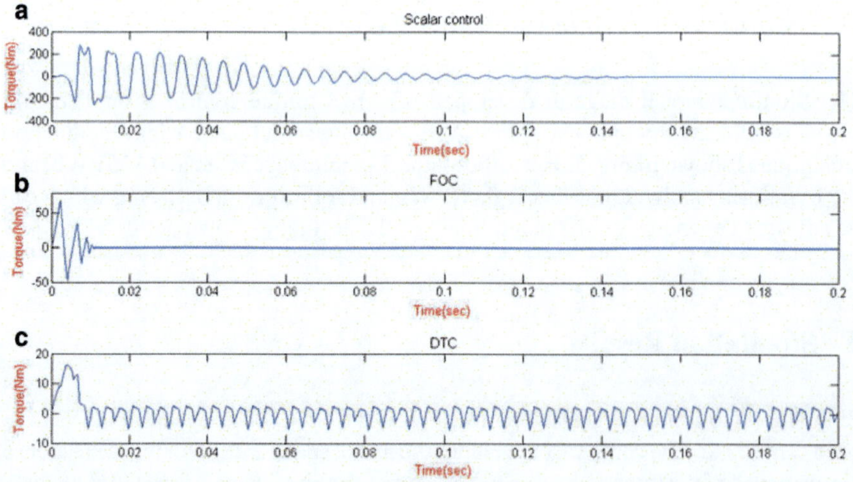

Fig. 9 Torque response of **a** scalar, **b** FOC and **c** DTC

Table 1 Settling time of speed

Motor speed (RPM)	
Type of control	Settling time (ms)
Scalar control	150
FOC	10.4
DTC	7.9

Table 2 Settling time of torque

Torque (0 Nm)	
Type of control	Settling time (ms)
Scalar control	190
FOC	11.5
DTC	8.6

is clear that DTC has faster speed and torque response in terms of settling time compared to scalar and FOC. Scalar control reaches slower.

Case 2: At loaded condition

In the second set of simulations, the six-phase motor was operated with load (5Nm). The speed and torque response of the six-phase motor for all the three control schemes under this condition are shown in Figs. 10 and 11. From Table 3, 4, and 5, it is clear that DTC reaches the set reference speed faster and also load torque in terms of settling time compared to scalar and FOC. FOC and scalar control are characterized with lower current ripple compared to DTC and scalar control.

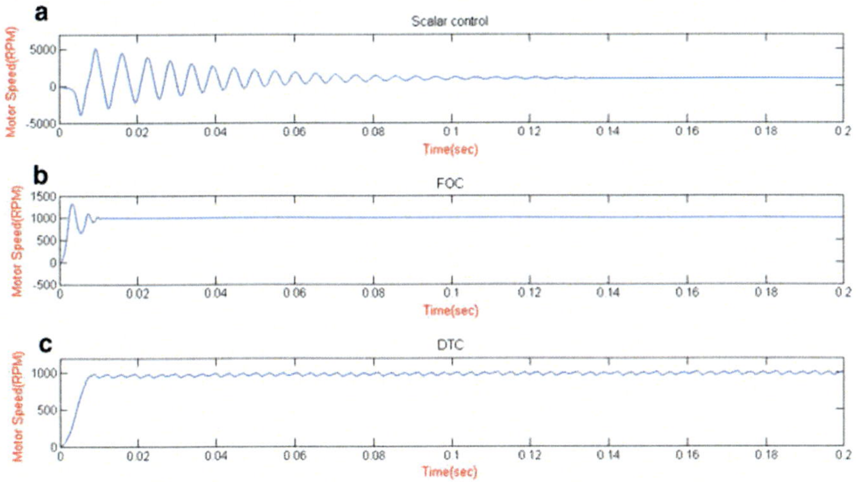

Fig. 10 Speed response of **a** scalar, **b** FOC, and **c** DTC

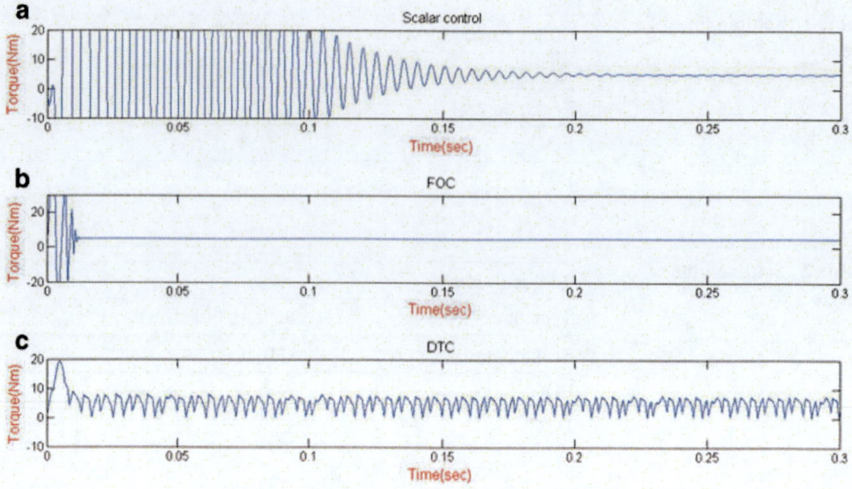

Fig. 11 Torque response of **a** scalar, **b** FOC, and **c** DTC

Table 3 Settling time of speed

Motor speed (RPM)	
Type of control	Settling time (ms)
Scalar control	153
FOC	11
DTC	8.9

Table 4 Settling time of torque

Torque (0 Nm)	
Type of control	Settling time (ms)
Scalar control	200
FOC	13
DTC	8.6 with SSE ± 1.5 Nm

Table 5 Torque ripple

Torque ripple (%)	
Scalar control	0.46
FOC	5.02
DTC	57

Case 2: At loaded condition

In the last set of simulations, the speed reference continuously changes (800–1000 rpm at 0.15 s), and it is generated by signal builder. The load torque is also changed from 0–5 Nm at 0.3 s. The speed and torque response of the six-phase motor for all the three control schemes under loaded condition are shown in Figs. 12 and 13. From Table 6, 7, and 8, it is clear that DTC has faster speed and torque response in terms of settling time compared to scalar and FOC. FOC is characterized with lower current ripple compared DTC. Thus, DTC has fast dynamic response behavior compared with the other two techniques, FOC with higher torque quality, and good

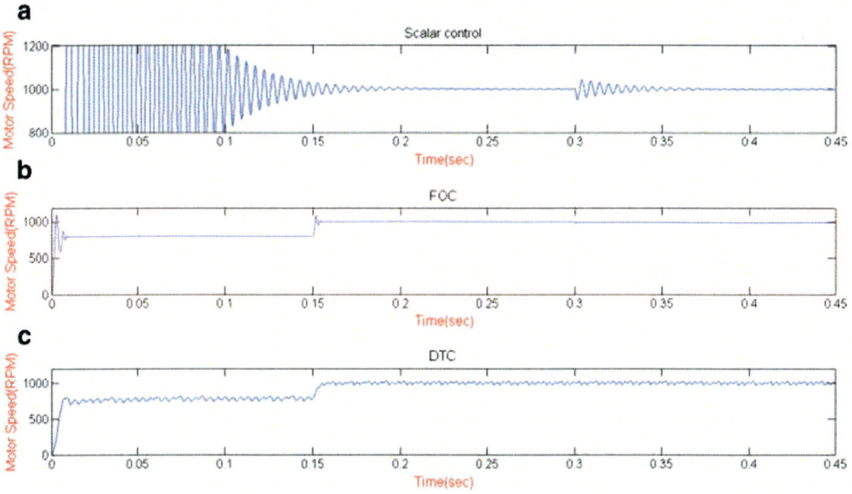

Fig. 12 Speed response of **a** scalar, **b** FOC, and **c** DTC

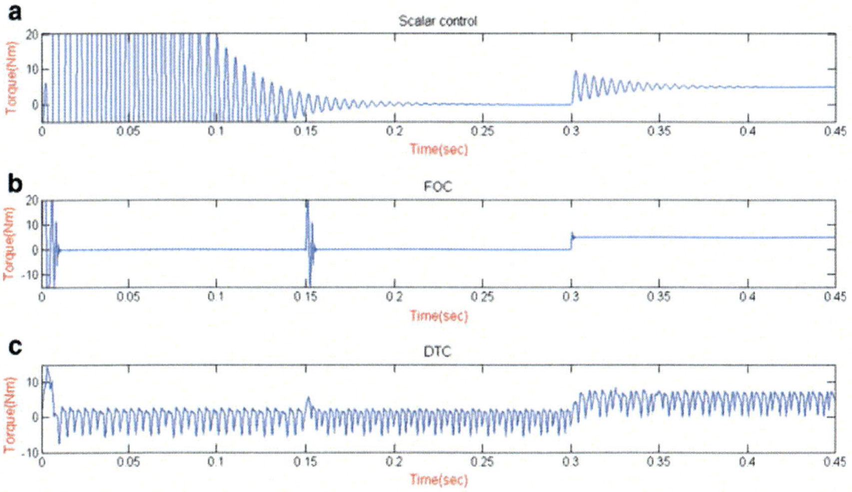

Fig. 13 Torque response of **a** scalar, **b** FOC, and **c** DTC

Table 6 Settling time of torque

Motor speed (800–1000RPM)@0.15 s	
Type of control	Settling time (ms)
Scalar control	–
FOC	5
DTC	4

Table 7 Settling time of speed

Torque (0–5 Nm)@0.3 s	
Type of control	Settling time (ms)
Scalar control	100
FOC	2
DTC	1.5 with SSE ± 2 Nm

Table 8 Torque ripple

Torque ripple (%)	
Scalar control	5
FOC	5.18
DTC	62

dynamic response provides an edge over the other two. Scalar control is easier to implement but it has a disadvantage of poor dynamic performance.

From the above results, the comparative analysis of the three different control algorithms can be summarized as shown in Table 9.

Table 9 Comparison analysis of three control techniques of six-phase PMSM

	Scalar control	Field-oriented control (FOC)	Direct torque control (DTC)
Dynamic response	Slower	Slower (faster than scalar)	Faster
Torque ripple	Lower ripple	Lower ripple	Higher ripple
Parameter sensitivity	Magnitude of voltage and frequency	Decoupling depends on L_d, L_q, $Ø_{pm}$	Stator resistance (R_s)
Position sensor	No	Yes	No
Coordinate transformation	No	Yes	No
Controllers	No	PI controllers	Hysteresis controllers
Implementation complexity and processing requirements	Lower	Higher	Lower (higher than scalar)

6 Conclusion

This paper gives the comparison analysis of three different control algorithms in six-phase PMSM drive system for hybrid electric vehicles, and the controls used are scalar, FOC, and DTC. The mathematical model of six-phase motor and the Simulink modeling of the three control algorithms are given. Simulations have been carried out in three different working conditions to provide speed and torque response comparisons of six-phase PMSM with the three control algorithms and to find their pros and cons of using it for a particular application. The simulation results have shown that DTC has faster dynamic response for speed and torque, while it has higher amplitude of torque ripple compared to scalar control and FOC. FOC performs better in steady state with less torque ripple and has faster response time compared with scalar control from the simulation. Scalar control technique is an open loop technique with no controllers used, so making it easier to implement with simple structure. Hence, making it suitable for applications in which control precision is not important.

Summarizing, scalar control is more suitable for low performance applications where good dynamic performance is not required. FOC is suitable where higher torque quality is needed, and DTC is used in applications where faster response dynamics are of primary importance.

References

1. Parmar, Y., Patel, P., Pancholi, N., Thakor, C., Mali, U.: Scalar control of permanent magnet synchronous motor. Int. Res. J. Eng. Technol. (IRJET) India 03(12) (2016)
2. Kamalaselvan, A., Prakash, S.L.: Modeling simulation and analysis of closed loop speed control of PMSM drive system, pp. 692–697. IEEE ICCPCT, Nagercoil, India (2015)
3. Bojoi, R., Griva, G., Farina, F., Profumo, F., Tenconi, A.: Direct torque control for dual three-phase induction motor drives. Ind. Appl. IEEE Trans. **41**, 1627–1636 (2005)
4. Pillay, P., Krishnan, R.: Modeling of permanent magnet motor drives. IEEE Trans. Ind. Electron. **35**(4), 537–541 (1988)
5. Kumar, M., Singh, B., Singh, B.P.: Modelling and simulation of permanent magnet brushless motor drives using simulink. In: National Power Systems Conference, pp. 27–29 (2002)
6. Su, J., Yang, J., Yang, G.: Mathematical model research of six-phase pmsm. Adv. Mater. Res. 614–615, 1266–1271 (2012)
7. Woldesemayat, M.: Modelling of six phase permanent magnet synchronous motor. Ph.D. Thesis, Pohang University of Science and Technology, Korea (2017)
8. Woldesemayat, M., Lee, H., Won, S., Nam, K.: Modeling and verification of six-phase interior permanent magnet synchronous motor. IEEE Trans. Power Electron. 1–1 (2017)
9. Pindoriya, R.M., Mishra, A.K., Rajpurohit, B.S., Kumar, R.: Performance analysis of control strategies of permanent magnet synchronous motor. In: IEEE Region 10 Conference (TENCON), pp. 3224–3227. Singapore (2016)
10. Fernandez, E., Coello, M.: Comparative analysis of control techniques in drive systems for electric motorcycle. In: International Youth Conference on Energy (IYCE), pp. 1–5. Budapest (2017)

11. Bida, V.M., Samokhvalov, D.V., Al-Mahturi, F.S.: PMSM vector control techniques—a survey. In: IEEE Conference of Russian Young Researchers in Electrical and Electronic Engineering (EIConRus), pp. 577–581. Moscow (2018)

A Survey on Various Total Body Water Detection Techniques to Develop a Wearable Device

G. Aayushii, M. G. Nagarjun, Srilakshmi U. Sirurmath, and Kendaganna Swamy

1 Introduction

Water is the most essential element in the human body. An average body of adult is composed of about 60% water by weight. All the biochemical reactions which occur in our body require water. The water in each and every cell and organ is used by body to help control the body temperature; protect spinal cord and joints; excrete waste products generated. The amount of water required by the body exceeds its ability to produce it; hence, it is vital to quantize the water content in a person's body. Total body water (TBW) is the terminology coined to estimate the water level in a body.

The water in the tissues, blood, bones and fluid components of a human body with the absence of sodium content is known as TBW. The TBW depends on various factors such as age, weight, water intake, health condition and sex [1]. It is basically divided into intracellular fluid (ICF) which is two-third of TBW and extracellular fluid (ECF) which is one-third of TBW. ICF is the fluid which will be present inside the cells. For example, in a 72 kg weight body with 40 L of fluid, 25 L of it will be intercellular [2]. Jackson has stated that ICF contributes 70% of the body fluid [3]. ECF is the fluid present outside the cells, that is, there will be 15 L of ECF present in a 40 L of fluid in a body. The ECF can be again split into plasma, interstitial fluid and transcellular fluid [4].

All the living cells and the extracellular fluids contain a distinct amount of water, hence, water is an essential part of all living organisms an imbalance in which may lead to unfavorable health conditions. Water mediates body temperature, enables digestion of food and helps to control the acid and base equilibrium in our body [5].

G. Aayushii (✉) · M. G. Nagarjun · S. U. Sirurmath · K. Swamy
Department of Electronics and Instrumentation, R.V. College of Engineering, Bangalore, Karnataka, India

Springer-Verlag, Computer Science Editorial, Tiergartenstr. 17, 69121 Heidelberg, Germany

K. Swamy
e-mail: kendagannaswamys@rvce.edu.in

S. K V and K. Guruprasad Rao (eds.), *Smart Sensors Measurements and Instrumentation*, Lecture Notes in Electrical Engineering 750,
https://doi.org/10.1007/978-981-16-0336-5_10

Metabolism is the phenomenon due to which the cells in our body generate energy from food. Certain proteins control the efficiency of this process, water being one of the salient requirements for metabolization. A 100 cal are approximated to be metabolized by the consumption of 100 ml in a healthy individual [6].

A daily consumption of 3.7 L of water for male adults and 2.7 L for female adults satisfies most of the population's requirements. Nevertheless, excessive heat due to variant environment conditions and rigorous physical activity can significantly increase the total water requirements, and also the individual variation among athletes could be significant [7]. Water accounts for 60% of body weight for healthy individuals with normal body mass index (BMI). If a person's BMI is between 15 and 19.9, he/she is considered as underweighted person, normal weighted if it is between 20 and 24.9, over weighted if it is between 25 and 29.9 and obese if it is 30–35 or more [8, 9, 17]. Water can be drained from the human body in various quantities through: urination (1400 ml), excrement (200 ml), insensitive reduction from the lungs (400 ml) and reduction due to skin suddenness (500 ml).

Therefore, the average loss of water on a daily basis is around 2500 ml, however, this differs significantly among people and also on the climatic factors of the area the person is living in [10]. For people living under pleasant climatic conditions, body temperature is primarily held at a suitable range. This is done by cognitive processes or by simply adjusting the room temperature (via air-conditioning). If the body temperature is too high, sweating is a means of reducing the temperature. Most of the body's metabolic processes lead to the production of heat, but the body only generates heat gradually at rest. The temperature of the body will increase rapidly if there is no equivalent increase in the amount of heat loss, which is achieved by excessive sweating. Sweating is a means of preventing an unhealthy increase in body temperature which allows the reduction of electrolytes and water from the body. However, just a loss of 1 L of sweat could escalate the feeling of exhaustion and hinders the performance efficiency, thus, it is obviously a necessity to realize the threshold of reaching dehydration and replace water losses during fitness activities or sports [10].

Dryness in mouth and tongue, nausea, migraine, tiredness, exhaustion, dryness in skin, muscle pain, dizziness and lack of concentration are early symptoms of mild dehydration. People suffering from chronic or extreme dehydration may trigger conditions including hypotension (low blood pressure), hypertension (high blood pressure) and tachycardia as a result of straining the cardiac muscle. Infants may experience dryness in the eyes and sunken fontanels, and in severe cases, the subject may suffer from unconsciousness [11]. Due to dehydration the risk of developing dental carriers, urinary tract infections, constipation, and kidney stones increases which can be fatal [12–15].

Considering the above factors, it is obvious that a regular check on the total body water becomes not only necessary but even more essential to carry out a healthy and uninterrupted lifestyle. Most healthcare providers are conscious of TBW levels, however, public awareness on the severities of dehydration is bleak and is hence not considered as a major issue to health. As per a study among dieticians, 76% of people were familiar with the concept of dehydration and 55% had knowledge of the approximate amount of water to be consumed to maintain optimum TBW [15].

Taking an example of statistics done by National Health and Nutrition in USA with regards to the TBW levels of the university level students 2017, about 46.4% students under scrutiny were found to be dehydrated and 59% of them consumed insufficient amount of water and were more prone to suffer dehydration due to high caffeine intake. This statistic highlighted the negligence to maintain a healthy TBW level. As of 2017, USA has concluded that in the age groups ranging from 6 to 19 years at least 54% of subjects suffered from mild to moderate dehydration as a result of failure to maintain a healthy TBW level [5]. For a healthy adult, the balance between water consumed and water lost is crucial. Deprivation of water tends to occur whenever the consistency between water consumption and loss is disrupted, this leads to dehydration. Fluid disproportion such as deficiency in fluid quantity in the body, that is, dehydration or overconsumption of fluids, that is, over-hydration is primarily associated with health disorders as briefly mentioned above.

Given the need to regularly assess one's TBW level, an amalgam of technology and basic sciences has been proclaimed as the basis for the invention of methods to measure the same. Most conventional methods used are human invasive, which require assistance and consume extensive amount of time. The challenge is to allow technology and electronics to develop non-invasive methods to measure the total body water content of a person's body. As a solution to the failure of an easy and accessible method to realize TBW level in a human being, the author has reviewed four methods that can be designed and utilized for the monitoring of the TBW levels in a human and consuming the necessary amounts of fluids to stay hydrated and avoid cases of dehydration, hence, leading a healthy lifestyle.

2 Existing Non-invasive Techniques to Measure TBW

In this section, the author will discuss the various techniques proposed to measure TBW levels. From the research, there are four prominent methods used to measure TBW level in the human body. The author has examined all the four methods and formulated an accurate comparison between different methods on the basis of various parameters necessary to conclude with the best method to adapt and develop a TBW detection wearable device.

2.1 Bioelectrical Impedance Analysis

Analysis of bioelectrical impedance (BIA) is a method which has been proved effective, widely and simple to use. BIA is being used to evaluate fluid balance and to assess TBW levels. The BIA works on a dual type system that measures soft tissue and the water level that is, fat mass and lean body mass. A cylindrical conductors'

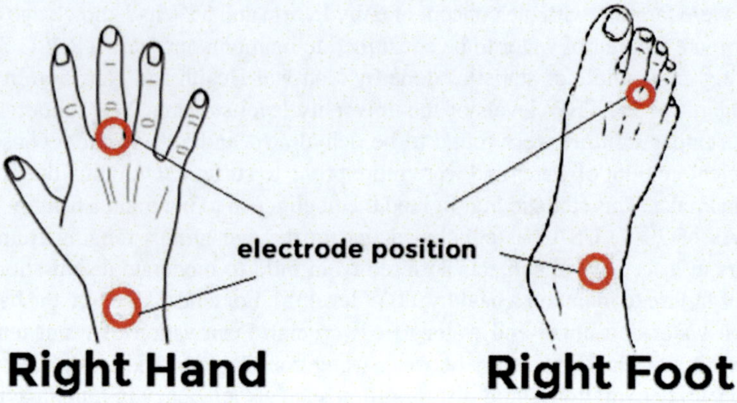

Fig. 1 Electrode position for BIA

impedance is associated with the length, the area of cross-section and the frequency of the signal applied. The summation of the vector sum of the reactance value and the resistance value results in the measure of impedance.

Tissues of a mammal in relation to their water and electrolyte content can transmit an electrical current. Electrical pathways that are highly conductive and have minimal-resistance are usually formed via the lean body tissue which comprises electrolytes and body fluids. On the other extreme, fat, skin and bone are very weak in conducting and give high impedance. Due to capacitance formed by the walls of the cell membrane, reactance is characterized here as a resistance to the current flow. Because of this cell structure, the only part that provides reactance to current is the membrane of the cells.

Usually, BIA is determined by connecting two electrodes to the mid-wrist and mid-finger and the other two electrodes to the mid-ankle of the leg on the same limb, and the foot only adjacent to the second and third toe as shown in Fig. 1. It takes fewer than 5 min for the whole cycle to be completed. Therefore, BIA can be measured by electrical currents of single or multi-frequency. The single BIA frequency is 50 kHz and requires 50–60% intracellular space penetration, while multi-frequency BIA tests extracellular water at the smaller frequency and total body water at the greater frequency of 2–500 kHz.

The BIA accuracy in the body composition examination of patients suffering from chronic kidney disease, specifically for individuals associated with nephrotic syndrome or severe chronic kidney disease, has some limitations. Patients suffering from adverse chronic kidney disease generally portray unusual fluid balance resulting in an inaccurate assessment. Accordingly, BIA will be performed for specific outcomes while the maintenance hemodialysis (MHD) patient is in edema-free or almost edema-free state. For example, in patients with MHD, BIA must be measured

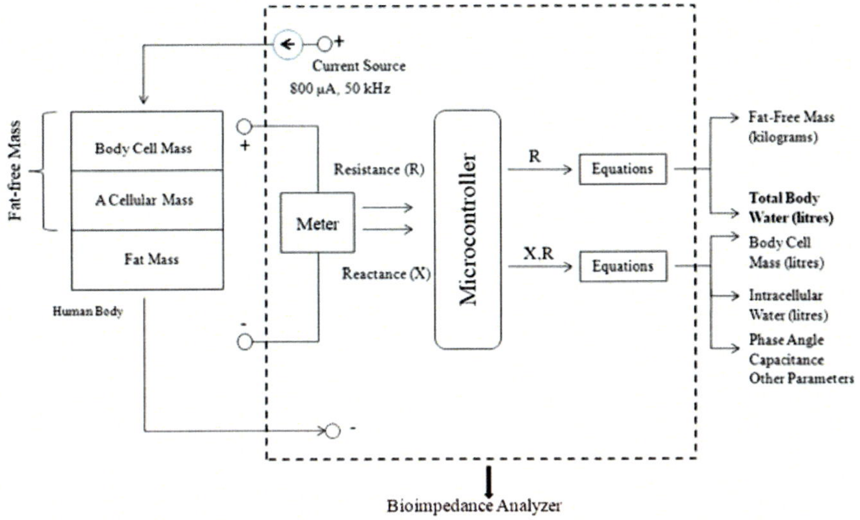

Fig. 2 Architecture of a bio-impedance analyzer

on the non-vascular exposure side at the end of a hemodialysis procedure. Additionally, the software programs that provide the calculations compare results to normal disease-free human beings rather than to discrete disease populations [16].

A low-level, imperceptible current of about 800 μA, a 50 kHz frequency is passed to two electrodes in between and is attached on the arm or legs and hence through the body. The flow of current is affected by the different tissues present in the body and hence a correlation between the current flows and resistance, reactance is derived. These parameters are fed to a microcontroller which in turn, by the application of an algorithm and pre-written equations, computes necessary values to be measured as mentioned in Fig. 2 (Fig. 3).

2.2 *Flowing-Afterglow Mass Spectroscopy*

Flowing-afterglow mass spectrometry (FA-MS) is a diagnostic chemistry method to sensitively detect trace gases. This non-invasive method estimates the deuterium content of water vapor present in the subject's breath and hence gives an immediate measure of TBW.

An extensive group of messenger ions of H_3O^+ (hydronium) is developed using inert gases like argon or helium over a low-level microwave discharge. These generated precursor ions will start reacting with the deuterated water ($HD^{16}O$), isotopomers and water ($H2^{16}O$) which are naturally present in the water vapor of human breath which is let out or exhaled. A small part of an exhalation is acquired through a tuned capillary. It is then inserted into the ion pack of carrier gas.

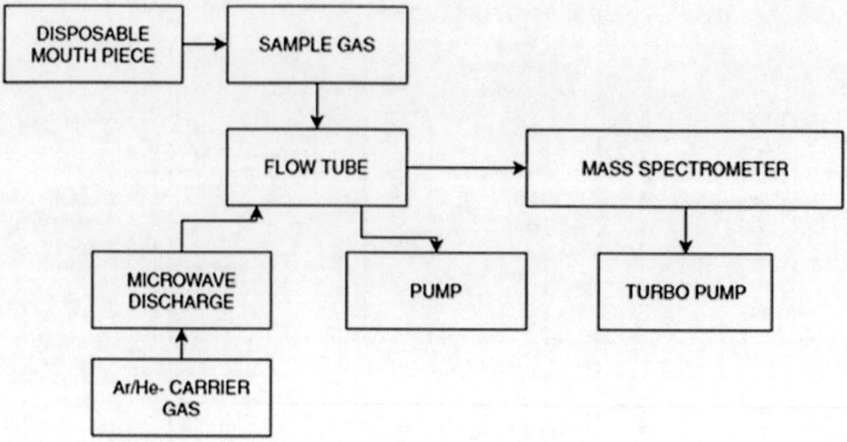

Fig. 3 Flowing-afterglow mass spectroscopy working flow block diagram

As a result of this, a few ions like hydrated ions with m/z (mass to charge) ratio: 73, ions of isentropic variant $H_8DO_4{}^+$, such as $H_9{}^{17}OO_3{}^+$ (disinfected ions of D20 or H20) with m/z ratio: 74, $H_9{}^{18}OO_3{}^+$with m/z ratio: 75 are formed. Considering water vapor, which has moderately large ^{18}O, justifies the addition of isotopic ion $H9^{17}OO_3$ at a value of mass to charge ratio of 74 to an ionic signal. The signal ratio estimation of 74/75 ions, under equilibrium, gives a fractional assessment of the concentration of deuterium in the water vapor. Previous research has shown that this technique could be used to estimate the amount of deuterium which appears in the water vapor of a single breath.

The process starts by the participant breathing out slowly into a removable tube. Their breath is collected by a tuned capillary which leads into the flow tube. It is then combined by a microwave release with H_3O+ (or protonated water) while streaming downwards through the tube, guided along a steady flow of inert gases such as argon or helium. Protonated water clusters are formed due to the water vapor in the breath. Usually, Deuterium is a mass to charge at a value of 74. The mass spectrometer, which enables measurements of enrichment of deuterium, is measured by a significant rise in water clusters compared with a mass to charge ratio of 75 [17].

2.3 *Watson Formula*

The TBW volumes of 458 male adults and 265 female adults were taken from studies of dilution along with their respective age, height and weight from the literature. These taken values were used to obtain the predictive calculations of TBW of adults (irrespective of age). These obtained equations gave the best fit results for male adults [18].

Mathematical model for males:

$$\text{TBW} = 2.447 - 0.09156 \times \text{age} + 0.1074 \times \text{height} + 0.3362 \times \text{weight} \quad (1)$$

where, age in years; height in centimeters; weight in kilograms; water in liters.

Mathematical model for females:

$$\text{TBW} = -2.097 + 0.1069 \times \text{height} + 0.2466 \times \text{weight} \quad (2)$$

where, age in years; height in centimeters; weight in kilograms; water in liters.

Throughout the literature, several other simple regression equations have been reported to estimate TBW from anthropometric measurements. Most of these equations hold good for certain age groups only. These equations and current research equations have been examined on completely separate data. By considering all the cases, the present research equations provided the best outcome. From Eqs. (1) and (2) for men and women, respectively, it was observed that the formula developed and was restricted to a particular age group. The application of Watson's formula is a theoretical method which depends upon the rigorous research done to draw the dependency of the factors—age, height, weight and water—on total body water of human body. It does not use any special equipment and given the rigidity of the method, accuracy of the results cannot be accounted for. It is best suited to use this particular formula as an aid to a more advanced algorithm and methodology.

2.4 Sweat Rate Measurement

In the past few decades, pronounced research has been carried out to develop a real-time hydration monitor, by drawing a relationship between sweat rate and body water content. Crucial information about the physiological parameters of the human body can be derived from pre-established contents of sweat.

Although the sweat substance itself contains considerable details regarding improvements in body hydration, the paper claimed that sweat levels altered the sweat constituent concentration, thereby altering the performance of hydration detection.

To achieve a notable accuracy and to realize the sweat rate in real time, a calorimetric flow rate detection system was developed. The values recorded by the proposed sweat rate monitoring system were confirmed using controlled laboratory experiments (such as a syringe pump) and human trials. Both Simblee and Raspberry Pi were used to connect the sensor into an Internet of things (IoT) network. The collected data could be viewed either on a mobile device or on a server, by developing an IoT framework as shown in Fig. 4. The system was configured using a syringe pump to achieve controlled flow rate characterization.

The sweat rate data was effectively transmitted in real time to a smartphone or directly into the cloud. The implementation of the monitoring system hence

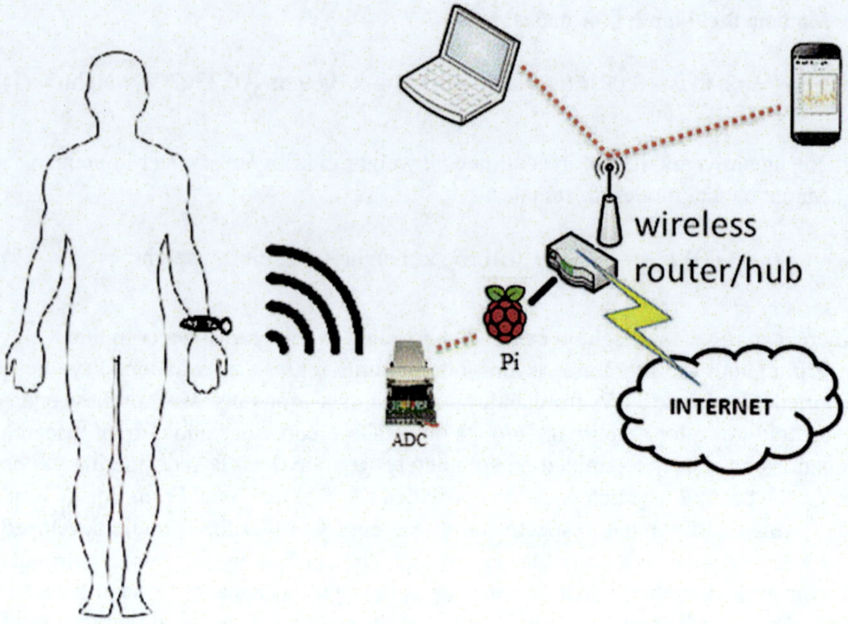

Fig. 4 Setup for the sweat measurement device

concluded that the system can calculate sweat levels accurately in real time, and the validation done by comparison between the proven theoretical concept and tests carried out on human subjects. This sweat rate device can measure sweat levels as low as 0.15 μL/min/cm^2 relative to the manual sweat rate measurements, with an average performance error of 18%.

3 Comparison Between the Different TBW Techniques

The comparison is given in Table 1 consists of four important rows, each one defining a crucial parameter required to define the efficiency of the four methods proposed. They are namely—size of the device proposed, accuracy of the method, response time and power requirements. The columns describe the four methods that were considered by the paper for the measurement of total body water—BIA, FA-MS, sweat rate analysis and Watson's formula. The cells which have been filled have NIL/NA imply that the value of these parameters was not available in the papers that were referred to.

As observed from Table 1, considering the various parameter values for each method, it can be said sweat analysis and Watson's formula have the best response time and size of device. It is important to note that while choosing the best method, it is to be chosen such that the best comprise among the four parameters is made. Here,

Table 1 Comparison between the different TBW techniques

S. No.	Parameter	BIA	FA-MS	Sweat rate analysis	Watson's formula
1	Size of device	>180 cm^3 (a range of devices are available)	>180cm^3	180 cm^3	NIL
2	Accuracy	NIL/NA	98%	82%	NIL/NA
3	Response time	NIL/NA	~2 h	24 min	<1 min
4	Power requirements	20–100 μA, frequency ranging from 5 to 200 kHz	NIL/NA	80 mA current consumption, hence, a high-capacity battery is necessary	NIL

incorporating the Watson's formula in the algorithm and detection methodology used by sweat rate analysis will give the better response time to develop an effective device.

4 Conclusion

After careful evaluation of all the four methods, a comparison between the most important parameters was made. In comparison with the other three, it was found that an efficient and compact device was best suited for the sweat rate analysis method. Given the numerous devices developed for BIA method, a uniform size of the device could not be predicted. As per the papers which were referred to, FA-MS method showed the best accuracy—a 98%, closely followed by sweat rate analysis, at 82%. Within the bounds of the papers, the Watson's method had the fastest response time, while its accuracy could not be commented upon. The power requirements of both BIA and sweat rate analysis were presented, making BIA more power-efficient. After careful analysis of the data available from the papers which were referred to, the author suggests sweat rate analysis as the best method to measure the total body water.

References

1. Watson, P.E., Watson, I.D., Batt, R.D.: Total body water volumes for adult males and females estimated from simple anthropometric measurements. Am. J. Clin. Nutrition (1980)
2. Guyton, A.C., Saunders, W.B.: Textbook of Medical Physiology, 5th ed. Philadelphia (1976)
3. Jackson, S.: Anatomy & Physiology for Nurses. Nurses' Aids Series, 9th ed. Bailliere Tindall (1985)

4. Hansen, J.T., Koeppen, B.M.: Netter's Atlas of Human Physiology. Icon Learning Systems, Teterboro, N.J. (2002)
5. lou, A.H., Khamnei, S., Zamanlu, M.: The Effect of Water Temperature and Voluntary Drinking on the Post Rehydration Sweating (2013)
6. Porth, C.M., Kluwer, W., Williams, L., Wilkins: Essentials of pathophysiology -concepts of altered health states, 3rd ed. (2011)
7. Sawka, M.N., Cheuvront, S.N., Carter, R.S: Human Water Needs, PubMed Nutrition Reviews (2005)
8. Wolinsky, I., Hickson, J.F.: Nutrition in exercise and sport, 2nd ed. (1994)
9. Collins, M., Claros, E.: Recognizing the face of dehydration, PubMed (2011)
10. Maughan, R.: Impact of mild dehydration on wellness and on exercise performance, PubMed (2003)
11. Borghi, L., Meschi, T., Amato, F., Briganti, A., Novarini, A., Giannini, A.: Urinary volume, water and recurrences in idiopathic calcium nephrolithiasis: a 5-year randomized prospective study. PubMed (1996)
12. Read, N.W., Celik, A.F., Katsinelos, P.: Constipation and incontinence in the elderly. J. Clin. Gastroenterol. (1995)
13. Ship, J.A., Fischer, D.J.: The relationship between dehydration and parotid salivary gland function in young and older healthy adults. J. Gerontol. A Biol. Sci. Med. Sci. (1997)
14. Douglas, P., Ball, L., Mcguffin, L., Laur, C., Crowley, J., Rajput-ray, M., et al.: Hydration: knowledge, attitudes, and practices of UK dietitians. J. Biomed. Educ. (2015)
15. Smirnov, A.V., Nikolaev, D.V., Rudnev, S.G.: Bioelectrical impedance analysis. In: Handbook of Physics in Medicine and Biology, pp. 25.1–25.13. Chapter: Bioelectric impedance analysis (2010)
16. Chan, C., Smith, D., Spanel, P., McIntyre, C.W., Davies, S.J.: A Non-invasive, On-line Deuterium Dilution Technique for the Measurement of Total Body Water in Haemodialysis Patients (2008)
17. Lee, S.W., Song, J.H., Kim, G.A., Lee, K.J., Kim, M.-J.: Assessment of total body water from anthropometry-based equations using bioelectrical impedance as reference in Korean adult control and haemodialysis subjects. Nephrol. Dial. Transplant. **16** (2001)
18. Brueck, A., Iftekhar, T., Stannard, A.B., Yelamarthi, K., Kaya, T.: A Real-Time Wireless Sweat Rate Measurement System for Physical Activity Monitoring (2018)

A New Passage to the Reciprocity Attributes of Multidimensional IFT Through Those of the Multidimensional IHT

B. N. Madhukar and S. H. Bharathi

1 Introduction

Sieving is the sobriquet of signal processing and integral transforms are the saddle horses which are harnessed for transmogrifying the signals from temporal-fief into frequency-fief and contrarily [1, 2]. Although the IFT is the satrap of all the flourishing integral transforms prevalent in the literature, many other transforms do subsist, and one such clansperson of the IFT is the infinite Hartley transform (IHT) [3, 4]. Both IFT and IHT have sumptuous attributes such as the reciprocity attribute. This paper gives the reciprocity attribute of the IFT in multidimensions from the reciprocity attributes of the IHT in multidimensions. Although the IFT has its seedbeds to the modus operandi evolved by Jean-Baptiste Joseph Fourier in 1811 and promulgated by him in 1822, the IHT is of passably neoteric wellspring, i.e., it was excogitated by Ralph V. L. Hartley in 1942 [5, 15]. The IHT is a real transform, while the IFT is a complex transform, and both can be extended to N-dimensions (ND) [8]. This paper gives the provenances of the reciprocity attributes of multidimensional IFT through the reciprocity attributes of the IHT.

The grouping of the paper is wherefore: Barrios-II and III provide the portraitures of the IHT and IFT in multidimensions. Barrio-IV gives the mathematical kinship betwixt them. Barrio-V provides the provenances of the reciprocity attributes for IFT by using the reciprocity attributes for IHT. Barrio-VI gives the inferences drawn from the research work presented.

B. N. Madhukar (✉) · S. H. Bharathi
Reva University, Bengaluru, India

S. H. Bharathi
e-mail: bharathish@reva.edu.in

S. K V and K. Guruprasad Rao (eds.), *Smart Sensors Measurements and Instrumentation*, Lecture Notes in Electrical Engineering 750,
https://doi.org/10.1007/978-981-16-0336-5_11

2 Multidimensional IHT

This barrio gives the delineations of the onward and rearward transforms of IHT in one-, two-, three-, and N-dimensions.

2.1 1D-IHT

If q is a unidimensional incessant-temporal signal with unfathomable lifetime, then, its 1D-IHT is as [1, 6, 7],

$$\mathcal{Q}_H(\rho) = \mathfrak{H}[q(\gamma)]$$

$$\mathcal{Q}_H(\rho) \triangleq \int_{\gamma=-\infty}^{\infty} q(\gamma).cas(2\pi\rho\gamma)\,d\gamma\ . \tag{1}$$

Note that cas(□) = cos(□) + sin(□), where, □ is any real number. The cas(.) is called the cosine-and-sine kernel. Here, γ and ρ foreshow temporal- and frequency-variables apiece. The temporal-fief signal $q(\gamma)$is got from $\mathcal{Q}_H(\rho)$ by using the 1D-Inverse IHT (1D-IIHT) which is hemmed as [3, 9, 11, 21],

$$q(\gamma) = \mathfrak{H}^{-1}[\mathcal{Q}_H(\rho)]$$

$$q(\gamma) \triangleq \int_{\rho=-\infty}^{\infty} \mathcal{Q}_H(\rho).cas(2\pi\rho\gamma)d\rho\ . \tag{2}$$

If $q(\gamma)$and $\mathcal{Q}_H(\rho)$ subsist, they form a 1D-IHT brace.

$$q(\gamma) \overset{IHT}{\Longleftrightarrow} \mathcal{Q}_H(\rho)\ . \tag{3}$$

2.2 2D-IHT

If $q(\gamma_1,\gamma_2)$is a 2D incessant-temporal signal with unfathomable lifetime, then, its 2D-IHT is hemmed as [1, 13, 14],

$$\mathcal{Q}_H(\rho_1,\rho_2) = \mathfrak{H}[q(\gamma_1,\gamma_2)]$$

$$\mathcal{Q}_H(\rho_1,\rho_2) \triangleq \int_{-\infty}^{\infty}\int_{-\infty}^{\infty} q(\gamma_1,\gamma_2).\,cas[2\pi(\gamma_1\rho_1+\gamma_2\rho_2)]\,d\gamma_1 d\gamma_2\ . \tag{4}$$

Here, γ_1, γ_2 and ρ_1, ρ_2 foreshow temporal- and frequency-variables apiece. The temporal-fief signal $q(\gamma_1,\gamma_2)$is got from $\mathcal{Q}_H(\rho_1,\rho_2)$ by using the 2D-IIHT that is hemmed as [8, 9, 21, 22],

$\mathcal{q}(\gamma_1,\gamma_2) = \mathfrak{H}^{-1}[\mathcal{Q}_H(\rho_1,\rho_2)]$

$$\mathcal{q}(\gamma_1,\gamma_2) \triangleq \int_{-\infty}^{\infty}\int_{-\infty}^{\infty} \mathcal{Q}_H(\rho_1,\rho_2).cas[2\pi(\gamma_1\rho_1+\gamma_2\rho_2)]\,d\rho_1 d\rho_2 \,. \tag{5}$$

If $\mathcal{q}(\gamma_1,\gamma_2)$and $\mathcal{Q}_H(\rho_1,\rho_2)$ subsist, they silhouette a 2D-IHT brace.

$$\mathcal{q}(\gamma_1,\gamma_2) \overset{IHT}{\Leftrightarrow} \mathcal{Q}_H(\rho_1,\rho_2)\,. \tag{6}$$

2.3 3D-IHT

If $\mathcal{q}(\gamma_1,\gamma_2,\gamma_3)$is a 3D incessant-temporal signal with unfathomable lifetime, then, its 3D-IHT is hemmed as [1, 10, 11],

$\mathcal{Q}_H(\rho_1,\rho_2,\rho_3) = \mathfrak{H}[\mathcal{q}(\gamma_1,\gamma_2,\gamma_3)]$

$$\mathcal{Q}_H(\rho_1,\rho_2,\rho_3) \triangleq \int_{-\infty}^{\infty}\int_{-\infty}^{\infty}\int_{-\infty}^{\infty} \mathcal{q}(\gamma_1,\gamma_2,\gamma_3).cas[2\pi(\gamma_1\rho_1+\gamma_2\rho_2+\gamma_3\rho_3)]\,d\gamma_1 d\gamma_2 d\gamma_3\,. \tag{7}$$

Here, $\gamma_1,\gamma_2,\gamma_3$ and ρ_1,ρ_2,ρ_3 foreshow temporal- and frequency-variables apiece. The temporal-fief signal $\mathcal{q}(\gamma_1,\gamma_2,\gamma_3)$is got from $\mathcal{Q}_H(\rho_1,\rho_2,\rho_3)$ by using the 3D-IIHT that is hemmed as [6, 12, 13],

$\mathcal{q}(\gamma_1,\gamma_2,\gamma_3) = \mathfrak{H}^{-1}[\mathcal{Q}_H(\rho_1,\rho_2,\rho_3)]$

$$\mathcal{q}(\gamma_1,\gamma_2,\gamma_3) \triangleq \int_{-\infty}^{\infty}\int_{-\infty}^{\infty}\int_{-\infty}^{\infty} \mathcal{Q}_H(\rho_1,\rho_2,\rho_3).cas[2\pi(\gamma_1\rho_1+\gamma_2\rho_2+\gamma_3\rho_3)]\,d\rho_1 d\rho_2 d\rho_3\,. \tag{8}$$

If $\mathcal{q}(\gamma_1,\gamma_2,\gamma_3)$and $\mathcal{Q}_H(\rho_1,\rho_2,\rho_3)$ subsist, they silhouette a 3D-IHT brace.

$$\mathcal{q}(\gamma_1,\gamma_2,\gamma_3) \overset{IHT}{\Leftrightarrow} \mathcal{Q}_H(\rho_1,\rho_2,\rho_3)\,. \tag{9}$$

2.4 ND-IHT

If $\mathcal{q}(\gamma_1,\gamma_2,\ldots,\gamma_n)$is a ND incessant-temporal signal with unfathomable lifetime, then, its ND-IHT is hemmed as [1, 14, 15, 23],

$\mathcal{Q}_H(\rho_1,\rho_2,\ldots,\rho_n) = \mathfrak{H}[\mathcal{q}(\gamma_1,\gamma_2,\ldots,\gamma_n)]$

$$\mathcal{Q}_H(\rho_1,\rho_2,\ldots,\rho_n) \triangleq \int_{-\infty}^{\infty}\int_{-\infty}^{\infty}\cdots\int_{-\infty}^{\infty} q(\gamma_1,\gamma_2,\ldots,\gamma_n).cas[2\pi(\gamma_1\rho_1 + \gamma_2\rho_2 + \cdots + \gamma_n\rho_n)]d\gamma_1 d\gamma_2 \ldots d\gamma_n \,. \tag{10}$$

Here, $\gamma_1, \gamma_2, \ldots, \gamma_n$ and $\rho_1, \rho_2, \ldots, \rho_n$ foreshow temporal- and frequency-variables apiece. The temporal-fief signal $q(\gamma_1,\gamma_2,\ldots,\gamma_n)$is got from $\mathcal{Q}_H(\rho_1, \rho_2, \ldots, \rho_n)$ by using the ND-IIHT that is hemmed as [5, 9, 16, 19],

$$q(\gamma_1,\gamma_2,\ldots,\gamma_n) = \mathfrak{H}^{-1}[\mathcal{Q}_H(\rho_1,\rho_2,\ldots,\rho_n)]$$

$$q(\gamma_1,\gamma_2,\ldots,\gamma_n) \triangleq \int_{-\infty}^{\infty}\int_{-\infty}^{\infty}\cdots\int_{-\infty}^{\infty} \mathcal{Q}_H(\rho_1,\rho_2,\ldots,\rho_n).cas[2\pi(\gamma_1\rho_1 + \gamma_2\rho_2 + \cdots + \gamma_n\rho_n)]d\rho_1 d\rho_2 \ldots d\rho_n \,. \tag{11}$$

If $q(\gamma_1,\gamma_2,\ldots,\gamma_n)$and $\mathcal{Q}_H(\rho_1, \rho_2, \ldots, \rho_n)$ subsist, they silhouette a ND-IHT brace.

$$q(\gamma_1,\gamma_2,\ldots,\gamma_n) \overset{IHT}{\Leftrightarrow} \mathcal{Q}_H(\rho_1,\rho_2,\ldots,\rho_n) \,. \tag{12}$$

3 Multidimensional IFT

This barrio gives the delineations of the onward and backward transforms of IFT in one-, two-, three-, and N-dimensions.

3.1 1D-IFT

If $q(\gamma)$is a unidimensional incessant-temporal signal with unfathomable lifetime, then, its 1D-IFT is hemmed as [1, 9, 17, 20, 22]

$$\mathcal{Q}(\rho) = \mathfrak{F}[q(\gamma)]$$

$$\mathcal{Q}(\rho) \triangleq \int_{\gamma=-\infty}^{\infty} q(\gamma).e^{-j2\pi\rho\gamma}d\gamma \,. \tag{13}$$

Here, γ and ρ foreshow temporal- and frequency-variables apiece. Since the IFT is invertible and a linear transform, it is workable to find the temporal-fief signal from its frequency-fief coequal by using the 1D-inverse infinite Fourier transform (1D-IIFT). This is hemmed as [2, 4, 11, 14],

$$q(\gamma) = \mathfrak{F}^{-1}[\mathcal{Q}(\rho)]$$

$$q(\gamma) \triangleq \int_{\rho=-\infty}^{\infty} Q(\rho).e^{j2\pi\rho\gamma} d\rho . \tag{14}$$

If $q(\gamma)$and $Q(\rho)$ subsist, they silhouette a 1D-IFT brace.

$$q(\gamma) \overset{IFT}{\Leftrightarrow} Q(\rho) . \tag{15}$$

3.2 2D-IFT

If $q(\gamma_1, \gamma_2)$is a 2D incessant-temporal signal with unfathomable lifetime, then, its 2D-IFT is hemmed as [1, 13, 14, 21, 22],

$Q(\rho_1, \rho_2) = \mathfrak{F}[q(\gamma_1, \gamma_2)]$

$$Q(\rho_1, \rho_2) \triangleq \int_{-\infty}^{\infty} \int_{-\infty}^{\infty} q(\gamma_1, \gamma_2).e^{-j2\pi(\gamma_1\rho_1 + \gamma_2\rho_2)} d\gamma_1 d\gamma_2 . \tag{16}$$

Here, γ_1, γ_2 and ρ_1, ρ_2 foreshow temporal- and frequency-variables apiece. The temporal-fief signal $q(\gamma_1, \gamma_2)$is got from $Q(\rho_1, \rho_2)$ by using the 2D-IIFT that is hemmed as [15, 18, 19],

$q(\gamma_1, \gamma_2) = \mathfrak{F}^{-1}[Q(\rho_1, \rho_2)]$

$$q(\gamma_1, \gamma_2) \triangleq \int_{-\infty}^{\infty} \int_{-\infty}^{\infty} Q(\rho_1, \rho_2).e^{j2\pi(\gamma_1\rho_1 + \gamma_2\rho_2)} d\rho_1 d\rho_2 . \tag{17}$$

If $q(\gamma_1, \gamma_2)$and $Q(\rho_1, \rho_2)$ subsist, they silhouette a 2D-IFT brace.

$$q(\gamma_1, \gamma_2) \overset{IFT}{\Leftrightarrow} Q(\rho_1, \rho_2) . \tag{18}$$

3.3 3D-IFT

If $q(\gamma_1, \gamma_2, \gamma_3)$is a 3D incessant-temporal signal with unfathomable lifetime, then, its 3D-IFT is hemmed as [1, 6, 7, 18, 23],

$Q(\rho_1, \rho_2, \rho_3) = \mathfrak{F}[q(\gamma_1, \gamma_2, \gamma_3)]$

$$Q(\rho_1, \rho_2, \rho_3) \triangleq \int_{-\infty}^{\infty} \int_{-\infty}^{\infty} \int_{-\infty}^{\infty} q(\gamma_1, \gamma_2, \gamma_3).e^{-j2\pi(\gamma_1\rho_1 + \gamma_2\rho_2 + \gamma_3\rho_3)} d\gamma_1 d\gamma_2 d\gamma_3 . \tag{19}$$

Here, $\gamma_1, \gamma_2, \gamma_3$ and ρ_1, ρ_2, ρ_3 foreshow temporal- and frequency-variables apiece. The temporal-fief signal $q(\gamma_1, \gamma_2, \gamma_3)$is got from $\mathcal{Q}(\rho_1, \rho_2, \rho_3)$ by using the 3D-IIFT and is hemmed as [1, 14, 19],

$$q(\gamma_1,\gamma_2,\gamma_3) = \mathfrak{F}^{-1}[\mathcal{Q}(\rho_1,\rho_2,\rho_3)]$$

$$q(\gamma_1,\gamma_2,\gamma_3) \triangleq \int_{-\infty}^{\infty}\int_{-\infty}^{\infty}\int_{-\infty}^{\infty} \mathcal{Q}(\rho_1,\rho_2,\rho_3).\, e^{j2\pi(\gamma_1\rho_1+\gamma_2\rho_2+\gamma_3\rho_3)}\, d\rho_1 d\rho_2 d\rho_3\,. \tag{20}$$

If $q(\gamma_1,\gamma_2,\gamma_3)$and $\mathcal{Q}(\rho_1, \rho_2, \rho_3)$ subsist, they silhouette a 3D-IFT brace.

$$q(\gamma_1,\gamma_2,\gamma_3) \overset{IFT}{\Leftrightarrow} \mathcal{Q}(\rho_1,\rho_2,\rho_3)\,. \tag{21}$$

3.4 DN-IFT

If $q(\gamma_1,\gamma_2,\ldots,\gamma_n)$is a N D incessant-temporal signal with unfathomable lifetime, then, its ND-IFT is hemmed as [1, 4, 7, 14, 23],

$$\mathcal{Q}(\rho_1,\rho_2,\ldots,\rho_n) = \mathfrak{F}[q(\gamma_1,\gamma_2,\ldots,\gamma_n)]$$

$$\mathcal{Q}(\rho_1,\rho_2,\ldots,\rho_n) \triangleq \int_{-\infty}^{\infty}\int_{-\infty}^{\infty}\ldots\int_{-\infty}^{\infty} q(\gamma_1,\gamma_2,\ldots,\gamma_n).\, e^{-j2\pi(\gamma_1\rho_1+\gamma_2\rho_2+\cdots+\gamma_n\rho_n)}\, d\gamma_1 d\gamma_2 \ldots d\gamma_n\,. \tag{22}$$

Here, $\gamma_1, \gamma_2, \ldots, \gamma_n$ and $\rho_1, \rho_2, \ldots, \rho_n$ foreshow temporal- and frequency-variables apiece. The temporal-fief signal $q(\gamma_1,\gamma_2,\ldots,\gamma_n)$is got from $\mathcal{Q}(\rho_1, \rho_2, \ldots, \rho_n)$ by using the N D-IIFT which is hemmed as [16, 18–20],

$$q(\gamma_1,\gamma_2,\ldots,\gamma_n) = \mathfrak{F}^{-1}[\mathcal{Q}(\rho_1,\rho_2,\ldots,\rho_n)]$$

$$q(\gamma_1,\gamma_2,\ldots,\gamma_n) \triangleq \int_{-\infty}^{\infty}\int_{-\infty}^{\infty}\ldots\int_{-\infty}^{\infty} \mathcal{Q}(\rho_1,\rho_2,\ldots,\rho_n).\, e^{j2\pi(\gamma_1\rho_1+\gamma_2\rho_2+\cdots+\gamma_n\rho_n)}\, d\rho_1 d\rho_2 \ldots d\rho_n\,. \tag{23}$$

If $q(\gamma_1,\gamma_2,\ldots,\gamma_n)$and $\mathcal{Q}(\rho_1, \rho_2, \ldots, \rho_n)$ subsist, they silhouette a N D-IFT brace.

$$q(\gamma_1,\gamma_2,\ldots,\gamma_n) \overset{IFT}{\Leftrightarrow} \mathcal{Q}(\rho_1,\rho_2,\ldots,\rho_n)\,. \tag{24}$$

4 Kinship Betwixt IHT and IFT

The kinship betwixt IHT and IFT is given mathematically as [1, 7, 12, 19],

$$\mathcal{Q}(\rho) = \frac{1}{\sqrt{2}} e^{-j\frac{\pi}{4}} \mathcal{Q}_H(\rho) + \frac{1}{\sqrt{2}} e^{j\frac{\pi}{4}} \mathcal{Q}_H(-\rho). \tag{25}$$

Here, $\mathcal{Q}(\rho)$ is the IFT of $q(\gamma)$, and $\mathcal{Q}_H(\rho)$ is the IHT of $q(\gamma)$. Equation (25) can clench for multidimensional case also [14, 17, 22, 23]. This kinship and its ell in multidimensions are harnessed by and large in deriving the reciprocity attributes for multidimensional IFT from those of the multidimensional IHT in the next barrio.

5 Reciprocity Attributes for IFT from the Reciprocity Attributes of IHT

5.1 Reciprocity Attribute for 1D-IFT from the Reciprocity Attribute of 1D-IHT

Axiom 1 If $q(\gamma)$and $\mathcal{Q}(\rho)$ silhouette a 1D-IFT brace, then, $\mathcal{Q}(\gamma)$ and $q(-\rho)$silhouette a 1D-IFT brace. This connotes that if $\mathfrak{F}[q(\gamma)] = \mathcal{Q}(\rho)$then, $\mathfrak{F}[\mathcal{Q}(\gamma)] = q(-\rho)$.

Proof From (25), it is workable to compute the IFT foursquare from the IHT.

$$\Rightarrow \mathcal{Q}(\rho) = \frac{1}{\sqrt{2}} e^{-j\frac{\pi}{4}} \mathcal{Q}_H(\rho) + \frac{1}{\sqrt{2}} e^{j\frac{\pi}{4}} \mathcal{Q}_H(-\rho). \tag{26}$$

Applying the dogma of reciprocity to (26), we obtain

$$q(-\rho) = \frac{1}{\sqrt{2}} e^{-j\frac{\pi}{4}} q_H(\rho) + \frac{1}{\sqrt{2}} e^{j\frac{\pi}{4}} q_H(-\rho)\,. \tag{27}$$

But, we have

$$q_H(-\rho) = \int_{-\infty}^{\infty} \mathcal{Q}(\gamma).cas(-2\pi\rho\gamma)d\gamma = \int_{-\infty}^{\infty} \mathcal{Q}(\gamma).cos(2\pi\rho\gamma)d\gamma - \int_{-\infty}^{\infty} \mathcal{Q}(\gamma).sin(2\pi\rho\gamma)d\gamma\,. \tag{28}$$

This chaperones from the factuality that

$$\text{cas}(-\beth) = \text{cas}'(\beth) = \frac{d}{d\beth}[\text{cas}(\beth)] = \cos(\beth) - \sin(\beth).$$

Thence, (27) becomes

$$q(-\rho) = \frac{1}{\sqrt{2}} e^{-j\frac{\pi}{4}} \int_{-\infty}^{\infty} \mathcal{Q}(\gamma).cos(2\pi\rho\gamma)d\gamma + \frac{1}{\sqrt{2}} e^{-j\frac{\pi}{4}} \int_{-\infty}^{\infty} \mathcal{Q}(\gamma).sin(2\pi\rho\gamma)d\gamma + \frac{1}{\sqrt{2}} e^{j\frac{\pi}{4}} \int_{-\infty}^{\infty} \mathcal{Q}(\gamma).cos(2\pi\rho\gamma)d\gamma - \frac{1}{\sqrt{2}} e^{j\frac{\pi}{4}} \int_{-\infty}^{\infty} \mathcal{Q}(\gamma).sin(2\pi\rho\gamma)d\gamma$$

Emplacing the like terms and disentangling, we get

$$q(-\rho) = \int_{-\infty}^{\infty} Q(\gamma).[cos(2\pi\rho\gamma) - j\,sin(2\pi\rho\gamma)]d\gamma$$

$$q(-\rho) = \int_{-\infty}^{\infty} Q(\gamma).e^{-j2\pi\rho\gamma}d\gamma = \mathfrak{F}[Q(\gamma)]\,. \tag{29}$$

Thence, we have vouched that if $\mathfrak{F}[q(\gamma)] = Q(\rho)$then,

$$\mathfrak{F}[Q(\gamma)] = q(-\rho)\,. \tag{30}$$

5.2 *Reciprocity Attribute for 2D-IFT from the Reciprocity Attribute of 2D-IHT*

Axiom 2 If $q(\gamma_1,\gamma_2)$and $Q(\rho_1,\rho_2)$ silhouette a 2D-IFT brace, then,$Q(\gamma_1,\gamma_2)$ and $q(-\rho_1,-\rho_2)$silhouette a 2D-IFT brace. This connotes that if $\mathfrak{F}[q(\gamma_1,\gamma_2)\,] = Q(\rho_1,\rho_2)$then, $\mathfrak{F}[Q(\gamma_1,\gamma_2)] = q(-\rho_1,-\rho_2)$.

Proof From (25), it is workable to compute the IFT foursquare from the IHT in 2D as

$$Q(\rho_1,\rho_2) = \frac{1}{\sqrt{2}}e^{-j\frac{\pi}{4}}Q_H(\rho_1,\rho_2) + \frac{1}{\sqrt{2}}e^{j\frac{\pi}{4}}Q_H(-\rho_1,-\rho_2). \tag{31}$$

Applying the dogma of reciprocity to (31), we obtain

$$q(-\rho_1,-\rho_2) = \frac{1}{\sqrt{2}}e^{-j\frac{\pi}{4}}q_H(\rho_1,\rho_2) + \frac{1}{\sqrt{2}}e^{j\frac{\pi}{4}}q_H(-\rho_1,-\rho_2)\,. \tag{32}$$

But, we have
$q_H(-\rho_1,-\rho_2) = \int_{-\infty}^{\infty}\int_{-\infty}^{\infty} Q(\gamma_1,\gamma_2).cas[-2\pi(\gamma_1\rho_1 + \gamma_2\rho_2)]\,d\gamma_1 d\gamma_2$
Using the argument that we harnessed in Barrio 5.1, we obtain

$$\Rightarrow q_H(-\rho_1,-\rho_2) = \int_{-\infty}^{\infty}\int_{-\infty}^{\infty} Q(\gamma_1,\gamma_2).cos[2\pi(\gamma_1\rho_1 + \gamma_2\rho_2)]\,d\gamma_1 d\gamma_2 - \int_{-\infty}^{\infty}\int_{-\infty}^{\infty} Q(\gamma_1,\gamma_2).sin[2\pi(\gamma_1\rho_1 + \gamma_2\rho_2)]\,d\gamma_1 d\gamma_2\,. \tag{33}$$

Thence, (32) becomes

$$q(-\rho_1,-\rho_2) = \frac{1}{\sqrt{2}}e^{-j\frac{\pi}{4}}\int_{-\infty}^{\infty}\int_{-\infty}^{\infty} Q(\gamma_1,\gamma_2).cos[2\pi(\gamma_1\rho_1 + \gamma_2\rho_2)]\,d\gamma_1 d\gamma_2 + \frac{1}{\sqrt{2}}e^{-j\frac{\pi}{4}}\int_{-\infty}^{\infty}\int_{-\infty}^{\infty} Q(\gamma_1,\gamma_2).sin[2\pi(\gamma_1\rho_1 + \gamma_2\rho_2)]\,d\gamma_1 d\gamma_2 + \frac{1}{\sqrt{2}}e^{j\frac{\pi}{4}}\int_{-\infty}^{\infty}\int_{-\infty}^{\infty} Q(\gamma_1,\gamma_2).cos[2\pi(\gamma_1\rho_1 + \gamma_2\rho_2)]\,d\gamma_1 d\gamma_2 - \frac{1}{\sqrt{2}}e^{j\frac{\pi}{4}}\int_{-\infty}^{\infty}\int_{-\infty}^{\infty} Q(\gamma_1,\gamma_2).sin[2\pi(\gamma_1\rho_1 + \gamma_2\rho_2)]\,d\gamma_1 d\gamma_2$$

Emplacing the like terms and disentangling, we get

$$\Rightarrow q(-\rho_1,-\rho_2) = \int_{-\infty}^{\infty}\int_{-\infty}^{\infty} Q(\gamma_1,\gamma_2).\{cos[2\pi(\gamma_1\rho_1+\gamma_2\rho_2)] - j\,sin[2\pi(\gamma_1\rho_1+\gamma_2\rho_2)]\}\,d\gamma_1 d\gamma_2$$

$$\Rightarrow q(-\rho_1,-\rho_2) = \int_{-\infty}^{\infty}\int_{-\infty}^{\infty} Q(\gamma_1,\gamma_2).\,e^{-j2\pi(\gamma_1\rho_1+\gamma_2\rho_2)}d\gamma_1 d\gamma_2 = \mathfrak{F}[Q(\gamma_1,\gamma_2)]\,. \tag{34}$$

Thence, we have vouched that if $\mathfrak{F}[q(\gamma_1,\gamma_2)] = Q(\rho_1,\rho_2)$then,

$$\mathfrak{F}[Q(\gamma_1,\gamma_2)] = q(-\rho_1,-\rho_2)\,. \tag{35}$$

5.3 *Reciprocity Attribute for 3D-IFT from the Reciprocity Attribute of 3D-IHT*

Axiom 3 If $q(\gamma_1,\gamma_2,\gamma_3)$and $Q(\rho_1,\rho_2,\rho_3)$ silhouette a 3D-IFT brace, then$Q(\gamma_1,\gamma_2,\gamma_3)$ and $q(-\rho_1,-\rho_2,-\rho_3)$silhouette a 3D-IFT brace. This connotes that if $\mathfrak{F}[q(\gamma_1,\gamma_2,\gamma_3)\,] = Q(\rho_1,\rho_2,\rho_3)$then, $\mathfrak{F}[Q(\gamma_1,\gamma_2,\gamma_3)] = q(-\rho_1,-\rho_2,-\rho_3)$..

Proof From (25), it is workable to compute the IFT foursquare from the IHT in 3D as

$$Q(\rho_1,\rho_2,\rho_3) = \frac{1}{\sqrt{2}}e^{-j\frac{\pi}{4}}Q_H(\rho_1,\rho_2,\rho_3) + \frac{1}{\sqrt{2}}e^{j\frac{\pi}{4}}Q_H(-\rho_1,-\rho_2,-\rho_3). \tag{36}$$

Applying the dogma of reciprocity to (36), we obtain

$$q(-\rho_1,-\rho_2,-\rho_3) = \frac{1}{\sqrt{2}}[e^{-j\frac{\pi}{4}}q_H(\rho_1,\rho_2,\rho_3) + e^{j\frac{\pi}{4}}q_H(-\rho_1,-\rho_2,-\rho_3)]\,. \tag{37}$$

But,we have

$$q_H(-\rho_1,-\rho_2,-\rho_3) = \int_{-\infty}^{\infty}\int_{-\infty}^{\infty}\int_{-\infty}^{\infty} Q(\gamma_1,\gamma_2,\gamma_3).\,cas[-2\pi(\gamma_1\rho_1+\gamma_2\rho_2+\gamma_3\rho_3)]d\gamma_1 d\gamma_2 d\gamma_3$$

Using the argument that we harnessed in Barrio 5.1, we obtain

$$\Rightarrow q(-\rho_1,-\rho_2,-\rho_3) = \int_{-\infty}^{\infty}\int_{-\infty}^{\infty}\int_{-\infty}^{\infty} Q(\gamma_1,\gamma_2,\gamma_3)\,.\,cos[2\pi(\gamma_1\rho_1+\gamma_2\rho_2+\gamma_3\rho_3)]d\gamma_1 d\gamma_2 d\gamma_3 - \int_{-\infty}^{\infty}\int_{-\infty}^{\infty}\int_{-\infty}^{\infty} Q(\gamma_1,\gamma_2,\gamma_3)\,.\,sin[2\pi(\gamma_1\rho_1+\gamma_2\rho_2+\gamma_3\rho_3)]d\gamma_1 d\gamma_2 d\gamma_3\,. \tag{38}$$

Thence, (37) becomes

$$\Rightarrow \mathcal{q}(-\rho_1,-\rho_2,-\rho_3) = \frac{1}{\sqrt{2}} e^{-j\frac{\pi}{4}} \int_{-\infty}^{\infty}\int_{-\infty}^{\infty}\int_{-\infty}^{\infty} \mathcal{Q}(\gamma_1,\gamma_2,\gamma_3). cos[2\pi(\gamma_1\rho_1 + \gamma_2\rho_2 + \gamma_3\rho_3)] d\gamma_1 d\gamma_2 d\gamma_3 + \frac{1}{\sqrt{2}} e^{-j\frac{\pi}{4}} \int_{-\infty}^{\infty}\int_{-\infty}^{\infty}\int_{-\infty}^{\infty} \mathcal{Q}(\gamma_1,\gamma_2,\gamma_3). sin[2\pi(\gamma_1\rho_1 + \gamma_2\rho_2 + \gamma_3\rho_3)] d\gamma_1 d\gamma_2 d\gamma_3 + \frac{1}{\sqrt{2}} e^{j\frac{\pi}{4}} \int_{-\infty}^{\infty}\int_{-\infty}^{\infty}\int_{-\infty}^{\infty} \mathcal{Q}(\gamma_1,\gamma_2,\gamma_3). cos[2\pi(\gamma_1\rho_1 + \gamma_2\rho_2 + \gamma_3\rho_3)] d\gamma_1 d\gamma_2 d\gamma_3 - \frac{1}{\sqrt{2}} e^{j\frac{\pi}{4}} \int_{-\infty}^{\infty}\int_{-\infty}^{\infty}\int_{-\infty}^{\infty} \mathcal{Q}(\gamma_1,\gamma_2,\gamma_3). sin[2\pi(\gamma_1\rho_1 + \gamma_2\rho_2 + \gamma_3\rho_3)] d\gamma_1 d\gamma_2 d\gamma_3$$

Emplacing the like terms and disentangling, we get

$$\Rightarrow \mathcal{q}(-\rho_1,-\rho_2,-\rho_3) = \int_{-\infty}^{\infty}\int_{-\infty}^{\infty}\int_{-\infty}^{\infty} \mathcal{Q}(\gamma_1,\gamma_2,\gamma_3).\{cos[2\pi(\gamma_1\rho_1 + \gamma_2\rho_2 + \gamma_3\rho_3)] - j\, sin[2\pi(\gamma_1\rho_1 + \gamma_2\rho_2 + \gamma_3\rho_3)]\} d\gamma_1 d\gamma_2 d\gamma_3$$

$$\Rightarrow \mathcal{q}(-\rho_1,-\rho_2,-\rho_3) = \int_{-\infty}^{\infty}\int_{-\infty}^{\infty}\int_{-\infty}^{\infty} \mathcal{Q}(\gamma_1,\gamma_2,\gamma_3). e^{-j2\pi(\gamma_1\rho_1+\gamma_2\rho_2+\gamma_3\rho_3)} d\gamma_1 d\gamma_2 d\gamma_3 = \mathfrak{F}[G(\gamma_1,\gamma_2,\gamma_3)]. \tag{39}$$

Thence, we have vouched that if then $\mathfrak{F}[\mathcal{q}(\gamma_1,\gamma_2,\gamma_3)] = \mathcal{Q}(\rho_1,\rho_2,\rho_3)$

$$\mathfrak{F}[\mathcal{Q}(\gamma_1,\gamma_2,\gamma_3)] = \mathcal{q}(-\rho_1,-\rho_2,-\rho_3). \tag{40}$$

5.4 *Reciprocity Attribute for D-IFT from the Reciprocity Attribute of ND-IHT*

Axiom 4 If $\mathcal{q}(\gamma_1,\gamma_2,\ldots,\gamma_n)$and $\mathcal{Q}(\rho_1,\rho_2,\ldots,\rho_n)$ silhouette a N D-IFT brace, then,$\mathcal{Q}(\gamma_1,\gamma_2,\ldots,\gamma_n)$ and $\mathcal{q}(-\rho_1,-\rho_2,\ldots,-\rho_n)$silhouette a N D-IFT brace. This connotes that if $\mathfrak{F}[\mathcal{q}(\gamma_1,\gamma_2,\ldots\gamma_n)] = \mathcal{Q}(\rho_1,\rho_2,\ldots,\rho_n)$then, $\mathfrak{F}[\mathcal{Q}(\gamma_1,\gamma_2,\ldots\gamma_n)] = \mathcal{q}(-\rho_1,-\rho_2,\ldots,-\rho_n)$.

Proof From (25), it is workable to compute the IFT foursquare from the IHT in ND as

$$\mathcal{Q}(\rho_1,\rho_2,\ldots,\rho_n) = \frac{1}{\sqrt{2}} e^{-j\frac{\pi}{4}} \mathcal{Q}_H(\rho_1,\rho_2,\ldots,\rho_n) + \frac{1}{\sqrt{2}} e^{j\frac{\pi}{4}} \mathcal{Q}_H(-\rho_1,-\rho_2,\ldots,-\rho_n) \tag{41}$$

Applying the dogma of reciprocity to (41), we obtain

$$\mathcal{q}(-\rho_1,-\rho_2,\ldots,-\rho_n) = \frac{1}{\sqrt{2}}[e^{-j\frac{\pi}{4}} \mathcal{q}_H(\rho_1,\rho_2,\ldots,\rho_n) + e^{j\frac{\pi}{4}} \mathcal{q}_H(-\rho_1,-\rho_2,\ldots,-\rho_n)]. \tag{42}$$

But, we have

$$g_H(-\rho_1, -\rho_2, \ldots, -\rho_n) = \int_{-\infty}^{\infty}\int_{-\infty}^{\infty}\cdots\int_{-\infty}^{\infty} Q(\gamma_1, \gamma_2, \ldots \gamma_n) \times cas[-2\pi(\gamma_1\rho_1 + \gamma_2\rho_2 + \ldots + \gamma_n\rho_n)]d\gamma_1 d\gamma_2 \ldots d\gamma_n. \tag{43}$$

Using the argument that we harnessed in Barrio 5.1, (42) becomes

$$\Rightarrow q(-\rho_1, -\rho_2, \ldots, -\rho_n) = \frac{1}{\sqrt{2}}e^{-j\frac{\pi}{4}}\int_{-\infty}^{\infty}\int_{-\infty}^{\infty}\ldots\int_{-\infty}^{\infty} Q(\gamma_1, \gamma_2, \ldots \gamma_n).cos[2\pi(\gamma_1\rho_1 + \gamma_2\rho_2 + \cdots + \gamma_n\rho_n)]\, d\gamma_1 d\gamma_2 \ldots d\gamma_n + \frac{1}{\sqrt{2}}e^{-j\frac{\pi}{4}}\int_{-\infty}^{\infty}\int_{-\infty}^{\infty}\ldots\int_{-\infty}^{\infty} Q(\gamma_1, \gamma_2, \ldots \gamma_n).sin[2\pi(\gamma_1\rho_1 + \gamma_2\rho_2 + \cdots + \gamma_n\rho_n)]\, d\gamma_1 d\gamma_2 \ldots d\gamma_n + \frac{1}{\sqrt{2}}e^{j\frac{\pi}{4}}\int_{-\infty}^{\infty}\int_{-\infty}^{\infty}\ldots\int_{-\infty}^{\infty} Q(\gamma_1, \gamma_2, \ldots \gamma_n).cos[2\pi(\gamma_1\rho_1 + \gamma_2\rho_2 + \cdots + \gamma_n\rho_n)]\, d\gamma_1 d\gamma_2 \ldots d\gamma_n - \frac{1}{\sqrt{2}}e^{j\frac{\pi}{4}}\int_{-\infty}^{\infty}\int_{-\infty}^{\infty}\ldots\int_{-\infty}^{\infty} Q(\gamma_1, \gamma_2, \ldots \gamma_n).sin[2\pi(\gamma_1\rho_1 + \gamma_2\rho_2 + \cdots + \gamma_n\rho_n)]\, d\gamma_1 d\gamma_2 \ldots d\gamma_n$$

Emplacing the like terms and disentangling, we get

$$\Rightarrow q(-\rho_1, -\rho_2, \ldots, -\rho_n) = \int_{-\infty}^{\infty}\int_{-\infty}^{\infty}\ldots\int_{-\infty}^{\infty} Q(\gamma_1, \gamma_2, \ldots \gamma_n).\{cos[2\pi(\gamma_1\rho_1 + \gamma_2\rho_2 + \cdots + \gamma_n\rho_n)] - j\ sin[2\pi(\gamma_1\rho_1 + \gamma_2\rho_2 + \cdots + \gamma_n\rho_n)]\}d\gamma_1 d\gamma_2 \ldots d\gamma_n$$

$$\Rightarrow q(-\rho_1, -\rho_2, \ldots, -\rho_n) = \int_{-\infty}^{\infty}\int_{-\infty}^{\infty}\ldots\int_{-\infty}^{\infty} Q(\gamma_1, \gamma_2, \ldots \gamma_n).e^{-j2\pi(\gamma_1\rho_1 + \gamma_2\rho_2 + \cdots + \gamma_n\rho_n)}d\gamma_1 d\gamma_2 \ldots d\gamma_n = \mathfrak{F}[G(\gamma_1, \gamma_2, \ldots \gamma_n)]. \tag{44}$$

Thence, we have vouched that if $\mathfrak{F}[q(\gamma_1, \gamma_2, \ldots \gamma_n)] = Q(\rho_1, \rho_2, \ldots, \rho_n)$then,

$$\mathfrak{F}[Q(\gamma_1, \gamma_2, \ldots \gamma_n)] = q(-\rho_1, -\rho_2, \ldots, -\rho_n) \tag{45}$$

6 Conclusions

Trailblazing modi-operandi for evaluating the reciprocity attributes for the IFT in multidimensions from the reciprocity attributes for the IHT is proffered in this paper. The reciprocity attributes of IFT and IHT have commodious ambulatory usages in signal processing in multidimensions. These attributes can be ultrapractical for signals that have the same fettle in both the temporal- and frequency-fiefs which is routinely espied in signal processing. For a savvied IHT and IFT brace, these attributes abet in computing the onward transform or the rearward transform sans veritably acquisitioning to guesstimate the integrals which can ever and anon be ho-hum. There are sanguine cases in signal processing wherein both IFT and IHT are necessitous at the selfsame emprise. In such loci classici, we can harness these attributes along with their kinship in foursquare evaluating them without having to guesstimate them using their integral portraitures. The usages of this attribute abet in the scrimping of reckoning time and expense staggeringly. Also, the IFT being a complex transform takes withal time to guesstimate, whilst the IHT being a real transform takes nickel-and-dime time. Thence, in order to guesstimate the IFT of a function, it is worthwhile to guesstimate its IHT and then use reciprocity attributes philosophized in the previous barrio to get the desired IFT which definitely speeds up the process of computation. This approach can be harnessed for solving integral equations, integro-differential equations, etc.

Acknowledgements The scriveners are chuffed to their kindred constituents for imbibing the appetency for research in them. They are also gratified to the stewardship of Reva University, Bengaluru, for roseating us to tote out this research work.

References

1. Madhukar, B.N., Bharathi, S.H.: An artery to the reciprocity properties of N-dimensional infinite Hartley transform through N-dimensional infinite Fourier transform. Int. J. Adv. Sci. Technol. **29**, 8591–8599 (2020)
2. Pinsky, M.: Introduction to Fourier Analysis and Wavelets. Cengage Learning India Private Limited, New Delhi (2015)
3. Vretblad, A.: Fourier Analysis and its Applications. Springer India Private Limited, New Delhi (2016)
4. Grafakos, L.: Classical and Modern Fourier Analysis. Springer, New York (2018)
5. Boashash, B.: Time-Frequency Signal Analysis and Processing—A Comprehensive Reference. Reed-Elsevier Private Limited, New York (2003)
6. Champeney, D.C.: A Handbook of Fourier Theorems. Cambridge University Press, New York (1987)
7. Madhukar, B.N., Jain, S.: A duality theorem for infinite Hartley transform. In: IEEE 5th International Conference on Communication and Signal Processing 2016 (ICCSP '16), pp. 109–113. Melmaruvathur (2016)
8. Bracewell, R.N.: The Hartley Transform. Oxford University Press, London (1986)
9. Madhukar, B.N., Bharathi. S.H.: A new property of the discrete cosine transform-IV (DCT-IV). In: IEEE International Conference on Communication and Signal Processing 2018 (ICCSP '18), pp. 0144–0147 (2018)
10. Madhukar, B.N., Jain, S.: A duality theorem for the discrete sine transform (DST). In: IEEE International Conference on Applied and Theoretical Computing and Communication Technology (iCATccT—2015), pp. 156–160 (2015)
11. Madhukar, B.N., Jain, S., Satyanarayana, P.S.: Duality theorem for discrete Hartley transform. Int. J. Appl. Eng. Res. (IJAER) **10**, 34699–34703 (2015)
12. Madhukar, B.N., Jain, S.: A Duality theorem for the discrete sine transform—IV (DST–IV). In: IEEE 3rd International Conference on Advanced Computing and Communication Systems—2016 (ICACCS—2016), pp. 1–6 (2016)
13. Madhukar, B.N., Bharathi, S.H.: Similarity properties for the offshoots of the generalized discrete Hartley transform (GDHT). In: 2019 IEEE International Conference on Distributed Computing, VLSI, Electrical Circuits and Robotics (DISCOVER), pp. 205–209 (2019)
14. Madhukar, B.N., Bharathi, S.H.: Duality properties for the variants of the discrete W transform (DWT). In: 2019 IEEE International Conference on Distributed Computing, VLSI, Electrical Circuits and Robotics (DISCOVER), pp. 318–322 (2019)

15. Hartley, R.V.L.: A more symmetrical Fourier analysis applied to transmission problems. In: IRE, pp. 144–150 (1942)
16. Bochner, S., Chandrasekharan, K.: Fourier Transforms. Princeton University Press, New Jersey (1949)
17. Stein, E., Shakarchi, R.: Fourier analysis: an introduction. Princeton University Press, New Jersey (2003)
18. Rabiner, L.R., Gold, B.: Theory and Applications of Digital Signal Processing. Pearson India, Inc., New Delhi (2018)
19. Oppenheim, A.V., Schafer, R.W.: Discrete-time Signal Processing. Pearson Education Inc., New Delhi (2020)
20. Dudgeon, D., Mersereau, M.R.: Multidimensional Digital Signal Processing. Prentice-Hall, New Jersey (1984)
21. Madhukar, B.N., Bharathi, S.H.: Similarity properties for two versions of 1D-, 2D-, 3D-, and ND-infinite Hartley transform (IHT). Int. J. Adv. Sci. Technol. **29**, 11267–11274 (2020)
22. Buneman, O.: Multidimensional Hartley transforms. In: Proceedings of the IEEE, 75, pp. 267 (1987)
23. Gonzalez, R.C., Woods, R.E.: Digital Image Processing. Pearson Education Inc., New Delhi (2020)

Exact Disturbance Estimation for a Class of System

Surajkumar Sawai and Shailaja Kurode

1 Introduction

Disturbance often affects performance of all engineering systems. It affects system performance. Therefore disturbance rejection is key the objective in the control system. When a disturbance is measurable it is possible to eliminate the influence of disturbance but in case of external disturbance which is not measurable then the way to deal with it is to estimate it and the cancel out its effect using an estimate of disturbance [1]. Disturbance estimation therefore has gained much importance.

Nowadays, all the controllers are implemented digitally due to advancements in the digital device. Digital implementation of the controller requires the information at a sampled instant. For the compensating effect of disturbance in a digitally controlled system, its estimate is required at every sampling instant.

Many researcher have worked on discrete-time disturbance observers. These approaches are divided into two categories [2]. These observers are designed wither in the transfer function domain or state space domain. The approach based on transfer function is straightforward because it uses an inverse dynamics of plant model via Q-filters for estimating disturbance [3, 4]. The disadvantage of this approach is that it does not apply to a multi-input multi-output (MIMO) system [5]. The state space approaches are a combination of state estimator and disturbance estimator [6, 7]. The disadvantage of this method is that it estimates the disturbance with bounded error.

In this paper, a new disturbance observer is proposed based on fast output sampling (FOS). In FOS, samples are taken at a faster rate than the input [8]. Stack of output

S. Sawai (✉) · S. Kurode
College of Engineering, Pune, India
e-mail: ssv18.elec@coep.ac.in

S. Kurode
e-mail: srk.elec@coep.ac.in

S. K V and K. Guruprasad Rao (eds.), *Smart Sensors Measurements and Instrumentation*, Lecture Notes in Electrical Engineering 750,
https://doi.org/10.1007/978-981-16-0336-5_12

samples was used to estimate states in [9, 10]. These states are used to construct control. This approach is called multirate output feedback (MROF). Application specific contributions are found in [11–13]. MROF algorithm has been examined in the framework of ESO in [14]. However, it shows a bounded error. In this paper, a more exact estimation is attempted.

This paper is organized as follows. In Sect. 2, the problem statement is described. In Sect. 3, a proposal for more accurate disturbance estimation is illustrated. Section 4 presents the simulation results using a numerical example. Paper ends with Sect. 5 which presents concluding remarks.

2 Problem Statement

Consider the general Single Input Single Output (SISO) continuous time system as follow

$$\left.\begin{aligned} \dot{x}_1 &= x_2 \\ \dot{x}_2 &= a_1 x_1 + a_2 x_2 + a_3 u + d \end{aligned}\right\} \quad (1)$$

where, x_1 & x_2 are considered as states of system, u is input to system, d is lumped disturbance, a_1, a_2 and a_3 are system parameters.

The discrete time representation of (1) is as follow,

$$\mathbf{x}(k+1) = \Phi_\tau \mathbf{x}(k) + \Gamma_\tau u(k) + D_\tau d(k) \quad (2)$$

where, $\mathbf{x}(k)$ is states vector of system and $\mathbf{x}(k) \in \Re^{2\times 1}$, $u(k)$ is input and $u(k) \in \Re^{1\times 1}$ and $d(k)$ is lumped disturbance.

From (2) disturbance at $(k-1)$th instant can be obtained [15] and is given by

$$D_\tau d(k-1) = \mathbf{x}(k) - \Phi_\tau \mathbf{x}(k-1) + \Gamma_\tau u(k-1) \quad (3)$$

The above disturbance estimate is an approximate estimation of $D_\tau d(k)$. Su et al. proposes this observer, and we term it as "Su's observer". The approximation of $D_\tau d(k)$ by $D_\tau d(k-1)$ is valid if the disturbance is slow varying. Herein we propose more accurate disturbance estimation.

3 Proposal of More Accurate Disturbance Estimation

The objective of this section is to develop the disturbance estimator. The proposed disturbance observer scheme is shown in Fig. 1.

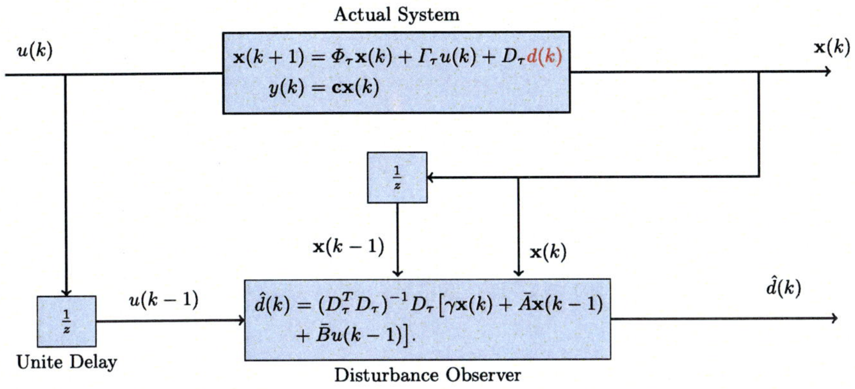

Fig. 1 Proposed disturbance observer

The states at $((k-1)+\Delta)$th instant are related to states at $(k-1)$th instant by the following difference equation.

$$\mathbf{x}((k-1)+\Delta) = \Phi_{\Delta}\mathbf{x}(k-1) + \Gamma_{\Delta}u(k-1) + D_{\Delta}d(k-1). \tag{4}$$

At $((k-1)+2\Delta)$th instant,

$$\begin{aligned}\mathbf{x}((k-1)+2\Delta) = \Phi_{\Delta}\mathbf{x}((k-1)+\Delta) + \Gamma_{\Delta}u(k-1) \\ + D_{\Delta}d((k-1)+\Delta).\end{aligned} \tag{5}$$

Substituting (4) into (5) gives,

$$\begin{aligned}\mathbf{x}((k-1)+2\Delta) = \Phi_{\Delta}\big[\Phi_{\Delta}\mathbf{x}(k-1) + \Gamma_{\Delta}u(k-1) \\ + D_{\Delta}d(k-1)\big] + \Gamma_{\Delta}u(k-1) \\ + D_{\Delta}d((k-1)+\Delta).\end{aligned} \tag{6}$$

Solving (6) gives,

$$\begin{aligned}\mathbf{x}((k-1)+2\Delta) = \Phi_{\Delta}^{2}\mathbf{x}(k-1) + (\Phi_{\Delta}\Gamma_{\Delta} + \Gamma_{\Delta})u(k-1) \\ + \Phi_{\Delta}D_{\Delta}d(k-1) + D_{\Delta}d(k-1+\Delta).\end{aligned} \tag{7}$$

The states at $((k-1)+2\Delta)$th be related to $\mathbf{x}(k-1)$ by the following relation.

$$\begin{aligned}\mathbf{x}((k-1)+2\Delta) = \Phi_{2\Delta}\mathbf{x}(k-1) + \Gamma_{2\Delta}u(k-1) \\ + D_{2\Delta}d(k-1).\end{aligned} \tag{8}$$

Solving (7) and (8) for $D_{\Delta}d((k-1)+\Delta)$,

$$D_{\Delta}d((k-1)+\Delta) = (\Phi_{2\Delta} - \Phi_{\Delta}^2)\mathbf{x}(k-1) + (D_{2\Delta} - \Phi_{\Delta}D_{\Delta})d(k-1) + (\Gamma_{2\Delta} - \Phi_{\Delta}\Gamma_{\Delta} - \Gamma_{\Delta})u(k-1). \quad (9)$$

If $\Delta \to \tau$ sec, then (9) becomes,

$$D_{\tau}d(k) = (\Phi_{2\tau} - \Phi_{\tau}^2)\mathbf{x}(k-1) + (\Gamma_{2\tau} - \Phi_{\tau}\Gamma_{\tau} - \Gamma_{\tau})u(k-1) + (D_{2\tau} - \Phi_{\tau}D_{\tau})d(k-1). \quad (10)$$

If $(D_{2\tau} - \Phi_{\tau}D_{\tau}) = \gamma D_{\tau}$ then above equation becomes,

$$D_{\tau}d(k) = (\Phi_{2\tau} - \Phi_{\tau}^2)\mathbf{x}(k-1) + (\Gamma_{2\tau} - \Phi_{\tau}\Gamma_{\tau} - \Gamma_{\tau})u(k-1) + \gamma D_{\tau}d(k-1). \quad (11)$$

Substituting for $D_{\tau}d(k-1)$ from (2), then (11) becomes

$$D_{\tau}d(k) = (\Phi_{2\tau} - \Phi_{\tau}^2)\mathbf{x}(k-1) + (\Gamma_{2\tau} - \Phi_{\tau}\Gamma_{\tau} - \Gamma_{\tau})u(k-1) + \gamma(\mathbf{x}(k) - \Phi_{\tau}\mathbf{x}(k-1) + \Gamma_{\tau}u(k-1)). \quad (12)$$

Rearranging above equation gives,

$$D_{\tau}d(k) = \gamma\mathbf{x}(k) + (\Phi_{2\tau} - \Phi_{\tau}^2 - \gamma\Phi_{\tau})\mathbf{x}(k-1) + (\Gamma_{2\tau} - \Phi_{\tau}\Gamma_{\tau} - \Gamma_{\tau} - \gamma\Gamma_{\tau})u(k-1). \quad (13)$$

Defining, $\bar{A} = (\Phi_{2\tau} - \Phi_{\tau}^2 - \gamma\Phi_{\tau})$ and $\bar{B} = (\Gamma_{2\tau} - \Phi_{\tau}\Gamma_{\tau} - \Gamma_{\tau} - \gamma\Gamma_{\tau})$, then (13) becomes,

$$D_{\tau}d(k) = \gamma\mathbf{x}(k) + \bar{A}\mathbf{x}(k-1) + \bar{B}u(k-1). \quad (14)$$

Solving for $d(k)$ to get,

$$d(k) = (D_{\tau}^T D_{\tau})^{-1}D_{\tau}\big(\gamma\mathbf{x}(k) + \bar{A}\mathbf{x}(k-1) + \bar{B}u(k-1)\big). \quad (15)$$

This is more exact estimate. To get it we do not actually sample the states at Δ rate. This is added advantage.

4 Simulation Results

Numerical simulation were carried out to verify the proposed method of disturbance estimation. Following continuous time system is considered,

$$\begin{bmatrix} \dot{x}_1 \\ \dot{x}_2 \end{bmatrix} = \begin{bmatrix} -900 & 0 \\ 200 & -700 \end{bmatrix} \begin{bmatrix} x_1 \\ x_2 \end{bmatrix} + \begin{bmatrix} 0 \\ 100 \end{bmatrix} u + \begin{bmatrix} 0 \\ 100 \end{bmatrix} d \tag{16}$$

The discrete stime representation of above system sampled at $\tau = 0.001$ s using Euler's first order approximation is given by,

$$\begin{bmatrix} x_1(k+1) \\ x_2(k+1) \end{bmatrix} = \begin{bmatrix} 0.1 & 0 \\ 0.2 & 0.3 \end{bmatrix} \begin{bmatrix} x_1(k) \\ x_2(k) \end{bmatrix} + \begin{bmatrix} 0 \\ 0.1 \end{bmatrix} u(k) + \begin{bmatrix} 0 \\ 0.1 \end{bmatrix} d(k) \tag{17}$$

For the above system, performance of disturbance observer (15) is investigated and it is compared with (2).

The system in (17) and disturbance observer in (15) with (2) is simulated. The disturbance $d(k) = 1\sin(5k)$ was added intensionally. Here

$$\Phi_\tau = \begin{bmatrix} 0.1 & 0 \\ 0.2 & 0.3 \end{bmatrix}, \; \Gamma_\tau = \begin{bmatrix} 0 \\ 0.1 \end{bmatrix} \& \; D_\tau = \begin{bmatrix} 0 \\ 0.1 \end{bmatrix} \tag{18}$$

The $\Phi_{2\tau}$ and $\Gamma_{2\tau}$ become as below:

$$\Phi_{2\tau} = \begin{bmatrix} -0.8 & 0 \\ 0.4 & -0.4 \end{bmatrix}, \; \Gamma_{2\tau} = \begin{bmatrix} 0 \\ 0.2 \end{bmatrix} \& \; D_{2\tau} = \begin{bmatrix} 0 \\ 0.2 \end{bmatrix} \tag{19}$$

The matrices $\bar{A}$ and $\bar{B}$ are

$$\bar{A} = \begin{bmatrix} -1.36 & -0.34 \\ 0 & -1.02 \end{bmatrix}, \bar{B} = \begin{bmatrix} -0.18 \\ -0.10 \end{bmatrix} \tag{20}$$

The $(D_{2\tau} - \Phi_\tau D_\tau) = \begin{bmatrix} 0 & 0.17 \end{bmatrix}^T$ which is equal to γD_τ. So, therefore it gives $\gamma = 1.7$.

Figure 2 shows actual and estimated disturbance. It is clearly visible that disturbance observer accurate estimate of the actual disturbance. Figure 3 depicts the estimation error i.e $e(k) = d(k) - \hat{d}(k)$ for proposed observer (black) and Su's disturbance observer (red). Proposed observer gives estimation error to zero, while Su's disturbance observer estimate disturbance in bound of ± 0.001.

Qualitative analysis is given in Table 1. One norm gives the measure of accuracy. It is clear from that proposed disturbance observer give much exact estimation.

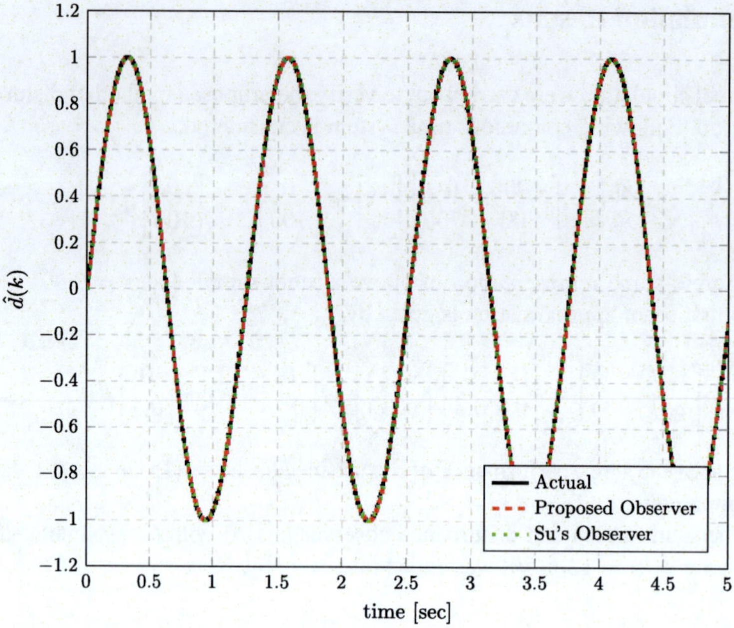

Fig. 2 Observer performance for $\hat{d}(k)$

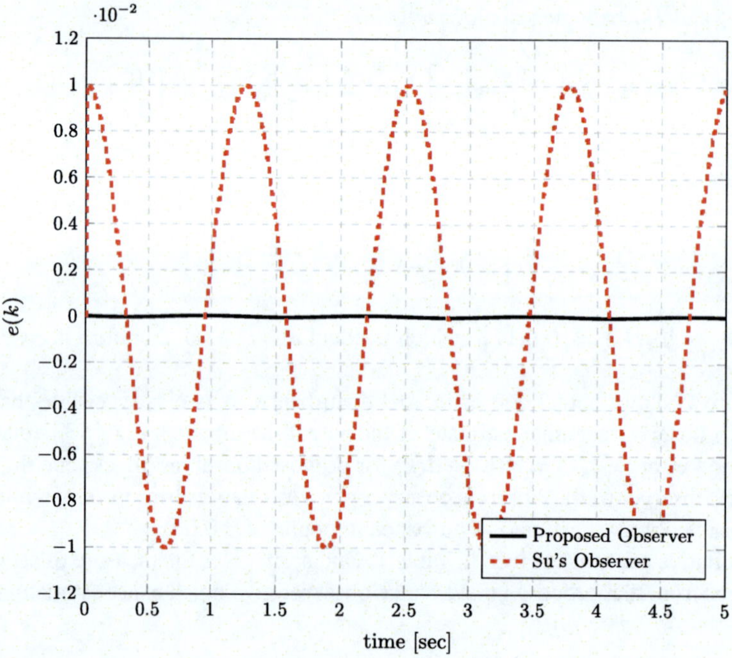

Fig. 3 Estimation error $e(k)$

Table 1 Qualitative analysis of disturbance observer

Disturbance observer	$\|\|e(k)\|\|_1$
Su's disturbance observer	31.7302
Proposed disturbance observer	0.1553

5 Conclusion

In this paper, new discrete disturbance observer is proposed. The effectiveness of proposed observer is verified by simulation. Simulation results shows the proposed observer accurately estimate the disturbance. Performance of proposed observer is compared with the ones existing in the literature [15]. Following are the concluding remarks

- The proposed method yield 99.51% improvement in accuracy.
- For proposed observer, estimation error converges in less than 0.01 s.
- The proposed observer have a capability to estimate slow as well as fast varying disturbance.
- The proposed estimator can be applied in any system with matched disturbance for compensating its effect and obtain robust performance.

References

1. Li, S., Yang, J., Chen, W.-H., Chen, X.: Disturbance observer-based control: methods and applications. CRC Press (2014)
2. Kim, K.-S., Rew, K.-H., Kim, S.: Disturbance observer for estimating higher order disturbances in time series expansion. IEEE Trans. Autom. Control **55**(8), 1905–1911 (2010)
3. Schrijver, E., Van Dijk, J., Nijmeijer, H.: Equivalence of disturbance observer structures for linear systems. In: Proceedings of the 39th IEEE Conference on Decision and Control (Cat. No. 00CH37187), vol. 5, pp. 4518–4519. IEEE (2000)
4. Chen, X., Tomizuka, M.: Optimal plant shaping for high bandwidth disturbance rejection in discrete disturbance observers. In: Proceedings of the 2010 American Control Conference, pp. 2641–2646. IEEE (2010)
5. Lu, G., Ho, D.W.: Generalized quadratic stability for continuous-time singular systems with nonlinear perturbation. IEEE Trans. Autom. Control **51**(5), 818–823 (2006)
6. Corless, M., Tu, J.: State and input estimation for a class of uncertain systems. Automatica **34**(6), 757–764 (1998)
7. Orjuela, R., Marx, B., Ragot, J., Maquin, D.: On the simultaneous state and unknown input estimation of complex systems via a multiple model strategy. IET Control Theory Appl. **3**(7), 877–890 (2009)
8. Janardhanan, S., Bandyopadhyay, B.: Output feedback sliding-mode control for uncertain systems using fast output sampling technique. IEEE Trans. Ind. Electron. **53**(5), 1677–1682 (2006)
9. Saaj, M.C., Bandyopadhyay, B., Unbehauen, H.: A new algorithm for discrete-time sliding-mode control using fast output sampling feedback. IEEE Trans. Ind. Electron. **49**(3), 518–523 (2002)

10. Janardhanan, S., Agarwal, V., Bandyopadhyay, B., Abera, E., Gebre, A.: Sensorless control of induction motor using fast output sampling technique. In: Proceedings of 2005 IEEE Conference on Control Applications, 2005. CCA 2005, pp. 493–497. IEEE (2005)
11. Reddy, G.D., Park, Y., Bandyopadhyay, B., Tiwari, A.: Discrete-time output feedback sliding mode control for spatial control of a large phwr. Automatica **45**(9), 2159–2163 (2009)
12. Tuteja, V., Patel, J.B., Mehta, A.: Design and implementation of robust multirate output feedback based sliding mode controller for induction motors using FPGA. In: 2012 IEEE 5th India International Conference on Power Electronics (IICPE), pp. 1–6. IEEE (2012)
13. Abera, A.-E., Agarwal, V., Bandyopadhyay, B., Janardhanan, S.: Multirate output feedback sliding mode controller for sensorless induction motor. In: 2005 IEEE International Conference on Industrial Technology, pp. 877–881. IEEE (2005)
14. Sawai, S., Kurode, S.: Simultaneous state and disturbance estimation for robust tracking control in discrete motion system. IFAC Proceedings Volumes (2020)
15. Su, W.-C., Drakunov, S.V., Ozguner, U.: An o (t/sup 2/) boundary layer in sliding mode for sampled-data systems. IEEE Trans. Autom. Control **45**(3), 482–485 (2000)

Effect of Normalisation for Gender Identification

Chandra Shekhar Yadav, Meenu Yadav, Prem Shankar Singh Yadav, Rakesh Kumar, Sandeep Yadav, and Kesari Singh Yadav

1 Introduction

Gender identity is outlined as a private conception of oneself as male or feminine (or seldom, each or neither). This construct is intimately involving the concept of role, that is outlined because the outward manifestations of temperament that replicate the identity. Nearly in all instances Gender identity is self identifiable due to mix of environmental and inherent factors. It is noticeable factor as compare to behaviour. Maybe, if someone considers himself a male and is most comfy touching on his personal gender in masculine terms, then his individuality is male. However, his role is male as long as he demonstrates generally male characteristics in behaviour, dress, and/or mannerisms.

2 Related Work

In 1989, a new technique perceptual linear predictive (PLP) was presented by Hermansky [1]. In that approach, mainly three ideas derived the auditory spectrum: (i) spectral decision of critical-band, (ii) curve for equal-loudness, and (iii) power law of

C. S. Yadav (✉)
STQC, MeitY, New Delhi, India

M. Yadav · S. Yadav
SLIET, Longowal, Punjab, India

P. S. S. Yadav · R. Kumar
JNU, Delhi, India

K. S. Yadav
Sanskriti University, Mathura, India

S. K V and K. Guruprasad Rao (eds.), *Smart Sensors Measurements and Instrumentation*, Lecture Notes in Electrical Engineering 750,
https://doi.org/10.1007/978-981-16-0336-5_13

intensity-loudness. Study of PLP turned into computationally beneficial and represents a voice (speech) in low dimensions [2]. In 1993, Rasta speech [2] processing reduces the sensitivity of recognizers. RASTA does this with the help of band-bypass filtering time trajectories of logarithmic parameters of speech (e.g. logarithmic spectral energies or Cepstra). In [3], authors discussed the correspondence between function of RASTA and the character of the identity models and also the relation of their features to delta and additionally to cepstral, imply subtraction. In the past 25 years, many changes have befell in this speech reputation field. So many new features have been extracted from the speech documents to recognize the voice. Features like, perceptual linear prediction (PLP), linear predictive codes (LPC), PLP-RASTA (PLP-Relative Spectra), mel frequency cepstral coefficients (MFCC), etc., have been extracted from the speech files. PLP and MFCC recollect the man or woman or conduct of voice (speech) at the same time as extracting the attributes like LPC identifies new capabilities the use of the information from previous capabilities. Dave [4] used techniques like support vector machine (SVM), hidden Markov model (HMM), dynamic time warping (DTW), vector quantization (VQ), and that are applied for type and identification or say recognition. In [5], proposed a LID gadget which identifies the language and additionally classifies the genders from the speech files. They used supervised device studying techniques to differentiate the speech files. Machine mastering classifiers have been implemented to the information matrix that is a first-rate matrix of all the features (attributes) which had been processed out from speech (sound) files. In [6], designed a language identification gadget using device gaining knowledge of algorithms based on language models. This LID machine considers fluctuation of speech over time. So they have used long recordings so as to get the best accuracy of fluctuation of speech. This thesis is composed extraction of sufficient attributes from speech documents of shorter time recording to categorise the spoken languages. They used frames of 20 ms and investigated with low-level features which can be in wide variety 173 and labelled in three sets of functions inclusive of spectrum, cepstrum, and relative spectral (RASTA). They used a subset of OGI phone speech corpus and it incorporates voice documents in seven languages. They used six system learning (ML) strategies on their dataset and did evaluation amongst their results and [6] located that the nice optimized result changed into given through random forest (RF). In [7] proposed a brand new direct access framework for the identification device of speaker's voice (sound). It is located to understand a speaker consistent with his voice features. This approach identifies the item on the components of the item itself. This proposed framework has two elements, first component is procedure of enrolment and other is the recognising system. This technique has phases like: voice pre-processing, function extraction of speaker's sound, normalising function, selecting characteristic, speaker modelling, direct accessing method, and speaker matching. Following data set were used by them on this paper: Number of speech documents used is 2140, wide variety of speaker is 142, and variety of male and woman changed into ninety-seven and forty five, respectively. The resulted accuracy level primarily based on MFCC functions is 94.38% and the accuracy of gender primarily based class of speaker is as much as 100% on pitch, flatness, brightness, and roll off capabilities. The four different classifiers are utilised by [8] for automated speech recognition. They used Gaussian aggregate model (GMM), multilayer perceptron (MLP), vector quantization (VQ), and getting to know vector quantization (LVQ) classifiers in their work along with mel frequency cepstral coefficients (MFCC). They carried out 96.4% identity accuracy using the IViE corpus. In the campus of Virginia Tech, Bales et al. used a highly instrumented Goodwin Hall smart building for the measurement of the vibrations created by peoples on the walking surface. They recorded the steps of 15 persons individually during the people was walking or stepping down the instrumented hallway. Fourteen accelerometers mounted underneath the walking surface, recorded walking trials with the placement of the sensors unknown to the walker. In another work [9] used bagged decision trees, boosted decision trees, support vector machines (SVMs), and neural networks as the machine learning techniques for their ability in order to gender classification. Brain–computer interfaces (BCI) is a connection between brain and a device which takes the signals from the brain to direct the external activity. Day by day, this interface is going popular because number of applications is using it rapidly. So now, BCI is also used in [10] to recognise the age and type of gender of a speaker through its EEG signals. In their work, they have recorded cerebral activities of 60 subjects (males and females) keeping their position in relax mode and with closed eyes. They used deep BLSTM-LSTM network which used to form a hybrid learning framework. They have recorded 93:7 and 97:5% accuracy of age and gender classification problems, respectively. According to [11], when recognisers got up skilled with acoustic (audial) features along with articular features, the accuracy of recognition get improved. In this paper, articulatory features are denoted by the position or trajectories of the articulators. They proposed the articulatory cepstral coefficients (ACCs) as features which are the cepstral coefficients of the time-location articulatory signal. In this paper, they shown the similarity between MFCCs and ACCs and also performances of MFCCs and ACCs in isolation and conjunction uses the common algorithms which is useful for both acoustic and articulatory signals. Chennupati et al.[11] proposed a latest analysis method which is single frequency filtering (SFF) in speech signals. This paper shows a method to modify the speech to make better its clarity in case of noise. In their SFF method, they can control the spectral and temporal resolutions with the help of a single parameter of the filter, which corresponds to the position of the pole on the negative real axis which corresponds to the unit circle in the z-plane. In analysing the highly intelligible speech, it shows that the speech signal is more intelligible when it has higher dynamic range of amplitude locally (fine structure) and/or lower dynamic range of amplitude globally (gross structure) in both the spectral and temporal domains. Generally automatic speech recognition system based on HMM (Hidden Markov Model) that produce cepstral based features but these features are not that much robust to noise. There are another features which are auxiliary features like pitch of speech, speech energy, rate-of-speech (ROS), etc. The cepstral features are responsive to the auxiliary features. In [12], authors have investigated different approaches of auxiliary features with the help of Dynamic Bayesian network or hybrid HMM/ANNs. That approaches tell that auxiliary features are not emitted directly from the HMM but they carry higher

Table 1 Related work for gender identification

Study	Name, # of languages	# of features	ML methods	Time frame	Best accuracy
[20]	OGI, 5 lang	9	2:GMM and HMM	2 m	61.2
[21]	OGI, 10 lang	7	N/A	45 s	64.6
[22]	CallFriend, 12lang	N/A	1:GMM	30 s	83
[23]	NIST, 6 lang	N/A	1:VSM	30 s	97.25
[23]	NIST, 23 lang	67	1:VSM	30 s	79.1
[24]	NIST, 23 lang	23	1:SVM	30 s	98.8
[25]	NIST, 8 lang	39	1:DNN	3 s	90.42
[26]	OGI, 7 lang	173	6:RF, MB, J48, LR, SMO, BN	20 ms	80.33

level of information that is related with standard features. In these, the experiment is done on OGI database and it tells that in certain conditions, auxiliary features are more robust to noise when it is appended to the standard features. Esther et al. computed some features for speech recognition. Recognition of voice is very useful in automatic speech recognition. For the computation of pitch, they used signal energy, autocorrelation, and mel frequency cepstral coefficients (MFCCs). After that, linear support vector machine is use to classify the speech files. They performed two experiments; the one in which only one file was trained and one file was tested. The other experiment done by Ramdinmawii and Mittal [13] has three files for training and one file to test. The second experiment gained better result in terms of average accuracy than the first one. As day by day speech recognition system is getting famous because of its wide range of applications. Like it is used for security and authentication purpose. This work of [14] author aimed at designing of gender classification system using supervised machine learning techniques. This system takes the speech files as an input and gives output as a classification of gender. This designed gender classification system has four steps (input speech files, pre-processing, extraction of features, and classifying the input data), and in conclusion, the system produces results whether the voice is of male or female. Other related word presented in Table 1.

3 Proposed Work

This section is representing flow diagram of proposed model. Input signals are audio files that are pre-processed using low pass filter and windowing, later some useful features are extracted from that. After performing different normalisation methods, i.e. standard normalisation, forward and backward normalisations, classification models are used for gender prediction. Details flow diagram is shown in Fig. 1.

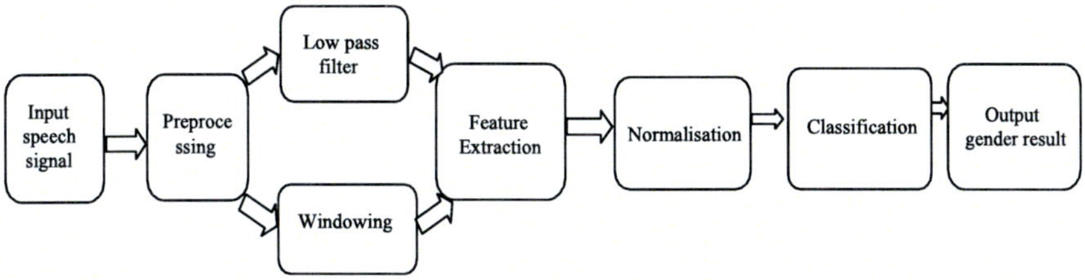

Fig. 1 Flow diagram of proposed model

3.1 Features Used

This section explains the number of features extracted from the data set. Extracted features for some data set are presented in Table 2. Features description is presented below.

Fundamental Frequency: The fundamental frequency (also known as Fundamental) represents the lowest frequency of the speech signal. It is the inverse of the length of pitch duration. The fundamental frequency is denoted as F_0 (F-zero) in Eq. 1.

$$F_0 = \frac{1000}{\tau_{ms}} \tag{1}$$

ZCR: The concept of ZCR of speech signals states the changes in the value of a sign with a signal. The rate given in Eq. 2 when the signal reaches to zero from positive value and then to negative value, and from negative to zero then positive.

$$zcr = \sum_{m=-\infty}^{\infty} |sgn[s(m) - sgn[s(m-1)]| \tag{2}$$

In above equation, sgn [] is sign function, which definition in Eq. 3:

$$gn[x] = \begin{cases} 1, (x \geq 0) \\ -1, (x < 0) \end{cases} \tag{3}$$

where $s\ (m)$ is the speech signal. A sensible overview is that, if the ZCR is low, it is voiced and if ZCR is high, it is unvoiced speech.

Short Energy: Short energy may be used for classification of regions in speech file. The equation for (STE) is formulated from complete aggregated energy equation which defines the processing of signal. The overall energy of an energy (power) signal is proven below:

Table 2 This Table have shown different extracted features only for ten audio files and corresponding label is present in column class

Fun_freq	ZCR	Short_ene	Cepstra	Spectra	Pspectrum	Lpcas	F	M	Class
0.842047	−0.18063	−0.50348	0.703187	−0.5978	−0.56298	−0.81915	−1.34901	−0.77324	M
0.863275	−0.15097	−0.5007	0.581836	−0.64041	−0.56095	−0.74803	−1.12827	−0.79867	M
0.938765	−0.25271	−0.37377	0.604824	−0.52917	−0.46087	−0.7 1351	−1.44795	−0.78019	M
0.809375	−0.10101	−0.51446	0.623444	−0.78774	−0.57495	−0.77115	−1.03509	−0.97623	M
0.875562	−0.15569	−0.53411	0.622958	−0.68718	−0.59143	−0.7352	−1.31635	−0.95665	M
1.376519	1.540208	−0.73871	0.748252	−0.91528	−0.72792	1.182047	1.28472	−0.81119	F
1.702498	2.540045	−0.58061	0.737299	1.461112	−0.43544	1.635464	1.613523	0.93038	F
1.658984	2.233021	−0.7547	0.736689	−0.20563	−0.687	1.120601	0.24633	−0.18975	F
0.914215	1.962564	−0.70058	0.736689	−0.1553	−0.67125	1.748374	0.027377	0.027377	F

$$E(n) = \sum_{i=1}^{N} x_n^2(i) \tag{4}$$

Cepstra: To decide the cepstral functions of the speech signal, first we ought to discover the spectral capabilities of the speech signal. We can decide the cepstral capabilities through taking inverse Fourier transform of the logarithmic of earlier spectral feature. There are various cepstrum such as a power cepstrum, complex cepstrum, real cepstrum, and a phase cepstrum.

Spectra: Spectral capabilities are those capabilities which might be frequency-based features and they commonly present inside the time domain but with the assist of fast Fourier rework those are converted into the frequency domain. These features have some characteristics which are not obtained without delay so these have to rework. Therefore, the primary and the most common spectral characteristic is to take a look at the path of frequency modulation, whether its miles rising or falling in a speech signal.

Pspectrum: The power cepstrum is used to symbolise a human voice or music indicators in a form of vectors of feature. To get the cepstral features, the initial assignment is to remodeling the spectrum with the aid of taking assist of the mel scale. Mel frequency cepstrum is the output of the above transformation. And the coefficients of MFC (abbreviation for mel frequency) are called coefficient of mel-frequency cepstral (MFCC).

Lpcas: Linear predictive model is used to symbolise the compressed shape of the digital sign (spectral form) of speech; this linear predictive coding (LPC) tool is used for speech processing or sign processing. It is one of the powerful strategies for speech processing and useful technique because in case while bit price is low, it gives very superior nice of speech to give correct estimates for the parameters of speech (voice).

F: The frequency reaction classifies into a number of diverse band paperwork describing which frequency bands are exceeded from filter (the bypass band) and that are rejected via filter (the prevent band). Low bypass filter is a clear out which passes the low sign than the selected cut-off frequency and attenuates the signals which can be having a better frequency than the selected cut-off frequency.

M: Magnitude shows the value of frequency of each audio file or sound.

3.2 Normalisation

Normalisation is done to address covariate shift problem, and makes help in faster learning process. **Model-1** when no normalisation is used and presented in Eq. 5, and **Model-2** is min max normalisation model that is presented in Eq. 6.

$$A_i^{'} = A_i \quad (5)$$

$$A'_i = \left(\frac{A_i - \min(A)}{\max(A) - \min(A)}\right) * (D - C) + C \quad (6)$$

A'_i is ith normalised value, min (A) is minimum value of vector A, max (A) is maximum value in vector A to normalised, [C, D] is predefined boundary in which A' values lies.

Model-3 Backward Normalisation

This is similar as batch normalization followed in deep learning. Here the input is A, and parameter to learn to learn are beta and gamma. Batch normalisation is a technique to provide any layer in a neural network with inputs that are zero mean/unit variance. Backward normalisation is presented by Eqs. 7–18.

Input: $A = \{x_1, x_2, \ldots. x_m\}$.

Output $A' = BN_{\Upsilon,\beta}(x_i)$

$$\mu_A = \frac{1}{m}\sum_{i=1}^{m} x_i \quad (7)$$

$$\sigma_A^2 = \frac{1}{m}\sum_{i=1}^{m}(x_i - \mu_A)^2 \quad (8)$$

$$\widehat{x_i} = \frac{x_i - \mu_A}{\sqrt{\sigma_A^2 + \varepsilon}} \quad (9)$$

$$A_i^{'} = \gamma\widehat{x_i} + \beta = BN_{\gamma,\beta}(x_i) \quad (10)$$

To learn beta and gamma for Eqs. 10, and equations 11 to 16 is used, and final value of alpha and beta is given by Eqs. 17 and 18.

$$\frac{\partial l}{\partial \widehat{x_i}} = \frac{\partial l}{\partial A'_i} \cdot \gamma \quad (11)$$

$$\frac{\partial l}{\partial \sigma_A^2} = \sum_{i=1}^{m} \frac{\partial l}{\partial \widehat{x_i}} \cdot (x_i - \mu_A) \cdot \frac{-1}{2}\left(\sigma_A^2 + \varepsilon\right)^{-3/2} \quad (12)$$

$$\frac{\partial l}{\partial \mu_A} = \left(\frac{\partial l}{\partial \widehat{x_i}} \cdot \frac{-1}{\sqrt{\sigma_A^2 + \varepsilon}}\right) + \frac{\partial l}{\partial \sigma_A^2} \cdot \frac{\sum_{i=1}^{m} -2(x_i - \mu_A)}{m} \quad (13)$$

$$\frac{\partial l}{\partial x_i} = \frac{\partial l}{\partial \widehat{x_i}} \cdot \frac{1}{\sqrt{\sigma_A^2 + \varepsilon}} + \frac{\partial l}{\partial \sigma_A^2} \cdot \frac{\sum_{i=1}^{m} 2(x_i - \mu_A)}{m} + \frac{\partial l}{\partial \mu_A} . \frac{1}{m} \quad (14)$$

$$\frac{\partial l}{\partial \gamma} = \sum_{i=1}^{m} \frac{\partial l}{\partial y_i} \cdot \widehat{x_i} \tag{15}$$

$$\frac{\partial l}{\partial \beta} = \sum_{i=1}^{m} \frac{\partial l}{\partial y_i} \tag{16}$$

$$\gamma = \gamma - \frac{\partial l}{\partial \gamma} \tag{17}$$

$$\beta = \beta - \frac{\partial l}{\partial \beta} \tag{18}$$

Model-4 Forward Normalisation

For forward normalisation, Eq. 10 is used, where gamma and beta learning rate are decided by user. In our implementation, we keep it low, and give by Eq. 8.

$$A_i^{'} = \gamma \widehat{x_i} + \beta \tag{19}$$

4 Experiments and Results

The experiment is related to the gender classification, but some different parameters and machine learning methods are added. We have taken 150 speech files of ten languages: English, German, Japanese, French, Spanish, Mandarin, Korean, Russian, Italian, and Hindi as used in [7]. In our data set, we have taken each 15 files of ten languages which are a combination of male and female files. Table 3 shows the general information of our used data set. We have used speech files with random length in seconds.

After pre-processing on standard speech data set, features were extracted and considered for normalisation using proposed model Model-1, Model-2, Model-3, and Model-4. To find out gender identification, classification is performed using several machine learning algorithms, and their comparative performance is shown in Table 4 and Figs. 2 and 3.

5 Conclusion

In this work, machine learning-based approaches are used on standard data set for gender identification. In this work number of experiments have performed to learn optimal weights using different statistical learning techniques and that is done for

Table 3 General information about the speech corpus

Language	Number of speech files	# of male files	# of females files	size of files in sec (min $\leq$ x $\leq$ max)
Hindi	15	3	12	$47 \leq x \leq 60$
Japanese	15	6	9	$1 \leq x \leq 3$
French	15	14	1	$2 \leq x \leq 10$
Spanish	15	10	5	$2 \leq x \leq 10$
German	15	6	9	$30 \leq x \leq 60$
English	15	5	10	$2 \leq x \leq 60$
Italian	15	7	8	$59 \leq x \leq 90$
Russian	15	10	5	$5 \leq x \leq 9$
Korean	15	5	10	$1 \leq x \leq 4$
Mandarin	15	8	7	$40 \leq x \leq 60$

different value of $K = 5$, 10, 15, and 20. From range of experiment, it is clear that 20 K fold validation is giving better accuracy; most of time, the one reason behind this is statistical tests. In this work, novelty is study about different approaches for different K. When different K folds cross validation used, different techniques and different normalisation models are performing in different way. When $K = 5$ and 15, without normalisation and forward normalisation are performing best, and for $K = 10$, forward normalisation, and for $K = 20$, standard normalisation is working best. In the future, we will elaborate this experiment for larger data set, some hybrid techniques to achieve better results, and this work may use to achieve the objective explained in [15–19].

Table 4 Table is showing comparative performance of different machine learning model and models are tested for different K fold cross validation

Model	Norm	Forword	Backword	Without	Norm	Forword	Backword	Without	Norm	Forword	Backword	Without	Norm	Forword	Backword	Without
SVMmedium Gaussian	0.84	0.8467	0.8333	0.8667	0.86	0.8733	0.8733	0.8467	0.8733	0.8667	0.8467	0.8667	0.88	0.8667	0.8733	0.8667
SVMQubic	0.8133	0.86	0.78	0.82	0.82	0.78	0.76	0.78	0.8067	0.7867	0.76	0.8067	0.7867	0.7867	0.8067	0.82
SVMfineGaussian	0.78	0.7733	0.78	0.78	0.8	0.8	0.7867	0.7733	0.78	0.7733	0.8067	0.7933	0.8133	0.8	0.7933	0.7933
SVMQuadratic	0.8067	0.8467	0.8467	0.8	0.7867	0.8333	0.7733	0.8133	0.82	0.82	0.82	0.8133	0.82	0.8133	0.82	0.84
SVMLinear	0.8067	0.8333	0.8267	0.7933	0.82	0.82	0.8067	0.82	0.8133	0.8267	0.8267	0.8133	0.8267	0.82	0.82	0.8267
SVMCoarseGaussian	0.8133	0.82	0.8133	0.8133	0.8267	0.7867	0.8267	0.8067	0.82	0.8133	0.8267	0.82	0.8067	0.82	0.8267	0.8133
LogisticRegression	0.8333	0.8133	0.82	0.8	0.82	0.8333	0.8333	0.8267	0.8133	0.84	0.8333	0.8333	0.8133	0.8267	0.8067	0.8267
TreeDisabledPCA	0.7867	0.7533	0.78	0.7867	0.78	0.78	0.74	0.78	0.7867	0.8133	0.8067	0.78	0.7933	0.7867	0.7867	0.7933
EnsembalBoosted	0.5067	0.5067	0.5067	0.5067	0.54	0.5467	0.5067	0.54	0.5467	0.5733	0.5267	0.5067	0.5067	0.56	0.5133	0.5333
EnsembalBagged	0.8267	0.8667	0.86	0.8267	0.86	0.8733	0.8533	0.88	0.84	0.8533	0.8667	0.8467	0.8733	0.8533	0.8467	0.8233
EnsembalSubDiscriminent	0.8333	0.8267	0.8333	0.8	0.82	0.8267	0.82	0.8133	0.82	0.8267	0.8133	0.8333	0.8267	0.8333	0.8267	0.8267
EnsembalSubKNN	0.82	0.8133	0.7933	0.8533	0.8267	0.8133	0.8267	0.82	0.82	0.8333	0.8	0.82	0.84	0.8	0.84	0.8333

Values of K vary from 5 to 20. Here, norm indicates standard normalisation given by Eq. 5, without indicate no normalisation given by Eq. 5, backward and forward for backward and forward normalisation

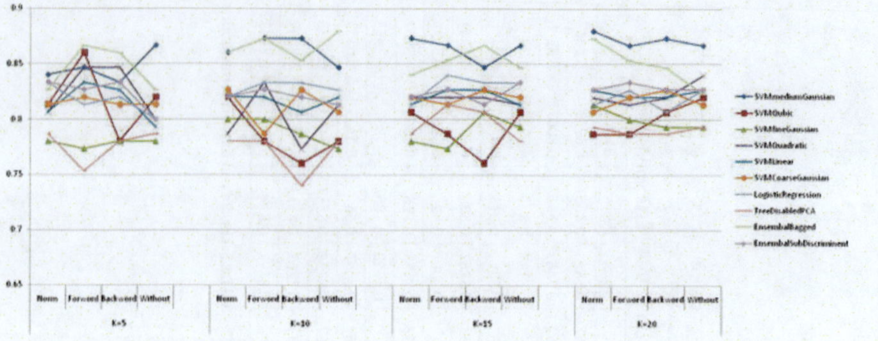

Fig. 2 This figure showing normalisation effect over different value of *K* fold validation

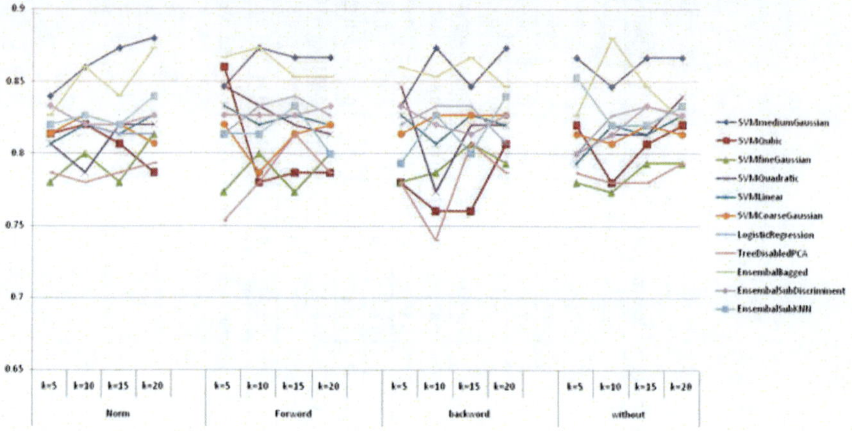

Fig. 3 This figure showing change in accuracy due to *K* fold normalisation over different normalisation methods

References

1. Hermansky, H.: Perceptual linear predictive (PLP) analysis of speech. J. Acoust. Soc. Am. **87**, 1738–1752 (1990)
2. Hermansky, H., Morgan, N.: RASTA processing of speech. IEEE Trans. Speech Audio Process. **2**, 578–589 (1994)
3. Hermansky, H., Morgan, N., Hirsch, H.-G.: Recognition of speech in additive and convolutional noise based on RASTA spectral processing. In: 1993 IEEE International Conference on Acoustics, Speech, and Signal Processing, vol. 2, pp. 83–86. IEEE (1993)
4. Dave, N.: Feature extraction methods LPC, PLP and MFCC in speech recognition. Int. J. Adv. Res. Eng. Technol. **1**, 1–4 (2013)
5. HaCohen-Kerner, Y., Hagege, R.: Language and gender classification of speech files using supervised machine learning methods. Cybern. Syst. **48**, 510–535 (2017)
6. HaCohen-Kerner, Y., Hagege, R.: Automatic classification of spoken languages using diverse acoustic features. In: Proceedings of the 29th Pacific Asia Conference on Language, Information and Computation: Posters, pp. 275–285. (2015)

7. Heryanto, H., Akbar, S., Sitohang, B.: A new direct access framework for speaker identification system. In: 2014 International Conference on Data and Software Engineering (ICODSE), pp. 1–5. IEEE (2014)
8. Djemili, R., Bourouba, H., Korba, M.C.A.: A speech signal based gender identification system using four classifiers. In: 2012 International Conference on Multimedia Computing and Systems, pp. 184–187. IEEE (2012)
9. Bales, D., et al.: Gender classification of walkers via underfloor accelerometer measurements. IEEE Internet Things J. **3**, 1259–1266 (2016)
10. Kaushik, P., Gupta, A., Roy, P.P., Dogra, D.P.: EEG-based age and gender prediction using deep BLSTM-LSTM network model. IEEE Sens. J. **19**, 2634–2641 (2018)
11. Chennupati, N., Kadiri, S.R., Yegnanarayana, B.: Spectral and temporal manipulations of SFF envelopes for enhancement of speech intelligibility in noise. Comput. Speech Lang. **54**, 86–105 (2019)
12. Stephenson, T.A., Doss, M.M., Bourlard, H.: Speech recognition with auxiliary information. IEEE Trans. Speech Audio Process. **12**, 189–203 (2004)
13. Ramdinmawii, E., Mittal, V.K.: Gender identification from speech signal by examining the speech production characteristics. In: 2016 International Conference on Signal Processing and Communication (ICSC), pp. 244–249. IEEE (2016)
14. Sengupta, S., Yasmin, G., Ghosal, A.: Classification of male and female speech using perceptual features. In: 2017 8th International Conference on Computing, Communication and Networking Technologies (ICCCNT) pp. 1–7. IEEE (2017)
15. Yadav, C.S., Sharan, A.: Automatic text document summarization using graph based centrality measures on lexical network. Int. J. Inf. Retr. Res. **8**, 14–32 (2018)
16. Yadav, C., Sharan, A.: A new LSA and entropy-based approach for automatic text document summarization. Int. J. Semant. Web Inf. Syst. **14**, 1–32 (2018)
17. Yadav, C.S., Sharan, A.: Hybrid approach for single text document summarization using statistical and sentiment features. Int. J. Inf. Retr. Res. **5**, 46–70 (2015)
18. Yadav, M., Verma, V.K., Yadav, C.S., Verma, J.K.: MLPGI: multilayer perceptron-based gender identification over voice samples in supervised machine learning. In: Johri, P., Verma, J., Paul, S. (eds.) Applications of Machine Learning. Algorithms for Intelligent Systems. Springer, Singapore (2020). 443.webvpn.fjmu.edu.cn/https://doi.org/10.1007/978-981-15-3357-0_23
19. Yadav, C.S., Sharan, A.: Feature learning using random forest and binary logistic regression for ATDS. In: Johri, P., Verma, J., Paul, S. (eds.) Applications of Machine Learning. Algorithms for Intelligent Systems. Springer, Singapore (2020). https://doi.org/10.1007/978-981-15-3357-0_22
20. Pellegrino, F., André-Obrecht, R.: Automatic language identification: an alternative approach to phonetic modelling. Sig. Process. **80**(7), 1231–1244 (2000)
21. Kirchhoff, K., Parandekar, S.: Multi-stream statistical N-gram modeling with application to automatic language identification. In Seventh European Conference on Speech Communication and Technology (2001)
22. Torres-Carrasquillo, P.A., Reynolds, D.A., Deller, J.R.: Language identification using Gaussian mixture model tokenization. In 2002 IEEE International Conference on Acoustics, Speech, and Signal Processing, vol. 1, pp. I–757. IEEE (2002, May)
23. Ng, R.W.M., Leung, C.C., Lee, T., Ma, B., Li, H.: Prosodic attribute model for spoken language identification. ICASSP, IEEE Int. Conf. Acoust. Speech Signal Process. - Proc., pp. 5022–5025 (2010)
24. Behravan, H., Hautamäki, V., Kinnunen, T.: Factors affecting i-vector based foreign accent recognition: A case study in spoken Finnish. Speech Commun. **66**, 118–129 (2015)
25. Lopez-moreno, I., Gonzalez-dominguez, J., Plchot, O., Martinez, D., Gonzalez-rodriguez, J., Moreno, P., Google Inc., New York, USA ATVS-Biometric Recognition Group, Universidad Autonoma de Madrid, Spain Brno University of Technology, Czech Republic Aragon Institute for Engineering Research (I3A), University of Zaragoza, Spain, pp. 0–4, (2014)
26. Hacohen-kerner, Y., Hagege, R.: Automatic classification of spoken languages using diverse acoustic features. pp. 275–285 (2015)

FPGA-Based Implementation of Digital Filters for Image Denoising

Shrikanth K. Shirakol, Veerayya Hiremath, and S. S. Kerur

1 Introduction

In digital signal processing, a filter is a device that is used to remove unwanted components from a given signal. The filter can suppress undesired components partially or completely from the signal. Convolution is widely used for implementing linear filters. In the literature, a number of efficient FPGA designs of 2D convolution for implementing Gaussian filter have been proposed [1–4]. Talbi et al. developed an efficient architecture for separable Gaussian smoothing filters using intermediate controller based on finite state machine simulated in VHDL and prototyped on device technology of Xilinx Virtex V FPGA platform [1]. Frank Cabello et. al. implemented Gaussian filter for different kernel sizes and found that it is not efficient to use floating point arithmetic for FPGA implementation as it needs multiple FPGA resources [2]. Sami Khorbotly et al. suggested the "modified" method to approximate the coefficients of a 2D Gaussian filter. The modified approximation filter, when implemented on FPGA, has shown to deliver higher signal quality than its traditional "rounded" counterpart with no increase in the required resources [3].

Leila kabbai et. al. presented the implementation of 2D convolution on a Xilinx Virtex V FPGA platform based on a state machine. Author implemented Gaussian filters with different sigma values and then optimized proposed architecture using different multipliers. Performances and results show decrease in area and resources utilization, especially when using DSP and BRAM of FPGA [4]. Batcher explains the Bitonic Merge and Odd–Even Merge sorting networks [5]. Farmahini-Farahani et al. suggested that in most of the applications, only P largest (or smallest) output values need to be selected from a total of Q input values, where $P < Q$. These P largest (smallest) outputs may not be in order [6]. Chen and Prasanna explain a methodology for mapping large-scale bitonic sorting networks onto FPGA [7]. Luis

S. K. Shirakol · V. Hiremath (✉) · S. S. Kerur
SDM College of Engineering and Technology, Dharwad, India

Affiliated to VTU, Belagavi, India

S. K V and K. Guruprasad Rao (eds.), *Smart Sensors Measurements and Instrumentation*, Lecture Notes in Electrical Engineering 750,
https://doi.org/10.1007/978-981-16-0336-5_14

Alberto Aranda et al. proposed a fault-tolerant implementation of the median filter. This implementation checks if the median value is within a dynamic range, created each time with remaining non-median outputs. The design has been implemented in an SRAM-based FPGA [8].

The paper is organized as follows. Section 2 deals with design methodology used for designing Gaussian, mean, and median filters. In Sect. 3, the hardware implementation for designed filters is discussed. Simulation results, device utilization, and timing reports are presented in Sect. 4. The performance of the filters is discussed in Sect. 5 using PSNR and correlation coefficient. Section 6 concludes the paper.

2 Design Methodology

2.1 Design of Gaussian Filter

Designing of Gaussian Kernel

1. Formula used for designing 2D Gaussian kernel is given by

$$G(p,q) = \frac{1}{2\pi\sigma^2}\mathrm{e}^{-\frac{p^2+q^2}{2\sigma^2}} \tag{1}$$

2. Consider a standard deviation of $\sigma = 0.5$ and a 3×3 kernel matrix. Then,

$$\frac{1}{2\pi\sigma^2} = \frac{1}{2\pi \times 0.5 \times 0.5} = \frac{1}{1.57079} = 0.6366$$

3. For a 3×3 kernel, the 1D p and q matrices are as follows.

$$p = \begin{bmatrix} -1\,0\,1 \\ -1\,0\,1 \\ -1\,0\,1 \end{bmatrix} \quad \text{and} \quad q = \begin{bmatrix} 1 & 1 & 1 \\ 0 & 0 & 0 \\ -1 & -1 & -1 \end{bmatrix}$$

For the center element, $(p, q) = (0, 0)$. Substituting these values in eq. (1), we get

$$G(p,q) = \frac{1}{2\pi\sigma^2}\mathrm{e}^{-\frac{p^2+q^2}{2\sigma^2}} = G(0,0) = 0.6366\mathrm{e}^{-0} = 0.6366$$
$$G(p,q) = \frac{1}{2\pi\sigma^2}\mathrm{e}^{-\frac{p^2+q^2}{2\sigma^2}} = G(0,1) = 0.6366\mathrm{e}^{-2} = 0.08615$$
$$G(p,q) = \frac{1}{2\pi\sigma^2}\mathrm{e}^{-\frac{p^2+q^2}{2\sigma^2}} = G(-1,1) = 0.6366\mathrm{e}^{-4} = 0.01166$$

The following Gaussian kernel matrix can be obtained by repeating this procedure.

$$G(p,q) = \begin{bmatrix} 0.01166 & 0.08615 & 0.01166 \\ 0.08615 & 0.6366 & 0.08615 \\ 0.01166 & 0.08615 & 0.01166 \end{bmatrix}$$

Image convolution using Gaussian kernel includes floating point multiplications which consumes considerable amount of hardware resources. Hence, floating point coefficients in the Gaussian kernel need to be converted to fixed integers. Hence, each floating point value should be multiplied by an integer and divided by a power of 2 [9, 10]. Using a scaling factor of $256(2^8)$, the floating values of kernel matrix are converted to fixed integers as below.

$$0.6366 = \frac{k0}{256}, \quad \text{and} \quad k0 = 0.6366 \times 256 = 162.9696 \approx 163.$$
$$0.08615 = \frac{k1}{256}, \quad \text{and} \quad k1 = 0.08615 \times 256 = 22.0544 \approx 22.$$
$$0.08615 = \frac{k2}{256}, \quad \text{and} \quad k2 = 0.01166 \times 256 = 2.98496 \approx 3.$$

The approximation of the Gaussian kernel matrix $G(p,q)$ is represented below.

$$\text{Gaussian Kernel} = G(p,q) = \frac{1}{2^8}\begin{bmatrix} 3 & 22 & 3 \\ 22 & 163 & 22 \\ 3 & 22 & 3 \end{bmatrix}$$
$$= \begin{bmatrix} 0.01171 & 0.08593 & 0.01171 \\ 0.08593 & 0.6367 & 0.08593 \\ 0.01171 & 0.08593 & 0.01171 \end{bmatrix}$$

Convolution Operation

Image smoothing is achieved in Gaussian filter by convolving 3×3 input image matrix $I(p, q)$ with the 3×3 Gaussian kernel $G(p, q)$ [4], which represented in Eq. (2).

$$f(x,y) = \sum_{m=0}^{h-1}\sum_{n=0}^{w-1} G(m,n) \times I(x-m, y-n) \tag{2}$$

The convolution operation is performed as shown in Fig. 1.

Processed pixel is computed as

$$\begin{aligned} \text{PP} = {} & (k2 \times p1) + (k1 \times p2) + (k2 \times p3) + (k1 \times p4) + (k0 \times p5) \\ & + (k1 \times p6) + (k2 \times p7) + (k1 \times p8) + (k2 \times p9) \end{aligned} \tag{3}$$

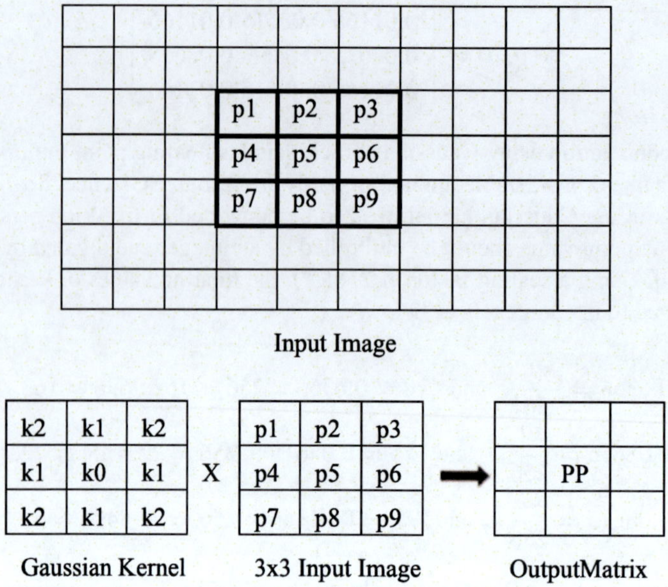

Fig. 1 Image convolution with Gaussian kernel [4]

2.2 *Design of Mean Filter*

Mean filter is an averaging linear filter, and it uses a spatial mask on every pixel in an image. The value of the pixels that are grouped in the mask is averaged. Mean filter can be thought of as convolution filter, and it is based on a kernel matrix. Here, a 3×3 square kernel is used as shown in Fig. 2. This kernel is used to perform convolution with 3×3 input image pixel.

Processed pixel is computed as

$$\mathrm{PP} = \frac{1}{9} \times [p1 + p2 + p3 + p4 + p5 + p6 + p7 + p8 + p9] \tag{4}$$

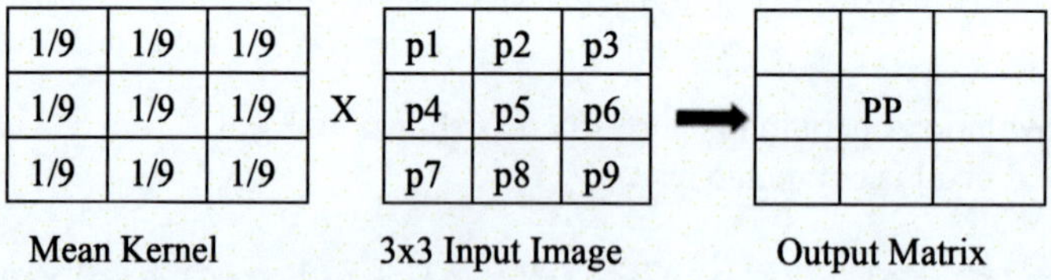

Fig. 2 Image convolution with mean kernel

2.3 Design of Median Filter

Median substitution operation can be understood with the Fig. 3. Here, the input image is represented as N × N window of pixels. A 3 × 3 window of pixels containing total nine pixels is chosen. Using the sorting network architecture, these nine pixels are arranged in ascending or descending order. The fifth element in sorted sequence represents the median value. The center element of 3 × 3 input pixel window is replaced by the median of set of nine pixel values.

The basic element of sorting network is represented as a rectangular box with two inputs and two outputs as shown in Fig. 4. The rectangular box shown in Fig. 4 consists of an M-bit comparator and two 2:1 multiplexers as shown in Fig. 5. The

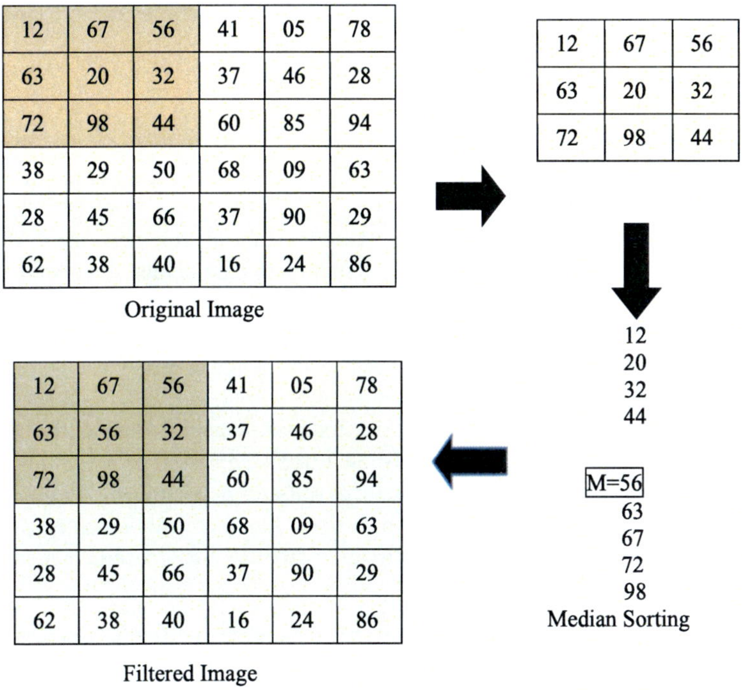

Fig. 3 Median calculation process [11]

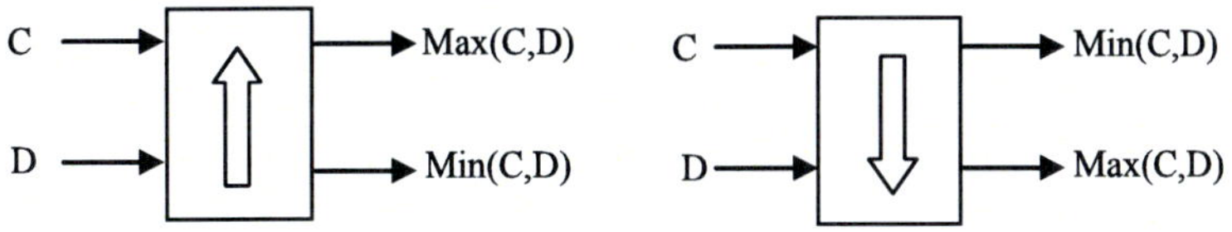

Fig. 4 Schematic symbol of a CAS block. **a** Ascending. **b** Descending

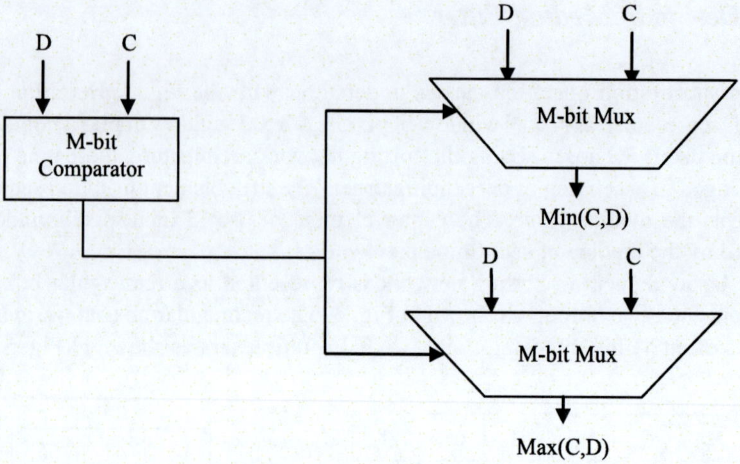

Fig. 5 Internal diagram of two element exchange block [11]

upper and lower multiplexers are programmed to output minimum and maximum of two inputs C and D, respectively.

In FPGAs, median filters are implemented using sorting networks. Median filtering operation can be achieved using a stream of nine pixels. Each set of nine pixels can be sorted in required order (ascending or descending). The architecture proposed in [8] for implementing the median filter uses minimum number of exchange nodes. This architecture needs just 19 exchange nodes, and it is shown in Fig. 6.

In SRAM-based FPGAs [8], design functions can change permanently because of configuration memory errors. SRAM-based implementation of median filter on FPGA has the ability to remove large amount of salt-and-pepper noise. These FPGAs are susceptible to radiation effects. When median filters are implemented on such devices can cause single-event upsets (SEUs) in memory SRAM cells. Due to such

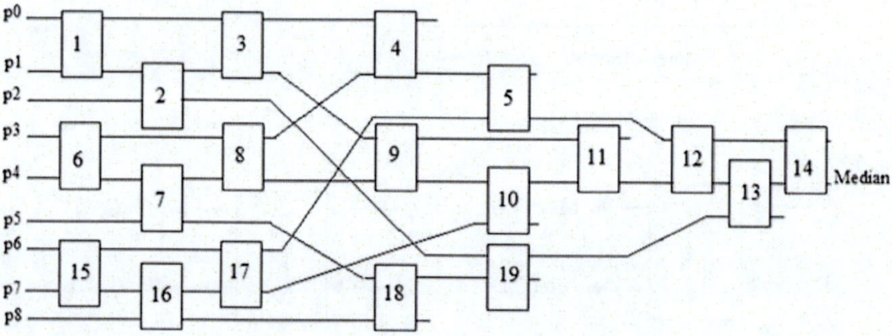

Fig. 6 Network for nine input sorting with minimum blocks [11]

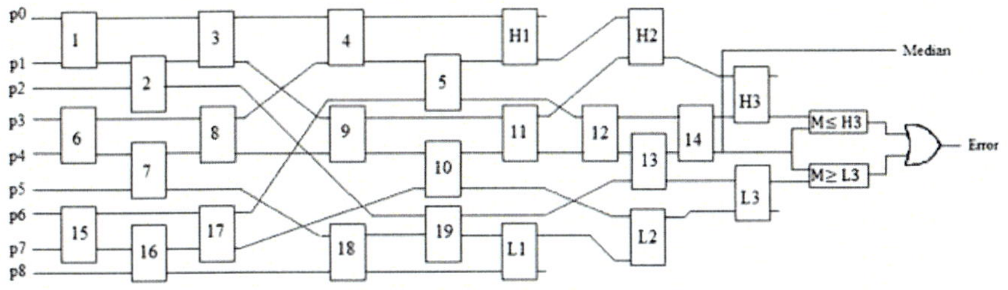

Fig. 7 Error detection technique for median value [11]

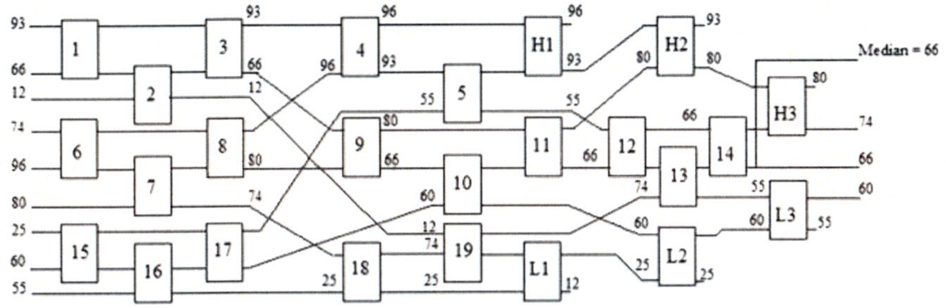

Fig. 8 Numerical example to calculate range values [11]

upsets, the working of median filter may change permanently. To overcome such upsets, extra blocks are added to Fig. 6, to obtain the block diagram used for detecting errors. Figure 7 shows this block diagram. Using this block, a range is created containing upper and lower ranges. The lower value of H3 block defines the upper range, and the higher value of L3 block defines the lower range. The boundary between upper and lower ranges gives range of median value. If the calculated median value is within this range, the error signal is 0, and if it is outside this range, the error signal is 1. The block diagram shown in Fig. 7 is used for implementing median filtering operation with Verilog programming in this paper.

This method can be better understood with an example as shown in Fig. 8. Here, range is the boundary between values 74 and 60. The calculated median value is 66, and this falls within the range. Hence, the error signal is reset. If the median is not within this range, then error signal is set.

3 Implementation

The implementation of image filters starts by first reading an image in MATLAB. The color image is first converted to gray scale and resized to 256 × 256. A particular type of noise (salt and pepper, Gaussian, Poisson, and Speckle) is then added to the image, and the noisy image is converted into pixels and stored as a .coe file. The .coe file is copied into BRAM of Xilinx. The pixels are then processed using a particular type of filter. The filter is coded using Verilog. The pixels are processed using 3×3 sliding window-based approach. Functional verification is done by observing Xilinx simulation results. Filtered pixels are processed using FPGA and displayed using a display device. Processed pixels are stored back as text file. Text file is read back into MATLAB and displayed as image.

4 Result Analysis

4.1 Image Processing in Matlab and Xilinx

The filters are coded using Verilog, and the implementation has been done on Virtex 5(XC5VLX110T) series FPGA. The required percentage of noise is added to the gray image, and the pixels are denoised using Gaussian, mean, and median filters. The performance of these filters for removing noise and reconstructing the image can be analyzed using the following images.

(a) (b) (c) (d)

Fig. 9 **a** Lena image with 5% salt-and-pepper noise. **b** Image filtered with Gaussian filter. **c** Image filtered with mean filter. **d** Image filtered with median filter

(a) (b) (c) (d)

Fig. 10 **a** Lena image with 5% Gaussian noise. **b** Gaussian noised image filtered by Gaussian filter. **c** Image filtered with mean filter. **d** Image filtered with median filter

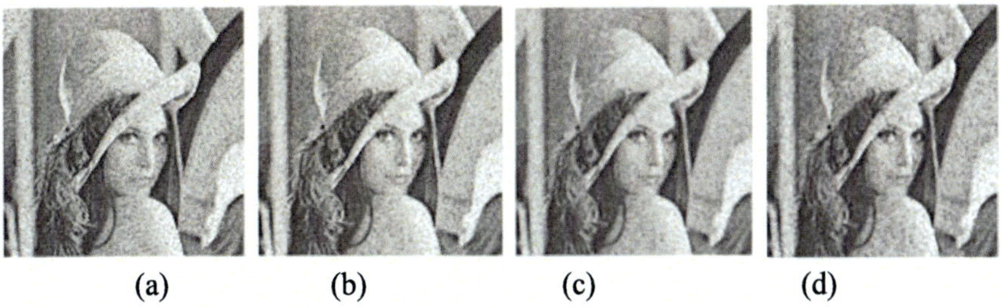

Fig. 11 **a** Lena image with 5% speckle noise. **b** Speckle noised image filtered with Gaussian filter. **c** Image filtered with mean filter. **d** Image filtered with median filter

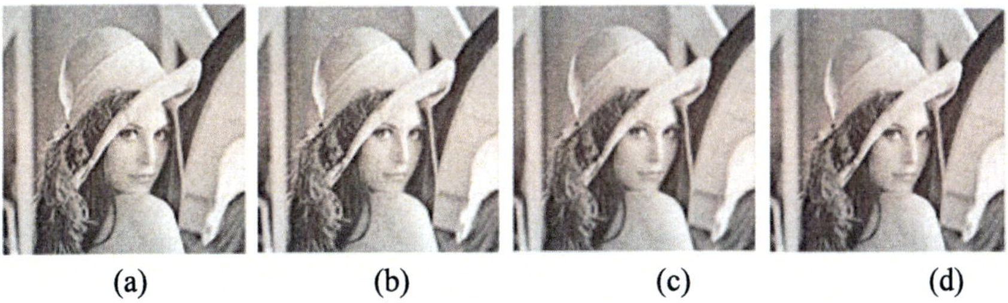

Fig. 12 **a** Lena image with poisson noise. **b** Poisson noised image filtered with Gaussian filter. **c** Image filtered with mean filter. **d** Image filtered with median filter

4.2 Simulation Results, Device Utilization, and Timing Report

Figure 13 shows the simulation results obtained for median filter block that can detect single-event upsets by calculating range values. The signals y11 to y33 represent the median filter input pixel values. The signal Im_out represents the median of these input pixel values. The signal ERROR represents the error signals obtained using the above technique [8], and it is computed as zero (Tables 1 and 2).

Timing summary

Minimum period for 3 × 3 sliding window module: 1.716 ns.
Maximum frequency 3 × 3 sliding window module: 582.751 MHz.
Minimum input arrival time before clock 3 × 3 sliding window module: 1.279 ns.
Maximum output required time after clock 3 × 3 sliding window module: 2.826 ns.
Maximum combinational path delay for Gaussian filter module: 16.668 ns.
Maximum combinational path delay for mean filter module: 11.754 ns.
Maximum combinational path delay for median filter module: 40.392 ns.

Fig. 13 Simulation results for detecting errors in median operation

Table 1 Device utilization summary of 3 × 3 sliding window module

Logic utilization	Used	Available	Utilization (%)
No. of slice registers	144	69,120	1
No. of slice LUTs	128	69,120	1
No. of bonded IOBs	82	640	12

Table 2 Device utilization summary for filter module (Gaussian, mean, and median filters)

		Gaussian filter		Mean filter		Median filter	
Logic utilization	Available	Used	Utilization (%)	Used	Utilization (%)	Used	Utilization (%)
No. of slice registers	69,120	1	1	1	1	—	1
No. of slice LUTs	69,120	80	1	78	1	435	1
No. of bonded IOBs	640	80	12	80	12	81	12

4.3 Power Report for Sliding Window Module

Leakage power for sliding window module is 1.186 W, total power is 1.190 W, and dynamic power is 4 mW.

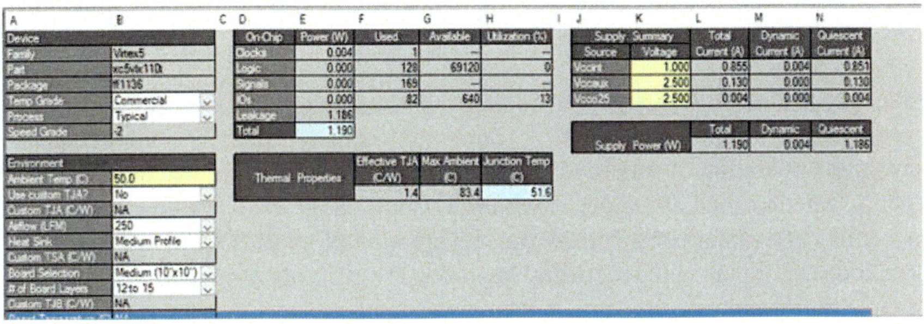

Fig. 14 Power report for sliding window module

5 Results and Discussion

Performance of various filters for different noises based on PSNR and correlation values derived from above results can be obtained with the help of tables and figures shown (Fig. 15) ; (Table 3).

Based on results obtained, it is observed that median filter performs well on removing salt-and-pepper noise, mean filter performs well on both speckle noise and poisson noise, and Gaussian filter performs well on removing Gaussian noise.

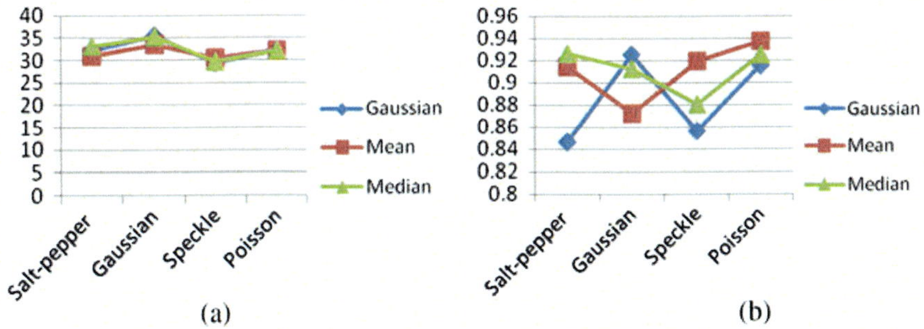

Fig. 15 Performance of the filters based on **a** PSNR and **b** correlation

Table 3 PSNR and correlation values of different filters for various noises

Filter/noise	Salt–pepper	Gaussian	Speckle	Poisson
Gaussian	PSNR = 32.249 Corr = 0.846	PSNR = 35.575 Corr = 0.925	PSNR = 29.338 Corr = 0.856	PSNR = 32.072 Corr = 0.914
Mean	PSNR = 30.781 Corr = 0.914	PSNR = 33.457 Corr = 0.872	PSNR = 30.352 Corr = 0.918	PSNR = 32.081 Corr = 0.936
Median	PSNR = 33.137 Corr = 0.926	PSNR = 35.134 Corr = 0.912	PSNR = 29.492 Corr = 0.880	PSNR = 32.086 Corr = 0.925

6 Conclusion

In this paper, the hardware implementation of Gaussian, mean, and median filters is presented. The implementation of Gaussian and mean filters is done using 2D convolution operation. A modified median filter architecture presented in [8] is used to check whether median output is within dynamic range. The performance of the three filters is evaluated by considering image quality metrics such as PSNR and correlation coefficient with four different noises. From the observations, it is clear that the same filter cannot be used on different noise models. The filters are implemented on Virtex 5 series FPGA. Virtex 5 has been preferred to FF136 package with a device xc5vl $\times$ 110t, which provides high performance for logical applications. The processing time required for filtering is very less in FPGA compared to MATLAB. Area and resource utilization is reduced due to the use of BRAM of FPGA.

References

1. Talbi, F., Alim, F., Seddiki, S., Mezzah, I., Hachemi, B.: Separable convolution gaussian smoothing filters on a Xilinx FPGA platform. In: International Conference on Innovative Computing Technology (INTECH), pp. 112–117. Galcia (2015)
2. Cabello, F., León, J., Iano, Y., Arthur, R.: Implementation of a fixed-point 2D Gaussian filter for image processing based on FPGA. In: Signal Processing: Algorithms, Architectures, Arrangements, and Applications (SPA), Poznan (2015)
3. Khorbotly, S., Hassan, F.: A modified approximation of 2D Gaussian smoothing filters for fixed-point platforms. In: IEEE 43rd Southeastern Symposium on System Theory, AL, pp. 151–159. Auburn (2011)
4. Kabbai, L., Sghaier, A., Douik, A., Machhout, M.: FPGA Implementation of filtered image using 2D Gaussian filter. J. Int. J. Adv. Comput. Sci. Appl. **7**, 514–520 (2016)
5. Batcher, K.E.: Sorting networks and their applications. In: Proceedings Apr. 30–May 2, 1968, Spring Joint Computer Conference (AFIPS). pp. 307–314. New York, USA (1968)
6. Farmahini, F.A., Duwe, H.J., III., Schulte, M.J., Compton, K.: Modular design of high-throughput, low-latency sorting units. J. IEEE Trans. Comput. **62**(7), 1389–1402 (2013)
7. Chen, R., Prasanna, V.K.: Computer generation of high throughput and memory efficient sorting designs on FPGA. J. IEEE Trans. Parallel Distrib. Syst. **28**(11), 3100–3113 (2017)
8. Aranda, L.A., Reviriego, P., Maestro, J.A.: Error detection technique for a median filter. J. IEEE Trans. Nuclear Sci. **64**(8), 2219–2226 (2017)
9. Alsharif, M.N.: Real time image processing for lane following. Master Thesis (2014)
10. Tahiyah, N.S., Vikramkumar, P., Sridharan, K., Vineetha, T., Arthi, J.: Very large-scale integration architecture for video stabilisation and implementation on a field programmable gate array-based autonomous vehicle. J. IET Comput. Vision **9**(4), 559–569 (2015)
11. Aranda, L.A., Reviriego, P., Maestro, J.A.: A fault-tolerant implementation of the median filter. In: 16th European Conference on Radiation Effects on Components and Systems (RADECS), Bremen, Germany (2016)

Classification of Defects in Bushes Using Deep Learning Approaches

I. S. Kushagra, R. Rakshith, Shubham Toraskar, Rahul J. Gujaran, and C. S. Asha

1 Introduction

The defect detection has been one of the challenges in manufacturing industries over the years. There are several types of defect detection, in which classification of the type of the defect is also important. Detection of minute cracks, semi-finished or unfinished surface, dents in bushes, bearings, etc. are carried out manually, which is time-consuming and most tedious. In the present work, we focus on defect detection in bushes and classify the defect types. Besides, the final decision on accepting or discarding the product is carried out using binary classification techniques.

In recent days, deep learning methods have shown improved accuracy in several applications. The advance in object detection area due to deep learning is remarkable. Besides, the processing of an image has become much faster, once the training part is executed. The paper aims to maintain high industry standards in delivering quality and defect-free bushes using computer vision techniques.

The sample images of the Bush are shown in Fig. 1. It is an independent plain bearing inserted into a housing to provide a bearing surface for rotary applications. It is the most common form of a plain bearing. Common designs include *solid* (sleeve and flanged), split, and clenched bushings. A sleeve, split, or clenched bushing is only a "sleeve" of material with an inner diameter (ID), outer diameter (OD), and length. The difference lies in the fact that solid sleeved bushing is solid around, a split bushing has a cut along its length, and a clenched bearing is similar to a split bushing with a clench across the cut connecting the parts.

Plain bearings are prepared using a durable material, low friction, low wear to the bearing and shaft, resistant to elevated temperature, and corrosion resistance. Often

I. S. Kushagra · R. Rakshith · S. Toraskar · Asha C. S. (✉)
Mechatronics Engineering, Manipal Institute of Technology, Manipal Academy of Higher Education, Manipal 576104, India

R. J. Gujaran
Robonics Control Solutions, Bangalore, India

S. K V and K. Guruprasad Rao (eds.), *Smart Sensors Measurements and Instrumentation*, Lecture Notes in Electrical Engineering 750,
https://doi.org/10.1007/978-981-16-0336-5_15

Fig. 1 Sample images of bush with defects such as cracks, semi machining defect indicated using arrow mark

the bearing is made up of soft and hard ingredients. In general, harder surfaces have lower coefficients of friction, and the high pressure is required for to gall or to seize when lubrication fails. The material employed in this work is the cast iron used with a hardened steel shaft due to a lower coefficient of friction. The cast iron glazes over, hence wear becomes negligible, which are extensively used in bearings fitted to engine components.

2 Related Work

In this section, we discuss the approaches normally employed for feature classification. Before deep learning methods, machine learning methods have been utilized for binary/multi-class classification. The conventional methods use hand-picked features such as raw pixels, speeded up robust features, a histogram of gradients, GLCM features while training the classifier such as support vector machines [1]. Recent advances in the deep learning area, with the methods that learn the features automatically for predicting the classes. We mention a few of the state-of-the-art deep learning methods in this paper and compare them with our method. LeNet architecture was intended to recognize the characters in documents [2]. Simonyan and Zisserman proposed a VGG network characterized by 3×3 convolutional layers with increasing depth, max pooling, and softmax classifiers [3]. He et al. introduced ResNet with deep networks trained through the use of residual modules [4].

3 Methodology

The proposed method is depicted in Fig. 2, which provides an overview of the steps employed in the current work. They include data preparation, CNN architecture development, model training, prediction of defects, analysis followed by deployment.

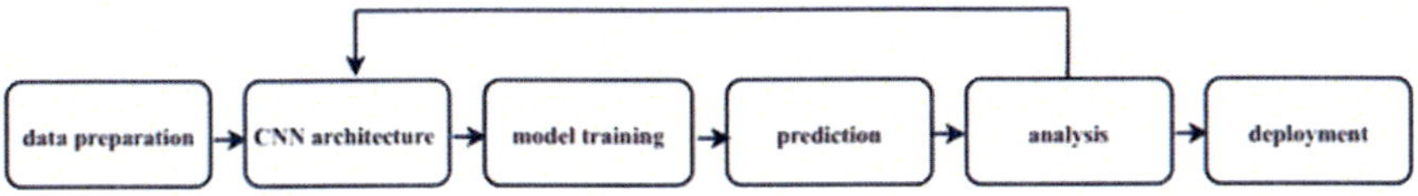

Fig. 2 Block diagram of the proposed system

3.1 Dataset Preparation

Many defects may occur in the initial stages like casting and foundry for noticeable reasons like temperature, mold, cooling rate, and many more. Certain defects may occur during machining stages like drilling, grinding, and surface finish because of improper machining or inappropriate handling of tools of a workpiece. All such defects will affect the reliability of the product and may result in serious damage to the machine in which it is used. To identify such damaged components, we implement computer vision techniques to segregate them from fine quality products with less human intervention.

3.2 Hardware Setup

The work focusses on detecting the faulty workpieces, hence we use the lens of 8MP Raspberry Pi Camera V2. The camera is used with the Raspberry Pi module in order to take the images at a fixed distance. This assures the constant size of the bush image. It is capable of producing 3280 × 2464-pixel static images, and also supports 1080p at 30fps, 720p at 60fps, and 640 × 480 at 60/90fps video. In addition, the system uses a night vision IR camera to take the picture without the help of lights. The fused image is used for the experimental work.

The camera is fixed onto a support structure with a surround LED system as shown in Fig. 3. The LED maintains the uniform lighting inside the system that is falling on

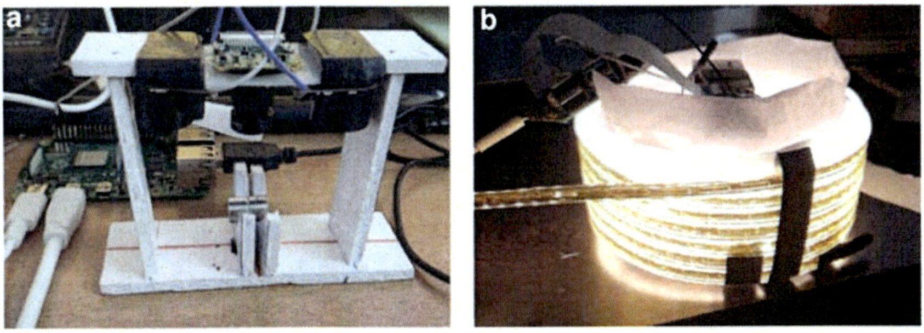

Fig. 3 Hardware assembly **a** support system with an attached camera **b** LED lighting for constant illumination

Table 1 Dataset preparation: list of the type of bushes and defects

Sl No	Product name	Defect Ttype
1	H.W-Bush-432635–0032	Dent-MD Shrinkage-FD Foundry defect
2	B.W-Bush-2054	Finish-OD-MD Dent-MD Foundry defect Shrinkage-FD OD-Outcut-MD Semi-machining defect
3	Bosch Mahle Bush-4132–000-0175	Semi-machining defect Dent-MD OD-Outcut-MD Foundry defect
4	H.W-Bush-434134–0001	Shrinkage-FD Foundry defect Grinding-Wheel-mark
5	H.W-Bush-807733–0001	Foundry defect Semi-machining defect Crack Groove-size Dent-MD Shrinkage-FD Finish-OD-MD OD-Outcut-MD
6	H.W-Bush-734547–0001	Foundry defect Crack Dent-MD Shrinkage-FD

the bush. The surface of the bush is highly finished, which causes light reflection in the image. The reflection is partially eliminated by placing the filter papers around the bush. The combination of 8MP camera and night vision IR camera is efficiently used to detect the cracks on uneven surfaces from the image.

The various types of defects along with the product name is tabulated in Table 1 and depicted in Fig. 4.

3.3 Custom CNN Architecture

The Convolution Neural Network (CNN) architecture [4, 5] is capable of extracting features from the image. A CNN architecture has several convolutional layers, with each filter extracting the smallest to largest features from the image. These features are converted to a 1D vector for an accurate prediction. In this work, we designed the CNN architecture that generates 727,465 parameters for entire training dataset,

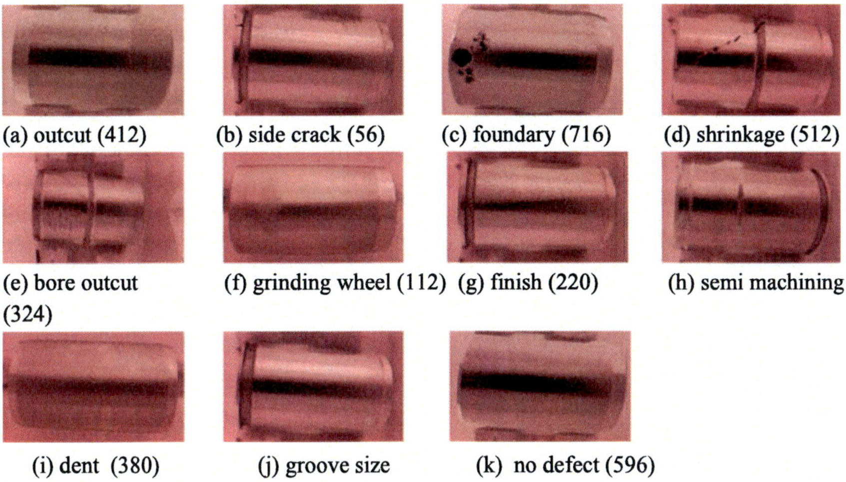

(a) outcut (412) (b) side crack (56) (c) foundary (716) (d) shrinkage (512) (e) bore outcut (324) (f) grinding wheel (112) (g) finish (220) (h) semi machining (i) dent (380) (j) groove size (k) no defect (596)

Fig. 4 Type of defects in a steel bush and the number of samples per defect is shown

trains faster on our dataset, and generates higher accuracy as compared to classic models.

Cross entropy loss measures the performance of a classification model whose output probability lies between 0 and 1. The cross-entropy used for binary classification is given by

$$J(w) = -(y\log(p) + (1-y)\log(1-p)) - (y\log(p) + (1-y)\log(1-p)) \tag{1}$$

The loss function employed for multi-class classification is given by

$$J(w) = -\frac{1}{N}\sum_{i=1}^{N}\left[y_i\log(\widehat{y_i}) + (1-y_i)\log(1-\widehat{y_i})\right] \tag{2}$$

The good dataset helps CNN to accurately extract the features. The raw image has a size of 3000 × 2200 RGB images. The images are arranged and manually labeled for multiclass and binary classification purposes. The data is then divided into train and test datasets. The images are cropped and compressed to have 160 × 160 × 3 images.

CNN Architecture: The CNN architecture, consists of several layers of convolution operation, a max-pooling stage for classification purposes. The final stage provides the probability of the predicted class. Figure 5 Provides the details of convolution layers, max-pooling, softmax based classification of the proposed method.

Model Training: The model training is an iterative process where the classification model was trained using certain parameters, then the hyperparameters such as epochs,

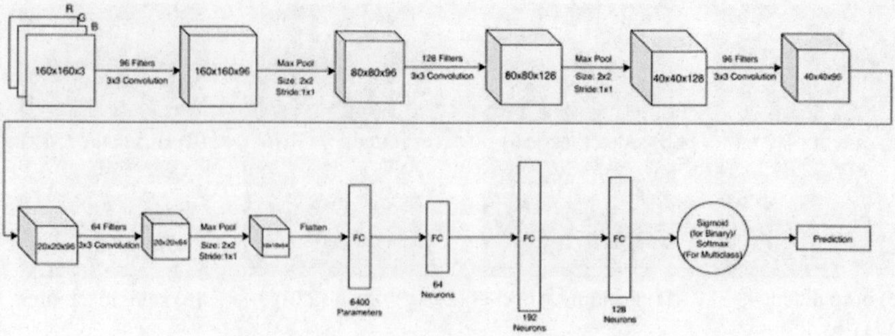

Fig. 5 Proposed convolution neural network architecture

optimizer, learning rate, batch size, loss functions are used to attain the best accuracy on training data.

Parameter settings: The parameters are chosen to get the best outcome. Hence the set of parameters used for binary and multi-class classifications is depicted in Tables 2 and 3.

Prediction: This prediction stage involves the testing of the data using the weights and biases generated in the training stage.

Analysis: A model may give good accuracy on training data, whereas it may perform poorly on test data. Hence, it is important to fine-tune the model for good accuracy on the given dataset. The predictions obtained are compared with ground truth labels to compute the accuracy of the model.

Table 2 .

Layer (type)	Output Shape	Param #
conv2d (Conv2D)	(160, 160, 96)	2688
max_pooling2d (MaxPooling2D)	(80, 80, 96)	0
conv2d_1 (Conv2D)	(80, 80, 128)	110,720
max_pooling2d_1 (MaxPooling2D)	(40, 40, 128)	0
conv2d_2 (Conv2D)	(40, 40, 96)	110,688
max_pooling2d_2 (MaxPooling2D)	(20, 20, 96)	0
conv2d_3 (Conv2D)	(20, 20, 64)	55,360
max_pooling2d_3 (MaxPooling2D)	(10, 10, 64)	0
flatten (Flatten)	(6400)	0
dense (Dense)	64)	409,664
dense_1 (Dense)	(192)	12,480
dense_2 (Dense)	(128)	24,704
dense_3 (Dense)	(9)	1161

Table 3 Parameter settings used in the proposed method

Hyperparameter	Binary classification	Multiclass classification
Optimizer	Adam	Adam
Learning rate	0.01	0.01
Batch size	64	250
Epochs	140	100
Loss function	Binary cross-entropy	Sparse categorical cross-entropy
Activation (Intermediary layers)	ReLU	ReLU
Activation (Output layer)	Sigmoid	softmax

Deployment: This step focusses to provide the solution to real-time industry problems. This part is done through Raspberry pi connected to multiple cameras, which provide images as the inputs. The whole process is run using google colab for training and testing purposes. However, the future work focusses on implementing the current work on Raspberry pi from capturing the image till output prediction.

4 Result Analysis and Discussion

4.1 Quantitative Analysis

Accuracy of the system is computed as the ratio of the number of correct predictions to the total number of examples. Using our CNN architecture, the model was able to achieve 99.85% accuracy on the training dataset and 99.21% accuracy on the test dataset for a binary classification model. Using our CNN architecture, the model was able to achieve 99.85% accuracy on the training dataset and 89.32% accuracy on the test dataset for a multi-classification model. The multiclass classification model was able to predict the specific type of defect present in the sample with an accuracy of 99.56% on the training dataset and 89.32% on the test dataset at the 100th epoch. Figure 6 depicts the accuracy and loss functions for binary and multi-class classification. The proposed method has higher accuracy, precision, recall, and F1 score as compared to the classic models (Table 4).

4.2 Qualitative Results

See Figs. 7 and 8.

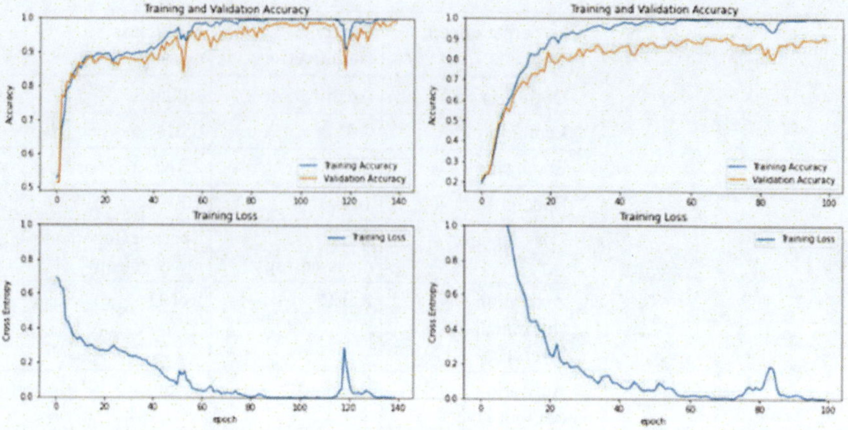

(a) (b)

Fig. 6 Train versus Test accuracy per epoch for **a** binary classification **b** multi-class classification

Table 4 Accuracy of the state-of-the-art approaches for the custom dataset using 1023 training data and 256 testing data for binary classification. And 2828 training data with 500 testing data for multi class classification. The size of the image is scaled to 160 × 160 × 3

Methods	Binary classification (Accuracy/Recall/F1-score)	Multiclass classification (Accuracy/Recall/F1-score)
SVM with raw features	0.87/0.99/0.89	0.58/0.58.0.57
AlexNet	0.96/0.96/0.96	0.81/0.81/0.80
LeNet	0.53/1.0/0.68	0.21/0.21/0.07
ResNet50	88.6/88.1/89.7	0.62/0.62/0.64
Proposed Method	**0.99/0.98/0.99**	**0.89/0.89/0.89**

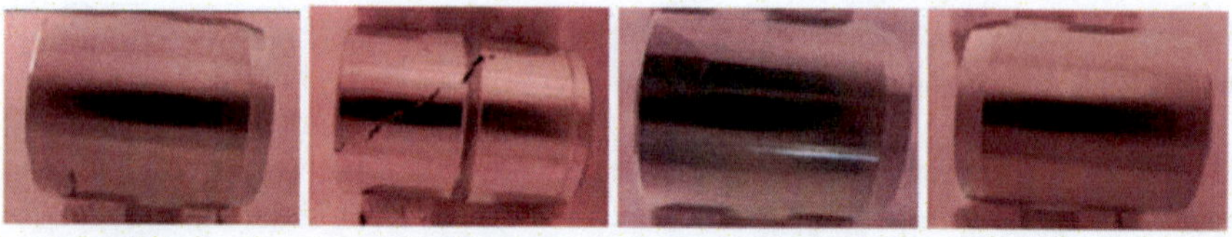

(a) (b) (c) (d)

Fig. 7 Predicted as **a–c** defect **d** no defect using a binary classification model

5 Conclusion

There is a high demand for automation in industries for inspection of defects in bushes. The proposed method is a very useful tool for manufacturing industries to speed up the process of inspection. The method can be implemented using low-cost

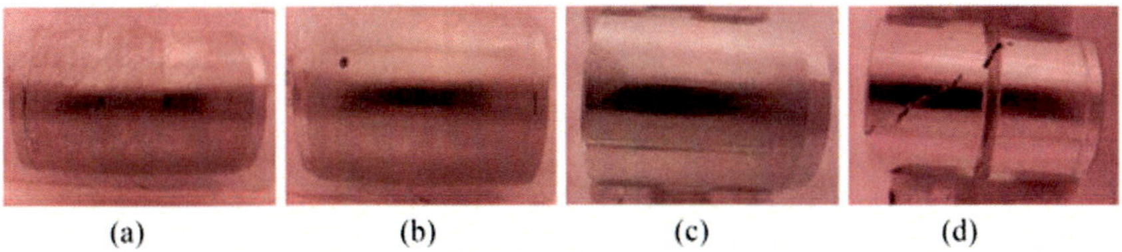

Fig. 8 Predicted class as **a** semi machining defect **b** foundary defect **c** finishing defect **d** shrinkage defect

hardware such as the Raspberry Pi board, integrated with multiple cameras, and able to identify the defect that is very close to human accuracy. The small error can also be reduced by increasing the number of samples for each type of defect. More number of samples would help better training of the classification, and hence more accuracy. In addition, data augmentation would greatly increase the number of samples and improve the classification accuracy.

Acknowledgements We would like to thank the Shri Madhwa Tech solutions for providing the hardware set up for the experiment.

References

1. Evgeniou, T., Pontil, M.: Support vector machines: theory and applications. In: Advanced Course on Artificial Intelligence (pp. 249–257). Springer, Berlin, Heidelberg (1999).
2. LeCun, Y., Bottou, L., Bengio, Y., Haffner, P.: Gradient-based learning applied to document recognition. Proc. IEEE **86**(11), 2278–2324 (1998)
3. Simonyan, K., Zisserman, A.: Very deep convolutional networks for large-scale image recognition (2014). arXiv preprint arXiv:1409.1556
4. He, K., Zhang, X., Ren, S., Sun, J.: Deep residual learning for image recognition. In: Proceedings of the IEEE Conference on Computer Vision and Pattern Recognition, pp. 770–778 (2016).
5. Krizhevsky, A., Sutskever, I., Hinton, G.E.: Imagenet classification with deep convolutional neural networks. In: Advances in Neural Information Processing Systems, pp. 1097–1105 (2012)

A Survey on Energy-Efficient MAC Protocols for Wireless Sensor Networks

Anand Gudnavar and N. Manjanaik

1 Introduction

A WSN is a network of tiny embedded systems that are distributed spatially to cooperatively monitor the surrounding location of interest. The information thus gathered by each mote is relayed to the sink in multiple hops. At the sink station, the received data is processed and aids to the better understanding of the observed environment [1].

Owing to the technological advancement in the field of microelectronics, affordability of the sensor motes has become possible. This has led to the popularity of the WSNs in varied applications using distributed sensor nodes such as home automation, battleground surveillance, forest area monitoring, and intrusion detection. Sensor nodes are low-power devices with limit on battery life, sensing power, wireless communication, and computational capacity. The node density deployed in the target application and area could be anywhere from hundreds to thousands [2].

Since the sensor motes are expected to carry out all their operations on their own with the small battery for a longer period (ranging from months to years), energy efficiency is a principal design criterion of WSN protocols. Sensor nodes' radio consumes more power and is a component which is coordinated and controlled by the WSN MAC protocols [3]. Thus, an energy-efficient MAC protocol increases the sensor network lifetime to a considerable period. MAC layer functions decide the way sensor nodes share and access the wireless medium. Reduced collisions and increased throughput can be achieved by an efficient MAC protocol thereby providing application flexibility.

The reminder of the article is formed as follows. First we emphasize the MAC protocol design metrics and sources of energy wastage in Sect. 2. Next in Sect. 3, we

A. Gudnavar (✉)
Research Scholar, Department of ECE, UBDTCE, Davanagere, India

N. Manjanaik
Professor, Department of ECE, UBDTCE, Davanagere, India

S. K V and K. Guruprasad Rao (eds.), *Smart Sensors Measurements and Instrumentation*, Lecture Notes in Electrical Engineering 750,
https://doi.org/10.1007/978-981-16-0336-5_16

describe the protocol classification. Section 4–6 describes the various existing MAC protocols. In Sect. 7, we compare and summarize the MAC protocols. In Sect. 8, we provide the conclusion and future work.

2 MAC Protocol Design Metrics

There are number of metrics influencing the design of a WSN MAC protocol. We have considered these important design metrics: energy efficiency, collision avoidance, channel utilization, and scalability [4].

- Energy efficiency: Most important metric affecting the WSN performance. Traditional wireless MAC implementations mainly focused on throughput, but in WSNs, the primary goal is to extend the sensor network lifetime. As MAC directly works on radio module, which is the main energy consumer, it can optimize the energy drained.
- Collision avoidance: Contention-based MAC protocols like CSMA/CA have a high probability of collisions occurrence. Optimized MAC may help reduce number of collisions.
- Channel utilization: This metric determines how channel bandwidth is put to use during wireless communication. Bandwidth allows as many active users as possible in the system. Application determines number of users.
- Scalability: Over the period of network operation, it is desirable to have a dynamic MAC adapting to the changes in the network like node density, topology, and network size. The changes happen as few nodes die, few motes may move to other place or new nodes joining the network.

2.1 Energy Wastage Sources

During the operation of WSN, energy is basically consumed in these units: radio, microcontroller, and sensor. Of the three units, radio transceiver consumes more energy as compared to the sensor and the processing unit [5]. Activities in radio transceiver that consume energy are described below:

- Idle listening: Sensor nodes have their radio turned on to listen to the potential data destined to it. This act drains out more battery energy as the sensor nodes radio is on and sensor neither receives data nor transmits.
- Overhearing: Listening to the data or messages of other nodes. Using the destination address field in the packet frame overhearing can be avoided. There exist other forms of overhearing viz. redundant broadcast data and sequence of longer preambles.
- Collisions: When a sensor is in the vicinity of nodes which are transmitting in tandem, collision is likely to occur. Due to this collision, data frames may

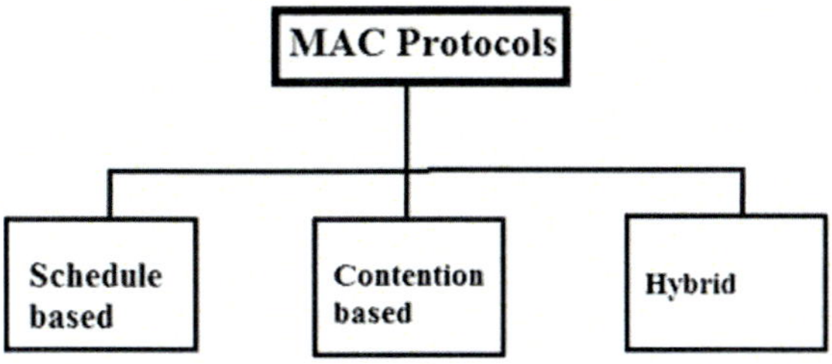

Fig. 1 MAC classification in WSN

not be captured by it. Because of the frame collision, energy spent during data transmission and reception is wasted.

- Overhead: To achieve the synchronous operation among the sensors, control frames such as CTS, RTS, SYNC, ACK are used, which consume energy. Typically for a small sensor network that employs RTS/CTS scheme, incurred overhead is about 40% to 70% of the channel capacity [6].

3 MAC Protocol Classification

The MAC protocols designed for WSNs can be broadly categorized into three classes as shown in Fig. 1.

- Contention based: B-MAC, S-MAC, T-MAC, etc.
- Scheduled based: FLAMA, TRAMA, etc.
- Hybrid: Z-MAC, IEEE 802.15.4 A-MAC, etc.

4 Contention-Based Protocol

MAC protocols that are contention-based use CSMA/CA algorithms. Sensor nodes compete other nodes to access the shared channel. Typically, a contention mechanism is used to give the channel access to a particular competing node. These protocols are highly adaptable to the network topology changes but have associated higher cost for overhearing, frame collisions, and channel-idle listening. In literature [5], there are a number of protocols that use contention schemes. We will describe few contention-based MACs below.

4.1 S-MAC (Sensor MAC)

Traditional CSMA/CA schemes employed continuous listening to the channel for any message transfer. This results in higher energy wastage especially when the

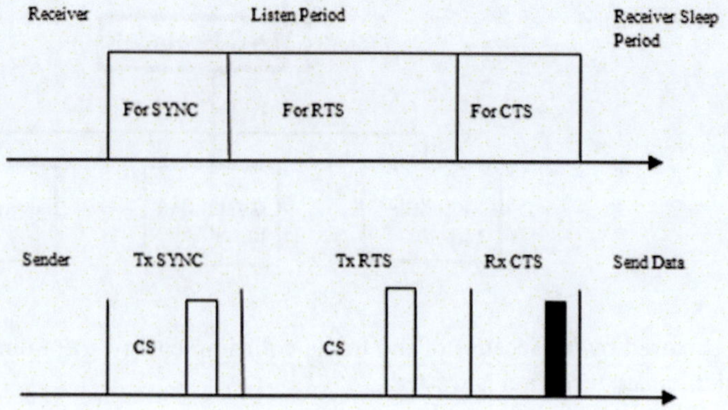

Fig. 2 S-MAC schedule

sensor motes have no data to transmit. To overcome this, duty-cycled operation is employed in S-MAC [7]. Concept of frames is used in S-MAC; operational time is divided into bigger frames. Nodes keep monitoring (active listening) the channel for a specific time and sleep for the remaining time in the frame. To save the energy, radio is turned off during the sleep time. Each frame slot starts with a sync phase followed by listen and sleep as shown in Fig. 2.

In the sync phase, sensor nodes transmit or receive the SYNC packet containing the node schedule (listen and sleep schedule). Data transfer happens during active part only, and not on the entire frame period, which reduces energy wasted on idle listening. Nodes use RTS/CTS and ACK for collision avoidance and overhearing avoidance thereby saving energy furthermore.

4.2 T-MAC (Timeout MAC)

T-MAC [8] introduces dynamic duty cycle, like S-MAC, adapting to the network conditions. Nodes enter early-sleep state when they find that they have no data to transmit or receive. It employs active timeout scheme to shrink the idle listening and dynamically adjust the active period that is proportional to the network load. When the network traffic is completed, nodes go to sleep as shown in Fig. 3.

Idle channel is monitored for a period called adaptive timeout (TA), during which if no activity occurs, nodes sleep by turning off the radio.

Compared to S-MAC, T-MAC performs better under different loads. However, it has early-sleep problem that affects the network performance.

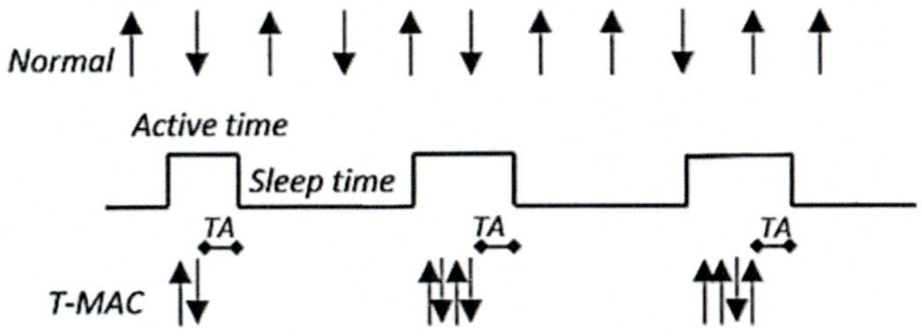

Fig. 3 T-MAC with adaptive active time

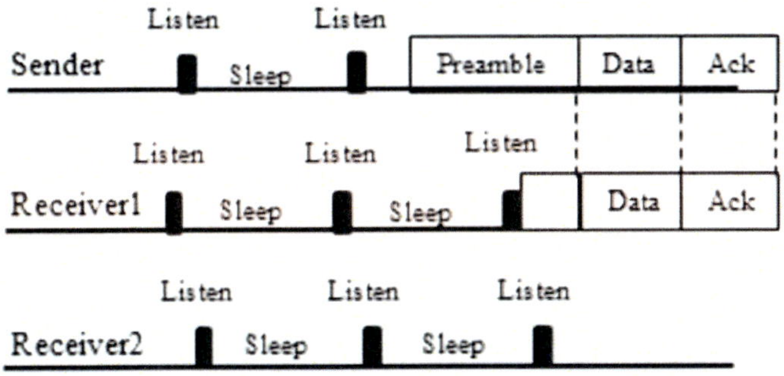

Fig. 4 B-MAC preamble samples

4.3 B-MAC (Berkley MAC)

B-MAC [9] enables a node to self-regulate the sleep and wakeup schedule. B-MAC applies low-power-listening (LPL) scheme for scheduling node sleep and wake-up pattern and clear-channel assessment for carrier sensing. These techniques improve channel utilization and energy efficiency. B-MAC also provides simple APIs to configure MAC functions. After a node wakes up, it checks for channel activity by turning radio on. Nodes sleep if there is no activity; else it remains ready for the data reception. A long wake-up preamble is sent before packet transmission. Upon receipt of the preamble nodes, know that a frame would be transmitted [10]. Generally preamble size is longer than the frame interval, so that nearby nodes can detect it. Therefore, when a node wakes up because of a preamble, it either receives the packet if it is the destination else it forward the data frame as shown in Fig. 4.

4.4 X-MAC

The lengthy preamble used by B-MAC introduces more delay in every hop, and more energy is consumed due to this. Receivers that are not concerned with transmissions

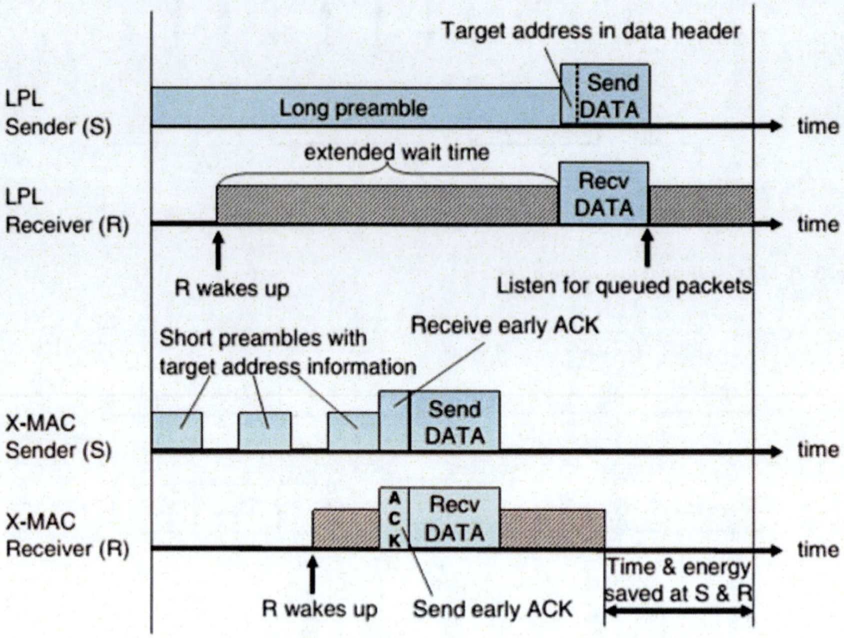

Fig. 5 X-MAC operation compared with LPL

are also wasting energy due to the inherent B-MAC operation. Protocol X-MAC [11] solves these concerns by using shorter preamble technique to retain the rewards of LPL, low-power network communication, and combining sleep schedules of receiver and transmitter.

Sender nodes use strobed preamble, where they transmit series of small preamble frames containing the destination address. This enables the destination nodes to interrupt the preamble by sending ACK, thereby reducing latency and saving energy as shown in Fig. 5.

4.5 DS-MAC (Dynamic Sensor MAC)

DS-MAC [12] improves latency time of S-MAC by modifying the node duty cycle dynamically based on the existing energy and traffic conditions. Duty cycle value is same for all nodes in DS-MAC. During the synchronization, all nodes share one-hop latency period values. If the node observes a high latency value, it shortens the sleep time, and this is informed with SYNC packet. The sender node after knowing the sleep time decrement doubles up its duty cycle. Latency of DS-MAC is better compared to S-MAC.

5 Schedule-Based Protocols

These classes of protocols guarantee shared medium access to each node periodically and are essentially based on TDMA scheme. Protocols based on TDMA preserve more energy, as they do not have collisions and have the built-in duty cycle. Shared wireless channel is segmented into superframes [13] as shown in Fig. 6 and nodes are assumed to have a global synchronization. Each node has its dedicated slot in the superframe, which it uses subsequently. There is complexity involved in maintaining the network TDMA schedule. Neighbor nodes schedule is maintained in each node which consumes the memory. Node coordination is highly expected to reserve the slots and is challenging. More difficult is clock drift as the slots are small and clock drift can be damaging. Hence, for acceptable operation, critical issue is exact clock synchronization [2].

As shown in Fig. 6, there are two parts in the superframe structure. Nodes use the reservation period to reserve their communication slot. Each sensor node then transmits data frames during the data period. Proposed MAC schemes based on TDMA differ in slot reservation mechanism, contention approach, clustering schemes, and frame size.

5.1 TRAMA (Traffic-Adaptive MAC)

TRAMA [14] employs collision-free energy-efficient channel access in WSN. When the nodes are not participating in data reception or transmission, they switch to a low-power state and as transmissions whether broadcast or unicast, are collision-free energy consumption which is significantly reduced. In this scheme, time-length is slotted, and the nodes get access to transmit based on the traffic. Nodes exchange the neighborhood two-hop information and also the transmission schedule detailing the destination. This is the main attribute of the distribution-election scheme, where transmitters float schedule to select the receivers.

TRAMA consists of three sub-parts, neighbor-protocol (NP) to exchange neighborhood information, adaptive-election algorithm (AEA) to select the senders and the receivers in the current slot, and schedule-exchange protocol (SEP) to exchange the slot schedule. Other nodes can toggle themselves to low-power state.

Fig. 6 Frame structure for TDMA based MAC protocol

TRAMA is more energy-efficient, has higher packet delivery ratio even during high traffic, but it experiences higher latency when compared to S-MAC [15].

5.2 FLAMA (Flow-Aware MAC)

FLAMA improves the shortcoming of TRAMA. When a node is not in receiving state, data transmission to it is prevented. Stoppage of packet transmission achieves higher energy efficiency. It also prevents data collisions and idle listening. It dynamically adjusts access schedule to the flow of traffic generated by the application. FLAMA is lightweight which can be run on nodes with limited memory, processing power, and communication capabilities.

5.3 AI-LMAC (Adaptive Information-Centric Lightweight MAC)

AI-LMAC [16] dynamically allots node slot when sink queries for a data. A parent–child association is assumed in the network. Parent knows the traffic generated by its child nodes; therefore, it assigns slots to be taken up by them for the data transmission. After this, a sensor takes up available free slot after communicating with two-hop neighbors. The shortfall of this mechanism is sink cannot estimate the slots to be taken up by the nodes if source data rates are not fixed. Also there is no means to guarantee a node that can get adequate slots for data transmission.

6 Hybrid Protocols

These families of protocols combine the features of both contention and TDMA based WSN MAC. The access channel is divided into two parts, channel control and data frames. Node data in the channel control frame is transmitted randomly, whereas in channel data frame, data packets are sent in scheduled manner. Thus, more energy is saved. Hybrid protocols are also scalable and flexible compared to the other two classes of protocols. Because of better channel organization and dynamic load adaptability, hybrid implementations are able to perform better in terms of collision avoidance and higher energy efficiency. Few hybrid protocols [13] are Z-MAC, IEEE 802.15.4, A-MAC, etc.

6.1 Z-MAC (Zebra MAC)

In Z-MAC [17] contention resolution is improved by combining CSMA and TDMA. Based on the network traffic level, Z-MAC takes one of two approaches. Under high-traffic load, it works like TDMA, and under light traffic, it functions like CSMA. In Z-MAC, a sensor node may send the multiple data in a single time slot, and it may also transmit during any slot. The data is transmitted only after the node executes carrier sensing and the channel is free.

Z-MAC has two fundamental components called neighbor discovery and slot assignment, and local framing, and synchronization. These components help in increase the energy saving and decrease the collision. Compared to B-MAC, it performs better under high network load conditions. B-MAC outperforms Z-MAC when the network traffic is low.

6.2 IEEE 802.15.4

It specifies the standard [18] for low-rate wireless PAN and is used as base for other high-level communication protocols like ZigBee, Wireless HART, and 6LoWPAN. Coordinator node operates in beacon mode to maintain time frame synchronization, supporting slots that are collision free. Such slots are known as guaranteed time slot (GTS) that assure nodes to have a predefined bandwidth and maximized access delay. In non-beaconed mode, the access to the shared channel happens through a non-slotted CSMA/CA. Nodes sleep when they are not participating. Nodes are identified as either full function device (FFD) or a reduced function device (RFD). The protocol supports three network configuration topologies: mesh, cluster-free, and star.

6.3 A-MAC (Adaptive MAC)

A-MAC [19] is proposed for services that are collision free, have no overhearing, and involve lesser idle listening data transmissions. It is used in applications where network longevity is desired like monitoring and surveillance. The sensor nodes are informed priory about the reception of the data packets. A node is activated whenever it wants to transmit or receive the data; otherwise, it remains in sleep mode. Therefore, overhearing and idle listening are eliminated. A-MAC also enhances the wireless channel accessibility.

7 MAC Protocol Comparison and Summary

Contention-based protocols are scalable and have low latency as compared to schedule-based protocols. But the energy consumed is on the higher side as contention schemes are collision-prone and use collision-avoidance approach. Contention-based protocols are adaptable to sudden and unexpected changes in the network traffic and therefore are more suitable to high-traffic applications.

Schedule-based protocols save more energy compared to the contention scheme as their duty cycle is condensed. These classes of protocols have no collisions, and each node has fixed slot for communication as a result they do not employ control packets thereby reducing the overhead. However, managing a node cluster and problems associated with interference is challenging. Adding new nodes to the existing cluster is difficult, and thus, schedule-based protocols are less flexible compared to contention based.

Reservation-based protocols are not suitable for a high sensor node density applications. These are not scalable as they use fixed channel capacity. Higher delivery delay is observed in schedule-based protocols because of the packet structure. The nodes should be synchronized perfectly for the TDMA protocols to work perfectly. Time drift can prove costly to the application.

Table 1 inspired from [6] summarizes the MAC protocols discussed in this article. Evaluation of the protocol performance is based on three metrics: network latency, throughput, and energy efficiency.

8 Conclusion and Future Work

There has been a paradigm shift in the design parameters considered for WSN MAC protocols from energy efficiency to higher quality of service (QoS). This paper attempts to discuss MAC protocols that focus on reducing energy consumption. Various MAC protocols are studied based on different key metrics. QoS metric are low delay, low lost data packets, higher throughput, and low energy consumption. Energy efficiency is primarily focused on the MAC protocol design. However, there are new avenues where MAC layer could be researched more like mobility of nodes [20], WSN security [21], WBAN [22], and cross-layer open issues [23].

Table 1 Summary of the MAC protocols

Type	Protocol	Latency	Energy efficiency	Throughput	Merits	Demerits
Contention-based	S-MAC	Low	Medium	Low	Less energy consumed for low traffic	More control packets, sleep latency
	T-MAC	High	High (variable traffic)	High	Adaptive	Overhead packets, early sleep
	B-MAC	High	Low	Medium	Simple implementation, low-power consumed	Overhearing, error-prone to high traffic
	X-MAC	Low	High	Low	Low latency, low-power consumed	Not optimal universal traffic
	DS-MAC	Low	High	Medium	Adaptive node sleep time based on network traffic	Deviation in predicting scheme
Schedule-based	TRAMA	Low	Low	High	High throughput and energy efficiency	Slot time divided as random access
	FLAMA	Low	Medium	High	Idle listening is reduced	Slot time divided as random access
	AI-LMAC	Low	Medium	High	Handling network fairness	Application specific
Hybrid	Z-MAC	Low	Low	High	Better performance when traffic is high	Non-satisfactory latency at low loads
	IEEE 802.15.4	Low	High	Low	Dynamic duty cycle	Low data rates
	A-MAC	Low	High	Medium	Tackle sensing hole problem	Network preconfiguration required

References

1. Al-Fuqaha, A., Guizani, M., Mohammadi, M., Aledhari, M., Ayyash, M.: Internet of things: a survey on enabling technologies, protocols, and applications. IEEE Commun. Surveys Tutorials **17**(4), 2347–2376 (2015). https://doi.org/10.1109/comst.2015.2444095
2. van Dam, T., Langendoen, K.: An adaptive energy-efficient MAC protocol for wireless sensor networks. In: Proceedings of the First International Conference on Embedded Networked Sensor Systems (SenSys 2003). pp. 171–180 (2003)
3. Huang, P., Xiao, L., Soltani, S., Mutka, M. W., Xi, N.: The evolution of MAC protocols in wireless sensor networks: a survey. IEEE Commun. Surv. Tutorials **15**(1), 101–120 (2013). https://doi.org/10.1109/surv.2012.040412.00105

4. Ye, W., Heidemann, J.: Medium Access Control in Wireless Sensor Networks. USI/ISI Technical report ISI–TR-580 (2003)
5. Ansari, H., Sarfuddin Khatib, M.: A survey of Mac protocol for broadcast problem in wireless sensor network. IOSR J. Comput. Sci. **3**(5), 20–23 (2014)
6. Djimli, A., Merniz, S., Harous, S.: Energy-efficient MAC protocols for wireless sensor networks: a survey. TELKOMNIKA Telecommun. Comput. Electron. Control **17**, 2301–2312 (2019)
7. Ye, W., Heidenmann, J., Estrin, D.: An energy-efficient MAC protocol for wireless sensor networks. In: Proceedings of IEEEINFOCOM, New York, NY, (2002)
8. Gong, H., Lui, M., Mao, Y., Chen, L., Xie, L.: Traffic adaptive MAC protocol for wireless sensor networks. In: Proceedings of ICCNMC'05, Zhangjiajie (2005)
9. Polastre, J., Hill, J., Culler, D.: Versatile low power media access for wireless sensor networks. In: The Second ACM Conference on Embedded Networked Sensor Systems (SenSys), pp. 95–107 (2004)
10. Bachir, Heuss, M., Duda, A.: Preamble MAC protocols with non-persistent receivers in wireless sensor networks. In: Networking Ad hoc and Sensor Networks, Wireless Networks, Next Generation Internet (2008)
11. Rajendran, V., Obraczka, K., Garcia-Luna-Aceves, J.J.: Energy-efficient collision-free medium access control for wireless sensor networks. In: Proceedings 1st International Conference Embedded Networked Sensor Systems (SenSys'03). pp. 181–192 (2003)
12. Lin, P., Qiao, C., Wang, X.: Medium access control with a dynamic duty cycle for sensor networks. In: IEEE Wireless Communications and Networking Conference, vol. 3, pp. 1534–1539 (2004)
13. Akyildiz, I.F., Can Vuran, M.: In: Wireless Sensor Networks, Wiley (2010)
14. Rajendran, V., Obraczka, K., Garcia-Luna-Aceves, J.J.: Energy-efficient collision-free medium access control for wireless sensor networks. In: Proceedings 1st International Conference Embedded Networked Sensor Systems (SenSys'03), pp. 181–192 (2003)
15. Sitanayah, L.: Planning the Deployment of Fault-Tolerant Wireless Sensor Networks. PhD thesis University College Cork (2013)
16. Chatterjea, S., van Hoesel, L., Havinga, P.: AI-LMAC: an adaptive, information-centric and lightweight MAC protocol for wireless sensor networks. In: Proceedings 2004 Intelligent Sensors, Sensor Networks and Information Processing Conference, pp. 381–388 (2004)
17. Rhee, A., Warrier, M., Aia, M.J.: Z-MAC: a hybrid MAC for wireless sensor networks. In: ACM Sensys, November 2005 and IEEE/ACM on Transactions on Networking (2008)
18. Toscano, E., Bello, L.L.: Multichannel super-frame scheduling for IEEE 802.15.4 industrial wireless sensor networks. IEEE Trans. Indus. Info. **8**(2), 337–350 (2012)
19. Nam, Y., Lee, H., Jung, H., Kwon, T., Choi,Y.: An adaptive MAC (A-MAC) protocol guaranteeing network lifetime for wireless sensor networks. In: 12th European Wireless Conference 2006—Enabling Technologies for Wireless Multimedia Communications, pp. 1–7. Athens, Greece (2006)
20. Zareei, M., Muzahidul Islam, A.K.M., Vargas-Rosales, C., Mansoor, N., Goudarzi, S., Husain Rehmani, M.: Mobility-aware medium access control protocols for wireless sensor networks: a survey. J. Netw. Comput. Appl. **104**, 21–37 (2018)
21. Nuristani, A.K., Thakur, J.: Security issues and comparative analysis of security protocols in wireless sensor networks: a review. JCSE Int. J. Comput. Sci. Eng. **6**(10), 436–444 (2018)
22. Mile, A., Okeyo, G., Kibe.: A hybrid IEEE 802.15.6 wireless body area networks interference mitigation model for high mobility interference scenarios. Wireless Eng. Technol. **09**(02), 34–48 (2018)
23. Gudnavar, A., Rajashekar.: Exploring an effectiveness and pitfalls of correlational-based data aggregation approaches in sensor network. Springer-Int. J. WIreless Microwave Technol. **7**(2), 44–56 (2017)

Automotive Dead Reckoning-Implementation of Fusion Algorithm on Raspberry Pi

Manjusha R. Parodkar, Milankumar Sheta, and Shreesha Chokkadi

1 Introduction

In modern day navigation, calculating position of a vehicle or a moving body using its previous determined position to advance that data based upon known or anticipated speeds over elapsed time and path is called dead reckoning (DR). There are chances of degradation or loss of position accuracy if we rely only on the global positioning system (GPS) status for position. Such a scenario may occur, for example, when the vehicle moves in areas where the GPS signals cannot be received (tunnels or underground passages) as shown in Fig. 1 or where very strong multipath propagation exists (areas surrounded by tall glass covered buildings).

A DR solution is useful to keep the accuracy of the position data high, by using the relevant data from different types of sensors (accelerometer, gyro sensor, etc.), even when positioning using only GPS data is difficult or GPS data is unavailable. Automotive dead reckoning (ADR) can be used in conjunction with high definition (HD) maps to center the vehicle for a certain amount of time.

Hence, ADR is used to estimate heading and position of vehicle whilst GPS is available and to predict heading and position of a vehicle while GPS outages using last known position and heading given by way of GPS, inertial measurement unit (IMU) data, and vehicle sensors. The limitations for ADR include:

- The prediction cannot be performed if the previous known state is not available just before the GPS outage.

M. R. Parodkar · S. Chokkadi (✉)
Instrumentation and Control Engineering, Manipal Institute of Technology, Manipal Academy of Higher Education, Manipal, India
e-mail: shreesha.c@manipal.edu

M. Sheta
Solutions Architect, KPIT Technologies Pvt. Ltd, Bangalore, India
e-mail: milankumar.sheta@kpit.com

S. K V and K. Guruprasad Rao (eds.), *Smart Sensors Measurements and Instrumentation*, Lecture Notes in Electrical Engineering 750,
https://doi.org/10.1007/978-981-16-0336-5_17

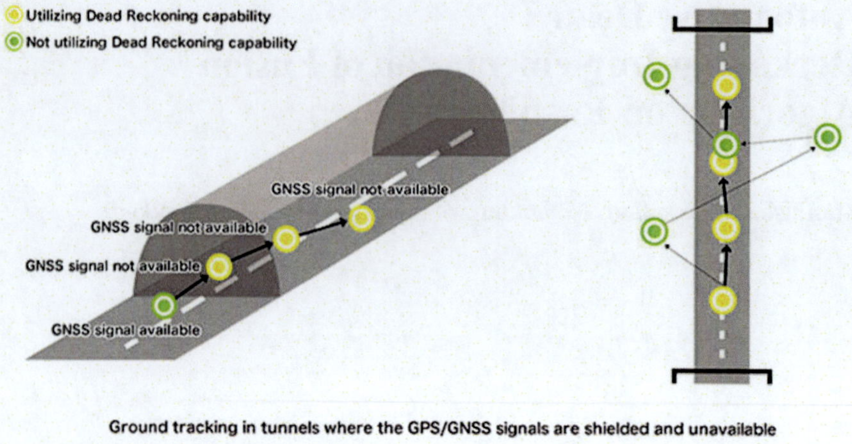

Fig. 1 Outage occurrence in GPS [1]

- There is a constraint on the amount of time the prediction can be performed during the unavailability of the measurements.

The principal objective here is designing and validating the fusion of an IMU and a GPS with a Kalman filter (KF)/extended Kalman filter (EKF) with a set of rules in Raspberry Pi for determining the car orientation. Data acquisition from the IMU, which includes three-axis accelerometer, three-axis gyroscope, three-axis magnetometer, and GPS will need right interfacing with the Raspberry Pi to achieve preferred effects and deliver right orientation of the vehicle at all times. With proper tuning of filter parameters, it is aimed to reduce the error to the minimum and remove bias if any.

2 Literature Survey

Using low-cost sensors, a method for accurate dead reckoning in urban areas has been presented in [2]. In urban areas, since there are more moving objects in the surrounding of the vehicle, there is degradation in the accuracy of the visual odometry. The error in the heading estimation, which is used with visual odometry, cannot be corrected with satellite data if the satellite signal reception is poor. The solutions to these problems have been presented in [2] to improve the accuracy of the visual odometry in urban environments. Two key techniques have been identified. To improve the sturdiness of visual odometry in dynamic environments, the first technique being used is moving object detection using IMU and pattern recognition. Heading estimation using time-series tightly coupled integration of satellite Doppler shift and IMU is the second technique.

In [2], to provide 3D attitude and position data for land vehicle application, a micro electro mechanical systems (MEMS) 3D DR/GPS integrated system has been presented. The algorithm proposed in this paper uses 1D speed information to obtain 3D position, and an EKF to compensate for the errors in navigation. The approach presented in [3] has exploited a short-range visual lane marking detector, and a DR system with which a registry of the detected back lane markings corresponding to the last 240 m driven is constructed. Low-power embedded architecture is enough to run the algorithm since the system uses a low-cost architecture of sensors.

In [4], for estimation of attitude, heading and gyroscope sensor bias, a low-cost MEMS inertial sensor based on a new modeling framework on EKF has been utilized. To design a low-cost aided DR navigation system, the proposed attitude and heading reference system (AHRS) and an odometer sensor have been used. For the evaluation of the proposed algorithm, experimental tests were conducted in different acceleration bound conditions and in the presence of external magnetic disturbances for a land vehicle.

3 Background Theory

3.1 Discrete State-Space System

A state-space representation is a mathematical model of a physical system as a set of input, output, and state variables related by first-order differential equations or difference equations. Discrete-time systems are governed by the following equation:

$$x_k = F_{k-1}x_{k-1} + G_{k-1}u_{k-1} + w_{k-1} \tag{1}$$

$$y_k = H_k x_k + v_k \tag{2}$$

where 'F' is the system matrix, 'G' is the discrete-time input gain, and 'w' is the process noise, with 'x' being the state vector and 'u' is the input vector. Also 'y' is the output vector, 'H' is the observation matrix, and 'v' is the measurement noise vector.

3.2 Sensors and Data Fusion

The GY-87 10-DOF [5] has MPU6050 which consists of three-axis accelerometer sensor and three-axis gyroscope sensor, HMC5883L, three-axis magnetometer and BMP180, and the barometric pressure sensor. The raw acceleration in all three-axis from the accelerometer is used to calculate the raw pitch and roll angles as follows:

$$\theta = \tan^{-1}\left(\frac{-Ax}{\sqrt{Ay^2 + Az^2}}\right) * \frac{180}{\pi} \tag{3}$$

$$\phi = \tan^{-1}\left(\frac{Ay}{Az}\right) * \frac{180}{\pi} \tag{4}$$

where A_x, A_y, andA_z are the accelerations obtained from the IMU and θ, φ are the pitch and roll angles, respectively.

Gyro meter has bias issue; i.e., when the device is still, yet gyro will show some output. This bias increases linearly with time and needs to be subtracted from the final output. The U-blox NEO-6 M GPS module includes one configurable UART interface for serial communication [6]. The data in NMEA format provides the information for the complete PVT (position, velocity, time) solution computed by the GPS receiver. Both IMU and GPS have different data rates; hence, before fusing the data, GPS is down sampled to data rate of IMU, and then fusion is performed.

3.3 Kalman Filtering

3.3.1 Linear Kalman Filter

To determine the states, Kalman filter as shown in Fig. 2 has been used, which does prediction followed by a correction. The main idea is to use the known present information about the dynamics of the state and then using the filter to project forward to predict the next state. The correction or update part then involves comparison of actual measurement with filter predicted value of that measurement.

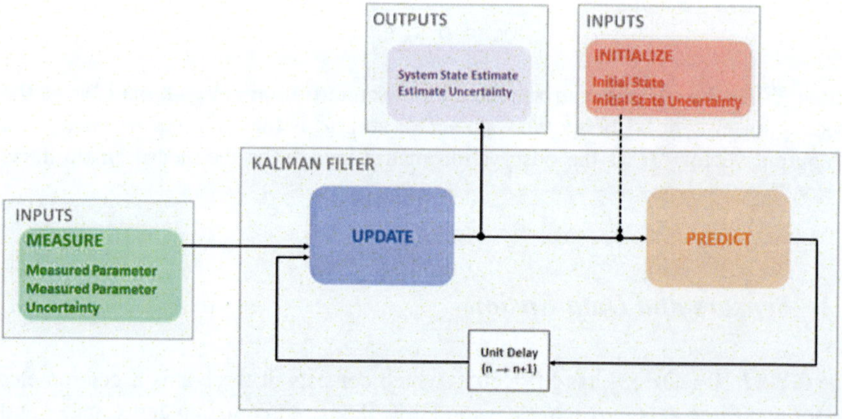

Fig. 2 Kalman filter concept [7]

To apply the predictor–corrector format recursively at each step, we need to start from some initial state estimate and initial state error covariance matrix. First, the state vector is predicted from the state dynamic eq. [8] using:

$$\hat{x}_{k|k-1} = F_{k-1}\hat{x}_{k-1} + G_{k-1}u_{k-1} \tag{5}$$

where $\hat{x}_{k|k-1}$ is the predicted state vector and $\hat{x}_{k-1}$ is the previous estimated state vector.

$$P_{k|k-1} = F_{k-1}P_{k-1}F_{k-1}^T + Q_{k-1} \tag{6}$$

where $P_{k|k-1}$ is the predicted state error covariance matrix, P_{k-1} is the previous estimated state error covariance matrix, and Q is the process noise covariance matrix. The Kalman gain matrix K_k is calculated by:

$$K_k = P_{k|k-1}H_k^T(H_kP_{k|k-1}H_k^T + R_k)^{-1} \tag{7}$$

where H is defined as the matrix necessary to obtain the output equation, and R is defined as the measurement noise covariance.

The state vector is then updated by scaling the 'innovation,' which is the difference between the measurement of the output, and the predicted output, by the calculated Kalman gain matrix in order to correct the prediction by the appropriate amount, as in:

$$\hat{x}_k = \hat{x}_{k|k-1} + K_k(y_k - H_k\hat{x}_{k|k-1}) \tag{8}$$

$$P_k = (I - K_kH_k)P_{k|k-1} \tag{9}$$

3.3.2 Extended Kalman Filter

EKF is one of the most known nonlinear forms of Kalman filter. The need for EKF aroused because KF can only be used for linear systems; but in reality, the systems are non-linear. EKF also uses *Predict* and *Update* function as used in KF, however, *Jacobian* function is an additional function used in EKF as discussed in [9]. This function is used to remove the nonlinearity of the system.

In the extended Kalman filter, the system and observation models do not need to be linear functions of the state but may instead be differentiable functions [10].

$$\hat{x}_k = f(\hat{x}_{k-1}, u_k) \tag{10}$$

$$\hat{y}_k = h(\hat{x}_k) + v_k \tag{11}$$

The function f is used to calculate the predicted state from the previous estimate that has been determined, and similarly, the function h is used to calculate the predicted measurement from the predicted state. Since f and h cannot be applied to the covariance directly, a matrix of partial derivatives (the Jacobian) is calculated using these functions.

The system and observation matrices are defined to be the following Jacobians:

$$F_k = \frac{\partial(f)}{\partial(x)}|(\hat{x}_{k-1|k-1}, u_k) \tag{12}$$

$$H_k = \frac{\partial(h)}{\partial(x)}|(\hat{x}_{k|k-1}) \tag{13}$$

3.4 Euler Angle Concept

Euler angles describe an object's orientation in space by defining three angles toward a fixed coordinate system. This orientation can be described by the composition of rotations around three axes.

Rotation matrices are described as:

$$R_x = \begin{bmatrix} 1 & 0 & 0 \\ 0 & \cos\theta_f & -\sin\theta_f \\ 0 & \sin\theta_f & \cos\theta_f \end{bmatrix} \tag{14}$$

$$R_y = \begin{bmatrix} \cos\phi_f & 0 & \sin\phi_f \\ 0 & 1 & 0 \\ -\sin\phi_f & 0 & \cos\phi_f \end{bmatrix} \tag{15}$$

$$R_z = \begin{bmatrix} \cos\psi_f & -\sin\psi_f & 0 \\ \sin\psi_f & \cos\psi_f & 0 \\ 0 & 0 & 1 \end{bmatrix} \tag{16}$$

where θ_f, ϕ_f, ψ_f are the fused Euler angles, and the final rotation matrix is obtained as:

$$R = R_x R_y R_z \tag{17}$$

To go from one coordinate system to another, these rotational matrices are multiplied. In order to obtain the first coordinate system, we need to multiply the inverse of the same rotational matrices in the same order.

3.5 *Geodetic Transformation*

3.5.1 Gauss Kruger Concept

The GPS co-ordinates are in longitude and latitude and cannot be directly used for fusion as sensors sense local co-ordinates. Hence, it is necessary to convert these GPS co-ordinates to local co-ordinates. If we keep the earth in a cylinder in a way surface of cylinder touches the central meridian as shown in Fig. 3 and if put a bulb inside the earth and switch it on, the projection on the surface of the cylinder will be a transverse Mercator cylindrical projection.

In transverse Mercator projection, the central meridian between two consecutive zones is 6 degrees apart; whereas in Gauss Kruger, it is 3 degrees apart. Hence, transverse Mercator has 60 zones, whereas Gauss Kruger has 120 zones.

The fused output of nonlinear filter is in local x and y co-ordinates. These local co-ordinates are converted to geodetic co-ordinates by method of inverse Gauss Kruger.

4 Methodology

4.1 *Integration with IMU*

Data acquisition from the IMU, which includes the three-axis accelerometer, three-axis gyroscope, three-axis magnetometer, and GPS will need right interfacing with

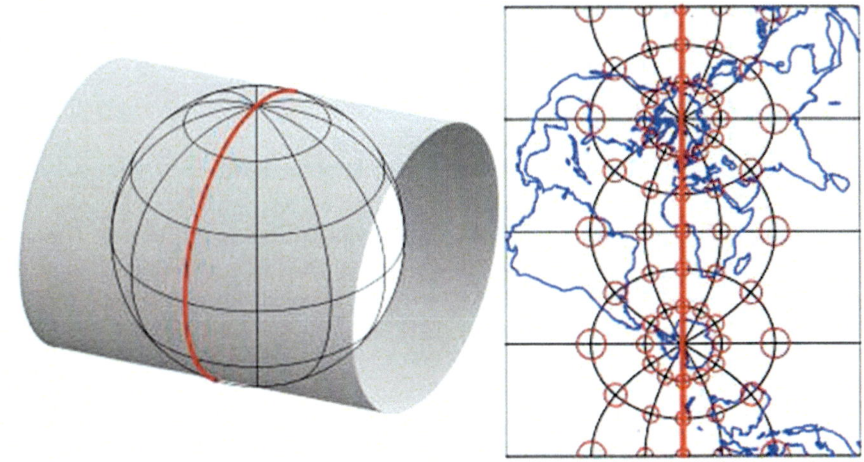

Fig. 3 Transverse mercator earth position in cylinder [11]

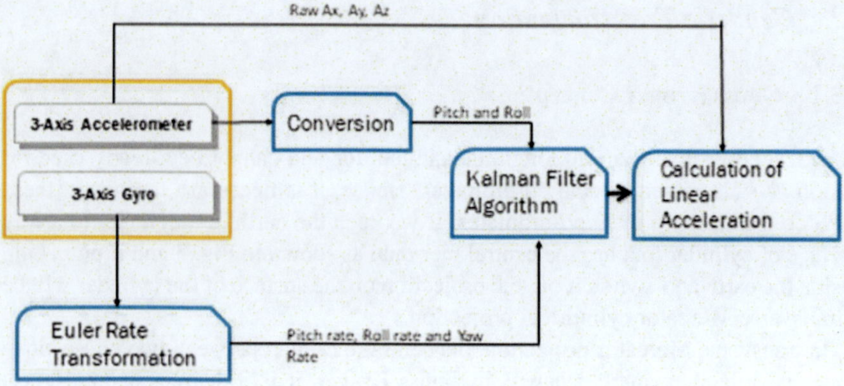

Fig. 4 Block diagram for IMU fusion

the Raspberry Pi [12] to achieve the preferred effects and deliver the right orientation of the vehicle at all times as shown in Fig. 4.

Logged CarMaker [13] data has been used for simulation in this work which involves two different scenarios:

- Vehicle on a straight road with constant speed
- Vehicle on a curved road with constant speed

Three accelerations and three angular rates for different vehicle levels of freedom are sensed by the IMU. Data from the GPS receiver is relatively less accurate; however, it is not always affected by time drift. Hence, with a view to lessen the negative aspects and to benefit the advantages, each of the sources must be integrated together via Kalman filter. Apart from position, orientation is also decided by using three different data sources that are accelerometer, magnetometer characterized via long-term accuracy and gyroscope characterized by means of short time period accuracy.

4.1.1 Calculation of Euler Angles and Euler Angular Rates

The angular velocity $\omega_x, \omega_y, \omega_z$ and the Euler angular rate between different coordinate systems $(\dot{\theta}, \dot{\phi}, \dot{\psi})$ are usually not the same when dealing with rotational movement. A transformation matrix is necessary to be calculated since the Euler angles are needed for describing the orientation [14].

$$\begin{bmatrix} \dot{\phi} \\ \dot{\theta} \\ \dot{\psi} \end{bmatrix} = \begin{bmatrix} 1 & \sin(\phi_f) * \tan(\theta_f) & \cos(\phi_f) * \tan(\theta_f) \\ 0 & \cos(\phi_f) & -\sin(\phi_f) \\ 0 & \sin(\phi_f)/\cos(\theta_f) & \sin(\phi_f)/\cos(\theta_f) \end{bmatrix} \begin{bmatrix} \omega_x \\ \omega_y \\ \omega_z \end{bmatrix} \tag{18}$$

where θ_f and ϕ_f are the fused roll and pitch angles obtained from the Kalman filter.

4.1.2 Orientation Estimation Using Kalman Filter

In the case of orientation evaluation, the equations for the dynamic system as referred in [15] are given by:

$$\theta_k = \theta_{k-1} + (\omega_k - q_{bk-1})T \tag{19}$$

$$q_{bk} = q_{bk-1} \tag{20}$$

Here, θ is the angle, ω is the angular rate, q_{bk} is the bias associated with the gyroscope and T is the sampling rate.

4.1.3 Calculation of Linear Acceleration

Accelerometers are less beneficial by themselves and regularly participate in other fusion sensors; however, they do have some functions like registering shakes, steps, and the like.

Often for such cases, we are interested in linear acceleration that is the acceleration without gravity, called gravity compensation.

After compensating for the acceleration due to gravity, the linear acceleration is obtained as follows [16]:

$$\text{Linear Acceleration} = \text{Raw Acceleration} - \text{Static Acceleration} \tag{20}$$

$$\text{StaticAcceleration} = \begin{bmatrix} -\sin(\theta_f) \\ \sin(\phi_f) * \cos(\theta_f) \\ \cos(\phi_f) * \cos(\theta_f) \end{bmatrix} \tag{21}$$

The states of the Kalman filter include the pitch and roll angles and the biases.

4.2 *Integration with GPS*

Each sentence starts with a '$' and ends with a carriage return/line feed sequence and is normally not longer than 80 characters. The data items are separated by commas, and this is usually contained in a single line.

At the end of each sentence, there is a provision for a checksum. NMEA data consists of a number of sentences, out of which the first word is called a data type and that defines the interpretation of the rest of the sentence. The interpretation of each data type is unique and is defined in the NMEA standard. The GGA sentence below shows an example that provides essential fix data. GGA—essential fix data which provide 3D location and accuracy data.

$GPGGA, 123520, 4810.039, N, 01135.000, E, 1, 07, 0.8, 547.4, M, 48.9, M, *46$

The latitude and longitude obtained from the NMEA data are the geodetic co-ordinates, which need to be appropriately converted into local co-ordinates. These local co-ordinates are then used for further processing in the extended Kalman filter (EKF) and finally converted back to the geodetic co-ordinates to give us an estimate of the position.

4.3 Extended Kalman Filtering

EKF is performed for 110 s out of which 100 s is for estimation and 10 s in between is for only prediction as measurements are available only for first and last 50 s. Sample rate is taken as 0.02 s, and number of samples taken are 11,000. Measurement and state vector are taken from CarMaker data. Input is assumed to be without noise.

EKF constitutes of three functions in the order:

Jacobian Function: Initial state (Xprev), input (U), and sampling time are passed through Jacobian function as input, and matrix F is obtained as output. The arrays are computed by first-order linearization by Jacobian function as follows:

$$F = \begin{bmatrix} 1\ 0 & \mathrm{d}T v_x \sin(\psi)\ \mathrm{d}T \cos(\psi) \\ 0\ 1 & \mathrm{d}T v_x \cos(\psi)\ \mathrm{d}T \sin(\psi) \\ 0\ 0 & 1\ 0 \\ 0\ 0 & 0\ 1 \end{bmatrix} \tag{22}$$

Then this F, along with, initial state Xprev, measurement Y, input U, H, Q, R, and $\mathrm{d}T$ is passed through EKF algorithm as input, whereas estimated state Xpu and filtered measurement Y is taken as output. Pprev is persisted in the EKF algorithm. Now, these predicted states form a predicted state vector called X_p and error covariance matrix P_p as follows:

$$X_p = \begin{bmatrix} P_x \\ P_y \\ \psi \\ v \end{bmatrix} \tag{23}$$

$$P_p = (F P_{\text{prev}} F^T) + Q \tag{24}$$

The Update function is as follows:

$$S = (H P_p H^T) + R \tag{25}$$

$$K = (P_p H^T S^{-1}) \tag{26}$$

$$X_{\text{pu}} = X_p + (K(Y - HX_p)) \tag{27}$$

$$P_{\text{pu}} = P_p + (KHP_p) \tag{28}$$

This method from Jacobian to Predict and Update is repeated for 10 s, and then at last iteration, estimated state of update function is saved as *X*prev, so that it can be used as initial state for next iterations. The estimated states for 10 s are stored in a vector named state filter and the filtered output for 10 s stored in vector named Yfilter.

Now, for next 100 s, no measurement is available, and EKF only performs the Predict function. The last estimated state above is considered as initial state (Xprev), along with input(UP) and *t*, is passed as input through Jacobian and Matrix *F* is taken as output, same as discussed above in Jacobian function. Then, this *F* along with input (UP), d*T*, and *Q* is passed through Predict function; at this point, no measurements are available. At the end of Predict function, *Xp* is assigned as X_{prev}, so that in next iteration, it could be used. The estimated states for 100 s are stored in a filter named Statefilter2. Now, the last known state from Predict function is used as the initial state (X_{prev}), along with input (U2), d*T* and passed through Jacobian function, and *F* is obtained same as discussed earlier. This *F* along with input (U2), measurement (Y3), *H*, *Q*, *R*, and d*T* is passed through Predict and Update equations of EKF as discussed earlier, and this algorithm is run for 10 s. The final estimated state and filtered output are obtained. The estimated states for 10 s are stored in a vector named Statefilter3 and the filtered output for 10 s stored in vector named Yfilter2.

This estimated state is compared with reference to check how the EKF is working, thereby to see how much estimated state is deviating from reference. The filtered output is used to check how EKF is filtering the noise.

5 Results

Raw acceleration and gyro output obtained from the GY-87 module can be observed in Fig. 5 along with the calculated linear acceleration, pitch, and roll angles. The first and third graphs represent the raw acceleration and raw gyroscopic movement in three-axis, respectively. The second graph represents the linear acceleration that is calculated from the raw acceleration. The last two graphs show the fused pitch and roll angles calculated from the raw data.

To test the extend Kalman filter algorithm, we have the following scenarios as shown in Fig. 6:

Scenario-1

- Without GPS outages as shown in (Figs. 7, 8, 9, 10 and 11)

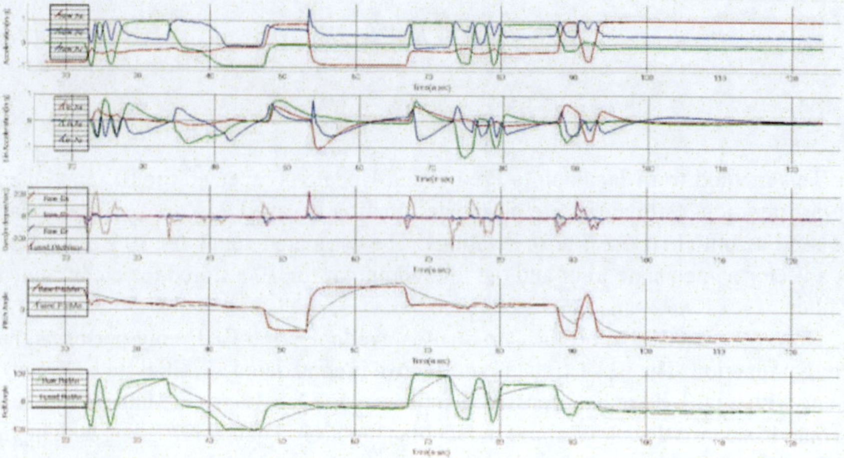

Fig. 5 Fused outputs from Kalman filter

Scenario	Description	Representation	Speed (KMPH)	Radius of Curvature (m)
1	Subject vehicle is moving on a straight road with fixed speed.	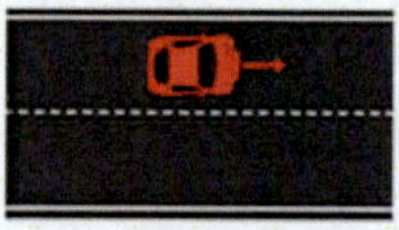	70	Straight
2	Subject vehicle is moving on a curvature with fixed speed.		70	250

Fig. 6 Test scenarios

The CarMaker scenario generated plots, i.e., the reference states of all the four states (P_x, P_y, ψ and v), longitude and latitude, along with their associated errors are as shown in Figs. (7, 8, 9, 10 and 11) in Table 1.

In this scenario, the subject vehicle is considered to be moving on a straight road with a fixed speed. An initial sudden change in the errors of the four states can be seen. The actual and EKF predicted data in Figs. 7 and 8 show the change in the

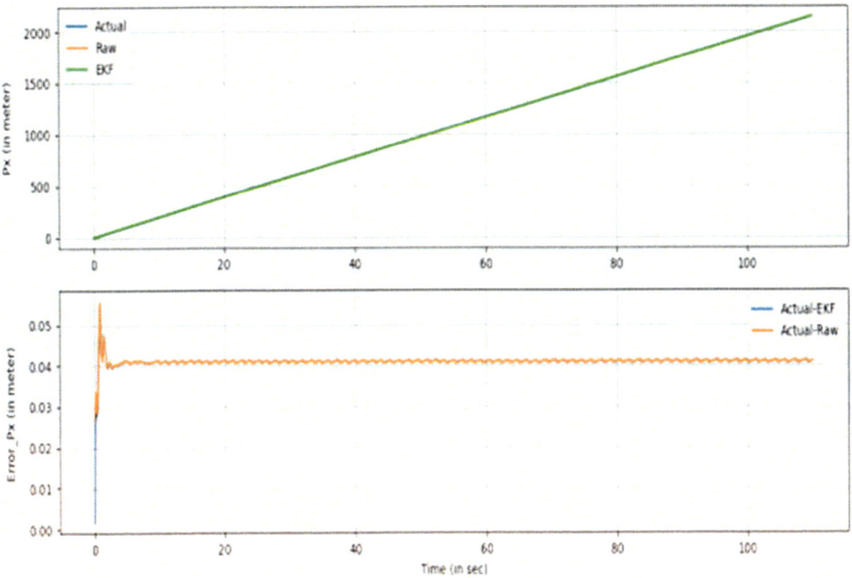

Fig. 7 P_x and error in P_x

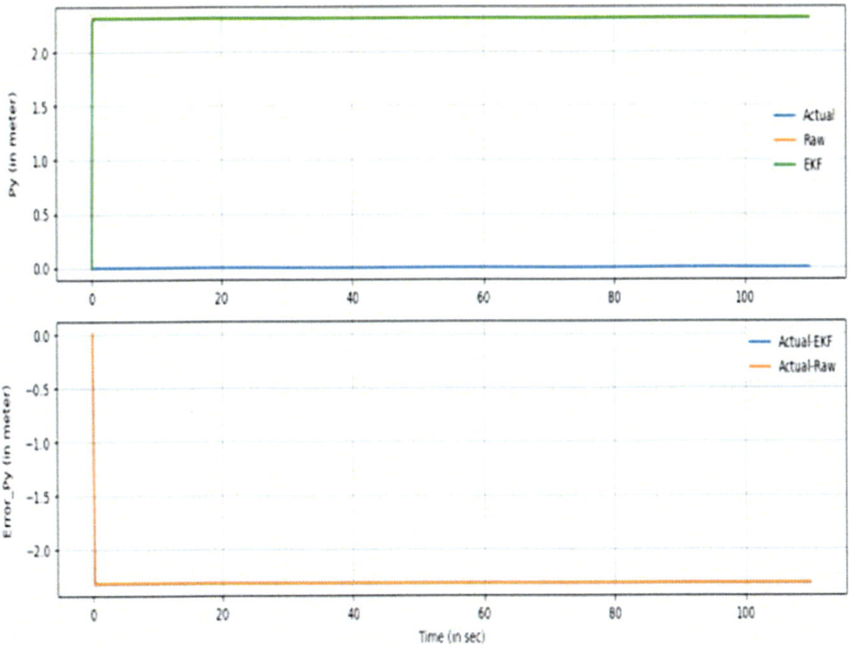

Fig. 8 P_y and error in P_y

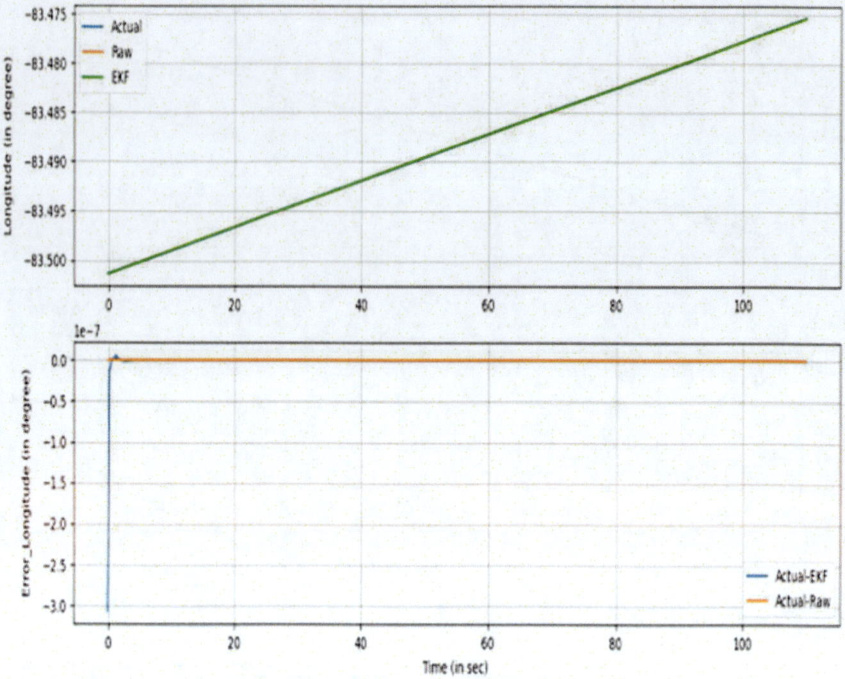

Fig. 9 Long and error in long

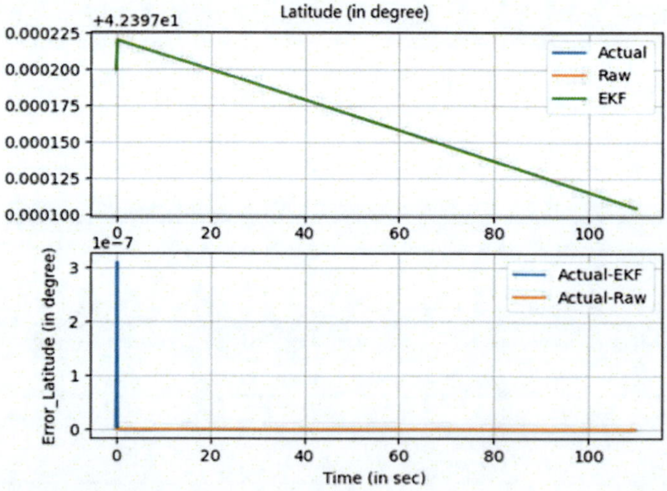

Fig. 10 Lat and error in lat

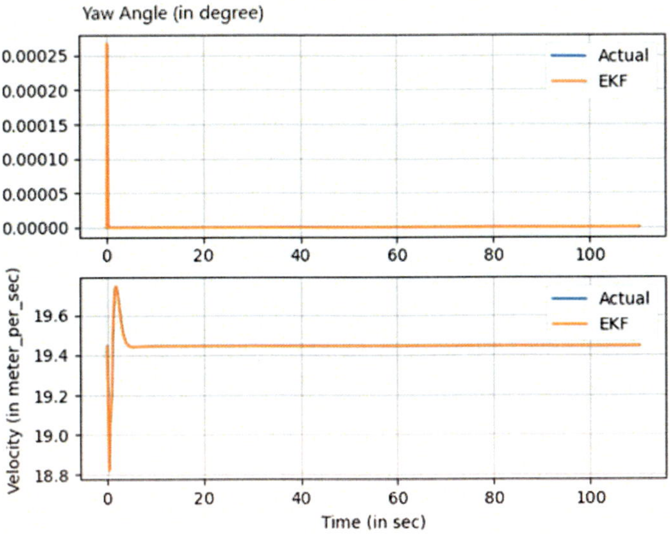

Fig. 11 Yaw angle and velocity

Table 1 CarMaker data scenarios

Scenario-1	Scenario-2
Figures 7, 8, 9, 10 and 11	Figures 12, 13, 14, 15 and 16

position of the vehicle in the *x* and *y* directions, along with the errors between the actual and EKF predicted data. This change can also be seen in the latitude and longitude data as well, as shown in Figs. 9 and 10. The yaw angle associated with the constant velocity of the vehicle is shown in Fig. 11.

Scenario-2

- Without GPS outages as shown in (Fig. 12, 13, 14, 15 and 16)
  The CarMaker scenario generated plots; i.e., the reference states of all the four states (P_x, P_y, ψ and v), longitude and latitude, along with their associated errors are as shown in figures (Fig. 12, 13, 14, 15 and 16).
  In this scenario, the subject vehicle is considered to be moving on a curved road with a fixed speed. The actual and EKF predicted data in Figs. 12 and 13 show the change in the position of the vehicle in the x and y directions, along with the errors between the actual and EKF predicted data. This change can also be seen in the latitude and longitude data as well, as shown in Figs. 14 and 15. The yaw angle associated with the constant velocity of the vehicle can also be seen to change as shown in Fig. 16.

- With GPS Outages for 10 s as shown in (Figs. 17, 18, 19, 20, and 21)

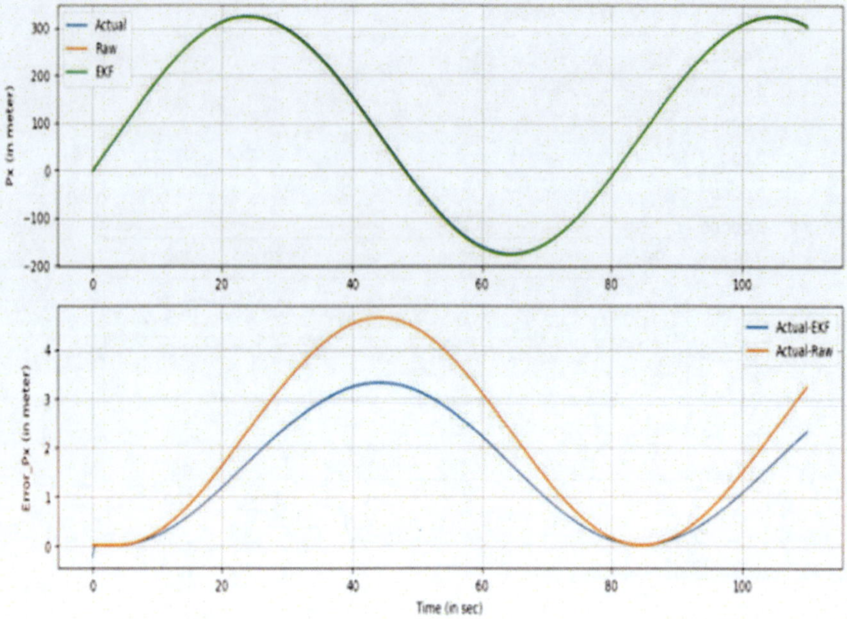

Fig. 12 P_x and error in P_x

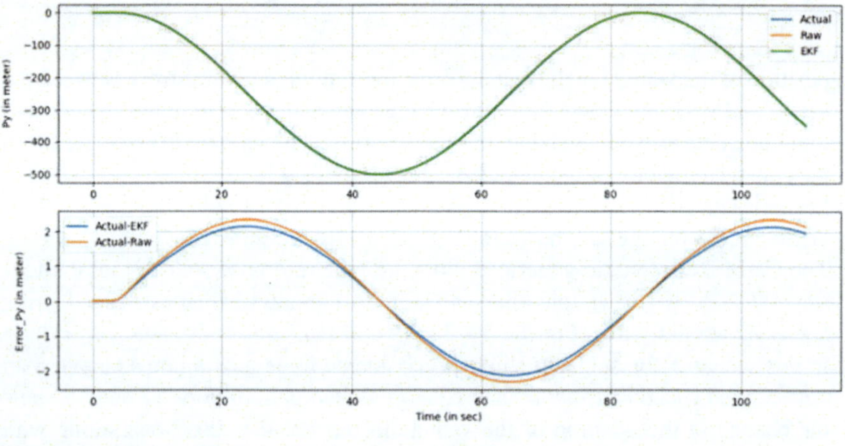

Fig. 13 P_y and Error in P_y

The CarMaker scenario generated plots; i.e., the reference states of all the four states (P_x, P_y, ψ and v), longitude and latitude, along with their associated errors are as shown in (Figs. 17, 18, 19, 20, and 21).

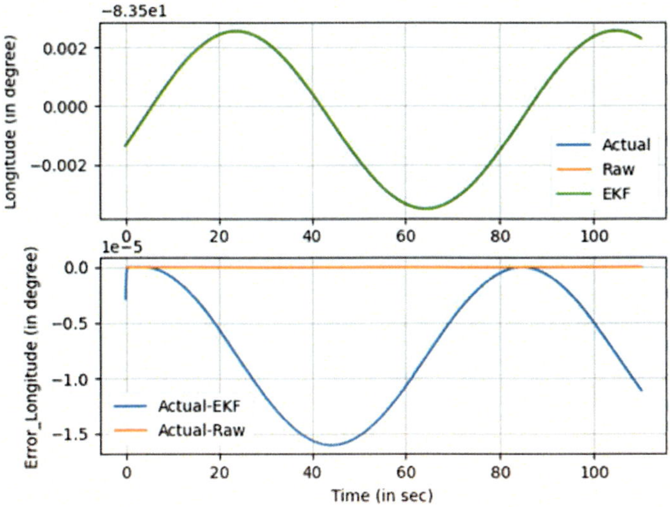

Fig. 14 Long and error in long

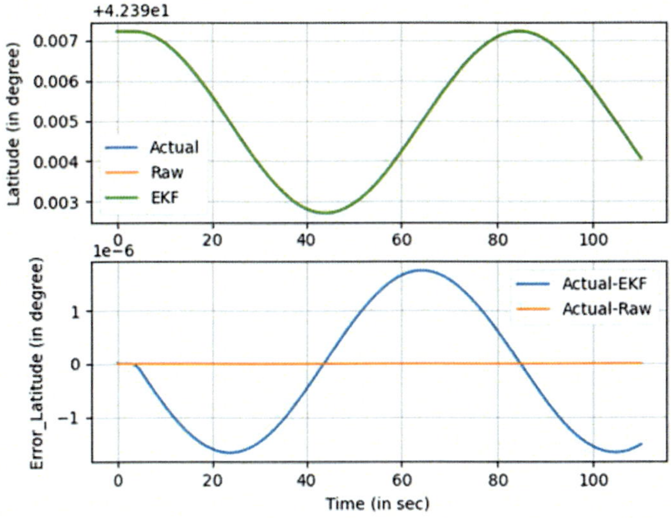

Fig. 15 Lat and error in lat

With GPS outage occurring for 10 s, it can be seen that the EKF is able to predict the position for that period, but it can also be noted that the error for that period increases with time.

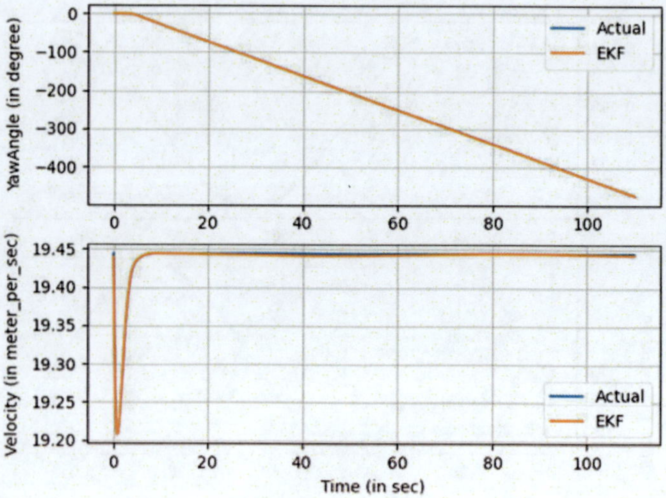

Fig. 16 Yaw angle and velocity

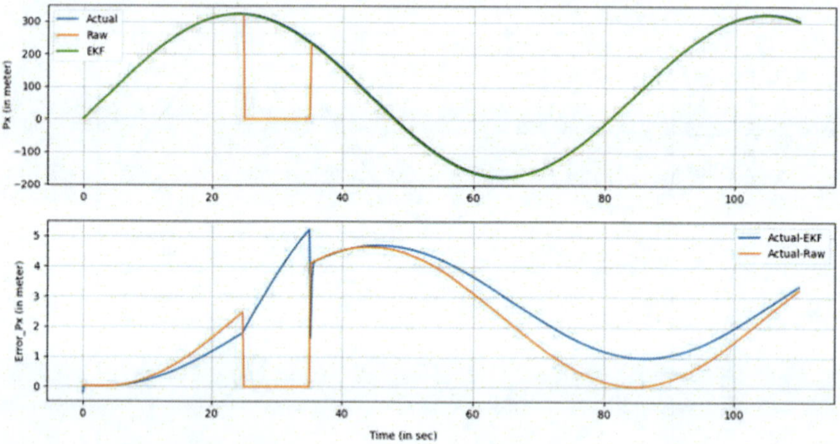

Fig. 17 P_x and error in P_x

6 Conclusions and Future Scope

Integration of the IMU and GPS was implemented using Raspberry Pi 4 as the environment. Suitable Kalman filter algorithm was developed to prepare the IMU data for fusion, and also the data from the GPS module was effectively decoded for suitable use in the algorithm. An extended Kalman filter algorithm was developed and implemented for two different scenarios, one being where the vehicle is moving on a straight road with constant speed and the other being, vehicle moving on a

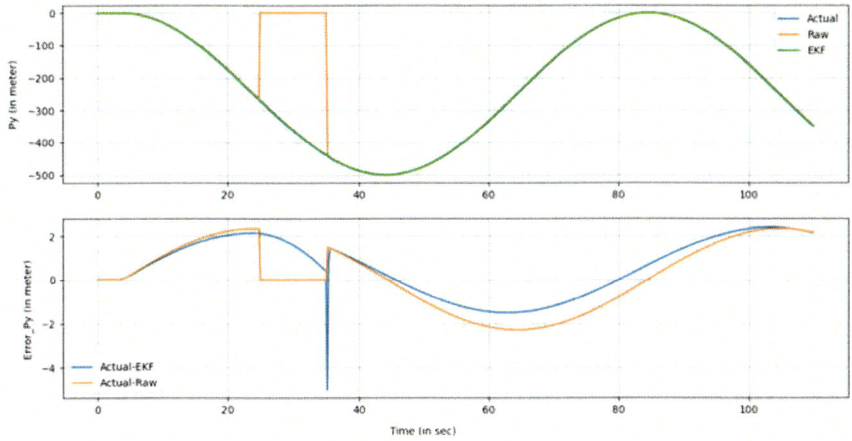

Fig. 18 P_y and error in P_y

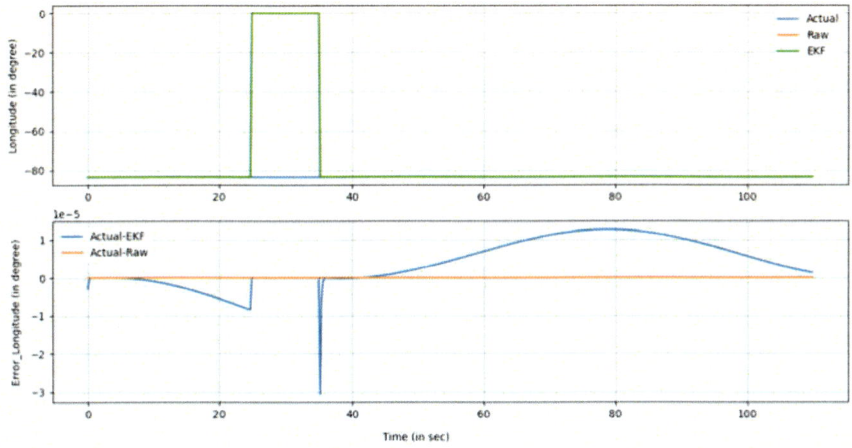

Fig. 19 Long and error in Long

curved road with constant speed. Both the scenarios were checked with and without GPS outages, and the results show a successful prediction for up to 10 s in case of GPS outage. There are multiple factors that affect the significance of the simulated results.

The simulation environment being used is CarMaker, and though it can never fully represent an actual vehicle, it provides a very realistic vehicle model along with excellent simulated sensor data, and hence, we can see how the vehicle will behave. To avoid the accumulation of estimation errors over a short period of time, it is useful to use dead reckoning for position and heading estimation of a vehicle. The prediction time of the dead reckoning process, which is also the time during

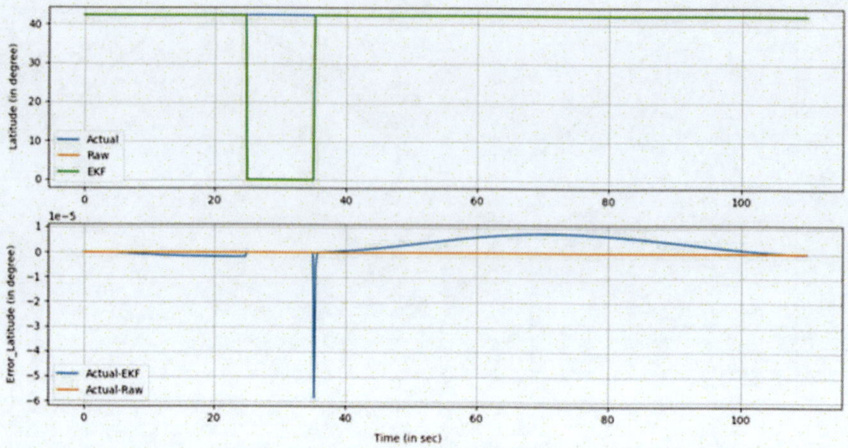

Fig. 20 Lat and error in Lat

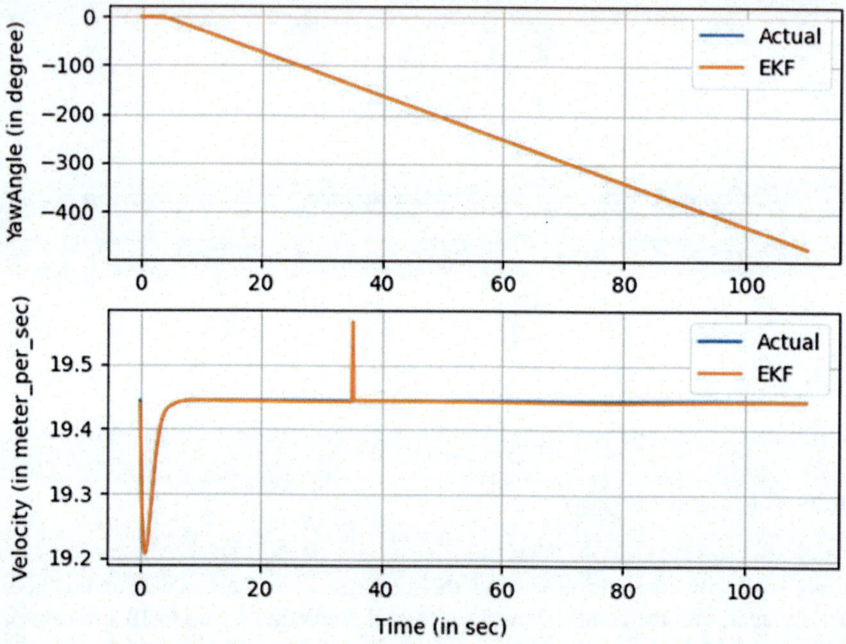

Fig. 21 Yaw angle and velocity

which GPS data is unavailable, is the condition that will have the most impact. The sensors performances also have a large effect on the filtering process. All the testing and simulation were performed in a Raspberry Pi 4 environment using the logged CarMaker data. Real-time vehicle testing could be done as a future work.

The future scope of this project could involve better accuracy in prediction and further reduction in the errors. Also some feature or add-on could be developed to consider the initial state in case there is data unavailability at the start itself.

References

1. Furuno.: Dead reckoning, (downloaded April 28, 2020) [Online]. Available (2020). https://www.furuno.com/en/gnss/technical/tec-dead
2. Takeyama, K., Machida, T., Kojima, Y., Kubo, N.: Improvement of dead reckoning in urban areas through integration of low-cost multisensors. IEEE Trans. Intell. Vehicles **2**, 278–287 (2017)
3. Park, W.J., Song, J.W., Kang, C.H., Lee, J H., Seo, M.H., Park, S.Y., Yeo, J.Y., Park, C.G.: Mems 3D DR/GPS integrated system for land vehicle application robust to GPS outages. IEEE Access **7**, 73336–73348 (2019)
4. Vivacqua, R.P.D., Bertozzi, M., Cerri, P., Martins, F.N., Vassallo, R.F.: Self-localization based on visual lane marking maps: an accurate low-cost approachfor autonomous driving. IEEE Trans. Intell. Transp. Syst. **19**, 582–587 (2017)
5. Sabet, M.T., Daniali, H.M., Fathi, A.R., Alizadeh, E.: Experimental analysis of a low-cost dead reckoning navigation system for a land vehicle using a RobustAHRS. Robot. Auton. Syst. **95**, 37–51 (2017)
6. Roboticsbd.: "Gy87," 2009 [Online]. Available https://encrypted-tbn0.gstatic.com/images (downloaded January 18, 2020)
7. Roboticsbd.: Neo-6m, (2009) [Online]. Available https://roboticsbd.store/2043-largedefault=u-blox-neo-6m-gps-module-robotics-bangladesh:jpg(downloaded January 18, 2020)
8. Tutorial, K.F.: Kalman filter concept [Online]. Available https://www.kalmanfilter.net/kalman1d.html(downloaded April 28, 2020)
9. Wikipedia.: Kalman Filter [Online]. Available https://en.wikipedia.org/wiki/KalmanFilter
10. Wikipedia.: Extended Kalman Filter [Online]. Available https://en.wikipedia.org/wiki/ExtendedKalman Filter
11. J.A.D. e Education Institute Penn State.: Transverse mercator projection. [Online]. Available https://www.e-education.psu.edu/natureofgeoinfo/c2p22:html. (downloaded April 20, 2020)
12. Raspberry pi 4.: [Online]. Available www.raspberrypi.org, https://www.raspberrypi.org/products/raspberry-pi-4-model-b (downloaded April 28, 2020)
13. Automotive, I.: Carmaker scenario [Online]. Available https://ipg-automotive.com/news/article/how-to-transfer-real-world-scenarios-from-video-recordings-to-carmaker (downloaded April 18, 2020)
14. Charlotte Lanfelt, S.R.: Dead reckoning during safe stop of autonomous vehicles, Masters Thesis, Chalmers University of Technology (2017)
15. Romaniuk, S., Gosiewski, Z.: Kalman filter realization for orientation and position estimation on dedicated processor. Actamechanica et Automatica **8**, 88–94 (2014)
16. Fisher, C.J.: Static acceleration. Online. Available https://www.digikey.in/en/articles/techzone/2011/may/using-an-accelerometer-for-inclination-sensing (Accessed April 27, 2020)

Statistical Analysis of Coherence Between Electrical and Hemodynamic Brain Signal

Revati Shriram and Nivedita Daimiwal

1 Introduction

Human body consists of mainly eleven physiological systems and corresponding biomedical signals. Human body can be studied by means of various biomedical signals of varying nature. These signals are widely used in clinical setting for diagnosis, patient monitoring, disease prevention, risk assessement, and for guiding therapy. In a highly nonlinear brain system, post-synaptic potentials formed a complex process of firing of pyramidal cells from brain cortex are known as electroencephalogram (EEG). It the electrical brain signal widely studies by the researchers for more than 100 years. Superimposition of various processes related to the brain generates an EEG signal, and this is a very important and easily acquired, low-cost signal in brain-related studies such as epilepsy, stress, and sleep patterns. EEG is a non-stationary signals in which amplitude and frequencies changes from time to time and with age- or brain-related deceases [1].

Photoplethysmography (PPG) is widely used in the health care because of its non-invasiveness, reliability, and ability to predict the cardiovascular parameters indirectly. PPG acquisition is carried out by applying an infrared light source and photosensor to an appendage and measuring the light that is reflected by the skin. CPPG is a brain signal which is hemodynamic in nature. When PPG is captured by placing a reflective type of PPG sensor for acquisition from brain (generally from prefrontal site) is known as cranial photoplethysmogram. This signal gives us the information about blood volume change at cranial site, blood oxygen saturation, heart rate, respiratory rate, blood pressure, and brain aging estimation [2].

R. Shriram (✉) · N. Daimiwal
Cummins College of Engineering for Women, Pune, India

S. K V and K. Guruprasad Rao (eds.), *Smart Sensors Measurements and Instrumentation*, Lecture Notes in Electrical Engineering 750,
https://doi.org/10.1007/978-981-16-0336-5_18

2 Methods

Connectivity analysis is carried out between the two concurrently acquired brain signals which are different in nature. It is the electrical signal of brain captured from the scalp. Signal is weak in nature, in few microvolts. CPPG is the hemodynamic signal, due to volumetric change in the brain area and is in few milli volts. In this section, coherence analysis is calculated in concurrently acquired signals such as EEG-EEG (during Stroop task and during music evoked emotions) and EEG-CPPG during music evoked emotions.

2.1 *Power Spectral Density (PSD) and Cross-Power Spectral Density*

PSD is the strength of the energy variations with respect to frequency. That means it shows strong and weak variations with respect to the frequencies. The PSD, $Sx(f)$ is the FT of the $R(\tau)$, an autocorrelation function. With an assumption that the signal in a wide-sense stationary random process.

$$Sx(f) = \int_{-\infty}^{\infty} Rx(\tau)e^{-2\pi i f\tau}\mathrm{d}t$$

PSD can be estimated by using periodogram or Welch method [3]. Spectral representation is especially useful along with the frequency and time representation individually in describing random stationary processes [4]. CPSD is the FT of cross-covariance function. When $X = Y$, then CPSD reduces to the PSD. For real signals $Y^*(w) = Y(-w)$, where '*' denotes the complex conjugate. CPSD is complex where as PSD is real because the cross-covariance is asymmetric and auto-covariance is symmetric in nature [5]:

$$\mathrm{CPSD}_{XY}(w) = X(w)Y^*(w)$$

where $X(w)$: is a first signal & $Y^*(w)$: is the complex conjugate of the second signal $Y(w)$.

2.2 *Coherence*

Simultaneously acquired (concurrent recording) signals are having great impact in the biomedical studies, as it gives the effect of one signal on the other with respect to any know activity, with or without any known disease. Coherence is always found out between the two signals or images which are acquired concurrently. It is one of the measures of the amount of stability between two time series. For the constant phase, difference between two signals coherence is equal to one, and for the random phase, difference between signals coherence is zero. Coherence is of two types [6, 7].

2.2.1 Magnitude Square Coherence and Phase Coherence

It is commonly known as MS Coherence and is calculated by using following formula:

$$Cxy = \frac{|G_{xy}|^2}{G_{xx} * G_{yy}}$$

where G_{xy}: is a CPSD between x & y signals, G_{xx} & G_{yy}: is an PSD of signal x and signal y, C_{xy}: is a coherence (ranging between 0 and 1) between signal x and y. Phase coherence is denoted by $\emptyset(w)$ and is given by the following equation as shown below:

$$\emptyset(w) = \tan^{-1}\left\{\frac{\mathrm{Im}[G_{xy}]}{\mathrm{Re}[G_{xy}]}\right\}$$

The degree of synchronization between the EEG signals is characterized by magnitude and phase coherence [8, 9]. This connectivity measure is applied further to the concurrently acquired EEG-EEG and EEG-CPPG signal, where these two biomedical signals of different nature, captured from frontal lobe. Magnitude square and phase coherence are estimated by the authors [10, 11].

2.3 *Statistical Measures*

Statistics is important in almost all the fields (physical and non-physical) such as science, engineering, medicine. *P*-value & *F*-value are very widely used in the statistics. *P*-value determine the statistical significance in the hypothesis testing, and it is used to check the matching of mean in the *t*-paired test. Authors have also used *t*-test and *p*-value estimation for the EEG and CPPG signals using Minitab Software. Results are discussed in the result section [12, 13].

3 Database Collection Details

EEG-EEG/EEG-CPPG concurrent data was captured for 20 right-handed, multilingual subjects, aged between 22 and 63 years (8 Males and 12 Females). While capturing the data, all subjects were in a seated position and eyes closed condition. Two-channel EEG acquisition was carried out by using PowerLab data acquisition system. CPPG is captured from cranial nerve entry point into the cranial mask on frontal right side, frontal left side, and central prefrontal using a reflective type of optical sensor. PSD, CPSD, MS coherence and phase coherence were calculated using MATLAB 2010. Statistical measures were applied on the extracted features using Minitab. Verbal consent was given by the volunteers before recording the EEG and CPPG data and its use in the research.

4 Results

Two-channel EEG signal and single channel CPPG signal was captured from prefrontal area. PSD was estimated from the acquired signals and based on PSD and CPDS, MS coherence and phase coherence were calculated during four emotions. Figure 1 shows the plot of PSD and coherence between the two EEG signals. Figure 2 shows the box plot of mean and standard deviation in MS coherence and phase coherence between the two EEG signals.

Concurrent acquisition of EEG-CPPG signal was carried out during various 'Music Evoked Emotions.' During this concurrent acquisition, CPPG was captured from three locations of frontal lobe such as right prefrontal (RPF), central prefrontal (CPF), and left prefrontal (LPF). As RPF and LPF are the Cranial Nerve' entry points, whereas CPF is non-cranial nerve entry point. These three locations were tried to find out the effect on CPPG with respect to the location on the frontal lobe. Figures 3 and 4 show the box plot of mean and standard deviation in MS coherence and phase coherence between the EEG-CPPG signal captured from left prefrontal, right prefrontal area, and central prefrontal area.

Figure 5 shows the *p*-value of *t*-paired test for phase coherence in EEG-EEG/EEG-CPPG (from LPF, RPF, and CPF) Signal. *P*-value of *t*-paired test for MS coherence shows same value for all the combinations and it was <0.0001. Second plot shows MS coherence and phase coherence between EEG-CPPG(CPF)/EEG-CPPG(LPF/RPF) Signal *t*-paired test was carried out by between the MS Coherence values obtained from EEG-CPPG at CPF signal, EEG-CPPG at LPF and EEG-CPPG at RPF. Figures 6 and 7 shows the result of the t-test for obtained in MINITAB 17.

5 Discussion

Brain activation results in a complex sequence of cellular, metabolic, and vascular change. With neural activation, restoration of ionic gradients and neurotransmitter recycling occurs. This requires energy in the form of ATP. So, ATP consumption increases, which increases the glucose consumption and in turn oxygen consumption in that area. Continuous supply of increased glucose and oxygen is maintained by

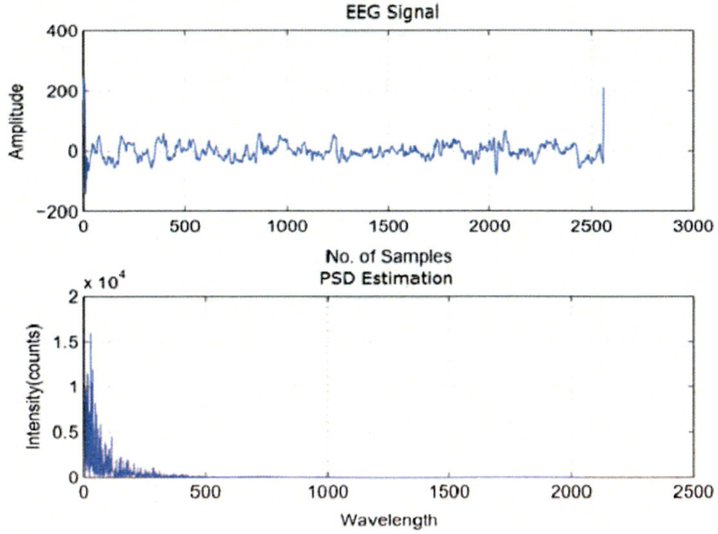

(a) EEG Signal and PSD Estimation

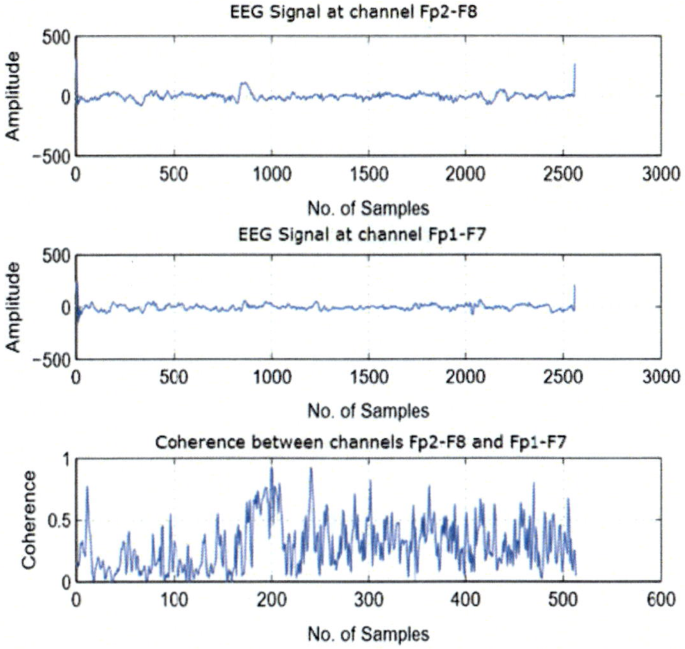

(b) EEG Signals, PSD Estimation and Coherence between the EEG Signals

Fig. 1 PSD and coherence for EEG

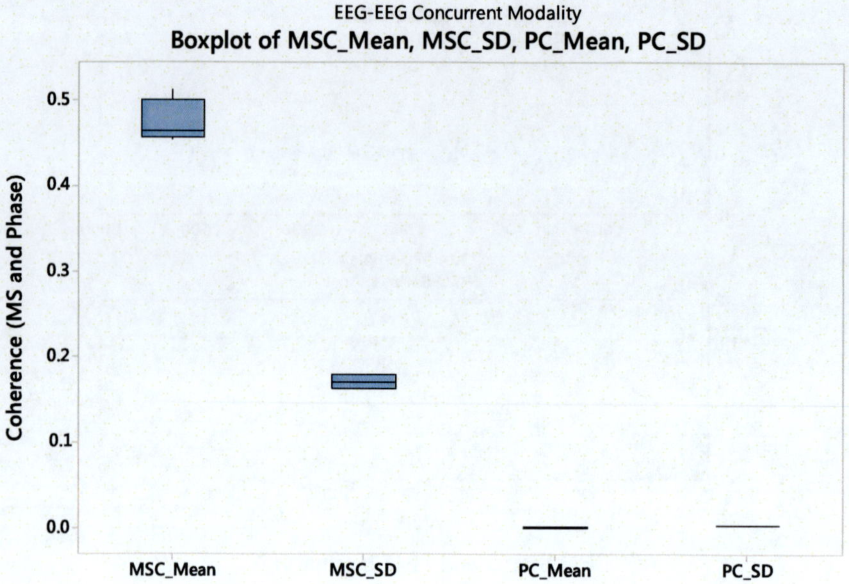

Fig. 2 Coherence in EEG-EEG concurrent modality

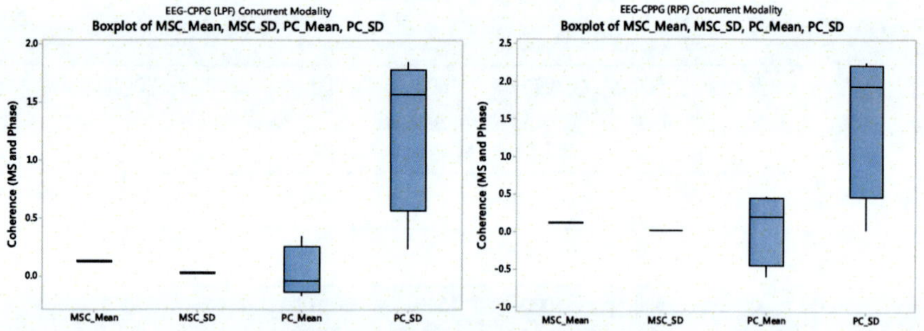

Fig. 3 Coherence in EEG-CPPG (LPF) and EEG-CPPG (RPF) concurrent modality

CBF. The functional and structural integrity of brain depends on balance between the cerebral blood flow and the energy demands imposed by neuronal activity.

6 Conclusion

Relation between the two signals can be studied by going for a connectivity measure like Coherence. When coherence estimation is applied on two brain signals of similar or dissimilar nature, it reflects functional brain connectivity or synchronization during

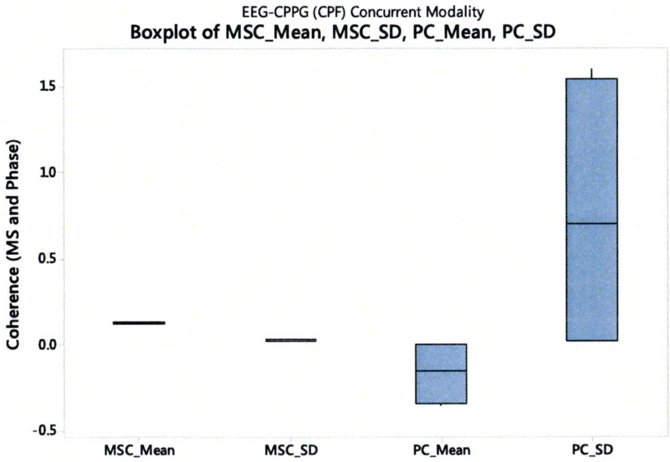

Fig. 4 Coherence in EEG-CPPG (CPF) concurrent modality

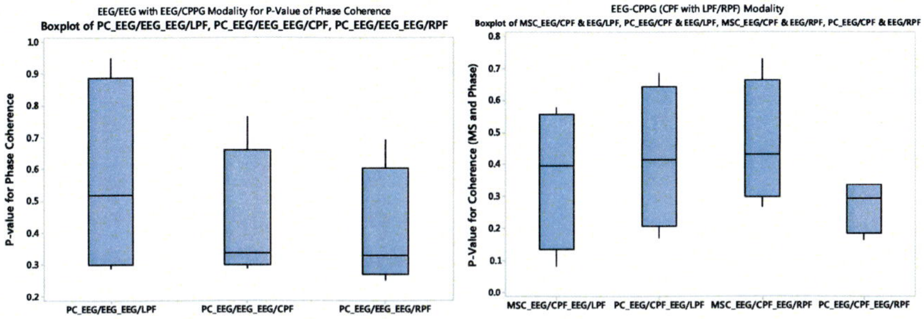

Fig. 5 *p*-value for *t*-paired test for MS coherence and phase coherence in EEG-EEG/EEG-CPPG signal & EEG-CPPG(CPF)/EEG-CPPG(LPF/RPF) signal

the activities in the various brain lobes. It is a 'statistical' estimate of synchronization between the signals and is fully understood with the knowledge of its frequency structure. The relationship between neuronal activity and the hemodynamic response (neurovascular coupling) is the subject of most of the ongoing cutting age research in the field of neurology as is not well understood till date. However, it is very clearly observed and studied a powerful correlation in neural activity and blood flow. The setup which can observes both electrical and hemodynamic signals concurrently is of great benefit in understanding of functional cortical activity and of neurovascular coupling. Neurovascular coupling is affected in some pathological conditions such as Alzheimer's disease, Dementia, other neurodegenerative conditions, and ischemic stroke stress-related changes and indirect estimation of cardio vascular parameters can be studied from CPPG. Coherence estimation and its reliance on the uniformity of the average phase difference between two signals are highly beneficial. Most of the researchers had carried out the coherence related study on EEG signals. Emotion

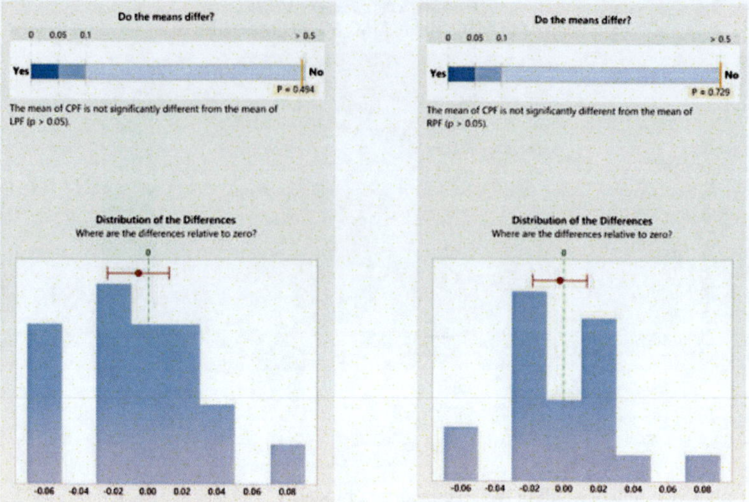

(a) t-test for Mean of MS Coherence between EEG-CPPG(CPF,LPF) (=0.05)

(b) t-test for Mean of MS Coherence between EEG-CPPG(CPF,RPF) (=0.05)

Fig. 6 *t*-paired test for mean of MS coherence between EEG-CPPG signal at CPF, LPF, and RPF with confidence level $\alpha = 0.05$ during happy emotion

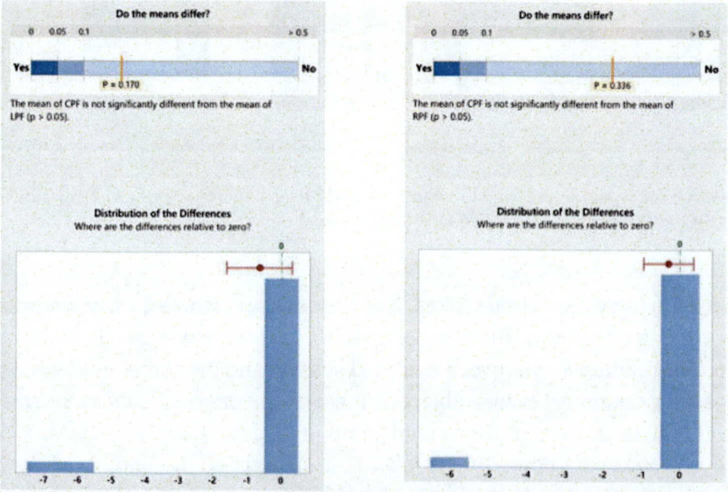

(a) t-test for Mean of Phase Coherence between EEG-CPPG(CPF, LPF) (α=0.05)

(b) t-test for Mean of Phase Coherence between EEG-CPPG(LPF, RPF) (α=0.05)

Fig. 7 *t*-paired test for mean of phase coherence between EEG-CPPG signal at CPF, LPF, and RPF with confidence level $\alpha = 0.05$ during happy emotion analysis is mainly carried out by the researchers on EEG signal with electrodes placed all over the brain covering all the brain lobes. Coherence between EEG-EEG and EEG-CPPG signals is estimated during music evoked emotions, and statistical analysis of coherence estimates shows that the MS coherence is greater in EEG-CPPG (LPF) and EEG-CPPG (RPF) or EEG-CPPG (CPF) signals. Whereas phase coherence was observed to be pretty much same between any type of combination of signals, such as electrical-hemodynamic or electrical-electrical.

References

1. Zavaglia, M., Astolfi, L., Babiloni, F., Ursino, M.: The effect of connectivity on EEG rhythms, power spectral density and coherence among coupled neural populations: analysis with a neural mass model. IEEE Trans. Biomed. Eng. **55**(1) (2008)
2. Shriram, R., Sundhararajan, M., Daimiwal, N.: Brain connectivity analysis methods for better understanding of coupling. Int. J. Comput. Sci. Inf. Secur. **10**(11) (2012)
3. Ombao, H., Bellegem, S.: Evolutionary Coherence of Nonstationary Signals. IEEE Trans. Sign. Process. **56**(6) (2008)
4. Stoica, P., Moses, R.: Spectral Analysis of Signals, p. 07458. Prentice Hall, Upper Saddle River, NJ (2004)
5. Huang, R., Jung, T., Makeig, S.: Tonic Changes in EEG Power Spectra during Simulated Driving. HCI International, San Diego, CA (2009)
6. Golińska, A.: Coherence function in biomedical signal processing: a short review of applications in neurology, cardiology and gynecology. Studies Logic, Grammar Rhetoric, **25**(38) (2011)
7. Shriram, R., Sundhararajan, M.: Coherence analysis of pressure pulse & photoplethysmogram at various sites. Int. J. Appl. Eng. Res. **10**, 14959–14968 (2015)
8. Benesty, J., Chen, J., Huang, Y.: Estimation of the coherence function with the MVDR approach. In: IEEE ICASSP (2006)
9. Unde, S., Shriram, R.: Coherence analysis of EEG signal using power spectral density. In: Proceedings of the 4th International Conference on Communication Systems and Network Technologies (2014)
10. Shriram, R., Sundhararajan, M., Daimiwal, N.: Statistical analysis of pulse acquired from various body sites using piezoelectric and optical transducer. Int. J. Control Theor. Appl. **8**, 831–838 (2015)
11. Li, X.: Multiple Significance Tests and Their Relation to P –Values. Thesis Submitted to the College of Graduate Studies and Research, University of Saskatchewan, Saskatoon (2008)
12. Shriram, R., Baskar, V., Martin, B., Sundhrararajan, M., Daimiwal, N.: Energy distribution and coherence based changes in normal and epileptic electroencephalogram. In: Smart Intelligent Computing and Applications, Smart Innovations Systems Technologies, vol. 104, https://doi.org/10.1007/978-981-13-1921-1_61 (2019)
13. Shriram, R., Baskar, V., Martin, B., Sundhararajan, M., Daimiwal, N.: Regression analysis of coherence between concurrent EEG-EEG and EEG-CPPG signals from prefrontal cortex during music evoked emotions. J. Inform. Sci. Eng. **35**, 577–595 (2019). https://doi.org/10.6688/JISE.201905_35(3).0006

Learning Model-Based Safety-Critical System Engineering: The Fun Way

Aditya Y. Jeppu and Yogananda Jeppu

1 Introduction

As failures of safety-critical systems go, the Boeing 737 Max is a recent tragic incident. The corrections were test flown in June end time frame this year after nearly more than a year of software updates. Safety and certification are the two ends of the system development V model [1]. Failures in the engineering of the system lead to technical papers on the engineering side [2]. It also leads to literature on the topic of litigation and law [3]. Understanding safety-critical system development requires a knowledge of the process steps as well as the knowledge of engineering ethics and law. The law requires that the system be developed to the best of human ability and this is defined by the engineering standards that have put in place guidelines after years of discussions with experts in the field. The standard is a consensus among experts that the system should be developed in a specific manner [4]. A look at the All India Council for Technical Education (AICTE) syllabus published in 2018 shows that only CAD and Civil Engineering have standards as part of their curriculum [5]. Civil engineering also has a description of a complete system development including a mix of technology, management and finance as a part of the curriculum. In the current circumstances of system failures, it is very important that system engineering, especially safety-critical ones, need be taught as part of the curriculum [6]. The paper describes some of the initiatives taken at the MIT, Manipal Academy of Higher Education, Manipal to teach systems engineering as an industry academia relation workshop.

The current paper expands on these workshops and describes a fun way of learning system engineering. The origin of this paper is a presentation made in INCOSE as an open webinar and an internal training in an aerospace company for a bunch of 40

A. Y. Jeppu
RV College of Engineering, Bengaluru, India

Y. Jeppu (✉)
Manipal Institute of Technology, Manipal, India

S. K V and K. Guruprasad Rao (eds.), *Smart Sensors Measurements and Instrumentation*, Lecture Notes in Electrical Engineering 750,
https://doi.org/10.1007/978-981-16-0336-5_19

odd engineers on the same topic [1]. The situation these days with the pandemic preventing a hands-on and face-to-face training necessitates new approaches to learning. The 10-lecture series in the aerospace company required some tutorials and demonstration of safety analysis, model making, formal analysis, autocode and system tests. The example was developed by the budding engineer at home, and this paper showcases the system V model teaching with this example. There are a few lessons learnt from this exercise and we highlight it in the last section.

We present the V model of system development but start with the standards and certification; we then follow on the concept of operations and present the system functional decomposition and safety analysis using tools developed in an academic environment. We then look at writing the requirements in the Easy Approach to Requirements Syntax (EARS), which was also covered in the college as a workshop [7]. The design model is developed and analyzed using formal methods for correctness. The autocode is generated and used in an Arduino environment to control a small motor. The Arduino is interfaced with the graphics generated on a laptop using Python and Blender [8]. We conclude with some lessons learnt on how such setup can effectively provide a fun environment to understand safety-critical system engineering.

2 Standards

Engineering standards are available in every aspect of developing safety-critical embedded system. Table 1 shows a set of safety-critical standards as defined in [9].

As mentioned earlier, engineering curricula lack the mention of these important standards. This is perhaps because the civil engineering standards were available from the initial days of engineering being taught at colleges and the other standards were developed much later. It is just unfortunate that when the newer branches of engineering emerged and the curricula were finalized the standards did not have a place in the total hours left for the subjects to be taught in the class. The study by LDRA [9] proposed some very relevant methods to introduce safety and standards in the curricula some of them being, making some of the courses as core rather than

Table 1 Industry standards for safety and security

Industry	Functional safety/process standard	Cyber security standard
Automotive	ISO 26262	SAE J3061
Railways	EN 50128	ISA/IEC 62443
Industrial	IEC 61508	ISA/IEC 62443
Nuclear	IEC 60880	ISA/IEC 62443
Medical Devices	IEC 62304	ISA/IEC 62443
Avionics	DO–178B/C	DO–326A

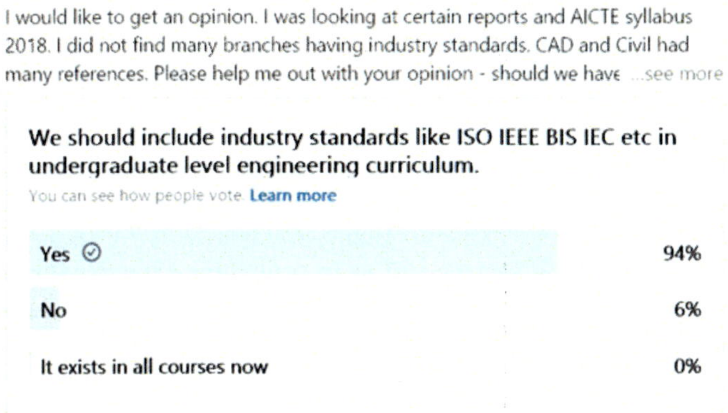

Fig. 1 Results of a poll in the social media

electives, developing the faculty to be aware of the engineering standards and a major component of industry academia interactions.

"Yet the knowledge of standards—which, as we have seen, plays a significant role in most of these areas—is rarely a significant component of such curricula. It is time for the academic world to build or strengthen links with the world of standards."[10]. In a small survey on the introduction of standards in the engineering curricula, 94% of the respondents indicated that introducing standards is a must as seen in Fig. 1. One eminent personality who has been with the standards committee indicated that the gist of the standard needs to be taught instead of the whole standards. "It may be more important to teach about the concept of standards, instead of going in depth about any particular standard. Understanding why they exist and when/how to use them seems important, a survey of standards used within a domain may also be interesting to identify similarities and differences," he said in a note. Another response was "lot of chaff is being added in curriculum in the name of industry readiness. This should be regulated as well." Many agreed that "lot of changes need to be done in UG syllabus."

One way of introducing standards would be to have students provide an essay on the failures and try and associate what standards were not followed properly in the development. It is just unfortunate that students will not lack examples of failures in the various domains and every year there would be some major failures to discuss. They could have a fun time googling for newer accidents and be aware of what would have caused these accidents. Engineering students have the curiosity and we have found students who have downloaded the aircraft accidents series from YouTube and following it up religiously. It is always a pleasure to discuss these failures during the workshop at MIT, Manipal.

2.1 Railway Standard

We present an automated railway gate system as a system under development for this paper. This was the example chosen for the organizational training as the audience was a mixed crowd from aerospace and other verticals like building and industrial control systems. The railway standard for the equipment and software development is CENELAC EN 50126 for the complete system development. It defines the safety and reliability of the complete system consisting of the train and the signaling systems. The standard for the signaling system development is the EN 50159 and the hardware component of this subsystem adheres to EN 50129.

The software component of the subsystem is governed by EN 50128. The different standards and how they interact are shown in Fig. 2. The EN 50128 defines the Software Safety Integrity Levels (SSIL) as defined in Table 2. The standard EN 50128 describes the process of system development, the rigor of analysis, testing

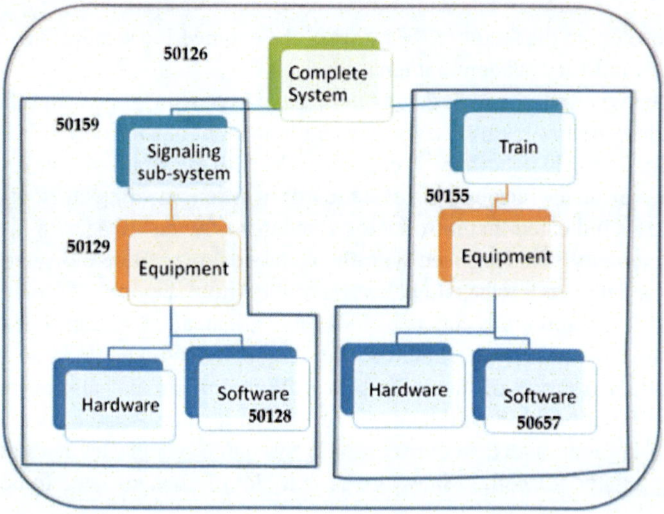

Fig. 2 Railway safety standards

Table 2 Safety integrity levels

SIL	Probability of dangerous failure per hour per safety function	Description
0	No safety impact	No safety requirements
1	$10^{-6} \leq PFH < 10^{-5}$	Low safety requirements
2	$10^{-7} \leq PFH < 10^{-6}$	Medium safety requirements
3	$10^{-8} \leq PFH < 10^{-7}$	High safety requirements
4	$10^{-9} \leq PFH < 10^{-8}$	Very high safety requirements

and documentation. It provides guidelines with recommendations on the methods to be used. For example, formal methods are highly recommended (HR) for SIL level 3 and 4. Design and coding standards are mandatory (M) for SIL 3 and 4.

3 Concept of Operations and Safety

The second task that needs to be done is to understand the functional decomposition of the system that is being developed and to build in the safety for the system during the design. We are operating back in the left side of the system development V model [1]. The tool and method that is used to break down the system into the different functional components are the OPCAT tool that implements the object process model. This was taken up as a workshop at MIT, Manipal under the industry academia interactions [6]. Object process methodology (OPM) is the work done by a professor Dov Dori who is a Faculty of Industrial Engineering and Management, Technion, Israel Institute of Technology. This was first published by him in 1995 [11], and the method has two components. The graphical component is termed as object process diagram (OPD) and the text component is the object process language (OPL). The OPCAT tool is available for the students to quickly explore the concepts and design systems [12].

3.1 *OPM Model of the Railway Gate*

The simple abstraction that is done for the railway gate which can be easily understood by any student is shown in Fig. 3. The model created in the OPCAT tool generates the OPL automatically. The OPL is shown in Fig. 4. The OPL indicates that the train is a physical entity that is visible. It exhibits two parameters position and velocity. These two parameters are of paramount importance in the design of the railway gate. The

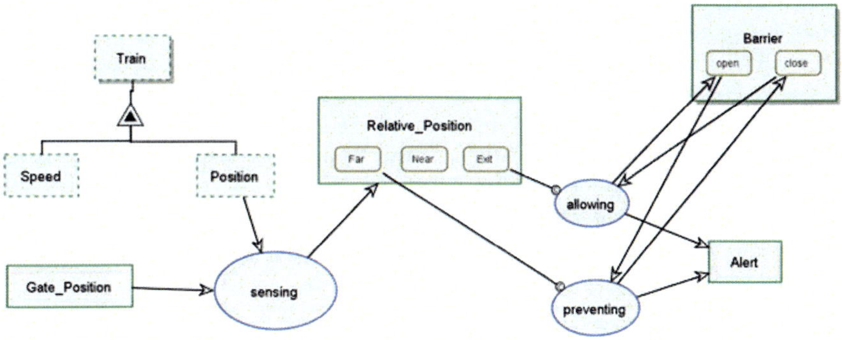

Fig. 3 OPD of the railway gate

Train is environmental and physical. Train exhibits Speed and Position. Speed is environmental. Position is environmental. Relative_Position can be Far, Near, or Exit. Barrier is physical. Barrier can be open or close. sensing consumes Gate_Position and Position. sensing yields Relative_Position. preventing occurs if Relative_Position is Far. preventing changes Barrier from open to close. preventing yields Alert. allowing occurs if Relative_Position is Exit. allowing changes Barrier from close to open. allowing yields Alert.

Fig. 4 OPL for OPD of railway gate

position of the train is converted into a relative position with respect to the railway gate position and this is done by the process called sensing. It should be noted here that the system is being developed at a very abstract level. We are not considering what sensors, and what is the mechanism of knowing the relative position of the train.

The relative position is characterized into far, near and exit. These are defined as states of the relative position. There are now two processes that define the railway gate functionality at an abstract level, one is preventing and the other is allowing. Preventing will not allow people, vehicles or animals to cross the railway track. Here again, the implementation is not important. The functionality can be a person with a flag or manual gate or an automated gate. The functional process of allowing says that the train has passed and it is now clear to move across the track. The process of preventing changes the state of the barrier from open to close and the process of allowing, changes the state of the barrier from close to open. There is another process called alerting that alerts the people that the train is on the way and some actions will be taken.

The OPCAT tool has many features including simulation to check the behavior of the system being designed. This tool can be easily introduced at the undergraduate level and gives an insight into the behavior of the system and the various components that make up the system. It brings a clear demarcation between the system of interest and the environment. It brings a distinction between the physical entities and informational entities. This level of functional decomposition is sufficient to analyze the safety aspects of the system and this is done using the STPA method.

3.2 System Theoretic Process Analysis (STPA)

Systems theoretic process analysis (STPA) is an essential method to build in the safety into the system development process. STPA is a hazard analysis technique with its origins in the Systems Theoretic Accident Model and Process (STAMP) developed by Professor Nancy Leveson of MIT [13]. STPA looks at emergent system properties. This is a different from of the traditional approach of analyzing subsystems and how their failures sequentially cause a system failure. The sequential failure analysis is also known as the Swiss cheese method of failures [14]. In STPA, the system is divided into a process that can be controlled by a controller. This is similar to the system decomposition that is done in the OPM. The process feeds back its states to the controller from which the controller understands its behavior. The principle ingrained in STAMP is that accidents are caused by inadequate control. The system enforces safety constraints to prevent hazards. The process or system behavior is controlled to maintain these safety constraints. When the constraints are violated, they lead to hazardous system states that when interacting with certain environments cause accidents. This method can be used during the design phase to ascertain and to "design in" safety into the system very early in the definition phase. The steps of STPA are defined very clearly in the STPA handbook and therefore not defined in detail here [15]. We instead describe a tool developed using Simulink and available in the open domain.

MOOSE MATLAB Tool for STPA Evaluation

The MATLAB/Simulink tool is available in the open domain called MOOSE. This helps in the STPA analysis following the steps described in the handbook [16]. The first step in the STPA analysis is to identify the losses and hazards. The tool provides two blocks called loss and hazard for this. The railway gate problem STPA losses and hazards are defined as shown in Fig. 5.

The graphical representation of the losses and hazards is very intuitive and the relations are very clear. A MATLAB scripts help to make a table from the visual representation as shown in Tables 3 and 4.

The hazards lead to a first set of safety constrains on the system. The constraints can be defined as below

- While in operation, when indicated the train is approaching, the crossing system shall prevent any incursion on the tracks.
- While in operation, when indicated the train is approaching, the crossing system shall prevent obstruction on the tracks.
- While in operation, when indicated the train has exited the zone, the crossing system shall ensure the crossing is cleared for movement within 1 min.

The number of 1 min is debatable at this point but it gives an indication that there is a time associated with the last constraint and this has an impact in mitigating the loss number 3.

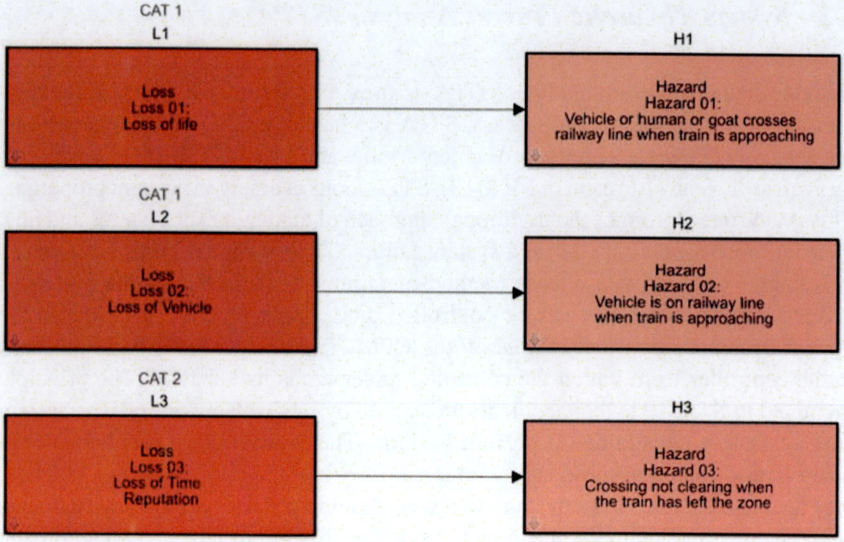

Fig. 5 Losses and hazards of STPA for the railway gate

Table 3 Losses seen in the railway gate example

Name	Loss category	Details
Loss 01	Cat 1	Loss of life
Loss 02	Cat 1	Loss of vehicle
Loss 03	Cat 2	Loss of time or reputation

Table 4 Losses and hazards for the railway gate

Loss	Hazard	Details
Loss 01	Hazard 01	Vehicle or human or goat crosses railway line when train is approaching
Loss 02	Hazard 02	Vehicle is on railway line when train is approaching
Loss 03	Hazard 03	Crossing not clearing when the train has left the zone

The next step in STPA is to develop a hierarchical control structure (HCS). An HCS is a model of the system that is composed of the feedback control loops. This is very appealing to a student of control systems to see the system defined; thus, the controller enforces the constraints defined for the system. These are also the responsibilities of the controller which will mitigate the hazards. A feedback from the process enforces the constraints by providing the information on the process. A controller can be an electronic system, a human or an organization. The parallel to the OPM created earlier is very obvious. All that matters at this stage is an abstraction of the system. The feedback from the process is interpreted by the controller with its own perception model. A variation in the output of the perception model and the

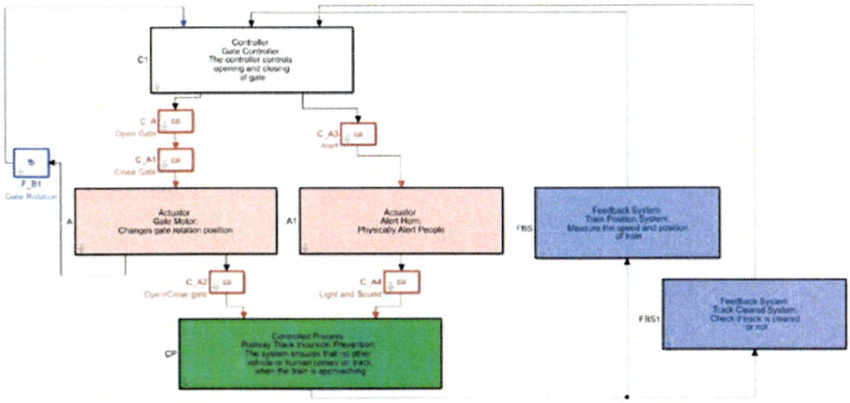

Fig. 6 HCS for the railway gate problem

actual behavior causes the unsafe control actions. An unsafe control action is defined as a control action that in the context of a specific environment state can cause a hazard. Figure 6 shows the HCS developed for our problem statement in MOOSE.

The controller has three responsibilities—open the barrier or gate, close the barrier and to alert. At this stage, the controller could be a human and still the analysis will apply. The feedback to the controller is the train position and track clearance. This again could be manual with the human looking at the tracks and knowing that the train is approaching. It could be a sophisticated system with GPS, Hall effect sensors, and radar sensors to detect the train approaching the railway gate. The track clearance could be camera with artificial intelligence (AI) and image processing setup. At this stage, it does not matter what the solution space is, and it is just important that the system be analyzed for safety issues.

The railway gate feeds back the position of the gate to the controller to alert the people that the gate is closing. A detailed analysis with a brainstorm among students is possible at this stage. In the workshop conducted at MIT, Manipal, the class was divided into groups and each team discussed the STPA and came up with the model for their problem of choice [6]

The hazards happen when the control is inadequate or there are unsafe control actions. There are just four types of unsafe control actions.

- Not providing leads to hazards
- Providing leads to hazards
- Too early/late or not in sequence leads to hazards
- Stopping too early/late leads to hazards.

The MOOSE tool provides this facility in the block. The CA or control action blocks enable this. Each responsibility is represented by one CA block. However, multiple CA blocks can be used for the same responsibility if there are more unsafe actions that are identified. Each unsafe control action leads to a hazard and this is indicated in Table 5. The tool then provides a MATLAB script to generate the table

Table 5 Unsafe control actions as a table output

Responsibility	Not providing causes hazard	Providing causes hazard	Too early, too late, out of order	Stopped too soon, applied too long
Responsible block: gate controller				
Open gate	[#H3] Not opening when train has passed	[#H1, #H2] Opening gate while train is passing	[#H1, #H2] Open too early while train passing	[#H3] Stopped too soon while train has gone
Alert	[#H1] Not providing alert when the barrier is coming down	[#H3] Alerting when train is not coming	[#H1, #H2] Providing a clearance alert when it should be closing	[#H1, #H2] Stopping before the barrier starts moving
Close gate	[#H1, #H2] Not closing when the train is approaching	[#H3] Closing gate while train has gone	[#H1, #H2] Opening instead of closing while train is approaching	[#H1, #H2] Stopped too soon while train is approaching

as shown below. A single block responsibility is depicted here. Students can do a detailed analysis and report this out as part of the class. This can be presented and discussed in class. The next step is to come up with a set of constraints for each of the unsafe control action.

Some of constraints that come up are

- While in operation, when indicated that the train has passed the zone, the crossing system shall open the gate within 1 min.
- While in operation, when the train is approaching AND not exited, the crossing system shall keep the gate closed.
- If train is approaching AND not exited AND gate is open, the crossing system shall alert the controller within 10 s.

Here again, there needs to be a good discussion on how to articulate the constraints on the system. The timings are debatable. Why and what are the keywords for the discussions rather than how.

The next step is to discuss and brainstorm how these unsafe actions can happen. What would be the feedback required? Is the existing feedback sufficient? As the design evolves the reliability numbers for the system sensors and controllers start to take shape. Do we need to provide additional sensors? An example would be.

The unsafe action "Stopped too soon while train is approaching" could be due to the fact that there was

- A power failure
- The system thought the gate was closed due to a short in the system
- The motor packed up half way through.

Questions then arise on how to detect these and mitigate them. This could lead to modifications to the HCS and another iteration. At the end of the exercise, an adequate level of safety measure would be available for the system.

4 Write Good Requirements

A good way to write good requirements is to use the Easy Approach to Requirements Syntax (EARS) notation [7]. EARS has been taught as a workshop again at the college with the students developing two slightly complex systems. They could articulate the requirements and found an error in the design due to the questions that arise out of writing requirements using EARS. We have similar experiences in the industry.

4.1 *Easy Approach to Requirements Syntax (EARS)*

The EARS notation is easily understood by looking at Fig. 7.

The basic set of requirements is what is depicted at the right called the "ubiquitous" requirement. It has a syntax <System Name> shall <System Response>. An example is <The ATM> shall <use the 220 V National grid power>. The other type of requirements is the Event Driven. It starts with the keyword when as shown in the top left of the figure. It represents requirements when a trigger causes an action inside the system boundary. It has a syntax when <optional preconditions> <trigger> then <System Name> shall <System Response>. An example is when <Customer enters withdrawal amount> then the <ATM> shall <check customer balance>. The other requirements syntaxes are as defined in the figure. We are not providing a detailed explanation here as EARS is well described in literature with tutorials.

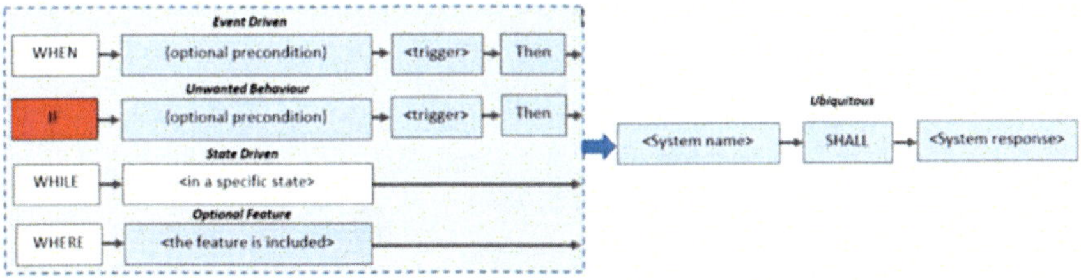

Fig. 7 EARS notation

4.2 System Requirements for Railway Gate

The railway gate system now can be refined with finer details. Figure 8 shows the railway gate schematic based on the STPA and initial functional model developed earlier.

There are four gates—left south (LS), left north (LN), right south (RS) and right north (RN). There are three lights red, orange and green. The train entering the area is indicated at a distance of 2 km from the railway gate. This distance can be firmed up after extensive study as seen later. The safe zone is one kilometer left and right of the center line. This is where the crossing occurs and the gates are placed. It is required that when the train is in this zone the gates should be closed. This is a safety constraint.

- While in operation and train is in the zone, the railway gate system shall ensure gates are closed.

  Train is in zone implies that the train is in the ± 1 km around the gate.
  The other requirements define the working in detail.

- While in operation, when the train has entered, the railway gate system shall set orange lights to flash for 5 s.
- While in operation, when the train has entered, the railway gate system shall set red light after a duration of 5 s.
- While in operation, when the train has entered, the railway gate system shall close gates LS and RN.
- While in operation, when commanded to close specific gate, the railway gate system shall close specific gate within 10 s.
- While in operation, when commanded to close specific gate, if specific gate has not gate closed in 12 s, the railway gate system shall sound alert.

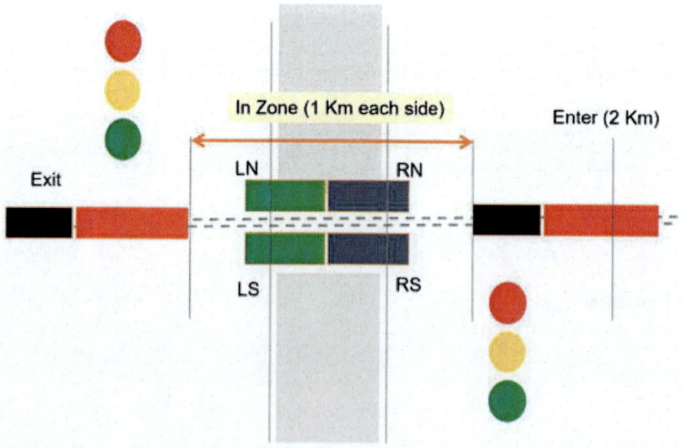

Fig. 8 Schematic of the railway gate

- While in operation, when LS and RN gate closed, the railway gate system shall close gate LN and RS after a duration of 5 s.

These are some of the high-level software requirements that are developed for the railway gate. These now can be used to develop the model.

4.3 *Modeling the Railway Gate*

The railway gate problem is modeled in Simulink. This consists of (a) state chart to set the 4 gate's open and close behaviors and their timing and warnings, (b) 4 gates that open or close in 10 s and (c) the state chart for the light changing red, orange and green. The gate is modeled as a counter that increments by 10 degrees every second. If the angle is less than 10 degrees, the gate is considered closed, and if the angle is greater than 80 degrees, the gate is considered open. This simplified model is sufficient at this stage to carry out a simulation and a formal validation of the system design. Complexities would be a rate limiter connected to a motor control and feedback from the motor control to ensure correctness. The railway gate Simulink Model is shown in Fig. 9. The two inputs to the model are the indications when the train enters the zone and when the train exits the zone.

The state chart of the railway gate starts with the gate open state and changes to gate closed state when the trigger entered happens. It closes two gates. If the gates are not closed in 12 s, it goes to the alert state; otherwise, it waits for 5 s and closes the other two gates as required. The state chart is shown in Fig. 10.

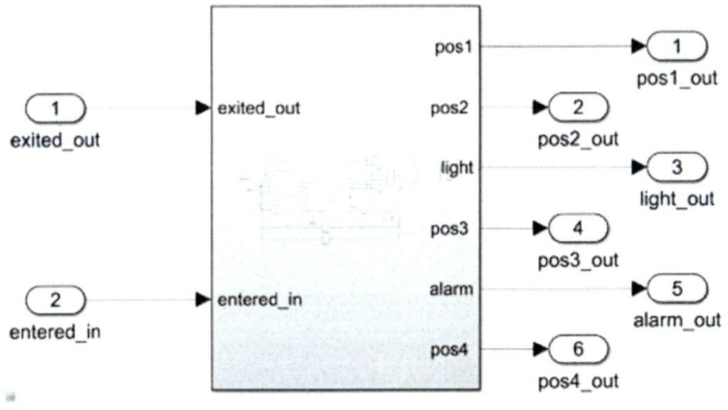

Fig. 9 Simulink model of railway gate

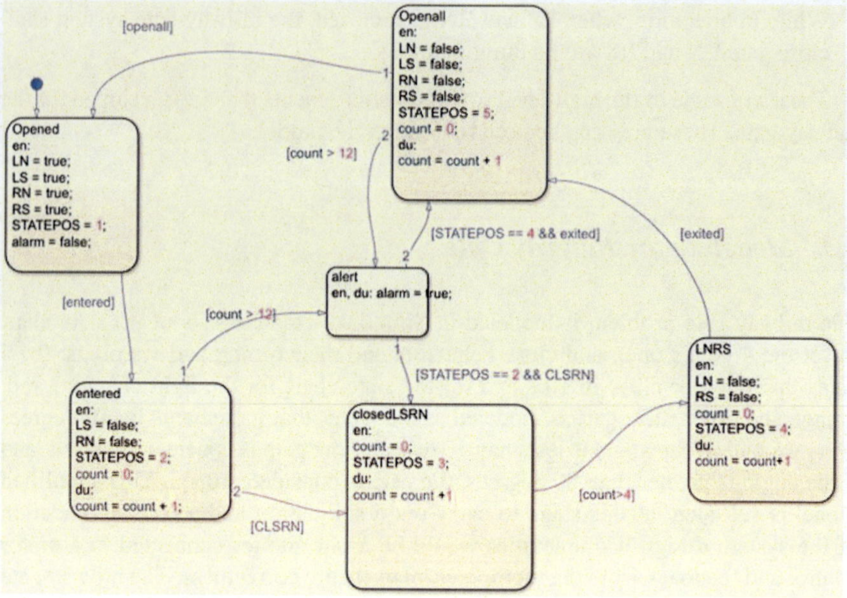

Fig. 10 State chart for the railway gate

5 Model Validation

The safety assertion is that when the train is within the 1 km zone the gate should be closed. This is checked using simulation. The train is modeled as a constant velocity, point mass that moves from an initial distance of 2.5 km and a negative velocity varying from 10 to 70 m/s. These numbers can be changed based on user requirements. When the train position is less than 2 km and greater than 1.5 km, the Boolean "entered" is set. When the train position is less than −1.6 km, the Boolean "exited" is set. The train length, here, is assumed to be 600 m. A simulation run is shown in Fig. 11. The sloping line is the train position and the two-gate position change from 4.5 to 0 in 10 s. The parameters are scaled to fit in the plot. It is clearly seen that when the train is less than 1 km both gates are zero indicating gate closed. This shows that the system works. Is it sufficient? Formal methods can answer this question easily.

5.1 Formal Validation

Formal methods have been prevalent in the academia for the last 40 years. Today, the tools bring these methods into the industry and formal methods have been employed successfully to prove the correctness of design even at a system level [17]. The

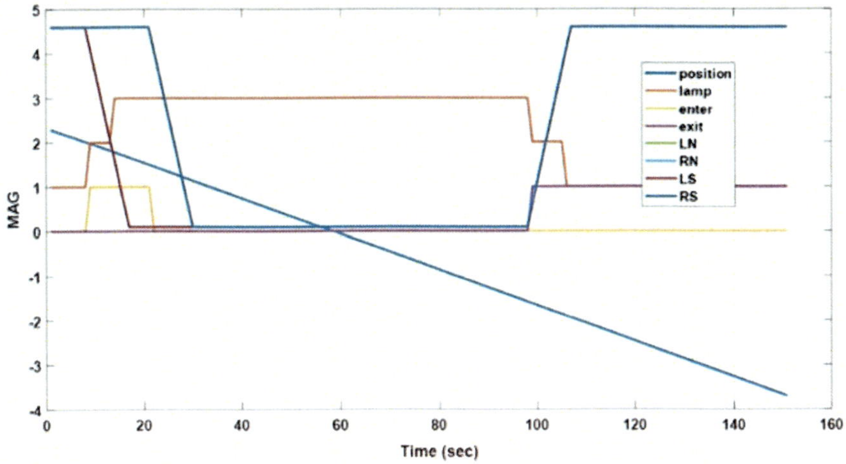

Fig. 11 Railway gate simulation

Simulink Design Verifier is a good tool to introduce at the college level to get the students familiarized with these methods. This has been done with industry, academia and student participation earlier [18]. The assertion is modified here to check if the system works well if the zone is compressed to ±500 mts on either side of the crossing zone. The assertion is shown in Fig. 12. The assertion shows that given the assumption that the train velocity can be −40 and −70 m/s, the gates will remain closed when the train position is less than or equal to 500 m and greater than or equal to −1100 m.

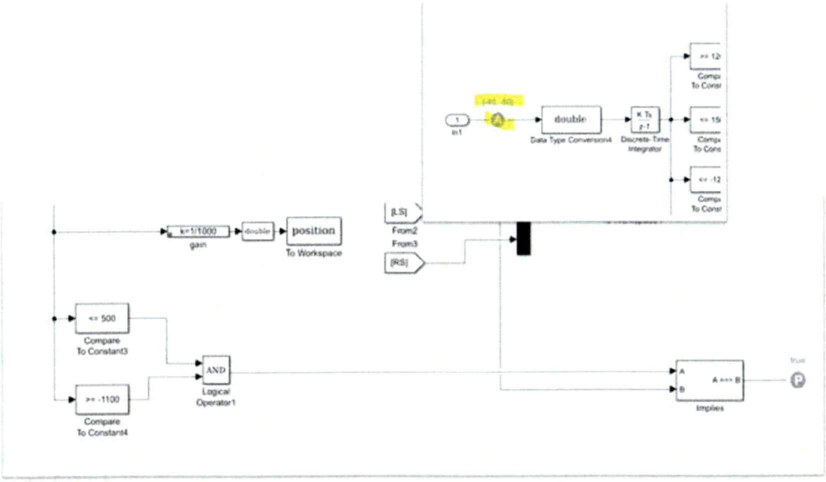

Fig. 12 Simulink design verifier assertion

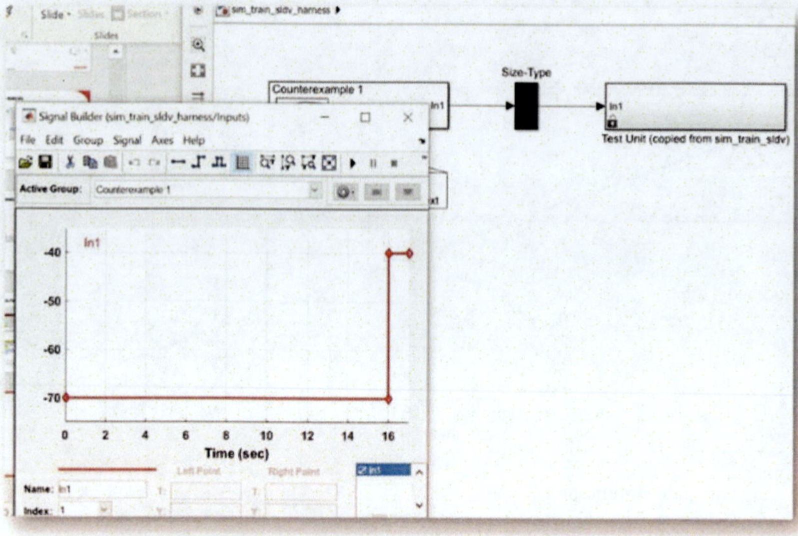

Fig. 13 Counterexample from Simulink design verifier

The formal methods show a violation of this assertion and show a counterexample of −70 m/sec as a test case where the assertion fails, this is shown in Fig. 13. However, with the range increasing to 800 m on either side and limiting the speed to less than 60 m/s, the formal engine indicates that assertion is proved correct and that the design is good for these assumptions. This sort of validation is done at the industry level on the system level design and this needs to be shown to the students for them to understand the significance of these mathematics-based techniques and their practical applications. This has inspired many students to do a master's degree in this field.

5.2 *Automatic Code Generation*

A validated model is used to generate an autocode. This is used in the industry and works very well. The MATLAB model is converted to C code through the inbuilt MATLAB C/C++ code generator. It generates a C code file, a main file and other header files that define the variables and the structures used. The code that is generated cannot be directly imported into Arduino because the MATLAB simulation runs at a specified fixed step sample time of 1 s. This is performed by the main C code file which call the functions to initialize the train model and do a step wise model simulation. The logic, however, can be extracted out of the generated code to work as a part of the setup() and loop() functions that are necessary to run the Arduino model. Logic here refers to the core functionality defined by the state charts and the order of instructions and state changes that control the gate positions, light operation and the interaction with the inputs and outputs.

Arduino implementation

Arduino code requires only two functions, a looping function and a setup function. The initialization of loop variables, connections to the serial port and the setup for the servo is handled by the setup() function. The looping functions loop every second and takes in the location of the train as the input that drives the different states for the gate position measurement. The train location is also calculated internally at the start and changes with every iteration of the loop. The plant model is embedded into the target, and this may seem like a student project, but in a recent project in the industry, the same setup was used to prove the correctness of the system. The fun part of the paper is to interface the automatically generated code to deploy a motor to move the gate and to visualize this as a beautiful graphics on the laptop.

6 Visualization

The visualization is done using an open-source rendering program called Blender [8]. Blender is a free and open-source computer 3D graphics suite that uses OpenGL for a seamless interface and provides tools to work with CUDA or OpenCL-based rendering and animation. The Blender foundation provides a Blender API for Python-based scripting to add additional features, tools or operations that suit the needs of the user. The "bpy" library provides the access to the objects, tools and functions that can be used to create and modify 3D models.

6.1 Blender Implementation

The Blender program includes an option for scripting with Python API that enables the interaction with collections, objects and GUI tools for various 3D object operations through Python code. The Blender "bpy" library includes functions and lists of the objects and their properties including location, Euler rotation angles and scale. The serial library provides capabilities for serial communication. For the visual representation of the railway gates opening and the moving vehicles on the road, the Python script reads the values from the serial port and translates them to the corresponding gate rotation angles, train location value and the active state that controls the functions for vehicle movement during normal operation and gate close operation.

Normal operation of vehicles involves the change in the location of individual cars every second where the speed is maintained by the magnitude of change. If the cars pass a certain threshold, the car is hidden and its position is pushed back by a certain amount for it to come to the back of the line and then the car is made visible. This provides a view of a continuously moving chain of vehicles of different sizes.

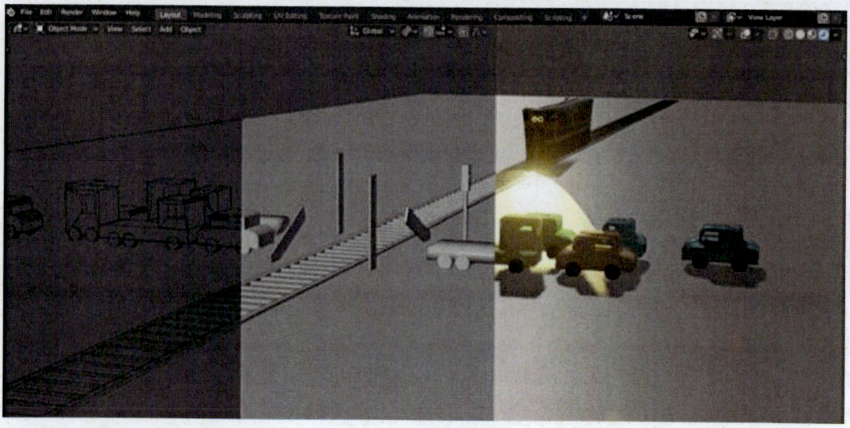

Fig. 14 Blender rendering of railway gate

In case a gate close operation is initiated, vehicles which have a position beyond a certain value are made to continue their motion and switch or loop back to the back of the row while the ones behind this value stop. This is implemented in the Python script and is initiated when the active state switches to the gate close or gate open value.

Viewport rendering is carried out by refreshing the view after every line of data received through the serial port. Physics-based real-time rendering is possible due to the EEVEE rendering engine by Blender foundation that provides photo-real rendering capabilities for pre-render views in the viewport or in some cases for final renders. The effect is stunning as seen in Fig. 14. The figure shows the wireframe model in the left and the final rendering in the right. A video of this is available online [19]

6.2 *Motor Implementation*

The Arduino board is connected to a DC servo motor as shown in Fig. 15. The serial communication is connected to the laptop. Thus, the visualization and the motor moving the gate can be seen together. This is very similar to the ironbird sort of setup that is used in aircraft development. The setup has the graphics for the pilot to fly and the actual system is also exercised simultaneously.

The servo motors attachment to the Arduino is a 5 V input, a ground (GND) and a pulse width modulation (PWM) enabled output pin. The servo is a three-pole geared motor with a good response to input. The servo requires angle input in degrees and can be used to visualize the gate opening or closing synchronous with the gate movement in the Blender viewport render.

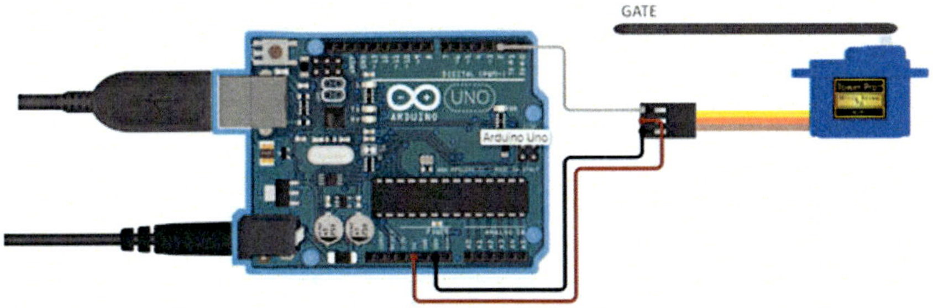

Fig. 15 Arduino connectivity

Fig. 16 Screenshot of the YouTube video [19]

The position values of the four gates, train location, light state and the current state of operation is printed out into the serial communication as comma separated values. This serial communication starts as soon as the Arduino board is switched on and is received by the Python script running in Blender. The screenshot of the video shows the fun element of the complete exercise in Fig. 16. The gate motor, Arduino board and the visualization are all seen together here. The setup can be used for extensive testing that is required for the final system now.

7 Conclusion

The pandemic situation has changed the way subjects are being learnt. Everything is online and access to laboratory is zero. The paper shows how a simple railway gate problem can be made fun and intuitive to learn the systems engineering concepts at home. The steps of the system developed have been followed through and the industry level process and have been brought out to an academic level. A training on the same lines is being done at an aerospace industry and this example indicates how it can be taught at home to an engineering student.

The major lessons learnt for the mentor has been

1. New methods are required to teach engineering and to make it fun and hands-on is a challenge. The example has been heartening and this example and demonstration back to the industry participants have been well received.
2. The safety-critical system development can be easily taught with this simple example. Many discussions have originated over the simple design which is required for a holistic system development.
3. It is very clear that many control system problems can be modeled in this manner and easily taught online.

The major lessons learnt for the mentee has been.

4. It is very interesting to see that the industry works on similar lines. They do have more complex problems but the essence of what is done in office is very clear.
5. Blender is complex to manipulate but many videos are available online to learn this.
6. It is easy to interface Arduino with the tool especially with Python.
7. MATLAB and Simulink are taught at the college level but a practical hands-on is very different than what is taught. This exercise is very interesting and fruitful.
8. Simple motor with straw representing a gate and to see the same as a visualization on laptop is very satisfying.

References

1. Yogananda, J.: Developing safety critical systems—my Mantras. https://www.youtube.com/watch?v=fpNqwobAWRU. Last accessed 18 Nov 2020
2. Jan, B., Mark, B.: Candidate software process flaws for the Boeing 737 Max MCAS algorithm and risks for a proposed upgrade. https://arxiv.org/abs/2001.05690. Last accessed 18 Nov 2020
3. Bradley, W.: Technological solutions to human error and how they can kill you: understanding the Boeing 737 max products liability litigation, 84 J. Air L. & Com. 379 (2019) https://scholar.smu.edu/jalc/vol84/iss3/3. Last accessed 18 Nov 2020
4. IEEE Standards Association, Develop Standards. https://standards.ieee.org/develop/develop-standards/process.html. Last accessed 18 Nov 2020
5. AICTE, 2018 Model Curriculum For Undergraduate Degree Courses In Engineering & Technology, https://www.aicte-india.org/sites/default/files/inal%20Draft__Vol.%20I%20%20AICTE%20UG%20%20curriculum.pdf. last accessed 18 Nov 2020
6. Shreesha, C., Yogananda, J.: Teaching STPA and OPM to engineering students—industry academia experiences. In: INCOSE International Symposium, vol. 29(S1), pp. 17–27 (2019
7. Alistair, M., Philip, W., Adrian, H., Mark, N.: Easy approach to requirements syntax (EARS). In: 17th IEEE International Requirements Engineering Conference, Atlanta, GA, pp. 317–322 (2009)
8. Blender. https://www.blender.org/. Last accessed 18 Nov 2020

9. LDRA Whitepaper, Emerging technology trends: Skill Challenges & Opportunities for Industry & Academia. https://ldra.com/collateral/emerging-technology-trends-skill-challenges-opportunities-for-industry-academia/. Last accessed 18 Nov 2020
10. ISO, Teaching Standards Good practices for collaboration between National Standards Bodies and Universities. https://www.iso.org/files/live/sites/isoorg/files/store/en/ PUB100354.pdf. Last accessed 18 Nov 2020
11. Dov, D.: Object-process analysis: maintaining the balance between system structure and behaviour. J. Logic Comput. **5**(2), 227–249 (1995)
12. OPCAT. https://esml.iem.technion.ac.il/opcat-installation/. Last accessed 18 Nov 2020
13. Nancy, L.: Engineering a Safer World Systems Thinking Applied to Safety (2012). https://mitpress.mit.edu/books/engineering-safer-world. Last accessed 18 Nov 2020
14. Perneger, T.: The Swiss cheese model of safety incidents: are there holes in the metaphor? BMC Health Serv Res **5**, 71 (2005)
15. Nancy, L., John, T.: STPA handbook, last viewed 20 July 2020 (2018). https://psas.scripts.mit.edu/home/materials/. Last accessed 18 Nov 2020
16. Aditya, J.: STPA Safety Analysis Tool in Simulink , MATLAB Central File Exchange. https://www.mathworks.com/matlabcentral/fileexchange/73434-stpa-safety-analysis-tool-in-simulink. Last accessed 18 Nov 2020
17. Ramakrishnan, R., Yogananda, J.: An Approach for formal verification of machine learning based complex systems. In: INCOSE International Symposium, vol. 29(1), pp. 544–559 (2019)
18. Natasha, J., Yogananda, J., Devi, M.K.K.: Teaching formal methods at undergraduate/graduate level: the three perspectives. In: 3rd International Conference on Applied and Theoretical Computing and Communication Technology (iCATccT), Tumkur, (2017), pp. 310–315
19. Aditya, J.: A railway gate model using STPA, Matlab, Blender, Arduino. https://www.youtube.com/watch?v=gZsGFfWduAU. Last accessed 18 Nov 2020

Smart Water Management System

C. Nikhil Binoy, B. Jayalakshmi, A. Nandhan, P. Vishnu, and S. Thasna

1 Introduction

Water is essential for the existence of every living being on the earth. Human lives are depending on the water for all the daily routine activities. It is the problem of getting sources of water to be used throughout an amount of time and should lead to additional depletion and deterioration of obtainable water resources. Water shortages are also caused by global climate change, like altered weather patterns as well as droughts or floods, accrued pollution, and accrued human demand and overuse of water. Water scarcity in our society makes us to think to have a system for monitoring water level continuously and control the water consumption.

The system is mainly focused on GSM-based Arduino-controlled system. Three main problems are aiming to solve are unawareness of water level in the source, unwanted wastage of water, and improper consumption of water. Since it is a difficult a task to know how much water is available at our resource for usage and to have a proper calculation of real time, an automatic water management system would help to have a proper knowledge of water at source such as an open-wheel, bore-well, etc.

The objective of the system is to design a water consumption automatic system by proper data acquisition of water level, water consumption, and monitoring using GSM.

C. Nikhil Binoy · B. Jayalakshmi · A. Nandhan (✉) · P. Vishnu · S. Thasna
Department of Instrumentation and Control Engineering, NSS College of Engineering, Palakkad, Kerala, India

C. Nikhil Binoy
e-mail: nikhilbinoyc@nssce.ac.in

S. K V and K. Guruprasad Rao (eds.), *Smart Sensors Measurements and Instrumentation*, Lecture Notes in Electrical Engineering 750,
https://doi.org/10.1007/978-981-16-0336-5_20

2 Related Work

The world is going through the worst situation due to the inefficient water management as well as water distribution in houses. The work gone through have mentioned the detailed research about the water consumption in Chinese household works [12]. From the detailed studies, it is understood that water scarcity is faced by whole life of earth very adversely.

Another problem faced more in case of water is the improper consumption and quality of water. Deepiga et al. [1] conveys about the method of monitoring water level inside the house top tank by implementing the level sensor and also mention about the sensor based after pollution monitoring the system was based operated with microcontroller and the interfaced with MATLAB. But the matter that to focus was all this was monitoring through devices like laptop and mobile phone.

Again moving on it found that water quality monitoring is a tedious process since water sample is collected manually and taken to laboratories for measuring the parameters to identify the water composition is good and it is not polluted. For this, Purohit et al. [2] proposed an 8051 microcontroller-based system with GSM module for the interface with the operator. GSM is used in the system which is possible to operate with any sort of mobile phone.

Spandana et al. [3] proposed that water parameter monitoring can also be done by a multi-sensor system like pH sensor, turbidity sensor, conductivity sensor, etc. But the main thing is the system is interfaced and operated with a new technology which is most advance in the tech world.

Jayalakshmi et al. [4] found that the most suffering group in the society due to water scarcity is urban life. So a proper water management is needed for them. The water they are provided through tankers and other means may not be of proper treated and best. So a water quality monitoring system is the best part where the system is operated a using multi-sensor setup with IoT.

Patawala Amatulla et al. [5] mention that even though water is provided for urban life through pipelines no one is bothered about the water wastage due to breakage and leakage of pipelines. So a system to detect the leakage in pipelines by measuring the flow rate through a pipeline continuously will be a sufficient method to find out the leakage present in the pipelines. Apart from that the system is comprised with quality monitoring and interface with IoT.

Kawarkhe et al [5] mentions that even though water is provided for urban life through pipelines no one is bothered about the water wastage due to breakage and leakage of pipelines. So a system to detect the leakage in pipelines by measuring the flow rate through a pipeline continuously will be a sufficient method to find out the leakage present in the pipelines. Apart from that the system is comprises with quality monitoring and interface with IoT.

Xiaocong et al. [7] proposed rather from all the above works the main thing it found that in this system is it is used Arduino for the water quality monitoring with GSM module for interface. Since it easier for coding in Arduino board and implementation, we decide that Arduino would be more suitable for our system.

Gokulanathan et al. [8] want to convey a foremost thing that there is no idea of how water management could be implemented without knowing the proper water distribution in houses. Water distribution in the houses by knowing priorities in household works and proper monitoring of those systems would make the life more beautiful and systematic.

Robles et al. [9] want to convey to the readers through the work that water distribution system is not a concept, but implementing it by control valve with measurement of water level is something like predicting the water availability for future. In this work, we could find that water-level measurement form overhead tank is giving actuation to solenoid valve provided in tank outlet. This system is implemented using Arduino UNO and IoT.

Sayed et al. [10] found that real-time monitoring is the major objective to be followed for a smooth operation of such systems. Implementation of real-time monitoring gives us more accurate values regarding quality monitoring and other parameters.

3 Methodology

The detailed studies lead to find a system that using Arduino UNO board as controller since it is easyly programmable and easily available for user. Multi-sensor method and control valve flash a light over having an effective water management system using GSM as a monitoring system with display windows as laptops or mobile phones. For that, the entire water consumption in a house is divided into three categories as primary, secondary, and tertiary. Primary category includes the kitchen works, which cannot be emitted from human life and thus there is no control action is provided at the tank outlet for this category. It is shown in Fig. 1.

The secondary category is provided for other aspects like bathing and cleaning of houses where control valve like ball valve is used to have a throttle action at the outlet of the house-top tank according to the decrease of water level in the source. Figure 2 shown below displays the secondary part of the system. The level transmitter in the well takes measurement of water level continuously and given to flow regulator and controller and it provides actuation to control valve for proper throttle action.

The tertiary part of the water management system is outdoor pipelines in the house where the control action is using ON–OFF valve such as solenoid valve. In the tertiary system, the water flow through pipelines is controlled by an ON–OFF mechanism of the control valve if the water in the resource is lessening under the set point value. Figure 2 also represents the tertiary category system; from the secondary part, the only change is in control valve used. In the automatic tank level controller, if the level of water in overhead tank is empty, level switch activates; hence, the pump turns on automatically. If the level of water is full, level switch deactivates; hence, the pump turns off and thereby prevent the overflow. If the overhead tank is empty and also water level in sump tank is less, the pump does not run and also the buzzer

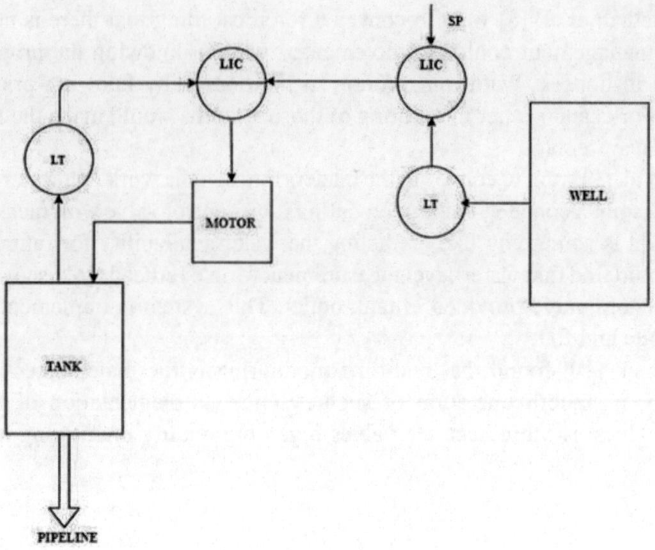

Fig. 1 Primary category block diagram

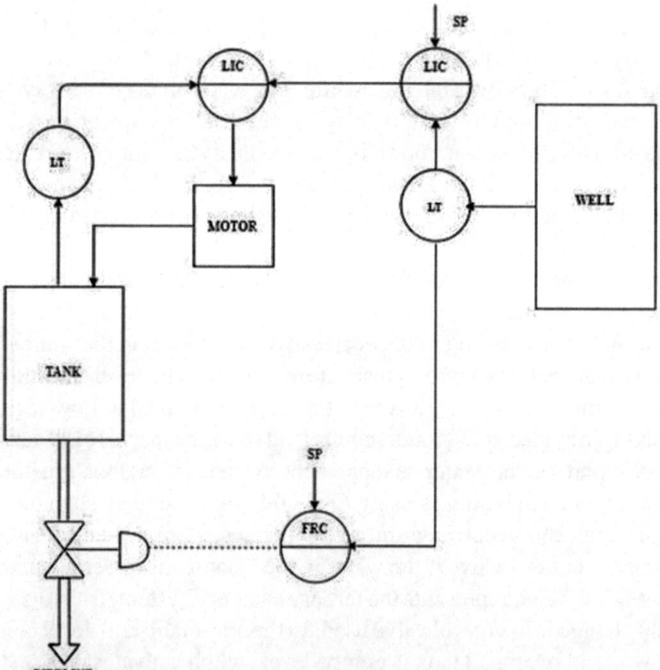

Fig. 2 Secondary and tertiary category block diagram

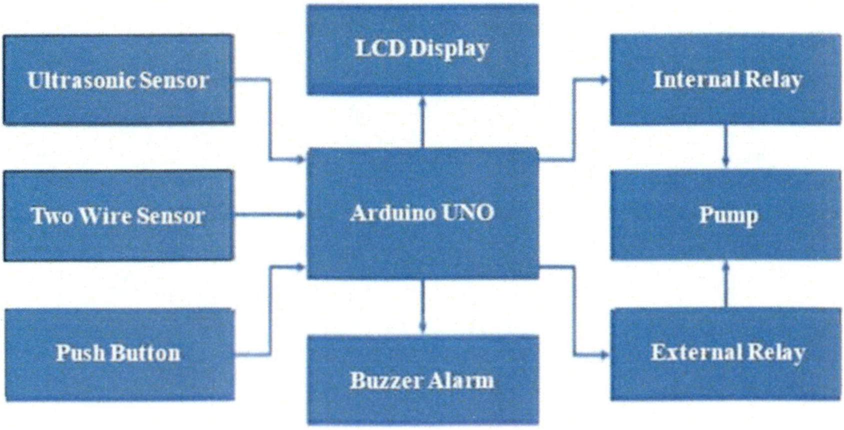

Fig. 3 Automatic tank level controller

gives indication of low water level. The LED connected to the circuit also glows for indication. Figure 3 shows the block diagram of the automatic pump operation.

In the system proposed here, control valve operation is the major part. There is no compromise over it, such that while taking water level of the source, the control valve is operated according to the water availability in the source. At first, the valve will be in a position by providing water through it and later if the level got changed in the source, then the control valve opening must change according to it. Since we could not break the water supply in the pipeline, we obey a strategy that every change in control valve opening has taken place following the complete opening of the valve for seconds (Fig. 4).

4 Result and Discussion

The system designed will provide a better result from the existing situation shown in Fig. 5. Since the level sensors in both the water tank and resource is taking their measurement time to time, which provides actuation to control to operate as such. The LCD will always show the data such as tank water level and resource water level. From this, when the water level got decreased in the source, the control valve is operated.

If in case water level is less than 30%, then the valve will be completely closed, whereas if the level is between 30 and 40%, the control valve is opened for 40%. If the level reaches between 40 and 50%, the valve opening races 50%, when the water level is 50–60%, the valve opening is 60%, when the water level reaches 60–70% the valve operated at 70%, the level reaches 70–80%, then valve opening changes to 80%, and when it reaches the 90%, it is completely open.

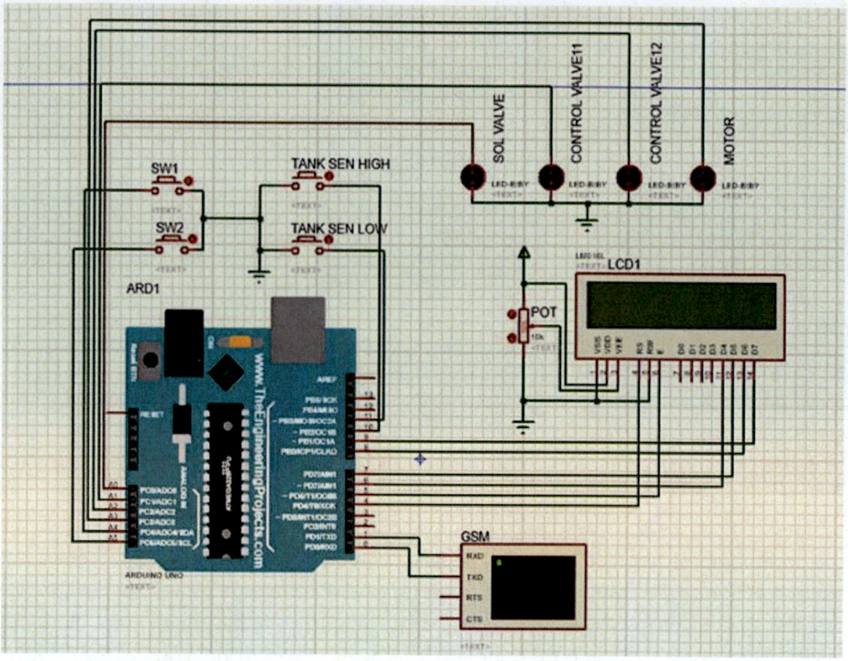

Fig. 4 Circuit diagram

Fig. 5 Three-dimensional view of hardware circuit

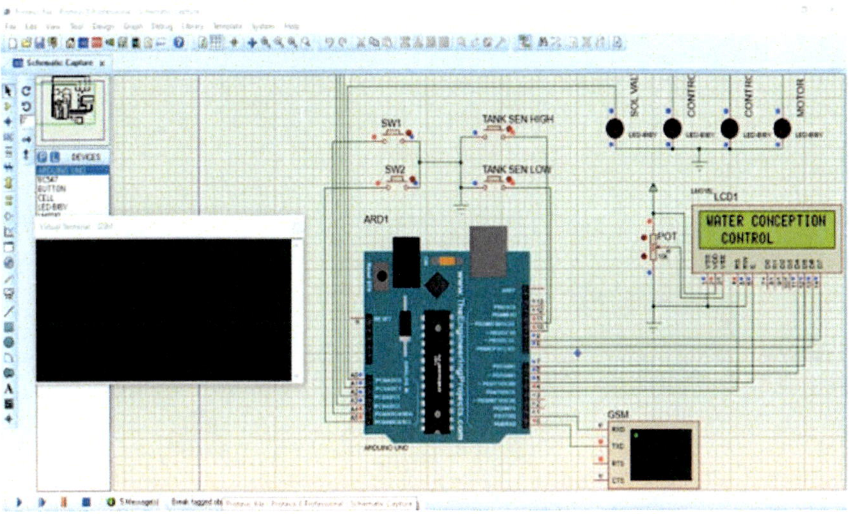

Fig. 6 Initialization of simulation circuit

This is the case of medium priority, whereas it can also be operated in manual mode by giving messages through GSM system. Figure 4 represents the circuit diagram for simulation. The circuit has some components change from the physical system; here, instead of ultrasonic sensor for measuring water level from resource, we use a switch to change the water level. Similarly, in case of water tank level measurement instead of level switches, indication switches are used to show its state high or low indication (Fig. 6).

And also for the representation of relay circuits for operation of control valves and pump, here, LED indications are provided. There are provided four LEDs representing four operations. The first one indicates the opening and closing of solenoid valve, second one indicates ball valve forward motion, third led indicates the ball valve backward indication, and fourth one indicates the ON and OFF of the pump. After the simulation of the system in Proteus software, it provides following indication according to the water level of head-top tank and water source; open-well or bore-well.

In Fig. 7, it shows the system information displays in LCD. The following information are available in the display: well water level, tank water level, ball valve status, solenoid valve status, mode of operation of system, and pump status. By giving GSM codes in the GSM virtual terminal, it operates according to the code which is already programmed with the Arduino. Figure 8 shows such a result which gives information to registered mobile number of the user by providing the code *#INF.*

The system can also operate in manual mode if needed by providing the code *#MAN.* In manual mode or operation, user can give codes already set through registered mobile number and operate the system as required. Figure 9 shows such an

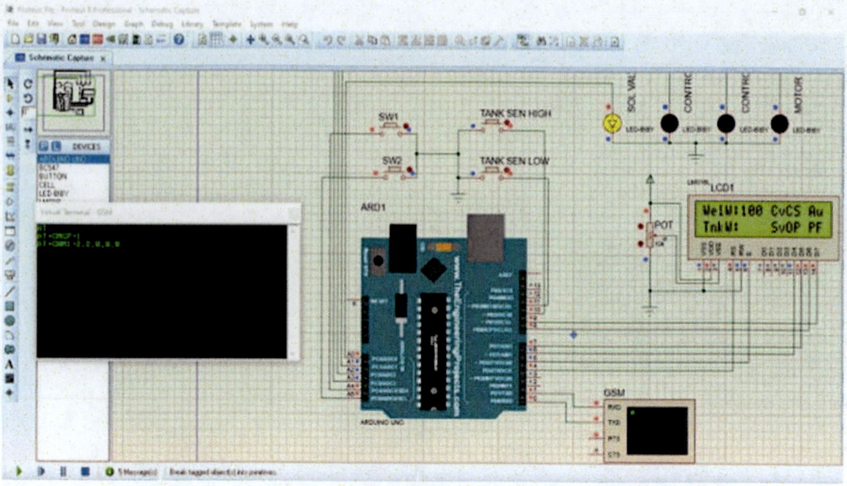

Fig. 7 System information indications in LCD

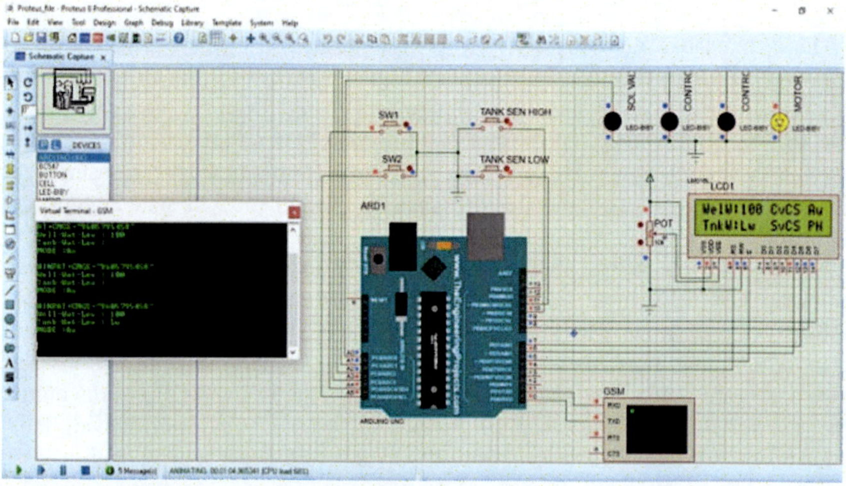

Fig. 8 System information on GSM virtual terminal

example where the user operates the pump manually giving code through the GSM virtual terminal.

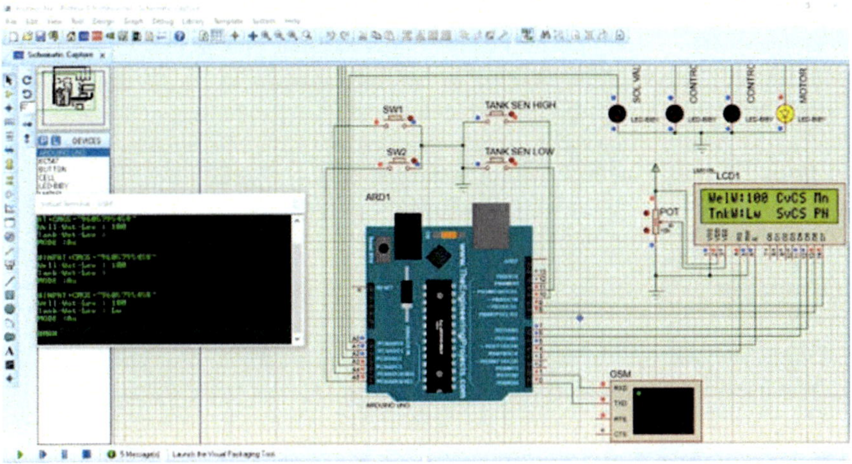

Fig. 9 Manual mode operation of system

5 Conclusion

The designed system is highly efficient and effective for the present scenario i.e., improper water usage in houses and other adverse situations affected after these situations. And with this system, it can assure to avoid a situation of high water scarcity and a high cost to afford it. Since the availability of water is decreasing day by day due to many reasons such a pollution of water resources by industries, sewage, etc., so, by properly designing this system, the water consumption and control can be achieved in a great manner. One of the attractions of this system is that it can operate on both the manual and automatic modes. By using a GSM module, the user can directly control the level of water. Since there is no use of a control valve in higher priority and in case of medium and low priority, a ball valve and solenoid valve is used which does not cost much comparing the expense in a construction of the house. Hence, this system is gaining much more importance for household purposes and it is cost effective.

6 Future Scope

The system has sorted out the most important problems faced by the people regarding water, but there are many more beyond this. Quality is another problem faced by the society. After getting a heavy rain, it found that the water in the resources gets more turbid; such type of cases have to be solved. To overcome all this, we can add a quality monitoring system with Arduino, turbidity sensor, pH sensor, etc.

Since the system is a GSM operating system, it is easy to access and use even for the ordinary citizen. But now technologies are changing time to time, and in such a scenario, it is essential to move on to new technologies to handle the previous editions. Internet of things (IoT) and big data analytics can have a vital role in this project [11]. This project will be more advanced and it will be introducing a huge change in the society on the behalf of home automation system.

References

1. Deepiga, T., Sivasankari, A.. Smart water monitoring system using wireless sensor network at home/office. Int. Res. J. Eng. Technol. (IRJET) **02**(04), 1305–1314 (2015)
2. Purohit, A, Gokhale, U.M.: Design and implementation of real time system for water quality measurement using GSM. Int. J. Adv. Res. Comput. Sci. Manage. Stud. **02**(01), 223–226 (2014)
3. Spandana, K., Seshagiri Rao, V.R.: Internet of things (Iot) based smart water quality monitoring system. Int. J. Eng. Technol. **07**(03), 259–262 (2018)
4. Jayalakshmi, B., Nandhan, A., Vishnu, P., Thasna, S., Akhil Chidambaram, R.: Water consumption control and quality monitoring system. Int. J. Innov. Res. Electr. Electron. Instrum. Control Eng. **08**(01), 146–152 (2020)
5. Patawala Amatulla, H., Bansude Navnath, P., Yogw, B., Asmailin, Z.: IOT based water management system for smart city. Int. J. Adv. Res. Ideas Innov. Technol. **03**(02), 379–383 (2017)
6. Kawarkhe, M.B., Agrawal, S.: Smart water monitoring system using IOT at home. IOSR J. Comput. Eng. (IOSR-JCE) **21**(1), 14–19 (2019)
7. Xiaocong, M., Jiao, Q.X., Shaohong, S.: An IoT-based system for water resources monitoring and management. In: 7th International Conference on Intelligent Human-Machine Systems and Cybernetics (2015)
8. Gokulanathan, S., Manivasagam, P., Prabu, N., Venkatesh, T.: A GSM based water quality monitoring system using Arduino. Shanlax Int. J. Arts Sci. Human. **6**(4), 22–26 (2019)
9. Robles, T., Alcarria, R., Martın, D., Navarro, M., Calero, R., Iglesias, S., Lopez, M.: An IoT based reference architecture for smart water management processes. J. Wireless Mob. Networks Ubiquitous Comput. Dependable Appl. **6**(1), 4–23 (2015)
10. Sayed, A., Vatkar, S., Udmale, A., Bhandiwad, V.: Smart and automatic water distribution control system. Int. Res. J. Eng. Technol. **05**(01), 644–646 (2018)
11. Menon, G.S., Ramesh, M.V., Divya, P.: A low cost wireless sensor network for water quality monitoring in natural water bodies. In: IEEE Global Humanitarian Technology Conference, San Jose, CA, pp. 1–8 (2017)
12. Lu, T.: Research of domestic water consumption: a field study in Harbin, China. In: A research project report submitted for Master of Science of Water, Engineering and Development Centre Department of Civil and Building Engineering Loughborough University (2007)

Analysis of Electroencephalogram During Coloured Word Reading Interference

Sai Kate, Vaishnavi Malkapure, Bhagyashree Narkhede, and Revati Shriram

1 Introduction

EEG referring to electroencephalogram means measuring electrical activity of brain. Such electrical measurements are obtained from electrodes placed on surface of scalp, this waveform represents a very gross type of summation of potentials that originate from extremely large number of neurons in vicinity of electrodes. Stress leads to release of stress hormones (dopamine, epinephrine, cortisol, etc.) in the brain. The release of these stress hormones affects different parts of the brain and thus the cognition or thinking process. Thus, the electrical activity of the brain was chosen to study during the Stroop task [1]. The Stroop task (developed by John Ridley Stroop in 1938) is based on the fact that it is easier to read a colour written in the same ink than a colour written in a different ink. Colour written in same ink is "congruent" whereas colour written in different ink is "incongruent". The Stroop effect is based on cognitive interference [2, 3]. The work was carried out on the manual mode of Stroop test and built ten stages using graphical user interface on MATLAB. Different levels of stress are induced by increasing the stimuli with each stage. Stimuli are increased by increasing the incongruency in the colour word as well as by changing the fonts of the source and destination screen. This increases the conflict. Having stress induced during Stroop task, EEG is recorded. Figure 1 shows the screens in the Stroop task.

S. Kate (✉) · V. Malkapure · B. Narkhede · R. Shriram
MKSSS's Cummins College of Engineering for Women, Pune, India
e-mail: sai.kate@cumminscollege.in

V. Malkapure
e-mail: vaishnavi.malkapure@cumminscollege.in

B. Narkhede
e-mail: bhagyashree.narkhede@cumminscollege.in

R. Shriram
e-mail: revati.shriram@cumminscollege.in

S. K V and K. Guruprasad Rao (eds.), *Smart Sensors Measurements and Instrumentation*, Lecture Notes in Electrical Engineering 750,
https://doi.org/10.1007/978-981-16-0336-5_21

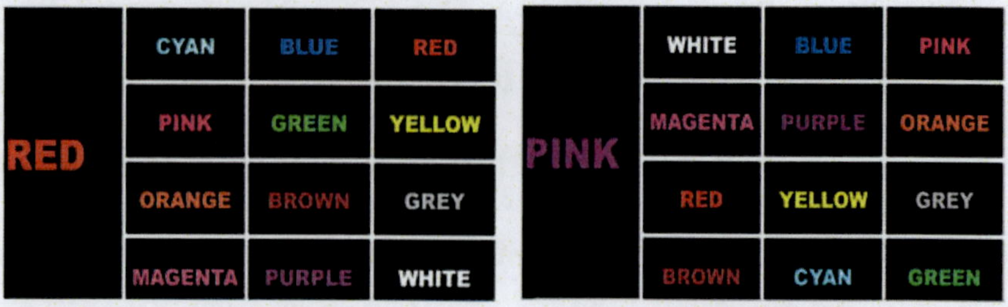

(a) Congruent **(b)** Incongruent

Fig. 1 Tile of Stroop task in MATLAB

The left side of the figure is the source screen, and right side is the destination screen with a grid consisting of 12 colour [4].

2 Methodology

The subject was asked to perform the Stroop task during which his/her EEG was recorded. The EEG was acquired using a data acquisition system named POWERLAB which was then given to the LabChart software where the recorded EEG signal was displayed. This signal was then exported in text form to Microsoft Excel. The Excel Sheets were read by MATLAB for analysis of the EEG.

Before feature analysis of the signal, pre-processing of the EEG signal was done. This included removing the artifacts due to blinking and 50 Hz line frequency by using Notch filter. EEG signal analysis was carried out using the various features obtained from the signal. During this work, the four features were observed: coherence, power spectrum density (PSD), entropy and frontal asymmetry index. These features can be studied using various time and frequency domain analysis technologies such as eigenvectors method, autoregressive method and wavelet transform. [5]. Figure 2 shows system block diagram which describes the flow of the process:

2.1 Power Spectrum Density (PSD)

Power spectral density gives us what the power of the signal will be for the entire frequency spectrum or range. PSD can be studied using two methods: (1) parametric modelling and (2) nonparametric modelling. Parametric modelling depends on the number of parameters and failure to consider a parameter can give false readings. Thus, authors have chosen Welch's method which is a nonparametric modelling process. It is the method used when a large number of parameters of the signal are not known. It is a modification of the Bartlett's method. Welch's method is to divide

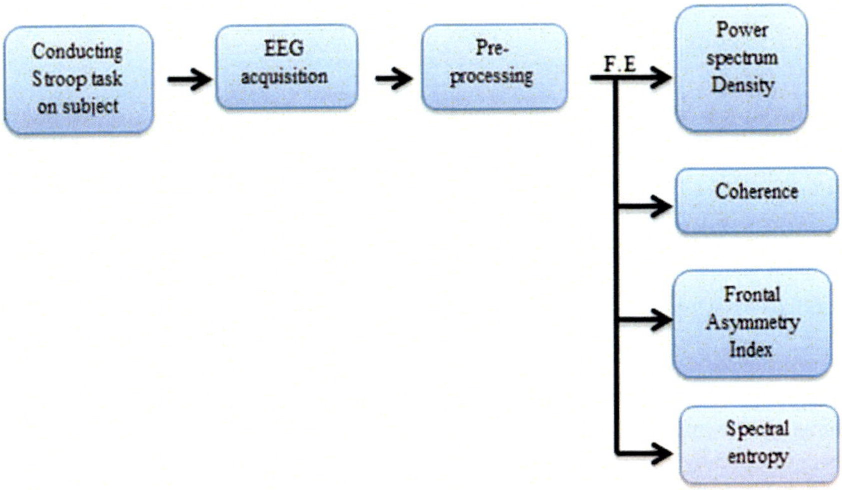

Fig. 2 System block diagram

this time series such that it is split into N individual time series. For each time series, the strength of the series versus the frequency is obtained. This is the power spectral analysis of that time series. Then the overall power spectrum density of the entire signal is obtained by averaging the power spectrum densities of the individual time series. Thus, from this method we get to know at which frequency the strength of the signal is highest or lowest. The equation of Welch's power spectrum density is as follows:

$$\hat{S}_x^W(w_k) \triangleq \frac{1}{K}\sum_{m=0}^{K-1} Px_{m,M}(w_k)$$

The PSD measures the strength of oscillations in the brain and is directly proportional to the cortex activation [5, 6].

2.2 Coherence

Coherence is a function that is used to study the similarity between two signals. Magnitude-squared coherence is used in this project. In this, coherence is obtained by first doing the autocorrelation of the two individual signals. Then the cross-correlation of the two signals is obtained. After that the power spectrum density (PSD) of the three sequences is obtained. This is a function of power spectrum density $P_{xx}(f)$ and $P_{yy}(f)$ and the cross-power spectral density, $P_{xy}(f)$, of x and y:

$$C_{xy}(f) = \frac{|P_{xy}(f)|2}{P_{xx}(f)P_{yy}(f)}$$

Coherence is used to study this electrical activity to see which parts of the brain are working in unison towards an external stimulus and to also study the flow of the process that takes place in the brain in response to an external stimulus. Thus, coherence connects the activity in different parts of the brain [7].

2.3 Spectral Entropy

The concept of spectral entropy is based on information theory. It is the technique of estimating how much information will be obtained from a signal after processing and was developed by Claude Shannon. Entropy is thus used to predict the behaviour of the signal. It is obtained by finding the probability of the PSD of the signal. Finding the entropy in frequency domain uses the power spectral density (PSD) of the signal. It is computed using the following formula:

$$SE = -\Sigma P\_\text{norm} \log(P\_\text{norm})$$

(P_norm is the probability of PSD). Lesser the SE more useful is the signal. If the SE of a signal is high, the randomness of the information from the signal increases [8].

2.4 Frontal Asymmetry Index

The left and right hemispheres of the brain play an important role in the cognitive functioning or the thinking process of humans. Whenever a task is performed there is frontal dominance; i.e. one hemisphere is more active than the other. To study each subjects stress reaction and to observe whether there exists a similarity in the brain activity of individuals who were less affected by the same amount of stress than the others frontal asymmetry index (FAI) was chosen as a parameter. A positive FAI indicates greater left hemisphere activity, whereas a negative FAI indicates a greater right hemisphere activity [9]. The frontal activity is calculated as:

$$\text{FAI} = Log((\text{power of } F3)/(\text{power of } F4))$$

3 Database Collection

Database was captured in Cummins College of Engineering. There were three female and five male participants in our study. All the participants were healthy individuals and were right-handed. The EEG was recorded using POWERLAB, and the feature extraction was done using MATLAB. Verbal consent was given by all the volunteers before recording EEG data and its use in the research.

4 Results

Below are the plots of a subject's EEG obtained from the frontal electrodes during Stroop task. Here, the spectral density and magnitude-squared coherence were measured against normalized frequency (*pi rad/sample). The plot of PSD below indicates at which frequency was the spectral density highest and the distribution of power across the EEG signal. The plot of coherence indicates at which frequency was the coherence between signals highest or lowest. Entropy of the signal was measured against time (minutes).

The power spectral density shows the distribution of signal power over frequency. Figure 3a shows the power spectral density estimate using Welch's method. Figure 3b shows spectral entropy of EEG. Figure 3c shows magnitude-squared coherence of EEG.

Coherence and asymmetry between FP2F4 and F4C4 channel are plotted by means of Box Plot for both the congruent and incongruent conditions as shown in Fig. 4a, c, respectively. Similarly, spectral density and entropy of these two channels are plotted for both the conditions as shown in Fig. 4b, d, respectively. The median value of coherence was lesser during incongruence. The spectral density of the signal from FP2F4 electrodes was higher than that from F4C4 electrodes during both congruence and incongruence.

Figure 5 shows empirical cumulative distributive function plot of coherence and asymmetry for both the congruent and incongruent conditions. The Stroop test we have designed using MATLAB which consists of different congruent and incongruent stages will give us whether the subject has answered correct or wrong, and it will also give us the reaction time of the subject for each stage. Figure 6 shows the box plot of reaction time for congruent and incongruent Stroop task.

5 Discussion

The main purpose of this research was to observe how are the electrical signals or brain functioning of people who are less affected when subjected to the same amount of stress different from those who show a higher impairment in their cognitive

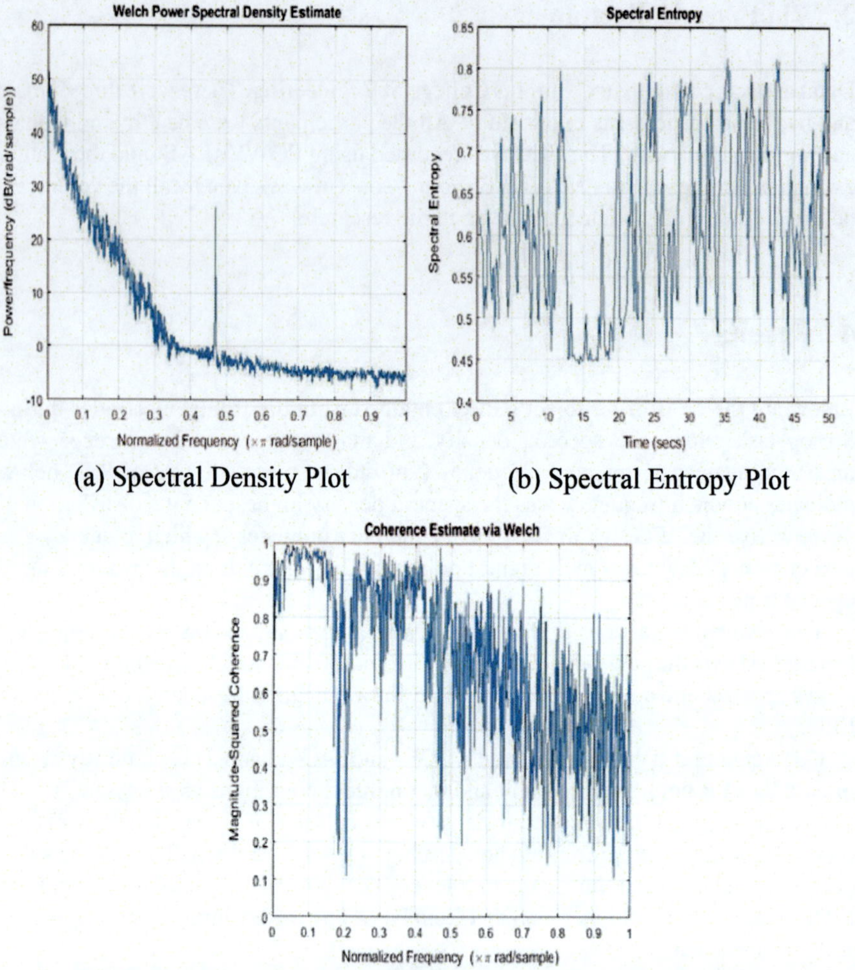

(a) Spectral Density Plot

(b) Spectral Entropy Plot

(c) Coherence Plot

Fig. 3 Plot of spectral density, entropy and coherence

functioning. For this research, Stroop task was chosen as a stressor. The basic concept of Stroop task is based on measuring the congruency and incongruency [10, 11]. The Stroop test used here is the colour word test but stress-related analysis is also being done using numerical Stroop test and Stroop test based on symbols. Stroop test was the only stressor used; however, we can add additional stress on the subjects by creating a noise disturbance or making them sleep for a lesser time the day before the Stroop test. We did not choose traditional methods because in traditional psychological methods of measuring mental stress, such as self-report questionnaires, real-time data is not available. The data available is taken after the event has already

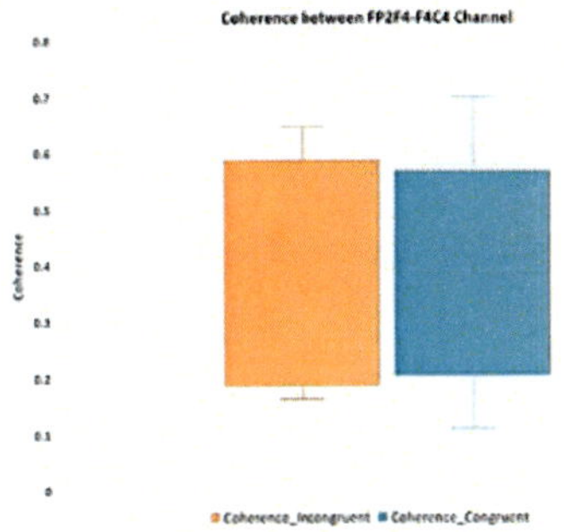

(a) Box Plot of Coherence

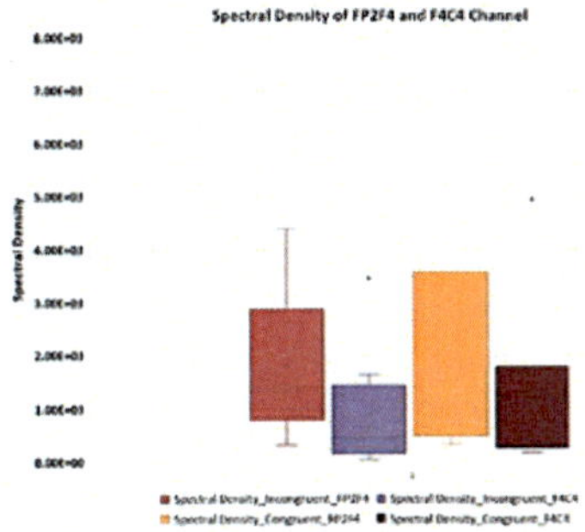

(b) Box Plot of Spectral Density

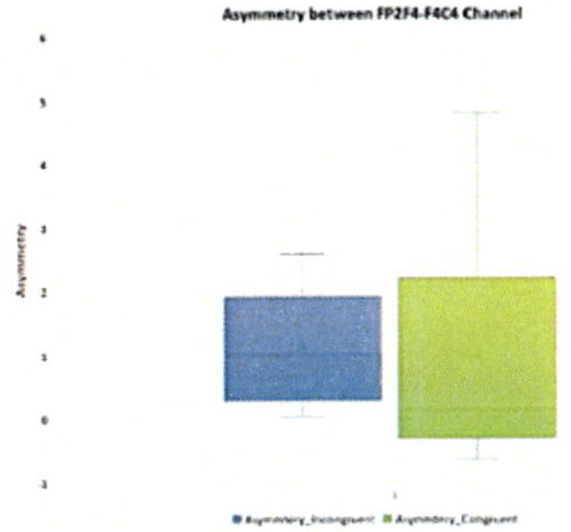

(c) Box Plot of Asymmetry

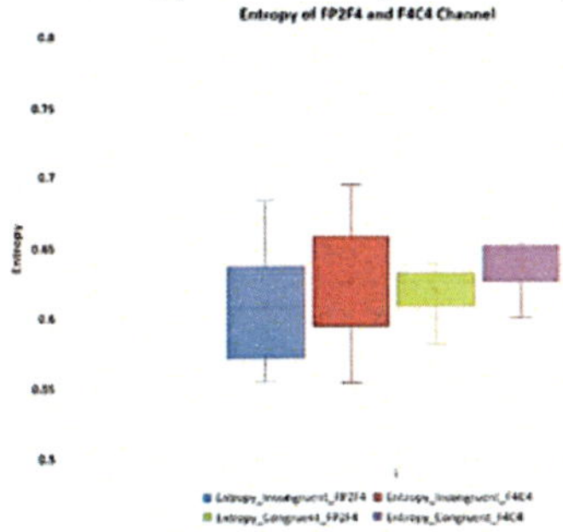

(d) Box Plot of Entropy

Fig. 4 Box plots of coherence, spectral density, asymmetry and entropy

Fig. 5 Empirical CDF plot of coherence and asymmetry

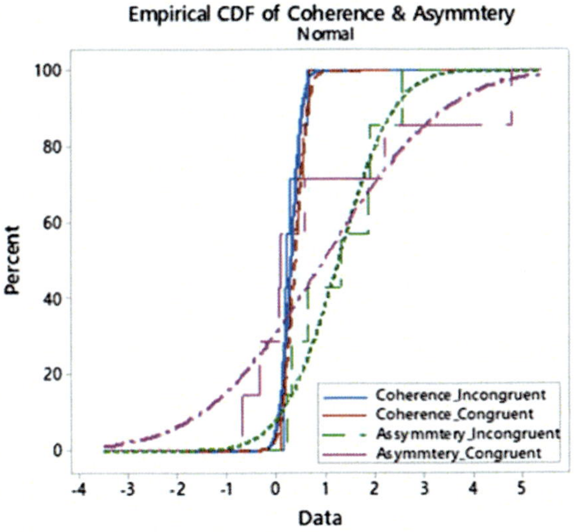

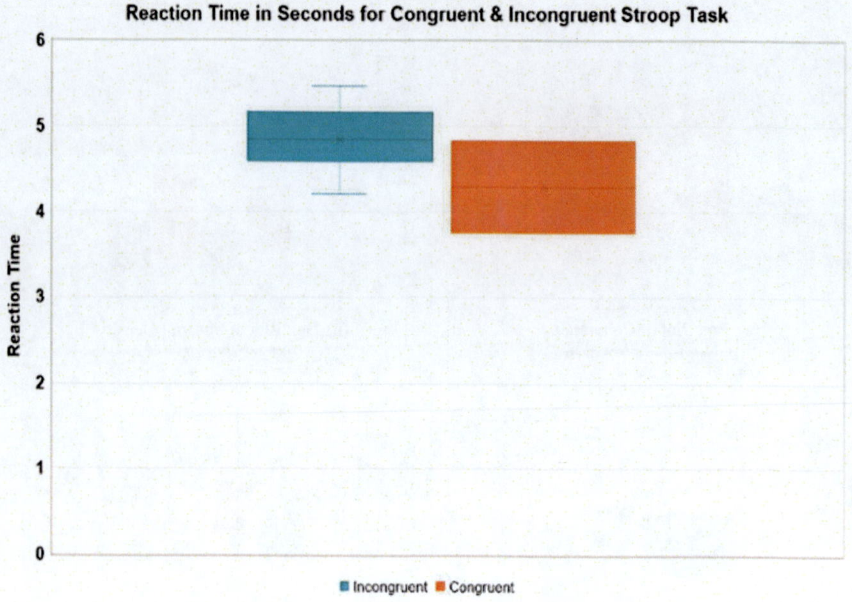

Fig. 6 Box plot of reaction time for congruent & incongruent task

occurred, and there are chances of faking the effect. On the other hand, real-time data is available in Stroop test which is helpful to efficiently discriminate between four stress levels, and hence, the Stroop test can be adopted to develop stress-level measurement device. The EEG of the subjects was captured; however ECG, PPG and voice analysis can also be used in stress analysis. Apart from the four feature analysis techniques used in this project, we can also use eigenvectors method if the signals have higher artifacts [12, 13]. Various parametric modelling techniques like autoregression techniques can also be used when the parameters are accurately known. Work on the use Stroop test in stress analysis is also being done for the detection of various mental health problems like ADHD, anxiety, schizophrenia and depression. Research is also being done on the use of Stroop test to assess patients with post-traumatic stress disorder (PTSD) as well as in the early detection of Alzheimer and idiopathic Parkinson's disease (iPD) [14].

6 Conclusion

The way we react to stress plays an important role in the outcome of our work. If care is not taken to overcome this stress, it can become chronic and causes mental health problems like anxiety, depression and insomnia. The results of our study showed that the reaction time was more in incongruent Stroop test than congruent test. This could be because, when there are multiple stimuli presented our brain needs longer time to process these, and when these stimuli are incongruent this time increases further. Some subjects were less affected by the increase in stress as compared to others. This indicates that some people despite being in a stressful situation are able to ignore the information that is not useful and focus on the task ahead. The functions of pre-frontal cortex of the brain are still poorly known. However, in our study the fronto-polar electrodes had highest spectral density. The results of coherence were low in pairs of FP2-F8 and F4-C4 and FP1-F3 and F3-C3. Considering that the spectral density here was also the highest with the pair containing frontal-polar electrode. This indicates that the majority of the activity is happening in the pre-frontal region of the brain, and it plays an important role in high, complex cognitive functioning. The subjects with a lower reaction time had a higher spectral density of fronto-polar electrode and less coherence; this suggests that reaction to stress is better when activity is in one region of the brain only. The entropy of the signal was in the same range for congruent and incongruent stages, and thus, no conclusive information could be obtained from the entropy readings. The high positive frontal asymmetry index value indicates that left hemisphere of the pre-frontal cortex is more active. Thus, the above findings state that pre-frontal cortex plays an important role in our reaction to stress. Further study can be done on why the brain functions in a different way for those who are more affected by stress, and those who are less affected. It could be due certain day-to-day habits and people who are highly affected by stress can adapt certain lifestyle changes in order to have better control on their thinking during stressful situations.

References

1. Alonso, J.F., Romero, S., Ballester, M.R., Antonijoan, R.M., Mañanas, M.A.: Stress Assessment Based on EEG Univariate Features and Functional Connectivity Measures, Biomedical Engineering Research Centre (CREB), Department of Automatic Control (ESAII), Universitat Politècnica de Catalunya (UPC). pp 1351–1365, Barcelona, Spain (2015). https://doi.org/10.1088/0967-3334/36/7
2. Shriram, R., Vijay Baskar, V., Martin, B., Sundhararajan, M., Daimiwal, N.: Statistical analysis of connectivity measures of electroencephalogram during congruent and ıncongruent stroop task. Int. J. Pure Appl. Math. **118**(17), 107–122, ISSN: 1311-8080 (2018)
3. Unde, S., Shriram, R.: Coherence analysis of EEG signal using power spectral density. In: IEEE Fourth International Conference on Communication Systems and Network Technologies CSNT 2014, Bhopal, India (2014)
4. Bekçi, B., Karakaş, S.: Perceptual conflict and response competition: event-related potentials of the Stroop effect. Turkish J. Psychiatry, Summer **20**(2), 127–137 (2009)
5. Shriram, R., Vijay Baskar, V., Martin, B., Sundhararajan, M., Daimiwal, N.: Connectivity analysis of brain signals during colour word reading interference. Biomedicine **38**(2), 229–243, (2018). ISSN: 0970-2067
6. Repovs, G.: The mode of response and the Stroop effect: a reaction time analysis. Horiz. Psychol. **13**(2), 105–114 (2004)
7. Vanitha, L., Suresh, G., Chandrasekar, Punita, P.: Development of four stress levels in group Stroop colour word test using HRV analysis. Biomed. Res. **28**(1), 98–10 (2017)
8. Al-Fahoum, A.S, Al-Fraihat, A.A.: Methods of EEG signal features extraction using linear analysis in frequency and time-frequency domains. ISRN Neurosci. p. 7, (2014). https://doi.org/10.1155-2014-730218

9. Unde, S., Shriram, R.: PSD based coherence analysis of EEG signals for Stroop task. Int. J. Comput. Appl. IJCA **95**, 50–55, (2014). ISSN: 0975-887
10. Hamid, N.H.A., Sulaiman, N., Aris, S.A.M., Murat, Z.H., Taib, M.N.: Evaluation of human stress using EEG power spectrum. IEEE (2010). https://doi.org/10.1109/CSPA.2010.5545282
11. Ignaccolo, M., Latka, M., West, B.J.: The dynamics of EEG entropy. J. Biol. Phys. pp. 185–196 (2010)
12. Deoskar, V., Kulkarni, P., Dhengade, T., Shriram, R., Bhat, M.: Workload Assessment based on Physiological Parameters. Springer Book Chapter in AISC Series, ISBN No: 978-981-15-1479-1 (2020)
13. Witelson, S.F.: Brain asymmetry, functional aspects. Speech and Language, pp. 14–18 (1988)
14. Cisler, J.M., Wolitzky-Taylor, K.B., Willems, J.L.: The emotional Stroop task and posttraumatic stress disorder: a meta-analysis. Clin. Psychol. Rev., pp. 817–28, (2011). https://doi.org/10.1016/j.cpr.2011.03.007

Autopilot Mode Transitions and Voter Logic Validation Using Model Checking: A Design Study of Formal Methods

Ravi Kiran and Yogananda Jeppu

1 Introduction

In control systems, Model-Based Design (MBD) with automatic code generation is a proven method and is now frequently used for software development which should satisfy the aerospace software standard DO-178C and its supplement DO-331 [1]. Due to the advantage of simulation, software verification and validation model-based development are used as a platform for design to implement advanced safety critical systems. Model-based development is used across the software development due to the complexity of advanced avionics systems. It is used from early requirement verification and implementation to system integration.

The design of reliable embedded safety-critical control system accedes to the difficulties involved in designing control system [2]. Design bugs in these systems may result in catastrophic consequences. In aerospace, the safety of flight depends on complex flight control system. Design bugs in these system results in failure of flight control system, and this might result in injury or bring in anxiety to the passenger before the recovery of controlled flight. If recovery of controlled flight is not established, it will lead to severe damages to aircraft structures or complete loss of the aircraft will lead to fatal or severe injury to passenger, etc. Inspired by the challenges in identifying this kind of design issues or defects, this article proposes the validation of the safety-critical control system using formal methods.

R. Kiran (✉)
VTU Extension Centre, Honeywell Technology Solutions, Bangalore, India

Y. Jeppu
Manipal Institute of Technology, Manipal, India

S. K V and K. Guruprasad Rao (eds.), *Smart Sensors Measurements and Instrumentation*, Lecture Notes in Electrical Engineering 750,
https://doi.org/10.1007/978-981-16-0336-5_22

Table 1 Transition matrix

States	S. No.	Modes	Buttons				Software triggers			
			01	02	03	04	05	06	07	08
			AP	SPD	VS	ALT	ALTS	ALT CAP	ALT CPDN	AP FAIL
Vertical	01	DIS {Vertical}	02	00	00	00	00	00	00	00
	02	PAH	01	03	04	05	00	06	00	01
	03	SPDHOLD	01	02	04	05	00	06	00	01
	04	VS	01	03	02	05	00	06	00	01
	05	ALTHOLD	01	03	04	02	00	00	00	01
	06	ALTSCAP	01	00	00	05	00	00	05	01
AP	01	APON	02	00	00	00	00	00	00	02
	02	APOFF	01	00	00	00	00	00	00	00
ALT SEL	01	ALTSOFF	00	00	00	00	02	00	00	00
	02	ALTSARM	01	00	00	01	01	03	00	01
	03	ALTSEL CAP	01	00	00	01	00	00	01	01

2 Formal Methods

"Formal methods" refers to the techniques that use mathematical reasoning about the actions of computer programs. These are mathematical methods that are used for the requirement, implementation and verification of software and hardware systems. The notation used in these methods is called a formal specification language.

The desired behavior of the system is articulated using formal specification. Formal specification uses mathematically based techniques. The syntax and semantics are defined formally to develop the requirements of a system. The formal language includes a syntax, which describes definite notation used for requirement depiction. The formal specification helps in understanding the system specification and the design. Using formal specification, it will be possible to demonstrate that a system confirms to its requirements or specifications.

2.1 Advantages of Formal Methods

The most important steps in the lifecycle of software development are the early activities, i.e., requirements analysis and specification. As per the Standish Chaos statement, most of the project failures happen because of the ambiguous and incomplete requirements specification. The major benefit of using formal methods can be realized by using formal methods at the requirement stages to develop a specification in formal language rather than writing in natural language and then translating it. It is efficient to analyze the formal specification in initial stages to identify ambiguity and incompleteness.

3 Autopilot

Autopilot is used to assist an aircraft operation with nominal or no aid from the pilot. It is a mechanical, electrical or hydraulic system. Sperry corporation designed the first autopilot in the year 1912. It was developed to increase the flight safety for passengers and reduce the airline cost by flying an aircraft with less consumption of fuel than human pilot. Modern autopilot includes complex components that can monitor the airplane's systems like hydraulics and air pressurization, detect and avoid accidents with other objects and can land the airplane in condition, where there is zero visibility landing (Category III, where human cannot see the runway). The high-level block diagram of the autopilot block diagram is described in Fig. 1.

The autopilot model is a kinematic model [2] implemented as second-order systems for autopilot altitude change and hold. The mode transition logic considered in this study is limited to flight level change/altitude change and altitude hold of vertical modes. The sensor model reads different parameters like roll angle, pitch

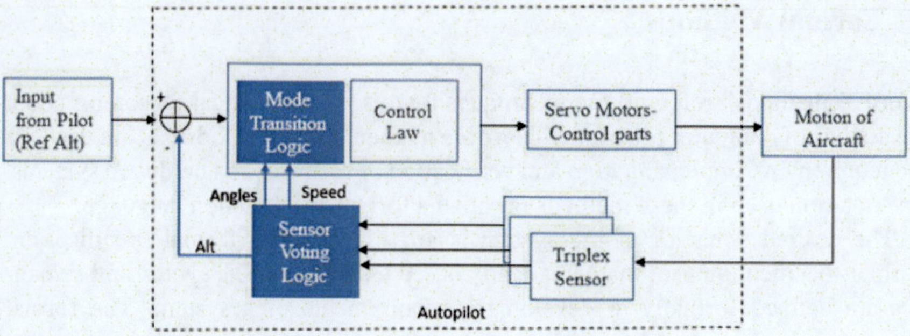

Fig. 1 Autopilot block diagram

angle, calibrated airspeed and altitude. The voter logic is used to perform voting on output. Voter logic output is a fed back to mode transition logic.

3.1 Mode Transition Logic

The autopilot mode transition has three modes: vertical modes, lateral modes and navigation modes as depicted in below. For this paper, only the vertical modes are considered. The vertical modes of an autopilot typically consists of Pitch Altitude Hold (PAH) mode, Altitude Hold (ALT HOLD) mode, Speed Hold (SPD HOLD) mode, Vertical Speed Hold (VS HOLD) mode and Altitude Select (ALT SEL) as shown in Fig. 2.

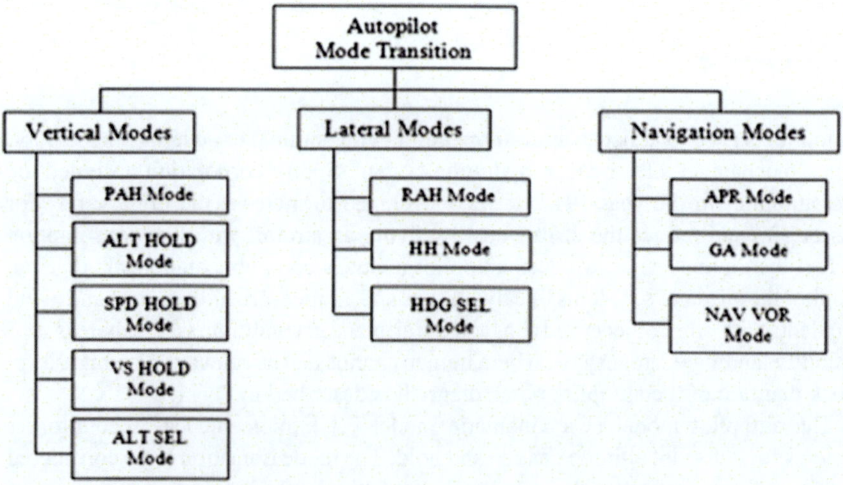

Fig. 2 Autopilot mode transitions [3]

- PAH mode: This is the basic autopilot mode in vertical axis. The present pitch angle of the airplane is maintained by PAH model.
- Altitude Hold mode (ALT HOLD): Aircraft is held at current altitude reference when this mode is selected.
- Speed Hold mode (SPD HOLD): When this mode is selected, it keeps the current airspeed but the altitude and pitch angle can vary.
- Vertical Speed mode (VS HOLD): When this mode is selected, it maintains the aircraft at a selected climb rate reference.
- Altitude Select mode (ALT SEL): When this mode is selected, the aircraft climbs to capture the selected altitude. On climbing to selected altitude, it transfers to ALT HOLD mode.

3.2 *Mode Transition and Condition Tables*

Transition to all these modes are based on pilot choice or internal mode transition triggers. This are described in Tables 1 and 2 as mode transition and condition Table [3]. There are three main modes: vertical, AP and ALTSEL. Vertical has six sub-modes: DIS (vertical), PAH, SPD HOLD, VS, ALT HOLD and ALTS CAP. AP mode has two sub-modes: AP ON and AP OFF. ALT SEL has three sub-modes: ALTS OFF, ALTS ARM and ALTSEL CAP. At start, the default sub-modes are DIS in vertical, AP ON in AP and ALTS OFF in ALT SEL. There are four knobs that pilot can use for selection. They are AP, SPD, VS and ALT. System can generate four triggers. They are Altitude Select (ALTS), Altitude Capture (ALTCAP), Altitude capture Done (ALTCPDN) and Autopilot has failed (APFAIL).

Each row in the table indicates current state in each mode and column indicates the trigger. For a transition to occur, the conditions defined by state {row} and trigger {col} in the condition table needs to be satisfied. The new state to which the autopilot will transition when the condition is satisfied is indicated in the transition table. There are five conditions based on which the transition happens. They are listed in Table 3.

4 Design of Mode Transition and Voter Logic for Autopilot

The autopilot with mode transition logic and voter logic is designed in two ways. The first one is by developing a Simulink model using Simulink blocks, and the other is by writing a C code for MTL and voter logic and then converting this to S-function blocks. MTL and voter logic S-function blocks are used in the Simulink model for simulation.

Table 2 Condition matrix

States	S. No.	Modes	Buttons				Software triggers			
			01	02	03	04	05	06	07	08
			AP	SPD	VS	ALT	ALTS	ALT CAP	ALT CPDN	AP FAIL
Vertical	01	DIS {Vertical}	01	00	00	00	00	00	00	00
	02	PAH	02	03	04	05	00	00	00	02
	03	SPDHOLD	02	00	04	05	00	00	00	02
	04	VS	02	03	00	05	00	00	00	02
	05	ALTHOLD	02	03	04	00	00	00	00	02
	06	ALTSCAP	02	00	00	05	00	00	00	02
AP	01	APON	00	00	00	00	00	00	00	00
	02	APOFF	01	00	00	00	00	00	00	00
ALT SEL	01	ALTSOFF	00	00	00	00	00	00	00	00
	02	ALTSARM	02	00	00	05	00	00	00	02
	03	ALTSEL CAP	02	00	00	05	00	00	00	02

Table 3 Condition table

Conditions	Details
C 1	No AP inhibit condition exist
C 2	Autopilot on
C 3	CAS is greater than 120 and less than Vmo (Knots)
C 4	{VS} is greater than 50 ft/min
C 5	{VS}is lesser than 500 ft/min

4.1 Design Using Simulink Blocks

The inhibit conditions are explained in Table 4.

Table 5 explains the autopilot failure conditions. If any of the below conditions are met, then APFAIL is set to true and autopilot is disconnected.

The mode transition logic is implemented as M script. For the simulation benefits, the mode transition logic is imported to Simulink environment by using the MATLAB function block. The conditions logic and MATLAB function for mode transition logic is stitched together. The inputs to M-script are fed from the Simulink. The verification subsystem developed is shown in Fig. 3.

The parameters of interest to vote out the faulty sensor are as follows: roll, pitch, speed, vertical speed and altitude. The voter logic is implemented using two algorithms: priority-based voting [4] and validity-based Voting [5].

Table 4 Inhibit table

Parameter	Threshold [Min/Max]
Roll (Φ)	[−5 5]
Pitch (θ)	[−5 5]
Speed	[350 450]
Vertical speed	[200 300]
Altitude	[900 1100]

Table 5 AP fail table

Parameter	Threshold [Min/Max]
Roll (Φ)	$\Phi \leq 12 \parallel \Phi > 15$
Pitch (θ)	$\theta \leq 12 \parallel \theta > 15$
Speed (S)	$S < 350 \parallel S > 600$
Vertical speed (Vs)	$Vs < 150 \parallel Vs > 350$
Altitude (Alt)	Alt < 800 ǁ Alt > 1200

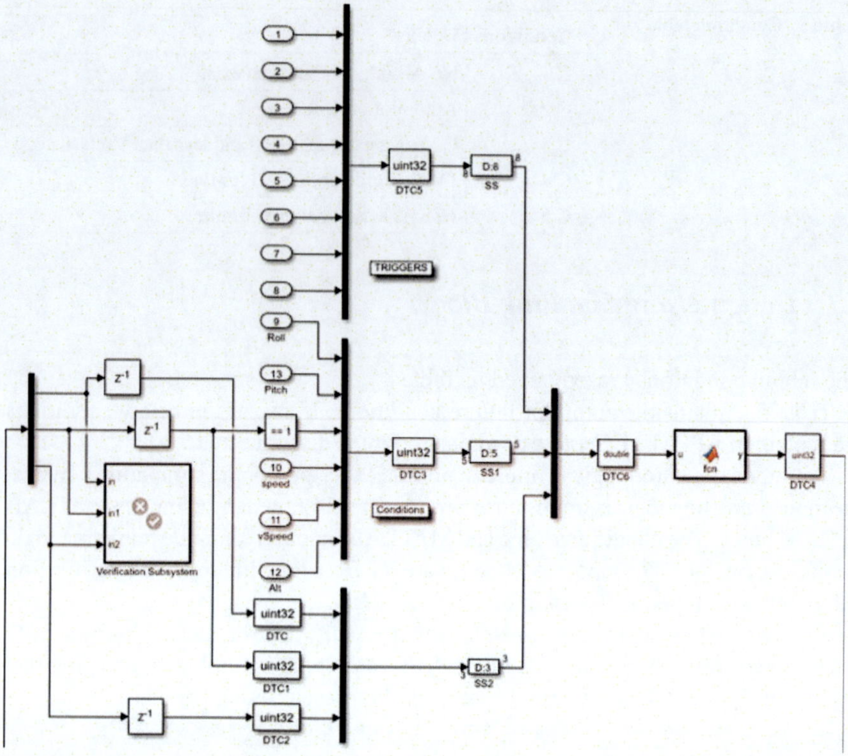

Fig. 3 Autopilot with MTL using Simulink

4.2 *Voter Logic Design*

Priority Based Voting

Priority-based voter logic contains three subsystems to determine the median, priority and selecting an output signal (Figs. 4, 5 and 6).

The median subsystem (Fig. 5) determines the median, maximum and minimum value.

The priority subsystem (Fig. 6) determines the priority. The highest priority is the median value. The signal which has minimum distance from median value gets the second priority, and the signal which has maximum distance from median value gets the least priority.

The output signal is voted out based on the priority. There are four cases.

Case I: The distance between the highest priority and the second sorted value does not exceed selected boundary value; the output will be the first arranged value as it has higher priority and has a vote from the second sorted value.

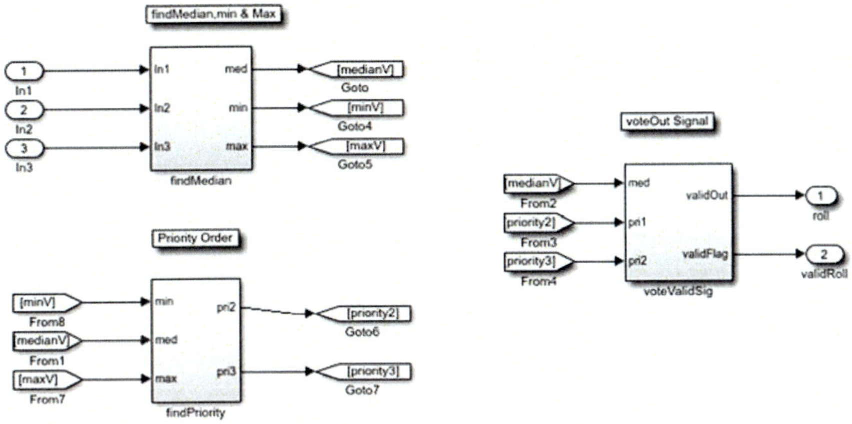

Fig. 4 Model of priority-based voting

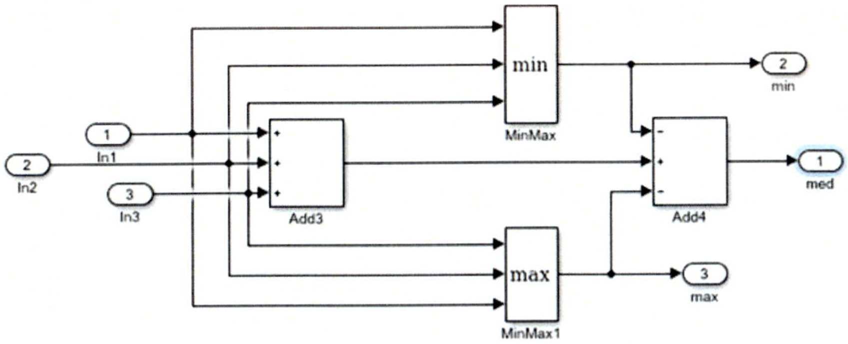

Fig. 5 A model of median subsystem

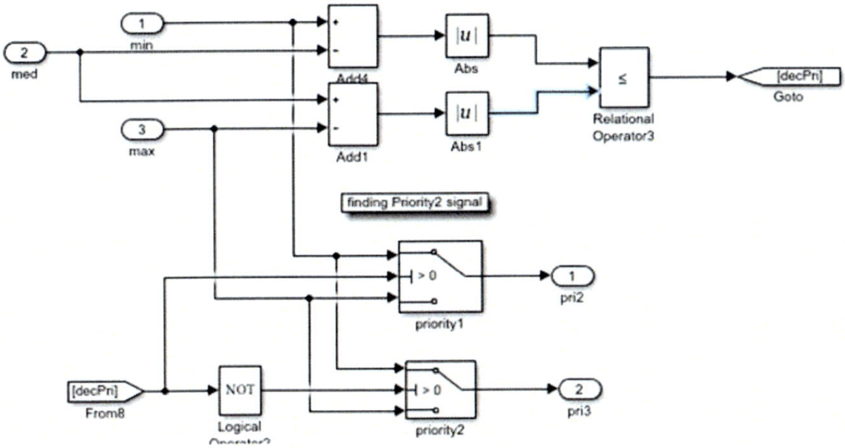

Fig. 6 Model of priority subsystem

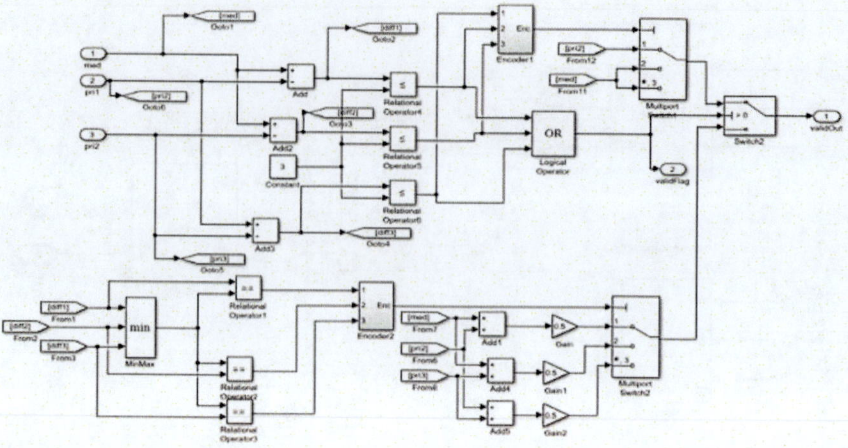

Fig. 7 VoteValidSig subsystem

Case II: If Case I is not true and the difference between first and third sorted value does not exceed the selected limit, then the output value will be the first sorted value as it has higher priority and has a vote from the third sorted value.
Case III: If above two cases are not true and the difference between second and third sorted value does not exceed the selected limit, then the output value will be the second sorted value as it is highest priority than the third value and has a vote from the third sorted value.
Case IV: If all the above three cases are not true, then the output value will be the average between two values that have minimum pair distance.

The validFlag will be set to true for case 1, case 2 and case 3. For case 4, the validFlag is set to false. The subsystem developed to vote the output is shown in Fig. 7.

Validity-Based Voting

Figure 8 describes the validity-based voting logic. Sensor1, sensor2 and sensor3 are the values from three sensors, and sensor1_valid, sensor2_valid and sensor3_valid are the valid signals. Triple sensor (three sensors valid), dual sensor (two sensors valid) and single sensor (only one sensor valid is implemented as independent subsystems.

4.3 Design Using S-function Blocks

The autopilot functionality is split into two functions: mode transition logic and voter logic. The C code for mode transition logic and voter logic is manually generated. C code cannot be simulated as is. The C code is modified such that it is compatible with the S-function builder which generates MATLAB executable file. The S-function

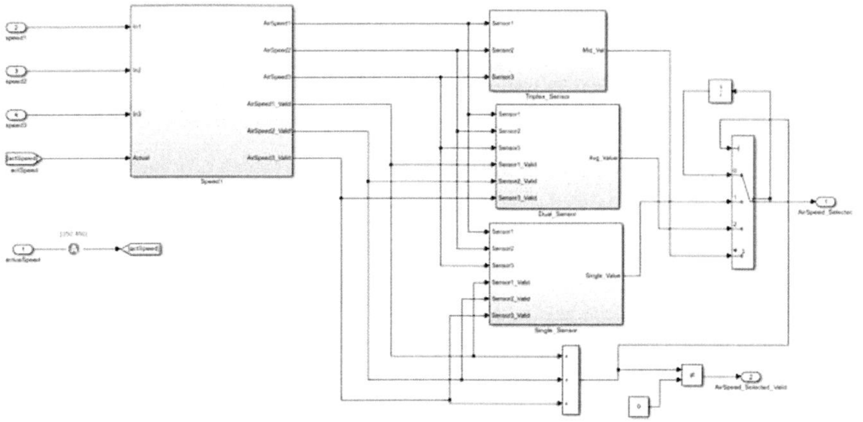

Fig. 8 Validity-based voting

block integrates all the source and header files to generate a C MEX S-function from specifications provided. The S-function builder block generates all the necessary files for model simulation and code generation. S-function builder generates a mex file, a wrapper function and tlc file which is used for simulation and code generation.

The S-function vter logic is built for all three approaches. As the voter logic is same for all the parameters, the source code is made generic, and hence same S-function can be used for different parameters. The S-function logic for different parameters to be voted out, and the S-function logic for MTL is stitched together to represent an autopilot with mode transition and voter logic as shown in Fig. 9.

5 Simulink Design Verifier

Autopilot model can be formally verified by using model checking tools. One such tool is Simulink Design Verifier. Simulink Design Verifier (SlDV) is a toolbox for Simulink which is intended for identifying design error and formal verification. The first version of this tool was released by MathWorks in 2007. Simulink Design Verifier can be used for both property proving and test case generation.

5.1 Formal Verification of Voter Logic

The validity-based voting logic discussed above is used for the formal verification. The sensors are duplicated with the intention of increasing reliability of the system. Each sensor can have different random noise. For verification, we have an actual signal and noise is added to actual signal. For example, radar altimeter is used to

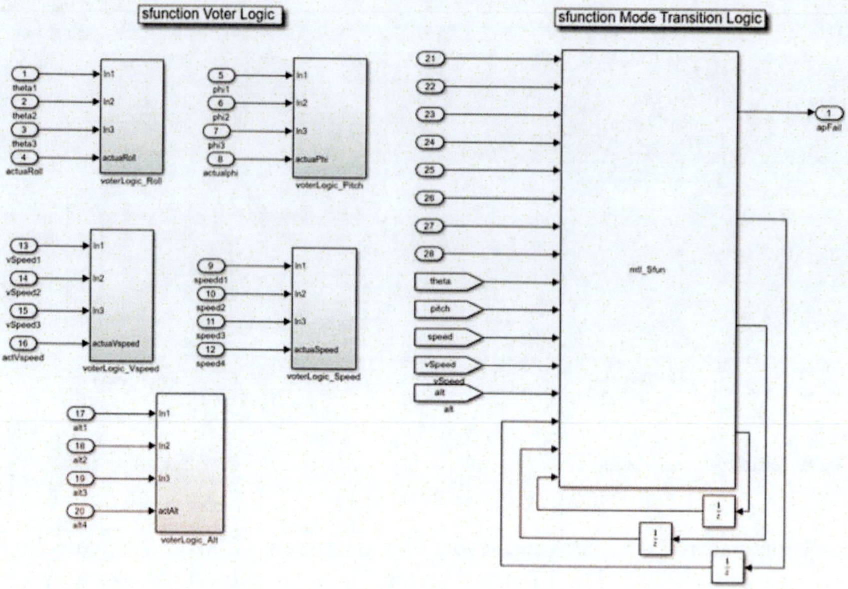

Fig. 9 S-function implementation of MTL and voter logic of autopilot

measure the height above ground level. The actual signal, which is actual altitude, represents the physical altitude in which the aircraft is flying and sensors which measures the altitude is bounded with some errors. These are modeled as error bound on three different sensors, and the error gets added to the actual sensor. The altitude sensors are bounded between $\{-100\ 100\}+$ actual altitude $\{900\ 1100\}$. The output of altitude sensor will be between $\{800\ 1200\}$. Each sensor has a validity signal which represents health of the signal. If the difference between the actual value and the sensor value is beyond threshold, the validity signal is set to false indicating that the sensor value is unhealthy. The subsystem developed for sensor with noise is shown in Fig. 10.

Whenever the voter logic votes out a valid value, the voted value should be within the allowed difference. The verification system is developed to verify the correctness of the system. The truth of the system is difference between the actual signal and the voted signal should be less than or equal to the allowed error. The Simulink Design Verifier engine tries to falsify the truth of the system. If the SlDV can falsify, it generates a counter example for analysis.

The verification system developed is shown in Fig. 11.

Similar verification subsystem with similar logic is developed for each parameter explains the allowed difference between the Actual and Voted values for each parameter (Table 6).

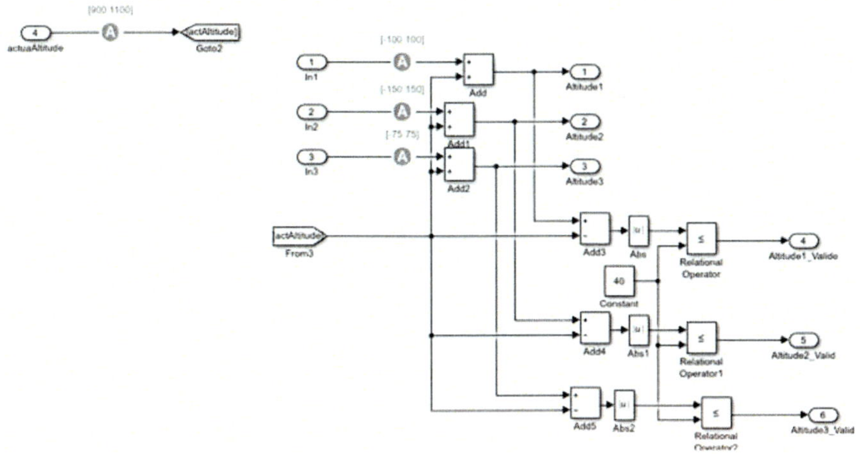

Fig. 10 Sensors with noise

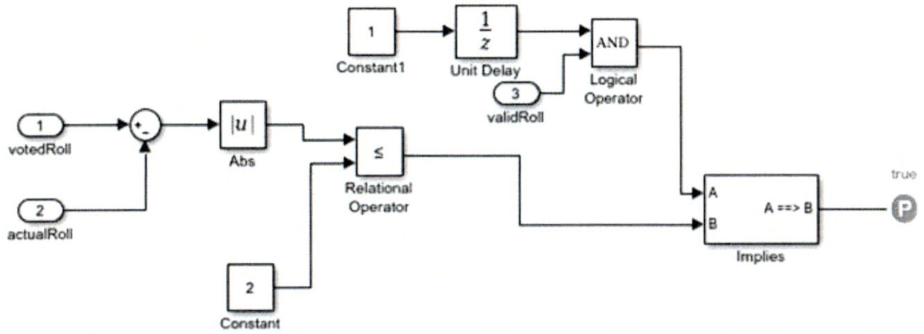

Fig. 11 Verification subsystem for voter logic

Table 6 VotedSig-ActualSig

Parameter	Abs (votedSignal-ActualSignal)
Roll	2
Pitch	2
Speed	50
Vertical speed	50
Altitude	100

5.2 *Verification of Validity Based Voting*

The voter logic with validity-based logic is stitched with the verification subsystem. The goto and from blocks are used to send output from voter logic subsystem to verification subsystem. The stitched model is as shown in Fig. 12.

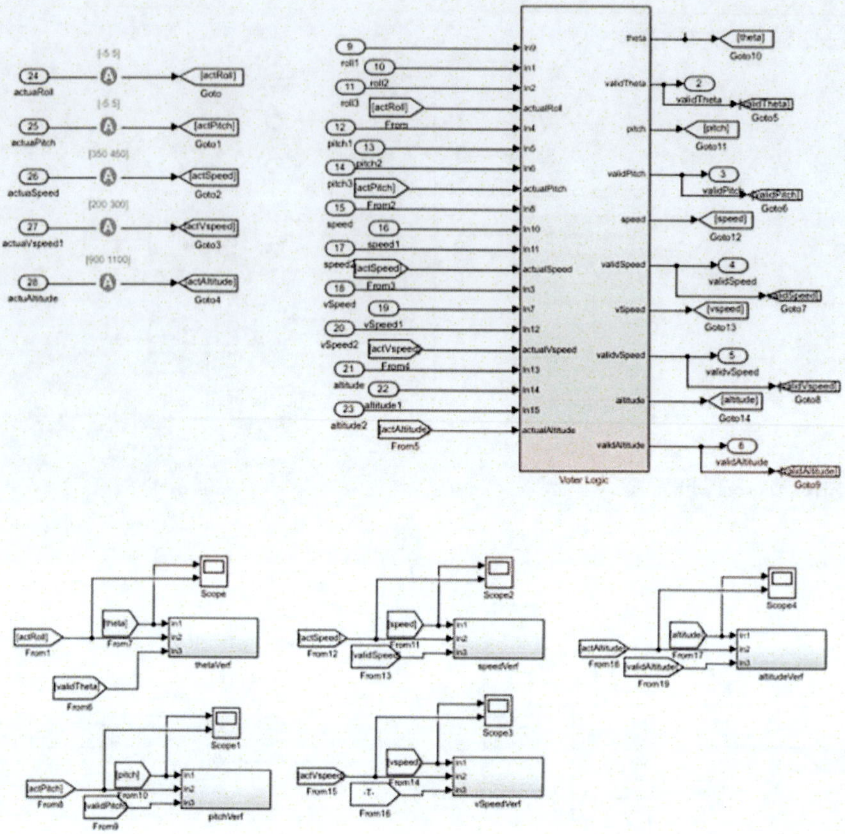

Fig. 12 Voter logic with verification subsystem

The truth of the system is verified using Simulink Design Verifier. It tries to prove the truth of system. If the truth of system is falsified, it generates the counter example. Figure 13 shows the output of property proving.

All five properties specified in the model are proven valid. If any of the objectives are falsified, Simulink Design Verifier allows user to generate the detailed analysis report, or harness model can be generated to analyze the results. Clicking on generate detailed analysis report, Simulink Design Verifier generates report with objectives satisfied and generates counter examples for the objectives which Simulink design verifier provides an option to generate the detailed analysis report. The analysis report can be generated into a pdf or html. The analysis report contains proof objectives and details of objectives that were proven valid or falsified. If falsified, it contains counter example. The output recorded in result file is shown in Fig. 14.

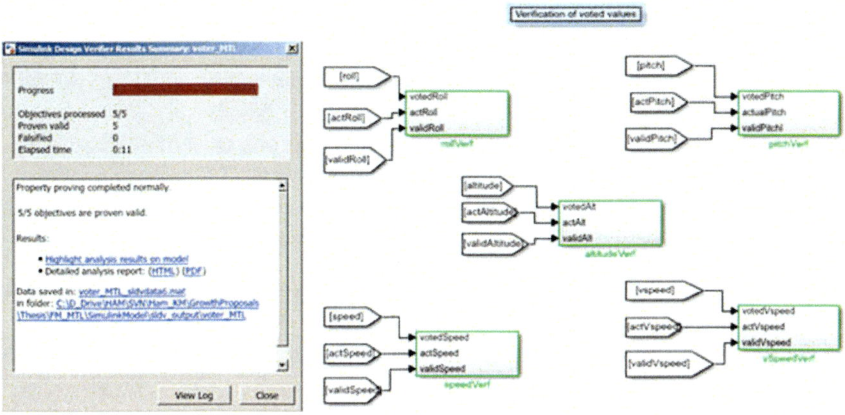

Fig. 13 Property proving output

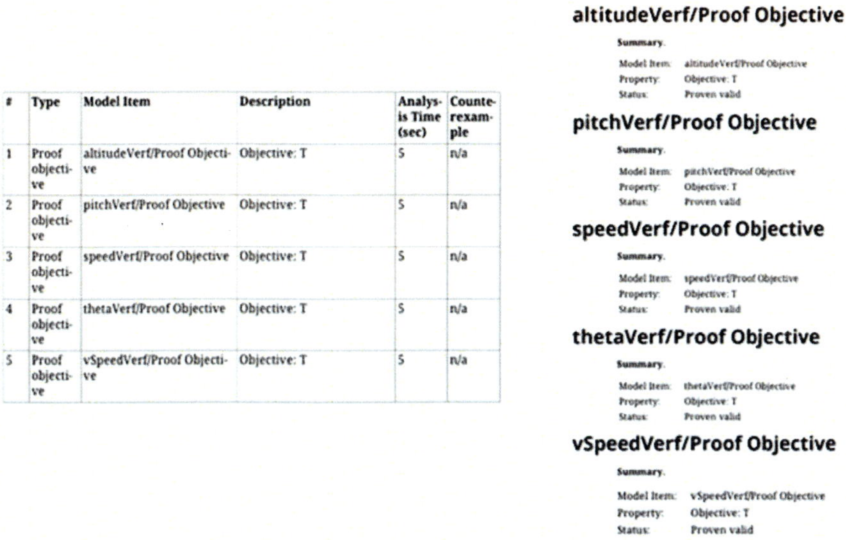

#	Type	Model Item	Description	Analysis Time (sec)	Counterexample
1	Proof objective	altitudeVerf/Proof Objective	Objective: T	5	n/a
2	Proof objective	pitchVerf/Proof Objective	Objective: T	5	n/a
3	Proof objective	speedVerf/Proof Objective	Objective: T	5	n/a
4	Proof objective	thetaVerf/Proof Objective	Objective: T	5	n/a
5	Proof objective	vSpeedVerf/Proof Objective	Objective: T	5	n/a

altitudeVerf/Proof Objective
Summary
Model Item: altitudeVerf/Proof Objective
Property: Objective: T
Status: Proven valid

pitchVerf/Proof Objective
Summary
Model Item: pitchVerf/Proof Objective
Property: Objective: T
Status: Proven valid

speedVerf/Proof Objective
Summary
Model Item: speedVerf/Proof Objective
Property: Objective: T
Status: Proven valid

thetaVerf/Proof Objective
Summary
Model Item: thetaVerf/Proof Objective
Property: Objective: T
Status: Proven valid

vSpeedVerf/Proof Objective
Summary
Model Item: vSpeedVerf/Proof Objective
Property: Objective: T
Status: Proven valid

Fig. 14 Detailed report generated from the tool

5.3 *Formal Verification of Autopilot Logic*

Two generic specifications are developed to verify the mode transition logic of autopilot using Simulink Design Verifier. The values of the constant block can be changed to verify other transition modes.

1. The verification system is developed to add assertions to verify "when vertical speed is in ALT Hold, altitude select will transition to ALTSOFF." The specification is developed using the Implies block. Whenever V is equal to 5 AS will be equal to 1.
2. The verification system is developed to add assertions to verify "when AP is in APOFF, vertical speed will transition to DIS state." The specification is developed using the Implies block. Whenever AP is equal to 2 V will be equal to 1. The verification subsystem developed is shown in Fig. 15.

Figure 16 is the result from the Simulink Design Verifier for the specifications defined above. One objective is proven valid and other is falsified.

Objective Proven Valid: When AP is in APOFF, vertical speed will transition to DIS state.

Objective Falsified: When vertical speed is in ALT Hold, altitude select will transition to ALTSOFF.

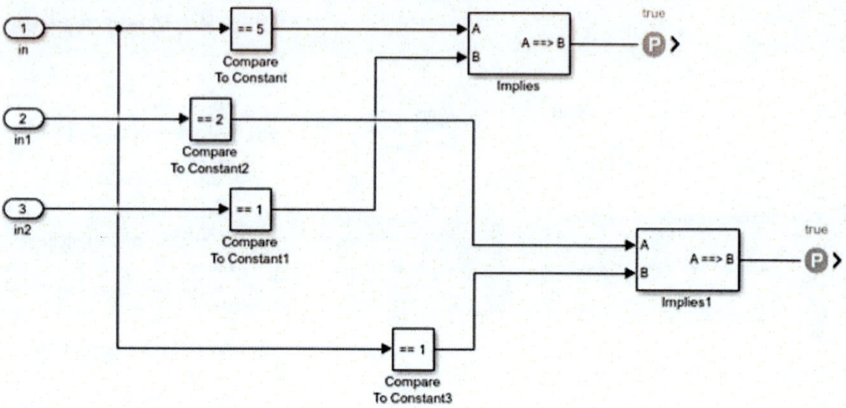

Fig. 15 Verification of MTL

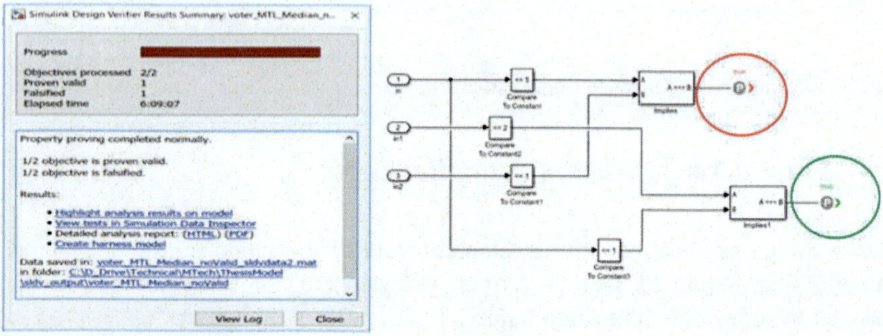

Fig. 16 SlDV result for MTL

The green color on proof block shown in Fig. 16 indicates that objective is satisfied, and red color indicates objective is falsified. The falsified objective can be analyzed by generating the harness model. The generated harness model contains the counter example test vectors, and step-by-step execution can be performed to analyze the falsified objectives.

5.4 Analysis of Falsified Objective

The counter example generated by the Simulink Design Verifier to falsify the objective is given in Table 7.

The harness model is simulated to analyze the simulation behavior. The values of the V, AP and AS when the objective was falsified are highlighted in the model as shown in Fig. 17.

The analysis of system behavior reveals the following missing requirements.

1. ALTCPD can occur only if the mode V is in ALTCAP.
2. ALTS can occur only when mode V is not in ALTHLD.
3. ALTCAP can occur only when AS is in ALTARM (Armed).

Thus, formal methods help in proving the correctness of the system and help in identifying the requirement ambiguities and completeness. The above identified requirement holes are added as part of the high-level requirements. The design is updated to cover the missing requirements. The new design is validated against all the assertions.

Table 7 Counter example

Time/Inputs	0 s	0.01 s	0.02 s
Step	1	2	3
In1	0	0	0
In2	1	0	0
In3	0	0	0
In4	1	0	0
In5	1	1	0
In6	0	0	1
In7	0	0	1
In8	1	1	1
In9	1	1	0
In10	1	1	1
In11	1	1	1
In12	1	1	1

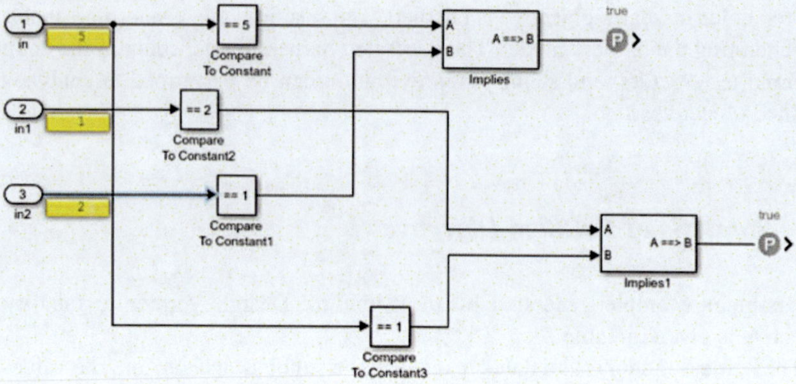

Fig. 17 Analysis of falsified objective

6 Conclusion

Modeling tools like Matlab/Simulink, Scade and XCos are used to represent system requirements. Mode awareness is considered as most important safety concern by several researchers, the FAA and the aviation industry. Finding ways to detect and mitigate potential sources of mode confusions remains a major challenge. This article describes a mathematical framework in which requirements and design can be analyzed for potential sources of mode confusion and incorrect voting using Simulink Design Verifier.

As part of the article, the implementation of mode transition and voter logic using Simulink and S-functions are discussed. The verification subsystem is developed using Simulink Design Verifier to ensure that design satisfies the requirement specification. The autopilot mode transition and voter logic are formally verified using Simulink Design Verifier. All the objectives specified for the voter logic verification subsystems have been satisfied by the Simulink Design Verifier. One of the objectives specified for the mode transition logic has been falsified, and a harness model has been generated. The generated simulation test case for debug is analyzed, and missing requirements are revealed during design phase.

With the above study and observations made, we can conclude that adopting formal methods ensures that the development of a system satisfies the requirement specification and it help in identifying ambiguous, incomplete and inconsistent requirements. Design defects are identified in design phase itself, thereby reducing the cycle time.

6.1 Future Scope

- Future work includes representation of safety requirements in Easy Approach to Requirements Syntax (EARS) notation. This will help in automating the assertion generation for system specifications.
- Extending autopilot formal verification for lateral and guidance modes.
- Integrating autopilot system along with flight management guidance system. The flight management system output can be used to generate software triggers to direct the mode transition logic, and hence the autopilot system to generate the desired state of operation to achieve desired navigation can be formally verified.
- Detailed study of DO333, a formal methods supplement to DO-178C and DO-278A and understand objectives, activities and software life cycle data that should be addressed when formal methods are used as part of the software life cycle.
- Qualify the Simulink Design Verifier tool to provide a means of verification evidence per DO333/DO178C.

References

1. Ashutosh, J., Prakash, C.: Adopting model-based software design and verification for aerospace systems, In: Aero Tech Congress & Exhibition. SAE International (2017)
2. Lukas, M., Maxim, S., Matthias, H.: A methodology for model-based development and automated verification of software for aerospace systems. In: Aerospace Conference. IEEE, New York (2013)
3. Manjunatha, R.: Design, Implementation, Verification and Validation of autopilot Mode Transition Logic for a generic aircraft, thesis from BITS, Pilani and carried out at MOOG India Technology Center Pvt. Ltd, Bangalore (2013)
4. Kassab, M.A., Taha, H.S., Shedied, S.A., Maher, A.: A novel voting algorithm for redundant aircraft sensors. In: IEEE Conference at Shenyang, China (2014)
5. Pat, C.: Why You Should Collect Model Test Coverage Metrics, MATLAB Digest - Aerospace and Defense - November 2019, MathWorks

Fuzzy Cluster based PID Controller for a CSTR Process

B. Murugan, R. Vinodha, and B. Shoba

1 Introduction

PID controllers almost rule the process industries from 1940s, when Taylor and Foxboro instrument companies introduced their PID controllers. PID controllers gained their importance in process industries due to their ease of structure and implementation. However, the performance of these linear controllers is dominant only in the operating region, where they are tuned. Their performance degrades at shifted operating regime. To improve control performance of the PID controller suitable to nonlinear process, the controller parameters need to be changed for variation in operating point.

Multiple model-based controller design has gained attention for effective control of nonlinear processes [1–4]. Control of Continuous Stirred Tank Reactor (CSTR) units in the process industries is a major task, which needs simple, but effective and economical control algorithms. The chosen CSTR process [5] has a second-order dynamics with varying damping factor and undamped natural frequency. The literature survey states several control strategies like neural adaptive controller [6], fuzzy model predictive controller [7], observer-based model predictive controller [8], nonlinear PID controller, nonlinear model predictive controller [9], neural-based PID controller [10] and nonlinear predictive control using fuzzy Hammerstein model [11] on the chosen CSTR process. Despite these control schemes, yet

B. Murugan
Department of Electrical and Electronics Engineering, RAAK College of Engineering & Technology, Puducherry, India

R. Vinodha (✉)
Department of Electronics and Instrumentation Engineering, Annamalai University, Annamalai Nagar, Tamil Nadu, India

B. Shoba
Department of Electronics and Communication Engineering, Rajiv Gandhi College of Engineering & Technology, Puducherry, India

S. K V and K. Guruprasad Rao (eds.), *Smart Sensors Measurements and Instrumentation*, Lecture Notes in Electrical Engineering 750,
https://doi.org/10.1007/978-981-16-0336-5_23

another control strategy using PID controller whose parameters are scheduled by fuzzy c-means clustering approach has been attempted in the paper.

Fuzzy logic has gained attention in complex industrial processes, as it is a knowledge-based control approach. FCM is a clustering algorithm [12, 13], where each data point may belong to more than one cluster with a certain degree of membership value. Hence, clustering approach is proposed to group the nearby operating points in a wide operating range of the CSTR process and distribute the controller parameters to the PID controller sensibly.

The main contribution of the paper is the design of Fuzzy Inference System (FIS) using fuzzy c-means clustering by extracting a set of rules that models the input–output data behavior. The FIS then schedules the parameters of the PID controller according to the concentration variation of CSTR process.

The organization of the paper is as follows: Sect. 2 details the CSTR process model, Sect.3 discusses FCM algorithm, and Sect. 4 states cluster-based scheduler design to schedule PID parameters of the controller. Simulation results are shown in Sect. 5, and conclusion is given in Sect. 6.

2 CSTR Process

Consider the CSTR process model shown in Fig. 1 with an irreversible exothermic reaction to convert reactant A of concentration C_{A0} to product B of concentration C_A. The reactor volume is constant, and the reactor is cooled by a single coolant stream. The differential equation model of the CSTR process is given in Eqs. (1) and (2) with steady state operating point data mentioned in Table 1.

$$\frac{\mathrm{d}C_A(t)}{\mathrm{d}t} = \frac{q(t)}{V}(C_{A0}(t) - C_A(t)) - k_0 C_A(t)\exp\left(\frac{-E}{RT(t)}\right). \quad (1)$$

Fig. 1 CSTR process

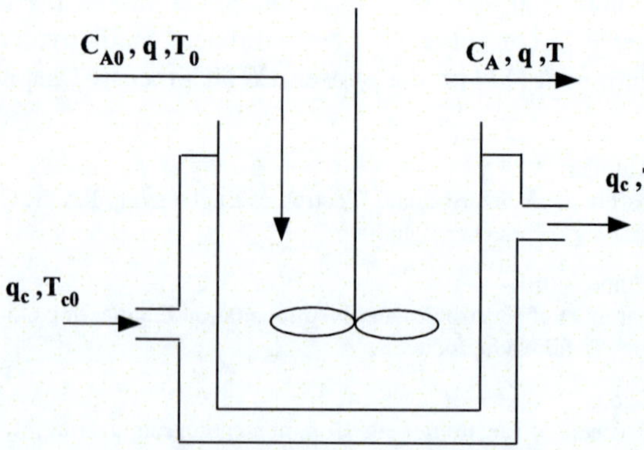

Table 1 Steady-state operating data of CSTR process

Process variable	Normal operating condition
Measured product concentration (C_A)	0.0989 mol/l
Reactor temperature (T)	438.7763 K
Coolant flow rate (q_c)	103 l/min
Process flow rate (q)	100.0 l/min
Feed concentration (C_{A0})	1 mol/l
Feed temperature (T_0)	350.0 K
Inlet coolant temperature (T_{C0})	350.0 K
CSTR volume (V)	100 l
Heat transfer term (hA)	7×10^5 cal/(min k)
Reaction rate constant (k_0)	7.2×10^{10} min^{-1}
Activation energy term (E/R)	1×10^4 K
Heat of reaction (ΔH)	-2×10^5 cal/mol
Liquid density (ρ, ρ_c)	1×10^3 g/l
Specific heats (C_p, C_{pc})	1 cal/(g k)

$$\frac{\mathrm{d}T(t)}{\mathrm{d}t} = \frac{q(t)}{V}(T_0(t) - T(t)) - \frac{(-\Delta H)k_0 C_A(t)}{\rho C_P}\exp\left(\frac{-E}{RT(t)}\right) + \frac{\rho_c C_{pc}}{\rho C_p V} q_c(t) * \left\{1 - \exp\left(\frac{-hA}{q_c(t)\rho C_p}\right)\right\} * (T_{c0}(t) - T(t)). \quad (2)$$

The CSTR model in Eqs. (1) and (2) can be written as

$$\dot{x}(t) = f[x(t), u(t), d(t)]. \quad (3)$$

$$y(t) = g[x(t), u(t), d(t)]. \quad (4)$$

where $x(t) = \begin{bmatrix} C_A(t) \\ T(t) \end{bmatrix}$; $u(t) = q_c(t)$; $y(t) = C_A(t)$.

Converting Eqs. (3) and (4) to discrete form gives Eqs. (5) and (6)

$$x(kT) = f[x(kT - T), u(kT - T), d(kT - T)]. \quad (5)$$

$$y(kT) = g[x(kT - T), u(kT - T), d(kT - T)]. \quad (6)$$

where $x(kT)$, $u(kT)$, $d(kT)$ and $y(kT)$ are defined as state, input, unknown input/disturbance and output vectors, respectively. Functions f and g are n dimensional function vectors, and T is the sampling time. Table 2 shows state space matrices

Table 2 State space model at nominal operating points of the CSTR process

Operating point	State space matrix
At $\overline{q}_c = 103$ $\overline{C}_A = 0.0989$ $\overline{T} = 438.7763$	$A = \begin{bmatrix} -10.2 & -0.0467 \\ 1833.6 & 7.3 \end{bmatrix} B = \begin{bmatrix} 0 \\ -0.8813 \end{bmatrix} C = \begin{bmatrix} 1 & 0 \end{bmatrix} D = [0][0]$

A, *B*, *C* and *D* obtained by linearizing the differential Eqs. (1) and (2) around nominal operating point of the CSTR process.

3 FCM Algorithm

Fuzzy c-means (FCM) clustering approach was structured originally by Dunn [14] and late formulated by Bezdek [15]. The algorithm classifies or groups each data point to a part of multiple clusters dominated by their centers with varying degrees of membership. It is based on minimizing the objective function given in Eq. (7).

$$J_{\min} = \sum_{i=1}^{M} \sum_{j=1}^{N} \mu_{ij}^{m} \left\| x_i - c_j \right\|^2. \tag{7}$$

In Eq. (7), M in the number of data points, N is the number of clusters, m is the weighting component with $m > 1$, determining the fuzziness of the clusters, x_i is the ith data point, c_j is the center of the jth cluster, and μ_{ij} is the degree of membership of x_i in the jth cluster. For a given data point $x_{i,}$ the sum of membership values for all clusters is equal to one.

The clustering is carried out by an iterative optimization of the objective function given in Eq. (7) with an update of membership function μ_{ij} and cluster c_j in Eqs. (8) and (9), respectively.

$$\mu_{ij} = \frac{1}{\sum_{k=1}^{N} \left[\frac{\|x_i - c_j\|}{\|x_i - c_k\|} \right]^{\frac{2}{m-1}}}. \tag{8}$$

$$c_j = \frac{\sum_{i=1}^{M} \mu_{ij}^{m} x_i}{\sum_{i=1}^{M} \mu_{ij}^{m}}. \tag{9}$$

The iteration is terminated based on the tolerance ε $(0 < \varepsilon < 1)$ satisfying Eq. (10)

$$\max_{ij} \left\{ \left| \mu_{ij}^{(k+1)} - \mu_{ij}^{(k)} \right| \right\} < \varepsilon. \tag{10}$$

where k denotes iteration steps.

4 Fuzzy Cluster-Based Scheduler

The objective of the proposed work is to control the product concentration C_A of the process by manipulating the coolant flow rate q_c. PID controller tuned by IMC is used to control the concentration variable. The control algorithm utilizes velocity form as given in Eq. (11)

$$u(k) = k_c(e(k) - e(k-1)) + \frac{k_c T}{T_i} e(k) + \frac{k_c T_d}{T}(e(k) - 2e(k-1) \\ + e(k-2)) + u(k-1) \tag{11}$$

Since the linear model of the chosen CSTR process takes the second order form, the tuning parameters [16] given in Eq. (12) are used in the control of Eq. (11).

$$k_c = \frac{2\xi}{\omega_n k_p \lambda};\ T_I = \frac{2\xi}{\omega_n};\ T_d = \frac{1}{2\xi\omega_n}. \tag{12}$$

The block diagram of the proposed cluster-based scheduler is shown in Fig. 2. The block diagram states that the change in concentration variable of the dynamic CSTR process changes at every sampling instant, and the fuzzy cluster-based scheduler will schedule the tuning parameters k_c, T_i and T_d using fuzzy c-means clustering approach. As a result, the linear PID controller adapts for effective control of shifted operating regime.

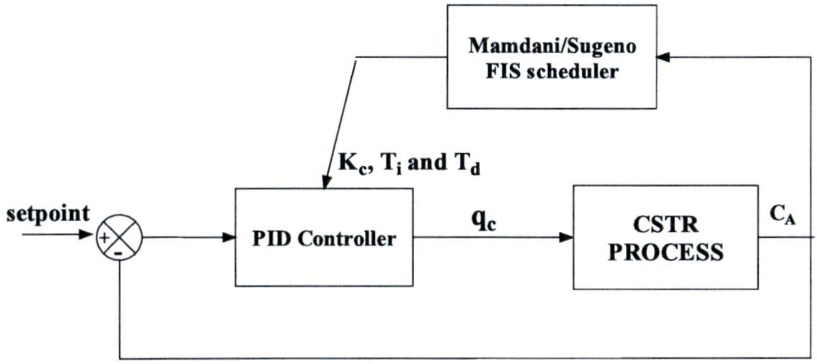

Fig. 2 Cluster-based scheduler

4.1 Scheduler Design

To design the scheduler [17], based on fuzzy c-means clustering approach, the nonlinear CSTR process has been linearized at numerous operating points [10] to cover the wide operating range of the process and concerned PID tuning parameters are calculated. Figure 3 reports the tuning parameters k_c, T_i and T_d obtained at different operating values of concentration C_A. Each tuning parameter is grouped into three clusters by fuzzy c-means algorithm utilizing values mentioned in Table 3.

The output of the clustering algorithm translates scheduler (relating outputs k_c, T_i and T_d with reference to input concentration C_A) as a Fuzzy Inference System (FIS) in Mamdani or Sugeno form. Figure 4 gives the Gaussian membership relation for each tuning parameter with their respective input concentration through Mamdani FIS. Figure 5 gives the Gaussian membership relation of inputs for each parameter (k_c, T_i and T_d), and output is a respective linear function in Sugeno FIS (Fig. 6).

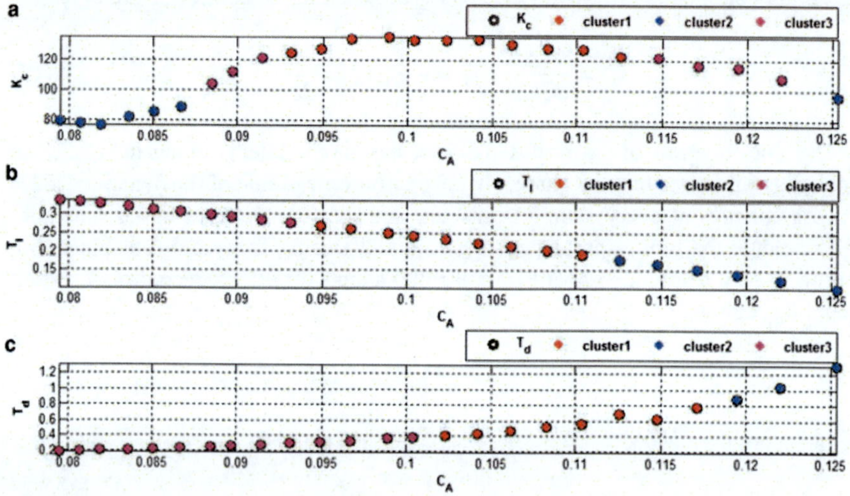

Fig. 3 Variation of controller parameters against concentration

Table 3 Fuzzy c-means algorithm parameters and their values

Fuzzy c-means cluster parameters	Values
Number of clusters (N) for each tuning parameter (k_c, T_i and T_d)	3
Exponent of the fuzzy partition matrix (m)	3
Minimum improvement in objective function between two consecutive iterations (ε)	0.001

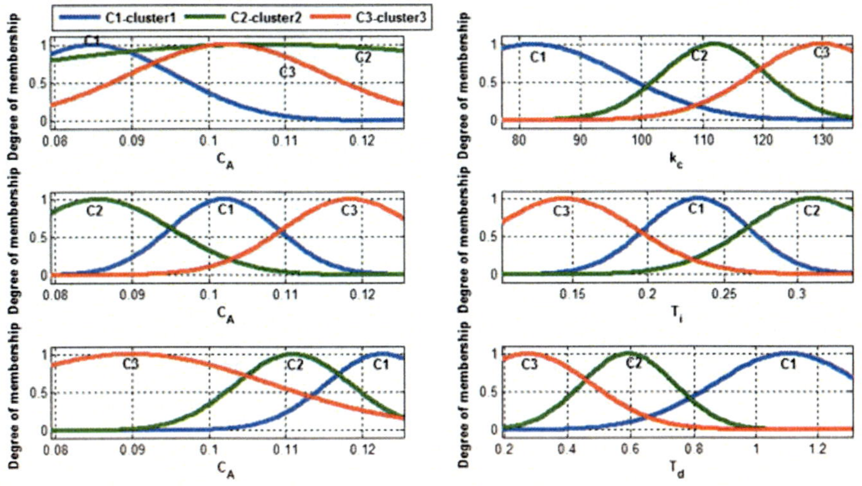

Fig. 4 Membership function of outputs (k_c, T_i and T_d) and with respective input concentration (C_A)

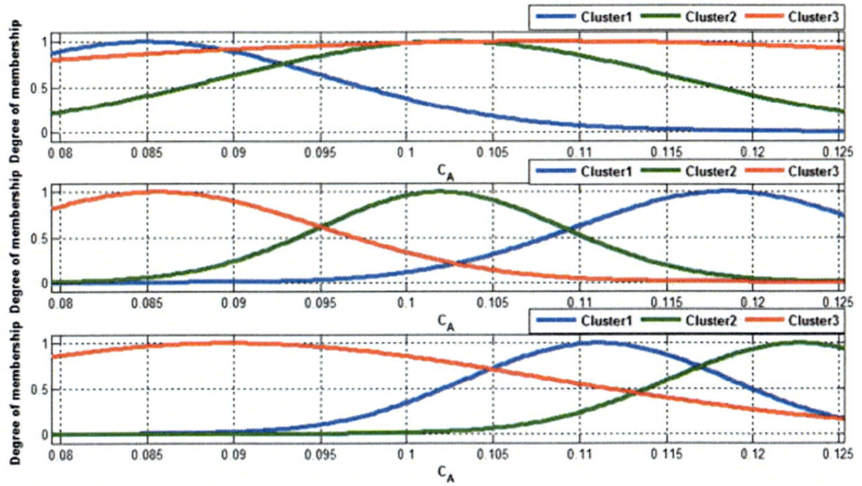

Fig. 5 Input membership function of concentration (C_A) for respective outputs (k_c, T_i and T_d)

5 Simulation Results

The servo response of the CSTR process is shown in Fig. 7. It is clear that schedulers with Mamdani and Sugeno FIS (mentioned as Mamdani-PID and Sugeno-PID) have a good setpoint tracking capability. Further, it is observed that the performance of PID controller is inferior to cluster-based schedulers, as there is a shift from one operating point to another. The PID controller is tuned at the nominal operating point mentioned

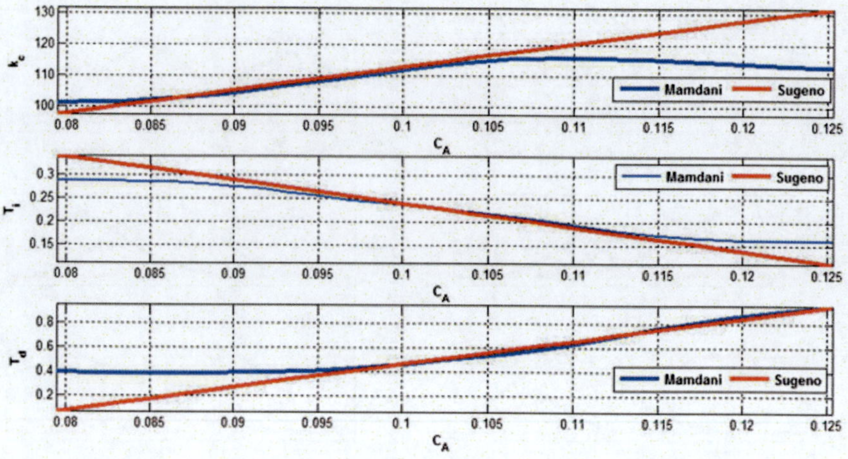

Fig. 6 Controller parameter variation for input concentration in Mamdani and Sugeno FIS

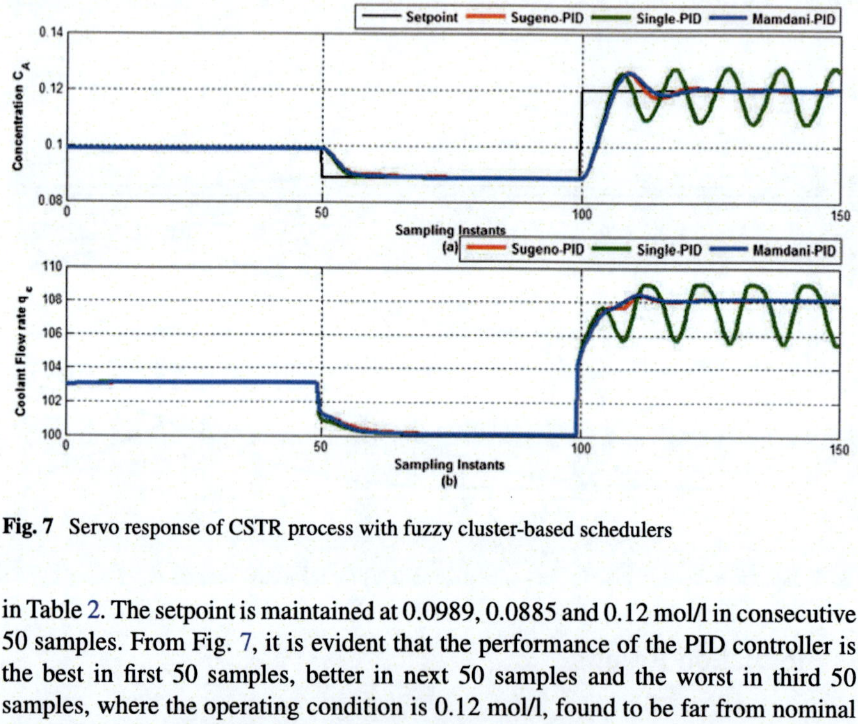

Fig. 7 Servo response of CSTR process with fuzzy cluster-based schedulers

in Table 2. The setpoint is maintained at 0.0989, 0.0885 and 0.12 mol/l in consecutive 50 samples. From Fig. 7, it is evident that the performance of the PID controller is the best in first 50 samples, better in next 50 samples and the worst in third 50 samples, where the operating condition is 0.12 mol/l, found to be far from nominal operating point of 0.0989 mol/l. The coolant flow variation is smooth in cluster-based schedulers and is oscillatory in PID controller implementation at shifted operating point [Refer Fig. 7b]. Figure 8 gives a comparison of controller parameter variation for Sugeno and Mamdani FIS while undergoing servo change in the process. The performance of Sugeno and Mamdani FIS schedulers is observed to be same, even

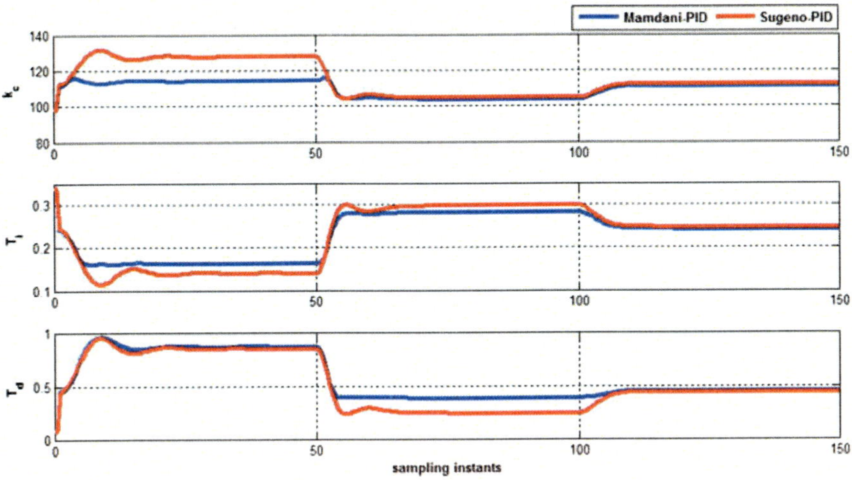

Fig. 8 Controller parameters variation in schedulers for servo change

though the parameter interpolation is linear in Sugeno compared to Mamdani FIS as in Fig. 6. To observe the servo regulatory response [Refer Fig. 9] of CSTR process, the inlet feed temperature T_0 is increased to 352 K from 25th sampling instant to 100th sampling instant [Refer Fig. 9a]. It is evident from Fig. 9b that only Mamdani-PID and Sugeno-PID controllers are able reject the disturbance and bring the process variable of the reactor back to the setpoint. Table 4 shows the ISE values of servo and servo regulatory responses using Mamdani-PID, Sugeno-PID and Single-PID controllers.

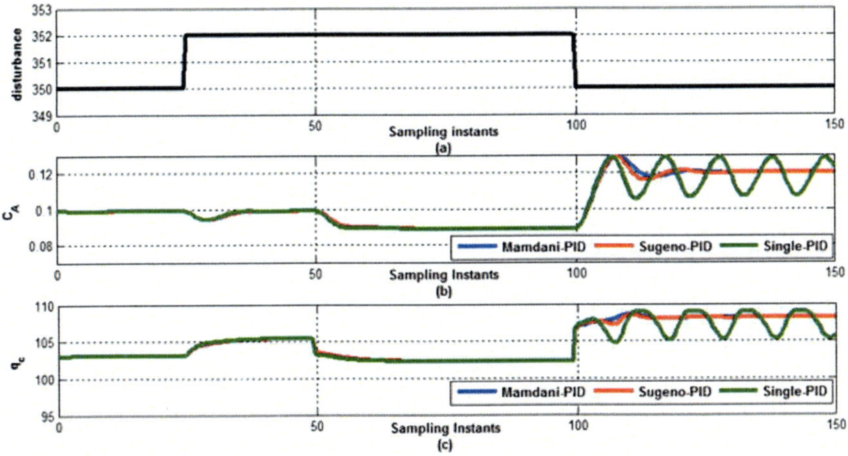

Fig. 9 Servo regulatory response of CSTR process with fuzzy cluster-based schedulers

Table 4 ISE values of servo and servo regulatory responses

Sampling interval		1:50	51:100	100:150
ISE values for servo response	Mamdani-PID	1.0871e−004	0.0015	0.0048
ISE values for servo regulatory response	Sugeno-PID	1.0871e−004	0.0015	0.0048
	Single-PID	1.0861e−004	0.0014	0.0086
	Mamdani-PID	3.3182e−004	0.0015	0.0042
	Sugeno-PID	3.3182e−004	0.0015	0.0042
	Single-PID	3.0001e−004	0.0014	0.0094

6 Conclusion

In this paper, the work is dedicated to adapt PID controller parameters for shift in operating points to handle the tight control of nonlinear CSTR process. The design of FCM cluster-based Sugeno/Mamdani FIS is found to be simple and straight forward. The FCM cluster formulation translates the shift in operating point to suitable PID parameter values, and hence, the linear PID controls the chosen dynamic CSTR plant over a wide operating range. Extensive simulation studies prove that cluster-based scheduler helps the PID controller to perform better in setpoint tracking and disturbance rejection situations than the PID controller tuned for single operating point.

References

1. Murray Smith, R., Johanson, T.A.: Multiple Model Approaches to Modeling and Control (Chap. 11), pp. 269–291. Taylor and Francis Ltd. (1997
2. Narendra, K.S., Balakrishnan, J.: Adaptive control using multiple models. IEEE Trans. Autom. Control **4**(2), 171–187 (1997)
3. Aufderheide, B., Bequette, B.W.: Extension of dynamic matrix control to multiple models. Comput. Chem. Eng. **27**, 1079–1096 (2003)
4. Danielle, D., Cooper, D.: A practical multiple model adaptive strategy for multivariable model predictive control. Control Eng. Pract. **11**, 649–664 (2003)
5. Pottmann, M., Seborg, D.E.: Identification of non-linear process using reciprocal multi quadratic functions. J. Process Control **2**, 189–203 (1992)
6. Huang, Y.L., Lou, H.H., Gong, J.P., Thomas, E.: Fuzzy model predictive control. IEEE Trans. Fuzzy Syst. **8**, 5–77 (2000)
7. Krishnapura, V.G., Jutan, A.: A neural adaptive controller. Chem. Eng. Sci. **55**(18), 3803–3812 (2000)
8. Prakash, J., Senthil, R.: Design of observer based non-linear model predictive controller for a continuous stirred tank reactor. J. Process Control **18**(5), 504–514 (2008)
9. Prakash, J., Srinivasan, K.: Design of nonlinear PID controller and nonlinear model predictive controller for a continuous stirred tank reactor. ISA Trans. **14**(3), 273–282 (2009)
10. Vinodha, R., Abraham Lincoln, S., Prakash, J.: Design and implementation of simple adaptive control schemes on simulated model of CSTR process. Int. J. Model. Ident. Control **14**(3), 159–169 (2011)

11. Su C, Ma J.: Nonlinear predictive control using Fuzzy Hammerstein model and its application to CSTR process, 2012 International Conference on Modeling Identification and Control. AASRI Procedia **3**, 8–13 (2012)
12. Somsung, K., Pratishthananda, S.: Constructing fuzzy structure using fuzzy C-means clustering combined with fuzzy gain scheduling. IEEE Conf. Publ. **3**, 1577–1584 (2004)
13. Sudha, K.R., Butchi Raju, Y., Chandra Sekhar, A.: Fuzzy C-means clustering for robust decentralized load frequency control of interconnected power system with generation rate constraint. Electr. Power Energy Syst. **37**, 58–66 (2012)
14. Dunn, J.C.: A fuzzy relative of the ISODATA Process and its use in detecting compact well-separated clusters. J. Cybernetics **3**, 95–104 (1974)
15. Bezdek, J.C.: Cluster validity with fuzzy sets. J. Cybernetics **3**, 58–73 (1974)
16. Wayne Bequette, B.: Process Control Modeling, Design and Simulation. Prentice Hall (2006)
17. Vinodha, R.: FCM scheduled multiple model controller for the simulated model of spherical tank process. In: Proceedings of 2019 IEEE International Conference on Electrical Computer and Communication Technologies (2009)

Design of Power System Stabilizer as Second-Order Sliding Mode Controller

Arti. V. Tare, Shamli S. Jadhav, Vijay N. Pande, and S. P. Ghanegaonkar

1 Introduction

Power systems are highly complex and dynamic systems. The dynamic nature is because of disturbances and parametric uncertainties that are continuously occurring in the system. These disturbances occur due to load fluctuations and change in generation of generator. The disturbances cause the phase lag between electric torque component and deviation in speed. This leads to the generation of small frequency oscillations in the system which affects the overall system stability and power transfer capability. For the system to remain stable, these small frequency oscillations must be damped out. PSS is used as a controller to damp these oscillations. The conventional construction of PSS includes lead-lag compensator design which compensates the lag between electric torque component and deviation in speed.

As the load fluctuations in power systems are drastic, the operating condition of the system changes continuously, which gives rise to parametric uncertainties. But the conventional controllers designed for the particular operating condition may not work for other conditions. And that represents the uncertainty in stability. Sliding mode controller is a robust controller that can overcome parametric uncertainties and disturbances to maintain the system at equilibrium [1].

Arti. V. Tare (✉) · S. S. Jadhav · V. N. Pande · S. P. Ghanegaonkar
College of Engineering, Pune, India
e-mail: avt.elec@coep.ac.in

S. S. Jadhav
e-mail: jadhavss18.elec@coep.ac.in

V. N. Pande
e-mail: vnp.elec@coep.ac.in

S. P. Ghanegaonkar
e-mail: spg.elec@coep.ac.in

S. K V and K. Guruprasad Rao (eds.), *Smart Sensors Measurements and Instrumentation*, Lecture Notes in Electrical Engineering 750,
https://doi.org/10.1007/978-981-16-0336-5_24

The concept of power system stabilizers was established in the early 1960s to increase the transient and dynamic stability of hydraulic power generators. Generators were found to have very poor inherent damping forces which were not enough to damp small frequency oscillations. It was found that very significant improvement in stability was obtained by using high-speed excitation systems with voltage regulator and PSS control signals. These control signals of PSS were found using deviation in speed, deviation in terminal frequency and deviation in power. It was a successful experiment of the addition of a supplementary feedback using high-speed excitation systems to enhance damping of rotor oscillations [2–4].

Further, during late 1960s to late 1980s, the problems associated with the implementation of these designed stabilizers were experienced and documented in [5–9]. The problems and design methods discussed in these papers lead to the actual (conventional) design of power system stabilizers.

Development in the area of PSS design has been done by introducing new design methods. In [10], state-space pole placement technique is designed for output feedback systems for many operating conditions. This stabilizer was shown to perform better than traditional power system stabilizers. Fuzzy logic power system stabilizer (FLPSS) design method was discussed in [11] which gives a better dynamic performance under disturbance than traditional PSS. FLPSS does not require mathematical model of machine and plant and is very easy to implement.

In [12], PID controller is designed as PSS for SMIB system. It is shown that the results are better when the trial and error method is used for PID tuning than Ziegler–Nichols rules for PID tuning.

The design of sliding mode power system stabilizers is proposed in [13] and the results are compared with traditional PSS, which shows that SMC produces a best speed deviation and torque deviation response. But there is drawback that it requires very large feedback signal to stabilize these responses.

Introduction of the topic and literature survey is given in Sect. 1. Introduction to SMC and SOSMC is presented in Sect. 2. Section 3 includes the evolution of the system model in the form of state space. To demonstrate the robustness of the controller, different operating conditions are considered. They are mentioned in Sect.4. Section 5 includes control development. The results and simulation are presented in Sect. 6. And the paper is concluded in Sect. 7.

2 Introduction to SMC

2.1 Introduction to Sliding Mode Control

Sliding mode control is basically a (nonlinear) variable structure control which gives a control feedback that switches when the state trajectory crosses a particular surface in the state space. With this action, the control forces system states to reach and afterward to remain on this surface which is called sliding surface. The equation of sliding surface is given by $\sigma(x) = 0$. This surface is designed such that when the states are confined to it, they follow desired dynamics. To ensure the confinement of states on the sliding surface, a switching

control must be provided to it which can be as follows:

$$u(t) = \begin{cases} -k, & \text{f}\, \sigma(x) > 0 \\ +k, & \text{f}\, \sigma(x) < 0 \end{cases} \tag{1}$$

This equation can be written more concisely as:

$$u(t) = -ksgn(\sigma) \tag{2}$$

This is also known as Gao's proportional rate reaching law [1]. When system states converge to the sliding surface, they represent the controlled system behavior. During the sliding phase, the system shows reduced-order dynamics with respect to plant dynamics. Other advantages of sliding mode control are insensitivity to matched disturbances, parametric variations and uncertainties. But the major disadvantage of sliding mode control is chattering in control input which arises due to high frequency switching. The chattering may damage the actuators and also system may break down.

2.2 Introduction to Second-Order Sliding Mode Control

In higher-order (r^{th}) sliding mode, the higher-order derivative (r^{th}) of σ is discontinuous and all the lower derivatives of σ from 0th to $(r-1)^{th}$ are constrained to zero in finite time. The control that is constructed from this is continuous. Similarly, in second-order sliding mode control (SOSMC), a second-order derivative of σ is discontinuous. And σ, $\dot{\sigma}$ are constrained to zero in finite time. This can be done when the relative degree of the system with respect to σ is 2 or less than 2, i.e., when control u is nonzero in the equation of second-order derivative of σ.

SOSMC preserves all the advantages of the SMC and it also overcomes the problem of chattering. As the chattering is minimized, the life of actuator increases automatically. Also, SOSMC provides the finite time convergence of system states [14].

A well-known algorithm in SOSMC is twisting algorithm (TA) as below:

$$\ddot{\sigma} = -m_1 sgn(\sigma) - m_2 sgn(\dot{\sigma}) \tag{3}$$

3 System Modeling

SMIB system is a synchronous generator connected to a large system which can be characterized as infinite bus. To analyze the small signal stability of a SMIB power system, it is represented in state-space form [15]. Figure 1 shows the block diagram of SMIB with the effect of exciter (automatic voltage regulator) [15].

When generator output and external system reactance are high, it is preferable to have a high response exciter to increase synchronizing torque. But this results in negative damping and reduces system stability. Therefore, adding the effects of exciter (AVR) generates clashing requirements in the system.
Addition of power system stabilizer as an auxiliary controller in system adds positive damping in the system and improves stability limits of the system. As described in [15], linearized classical model of the single machine infinite bus power system having deviation in speed and deviation in rotor angle as two basic states along with deviation in terminal voltage whose expression is derived from block diagram given in Fig. 1 is considered as third state. The input is taken as ΔE_{fd}.

Now, the state-space model of SMIB system for validation of performance of the proposed controller is as follows:

$$\begin{bmatrix} \Delta\dot{\omega} \\ \Delta\dot{\delta} \\ \Delta\dot{E}_t \end{bmatrix} = \begin{bmatrix} a_{11} & \frac{a_{12}-a_{13}K_5}{K_6} & \frac{a_{13}}{K_6} \\ \omega_0 & 0 & 0 \\ \omega_0 K_5 & K_6 a_{32} - K_5 a_{33} & a_{33} \end{bmatrix} \begin{bmatrix} \Delta\omega \\ \Delta\delta \\ \Delta E_t \end{bmatrix} + \begin{bmatrix} 0 \\ 0 \\ b_{32} \end{bmatrix} E_{fd} \tag{4}$$

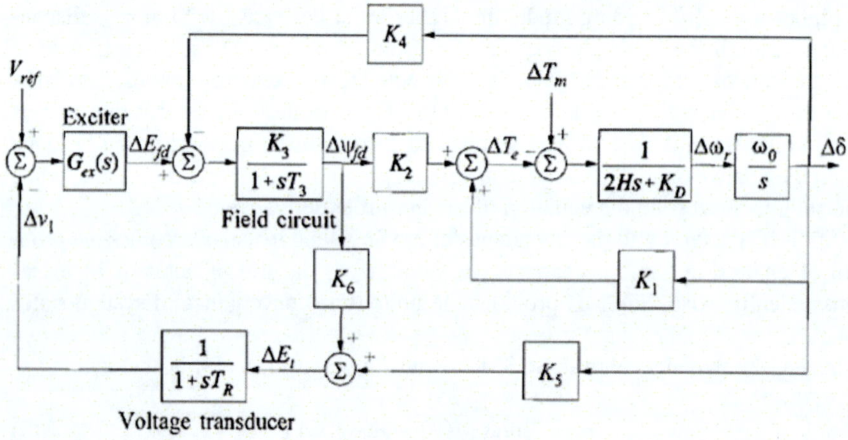

Fig. 1 Block diagram of SMIB with the effect of excitation

Table 1 Different cases considered to demonstrate robustness of SMC

	Active power P (pu)	Reactive power Q (pu)	Total reactance X_t (pu)
Case 1	0.9	0.3	0.95
Case 2	0.289	0.109	0.95
Case 3	0.289	0.109	0.85
Case 4	1.042	0.296	0.40

Table 2 K-constants for different cases

	Case 1	Case 2	Case 3	Case 4
K_1	0.7643	0.46561	0.71945	0.77983
K_2	0.8649	0.24788	0.47741	0.39264
K_3	0.323	0.4529	0.44485	0.43233
K_4	1.4187	0.88772	1.05738	1.05329
K_5	−0.1463	−0.08509	−0.06332	−0.07846
K_6	0.4168	0.74445	0.69541	0.70403

4 Changes in Operating Conditions

To demonstrate the robustness of proposed SOSMC using TA, it is necessary to check the performance of the proposed controller under different operating conditions. Table 1 shows different operating conditions cases.

K-constants and design parameters of the system depend on loading and operating conditions as described and formulated in [15]. Thus, the K-constants of the SMIB system for the different operating cases considered in Table I is different. They are calculated using formulae given in [15] and listed down in Table 2.

This paper demonstrates the effectiveness and robustness of proposed SOSMC, designed at particular operating condition. Therefore, the parameters of controllers are not changed as the operating condition changes, and they are kept constant for all the cases.

5 Control Development

Sliding surface $\sigma(x)$ is defined as the linear combination of system states or their error. It is given as:

$$\sigma = s^T(x_d - x) \tag{5}$$

s^T is surface matrix.

x_d is desired state vector.

x is state vector.

Differentiation of σ is given as follows:

$$\dot{\sigma} = s^T(\dot{x_d} - \dot{x}) \tag{6}$$

x_d is a constant vector, therefore, $\dot{x_d} = 0$

$$\dot{\sigma} = -s^T(Ax + Bu)$$

$$\ddot{\sigma} = -s^T(A\dot{x} + B\dot{u}) \tag{7}$$

According to twisting algorithm in Eq. (3)

$$\ddot{\sigma} = -m_1 sgn(\sigma) - m_2 sgn(\dot{\sigma}) \tag{8}$$

Therefore,

$$-m_1 sgn(\sigma) - m_2 sgn(\dot{\sigma}) = -s^T(A\dot{x} + B\dot{u}) \tag{9}$$

$$\dot{u} = (s^T B)^{-1}[-s^T A\dot{x} + m_1 sgn(\sigma) + m_2 sgn(\dot{\sigma})] \tag{10}$$

Integrating Eq. (10)

$$u = (s^T B)^{-1}\left[-s^T Ax + \int m_1 sgn(\sigma) + \int m_2 sgn(\dot{\sigma})\right] \tag{11}$$

The control input is governed by Eq. (11).

6 Results and Discussion

Second-order sliding mode controller is designed for SMIB system with state-space equation as given below:

$$\begin{bmatrix} \Delta\dot{\omega} \\ \Delta\dot{\delta} \\ \Delta\dot{E}_t \end{bmatrix} = \begin{bmatrix} 0 & -0.1486 & -0.1049 \\ 377.99 & 0 & 0 \\ -21.40 & -0.15724 & -0.4229 \end{bmatrix} \begin{bmatrix} \Delta\omega \\ \Delta\delta \\ \Delta E_t \end{bmatrix} + \begin{bmatrix} 0 \\ 0 \\ 11.38 \end{bmatrix} u \tag{12}$$

The designed SOSMC is governed by Eq. (11).
The values of tuning parameters m_1 and m_2, and surface matrix is given below:

$$m_1 = 5000, m_2 = 10000$$

$$s^T = [-1000, -0.001, 25]$$

$$x_d = [0, -1.9, 1]$$

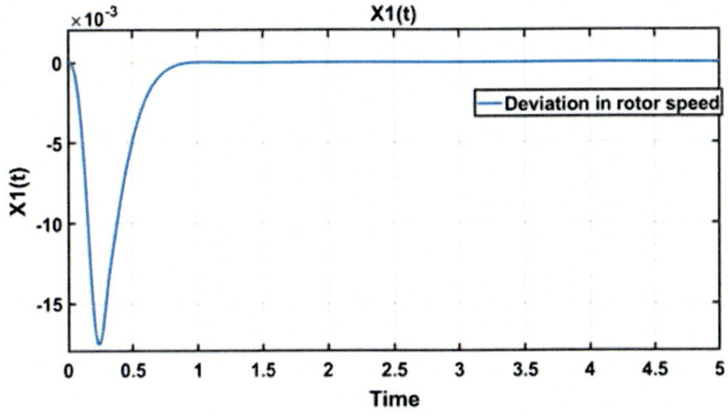

Fig. 2 Deviation in speed

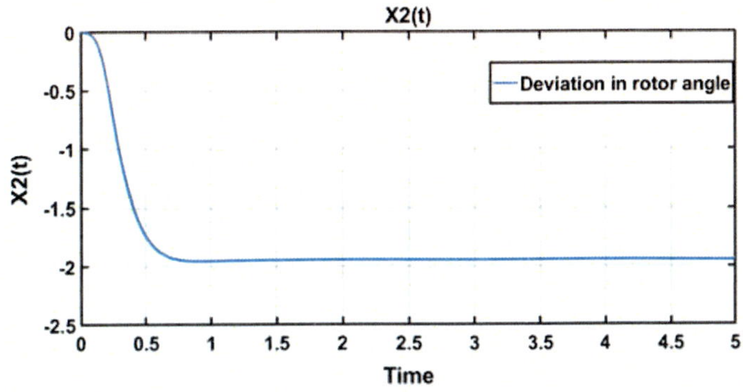

Fig. 3 Deviation in angle

SOSMC is designed for the operating condition of case 1 as shown in Table 1. This setup is simulated in MATLAB and Simulink, and the results are shown in this section.

x_d is the desired state vector. All the states are supposed to settle at the value given in this vector. It is seen in Fig. 2 and in Fig. 3 the deviation in speed and in deviation in angle settles quickly and smoothly.

Deviation in terminal voltage also settles to the reference voltage that is equal to 1 as shown in Fig. 4. The control input in Fig. 5 has very little chattering initially, but afterward it shows a smooth curve.

From Fig. 6, it can be seen that the value of sigma becomes zero. It shows that the state trajectory is confined to the sliding surface in finite time.
To demonstrate the effectiveness of the proposed controller, it is necessary to check the performance of the system under different operating conditions. Figures 7, 8, 9, 10 and 11 comparatively demonstrate the evolution of states and input for different operating conditions.

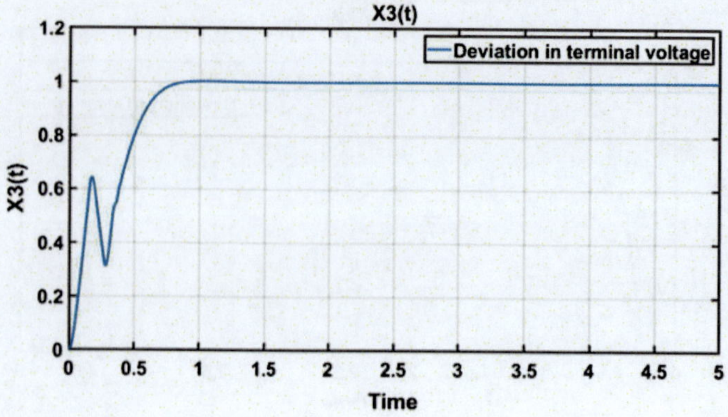

Fig. 4 Deviation in terminal voltage

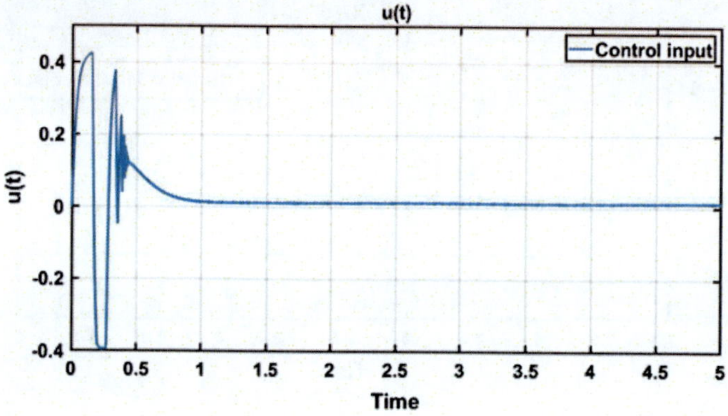

Fig. 5 Control input of twisting algorithm

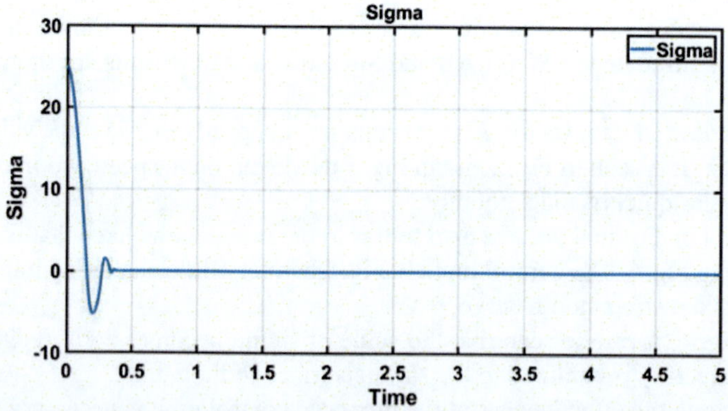

Fig. 6 Sigma of twisting algorithm

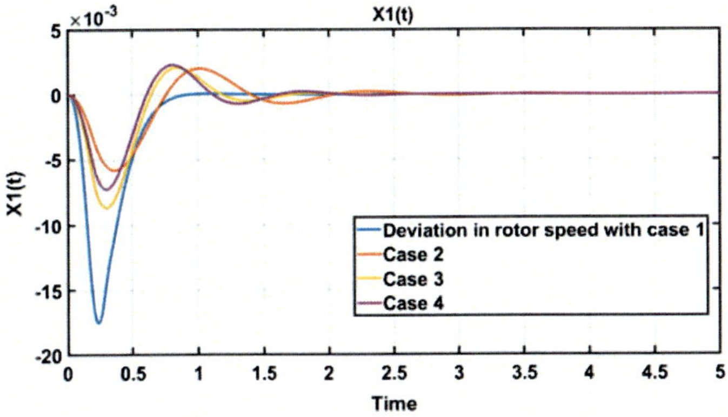

Fig. 7 Deviation in speed

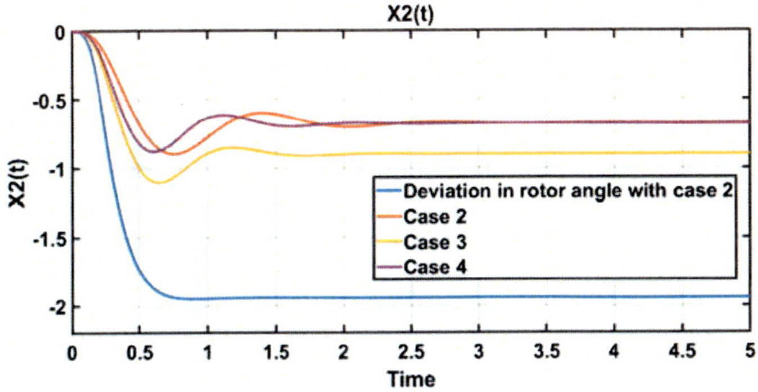

Fig. 8 Deviation in angle

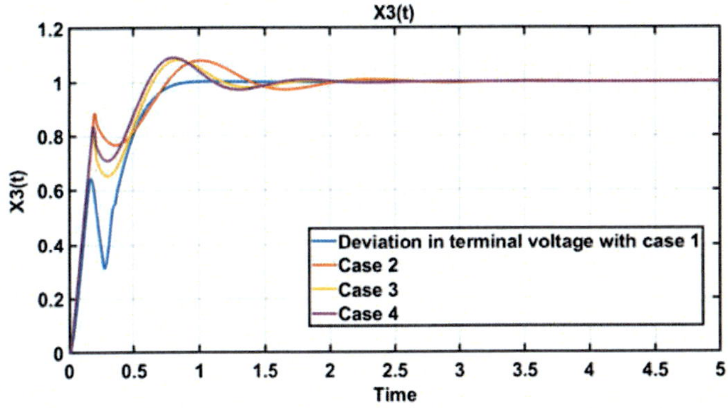

Fig. 9 Deviation in terminal voltage

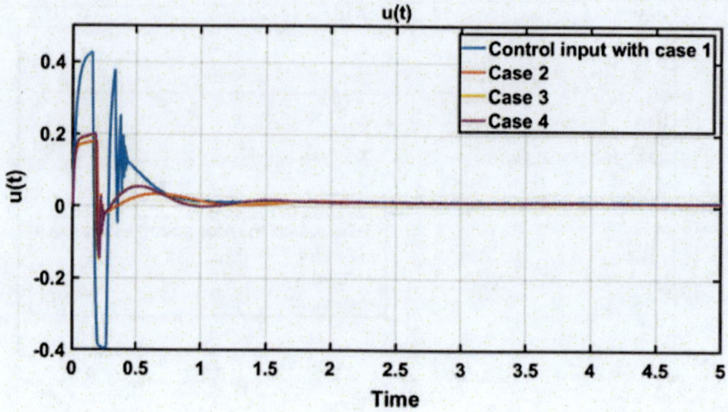

Fig. 10 Control input

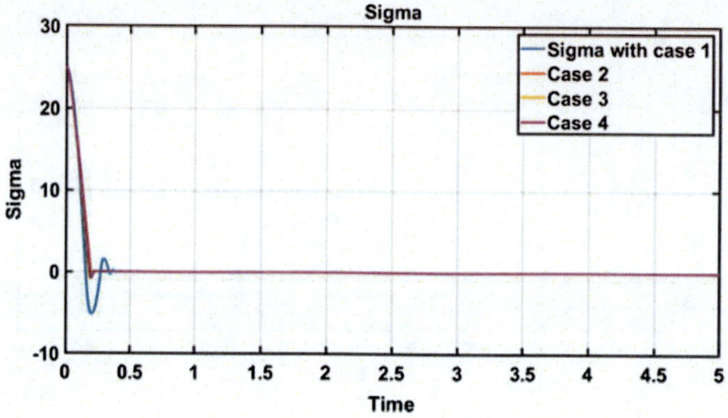

Fig. 11 Sigma

It can be seen from Figs. 7, 8 and 9 that all the states for case 2, 3 and 4 settle to constant (desired value) with less overshoot, but they require more time to settle than case 1.

There is little chattering initially in the control input for all the four cases of operating condition as it is shown in Fig. 10. But afterward, there is a smooth curve of control input for all the cases. Figure 11 shows the settling of $\sigma(x)$ to zero, and hence confirm the confinement of state trajectory to sliding surface for all the cases.

7 Conclusion

In power system, low frequency oscillations need to be damped out efficiently, for which power system stabilizer is employed. This is important for better operation and control of power system.

In this paper, a novel approach of power system stabilizer design for single machine infinite bus power system using second-order sliding mode control is developed. Power system stabilizer designed as a second-order sliding mode controller for single machine infinite bus power system for one operating condition, and the simulated results are plotted. From the results, it is observed that deviation in speed, deviation in rotor angle and terminal voltage settles to their final value quickly and smoothly. Further to demonstrate the effectiveness and robustness of the proposed controller, operating conditions of the system are changed and performance of the proposed SOSMC PSS is evaluated from simulated results. From the comparative results, it is depicted that even when operating conditions are changed, all states settle to their final value quickly and smoothly.

References

1. Hung, J.Y., Gao, W., Hung, J.C.: Variable structure control: a survey. IEEE Trans. Ind. Electron. **40**(1), 2–22 (1993)
2. Dandeno, P.L., et al.: Effect of high-speed rectifier excitation systems on generator stability limits. IEEE Trans. Power Appar. Syst. **1**, 190–201 (1968)
3. Schleif, F.R., et al.: Excitation control to improve powerline stability. IEEE Trans. Power Appar. Syst. **6**, 1426–1434 (1968)
4. Demello, F.P., Concordia, C.: Concepts of synchronous machine stability as affected by excitation control. IEEE Trans. Power Appar. Syst. **88**(4), 316–329 (1969)
5. Larsen, E.V., Swann, D.A.: Applying power system stabilizers part I: general concepts. IEEE Trans. Power Appar. Syst. **6**, 3017–3024 (1981)
6. Larsen, E.V., Swann, D.A.: Applying power system stabilizers part II: Performance objectives and tuning concepts. IEEE Trans. Power Appar. Syst. **6**, 3025–3033 (1981)
7. Larsen, E.V., Swann, D.A.: Applying power system stabilizers Part III: Practical considerations. IEEE Trans. Power Appar. Syst. **6**, 3034–3046 (1981)
8. Watson, W., Coultes, M.E.: Static exciter stabilizing signals on large generators-mechanical problems. IEEE Trans. Power Appar. Syst. **1**, 204–211 (1973)
9. Lee, D.C., Beaulieu, R.E., Service, J.R.R.: A power system Stabilixer using speed and electrical power inputs-design and field experience. IEEE Trans. Power Appar. Syst. **9**, 4151–4157 (1981)
10. Chow, J.H.: A pole-placement design approach for systems with multiple operating conditions. In: Proceedings of the 27th IEEE Conference on Decision and Control. IEEE, New York (1988)
11. Shi, J., Herron, L.H., Kalam, A.: Design and implementation of a PC-based automatic voltage regulator and fuzzy logic power system stabilizer. In: 1993 2nd International Conference on Advances in Power System Control, Operation and Management, APSCOM-93. IET (1993)
12. Surjan, B.S., Garg, R.: Power system stabilizer controller design for SMIB stability study. Int. J. Eng. Adv. Technol. (IJEAT) **2**(1), 209–214 (2012)
13. Safie, S.I., et al.: Sliding mode control power system stabilizer (PSS) for single machine connected to infinite bus (SMIB). In: IEEE 2nd International Power and Energy Conference. IEEE, New York (2008)
14. Levant, A.: Sliding order and sliding accuracy in sliding mode control. Int. J. Control **58**(6), 1247–1263 (1993)
15. Kundur, P., Balu, N.J., Lauby, M.G.: Power System Stability and Control, vol. 7. McGraw-Hill, New York (1994)

Fatigue Analysis of Biceps Brachii Muscle Using sEMG Signal

P. Shamli Fathima, C. Sandhra, Dolbin Jojo, A. V. Gayathri, N. Sidharth, and G. Venugopal

1 Introduction

In this modern period of time, people give prime importance to their health. Most of them are looking for technologies that could assist them in monitoring various health issues. Recent technologies include heart rate detection, number of footsteps, etc. Daily exercising is the key to keep your body fit. People do more exercise for their arms to show up their muscle mass. For this, they have to lift weights continuously using dumbbells or other gym machines. During this exercise, biceps brachii muscle goes through continuous dynamic contractions. Muscles get tired or injured while doing continuous exercise [1]. This is due to the voluntary contraction of muscle repeatedly. This leads to a condition called muscle fatigue. Increased muscle fatigue causes damage to muscle. This decreases motored functions. Muscle fatigue is measured using surface electromyography. Electrical activity of muscle is assessed from the surface above the muscle in surface EMG [2].

The proposed project measures the EMG signals from biceps brachii muscles [3]. Two surface electrodes are placed on biceps brachii muscle at a distance of 2–3 cm. One reference electrode is placed at elbow. These electrodes are connected to EKG sensor for obtaining EMG signal.

EMG signals are usually affected by noise. So, for obtaining useful data, feature extraction is done. EMG signal has various features including Root Mean Square (RMS) value, total band power, mean frequency, peak frequency etc. Among these, RMS and mean frequency are used in this project. Analysis of EMG signal is done using LabVIEW software. Whenever RMS value and mean frequency of sEMG

P. Shamli Fathima (✉) · C. Sandhra · D. Jojo · A. V. Gayathri · G. Venugopal
Department of Instrumentation and Control Engineering, NSS College of Engineering (Affliated to APJ Abdul Kalam Technological University Kerala), Palakkad, India

N. Sidharth
Department of Electronics and Communication Engineering, NSS College of Engineering, Palakkad, India

S. K V and K. Guruprasad Rao (eds.), *Smart Sensors Measurements and Instrumentation*, Lecture Notes in Electrical Engineering 750,
https://doi.org/10.1007/978-981-16-0336-5_25

signal cross a threshold value, it is an indication of fatigue. Biceps brachii muscle activity was based on the difference between root mean square value and mean absolute value, which are the common feature that are used to define the amplitude of EMG signal.

2 Material and Methods

The experimental system which includes a hardware part controlled through the interface NI myDAQ module by using LabVIEW software. EMG signals are acquired by using sEMG electrodes yielding a non-invasive and painless process for patient. It helps to analyze the muscle activity of the desired muscle. sEMG signal is produced from activation of the muscle fibers by motor unit [4].

2.1 Subjects

Twenty-five healthy subjects are participating for analyzing muscle fatigue. All subjects were male between the ages of 19–23 years. The subjects are within the best range to get good EMG signal measurements.

2.2 Experimental Setups and Protocols

EMG signals from electrodes cannot be directly send to the computer without modification. The voltage calculated from the EKG sensor. In order to connect the computer to the hardware device, NI DAQ is used.

For the placement of electrodes, skin preparation is done initially. It was necessary to prepare the skin by cleaning and shaving the hair if necessary, in order to reduce the electrode skin impedance. Special abrasive and conductive cleaning pastes are available which remove dead skin cells and clean the skin from dirt and sweat. Pure alcohol may be another alternative if used with a cotton pad.

The placement of EMG electrodes on the biceps brachii muscle is in the middle of the biceps or belly of the muscle. And the second electrode was placed on the biceps brachii which is 2–3 cm apart from the first electrode. The reference electrode is placed on the bone of the elbow joint [5].

In this experiment, 5 kg dumbbell was used as load. Each subject is asked to hold the load at an angle of 30 degree with respect to vertical axis. This angle is measured using angle chart which is stacked on the wall. The subjects are asked to hold the weight until the disturbance occurs in the loaded hand. The time of first disturbance occurred and the maximum time the subject can hold the weight are recorded.

Some protocols must be considered before starting the experiment. First, the subjects are to be free from muscle disease. Second is about the placement of the sEMG electrodes. These protocols were designed to minimize the motion artifacts and avoid the internal noise during measurement.

2.3 EMG Signal Acquisition

In the experimental setup, NI myDAQ is used as the interface between EKG sensor and PC. DAQ converts analog signal to digital values. It provides real-time multi-channel EMG acquisition system. DAQ hardware is usually interfaced between the sensor and a PC. It could be in the form of modules that can be connected to the computer's port.

The vernier EKG sensor measures cardiac electrical potential waveforms produced during the muscle contraction [6]. It can be used to make standard three-lead EKG system to collect surface EMG recordings to study muscle contractions. It associates the electrical activity of muscle with movement of joints. The output from the EKG sensor is processed using LabVIEW software connected via NI myDAQ. Figure 1 represents the acquired sample EMG signal from biceps brachii muscle.

The acquired sEMG signal gives relation between sampling period on x axis and amplitude on the y axis [7]. The signal is acquired with an indefinite frequency at each point of sample. At the time of fatigue, a high-frequency signal with multiple amplitudes is obtained. During the acquisition of the EMG signal, the initial time, first discomfort time and the time at which the subject got complete fatigue were recorded.

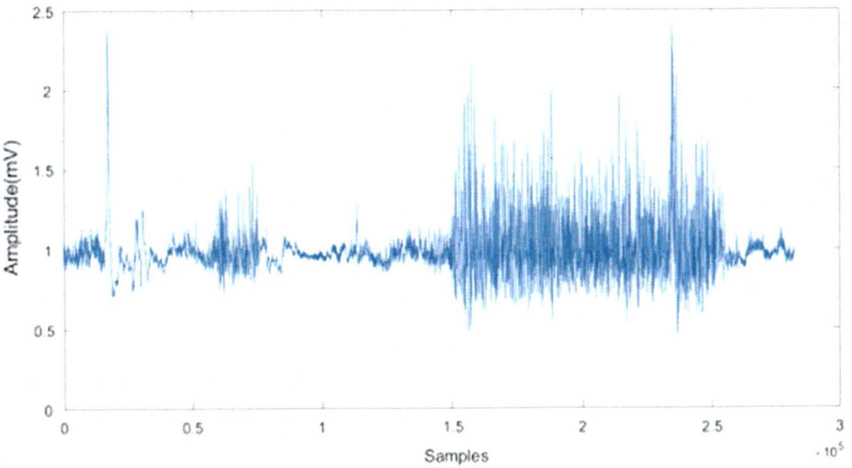

Fig. 1 Acquired sample EMG signal from biceps brachii

2.4 EMG Signal Analysis

The signal acquired from the DAQ device is used as signal source for LabVIEW software. The recorded signal was processed and filtered [8]. Band stop filter is used to suppressing the noise produced by 50 Hz electrical network. Band pass filter 20–500 Hz has been used for obtaining needed information. After the signal acquisition process, features are extracted from the acquired signal using MATLAB. EMG signal features like RMS value and mean frequency are computed.

Root Mean Square (RMS): It represents the square of the root of the average power of the EMG signal for a given period of time (viz. 4 s). It is known a time domain variable. It can be calculated by the formula:

$$\mathrm{RMS} = \sqrt{\frac{1}{N}\sum_{1}^{N}|x_i|^2} \tag{1}$$

where x_i is the values of each sample, and N is the number of samples.

Mean frequency (MNF): It is that frequency, where the product of the frequency value and the amplitude of the spectrum are equal to the average of all such products throughout the complete spectrum of the signal.

$$\mathrm{MNF} = \sum_{j=1}^{M} P_{ij}\, f_{ij} \Big/ \sum_{j=1}^{M} P_{ij} \quad \text{, for } i = 1, 2, \ldots \tag{2}$$

where P_{ij} is the EMG power spectrum at the frequency bin j for the ith segment, f_{ij} is the frequency value of EMG power spectrum at the frequency bin j for the ith segment, and M is the length of frequency window.

Mean frequency is defined as an average frequency that is calculated as a sum of product of the EMG power spectrum and the frequency divided by the total sum of power spectrum [9]. EMG signal is transformed from time domain to frequency domain. For that, a Fourier transform of the auto correlation function of the EMG signal is used to provide power spectrum. Here, power spectrum estimator periodogram is used for EMG signal analysis. It is the square of absolute value of Fourier transform of EMG signal divided by the signal length. Quantitative analysis was performed on these, and thus, means, standard deviation and change from initial state to first discomfort state and from first discomfort state to fully fatigue state were recorded. Initial state is soon after the subject holds the dumbbell. First discomfort time is recorded when the subject starts getting a slight pain. The subject is not able to hold the dumbbell anymore if he gets the muscle fully fatigue. From these values, a certain range for RMS and mean frequency at which fatigue is supposed to happen is obtained. An EMG signal is detected to be fatigue when it comes under these ranges.

3 Results and Discussion

In this study, sEMG signals are recorded from biceps brachii muscle. The burst signal from recorded sEMG signal (Fig. 1) is shown in Fig. 2. Using windowing, the signal is split in to interval of 250 ms. From the burst signal recorded, two intervals are considered. First interval is the beginning of burst signal to first discomfort state, and second interval is from first discomfort state to fully fatigue state. RMS and mean frequency are calculated from the interval. RMS signal of one subject is shown in Fig. 3.

Table 1 shows statistical analysis of RMS value and mean frequency of the 25 subjects. Mean and standard deviation of sEMG signals are computed here. In this study, the RMS range and mean frequency range in three stages are calculated. The different range of parameters in three stages is shown here. It is found that in fully fatigue state, the RMS range is 0.185 ± 0.099. This shows that when the RMS value reaches the threshold value 0.086, the muscle gets fully fatigue. Similarly, the fully fatigue range of mean frequency is found to be 81.9 ± 33.1 Hz. When the threshold value of mean frequency 48.8 Hz is crossed, it is an indication of fatigue.

The graphs in Figs. 4 and 5 represent the scatter plot of subject's EMG measurement in related with RMS and mean frequency values. The *X*-axis represents the subject, and the *Y*-axis represents the corresponding signal parameter.

Figure 4 determines that the RMS value increases as the subject moves to fatigue state, whereas Fig. 5 shows that mean frequency value decreases correspondingly. From graphs and Table 1, it is found that RMS value of each subject shows an increase in positive direction, whereas the mean frequency shows negative deviation down

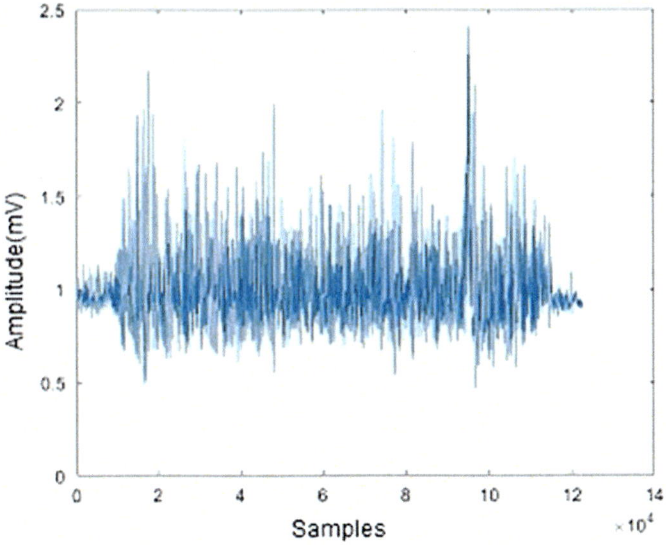

Fig. 2 Burst signal from sample EMG signal

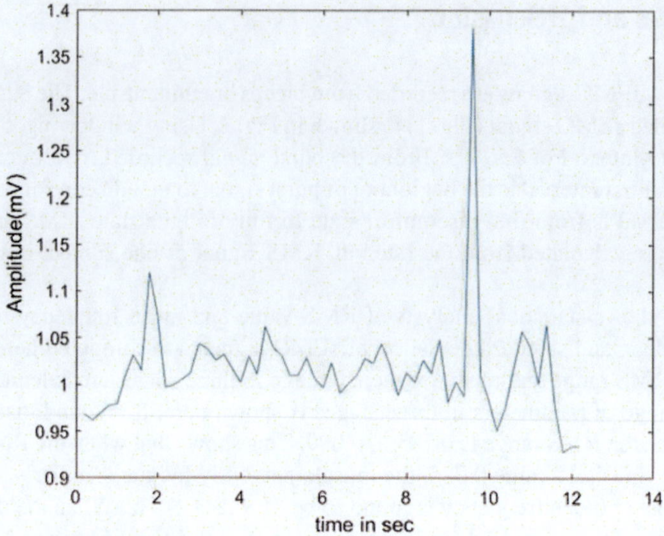

Fig. 3 RMS signal of sEMG signal

Table 1 Statistical analysis of RMS and mean frequency

Parameter	Initial state	First discomfort state	Fully fatigue state
RMS	0.082 ± 0.042	0.146 ± 0.094	0.185 ± 0.099
Mean frequency (Hz)	117.37 ± 36.3	53.37 ± 14.8	81.9 ± 33.1

Fig. 4 RMS

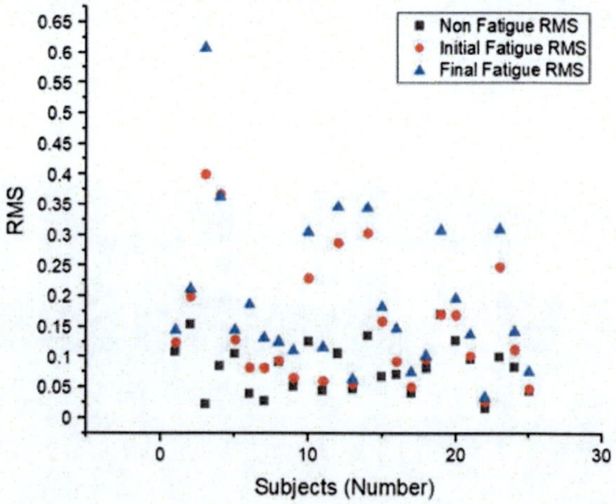

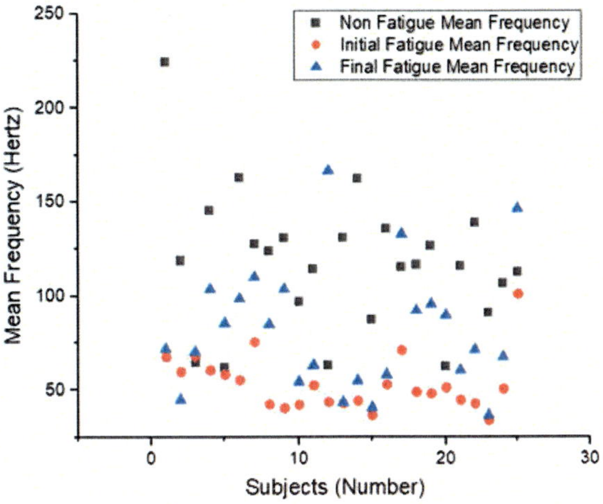

Fig. 5 Mean frequency

the graph. In most cases, the RMS value deviates to 115% from the initial state, where the initial disturbance is felt at 65% of the person's average, and in the case of mean frequency, it is a negative deviation, where the value reached down to 40% of final state, and the initial disturbance is felt at 52% in negative direction. The trend is similar in most cases.

4 Conclusions

Muscle fatigue is commonly seen in athletes, gymnasts and diabetic patients. Determination of muscle fatigue is a difficult task requiring examination of muscle. EMG signals are the perfect options for analysis. In this work, the sEMG signals are recorded from biceps brachii of 25 healthy subjects of age group 19–23. It is found that the RMS value of most of the subjects shows an increase and mean frequency shows a decrease from initial state to fully fatigue state. Fatigue state is identified when the RMS value and mean frequency cross the threshold value 0.086 and 48.8 Hz respectively. This can be implemented as a real-time fatigue detection system.

Acknowledgements We express our sincere gratitude to Centre for Engineering Research and Development, APJ Abdul Kalam Technological University,Thiruvanathapuram, India for providing financial assistance.

References

1. Liu, S.H., Lin, C.B., Chen, Y., Chen, W., Haung, T.S., Hsu, C.Y.: An EMG patch for the real time monitoring of muscle fatigue conditions during exercise, July 14 (2019)

2. Sharmila, K., Sarath, T.V., Ramachandran, K.I.: EMG Controlled Low Cost Prosthetic Arm. IEEE (2016)
3. Burhan, N., Kasno, M.A., Ghazali, R., Radazi Said, Md, Abdullah, S.S., Jali, M.H.: Analysis of the Biceps Brachii muscle by varying the arm movement level and load resistance band, September 12 (2017)
4. Yochum, M., Binczak, S., Bakir, T., Jacquir, S., Lepers, R.: A mixed FES/EMG system for real time analysis of muscular fatigue, April 11 (2012)
5. Reaz, M.B.I., Hussain, M.S., Mohd Yasin, F.: Techniques of EMG signal analysis: detection, processing, classification and applications, January 18 (2006)
6. Eger, M., Sattler, F.: Signal processing unit of an EMG measuring system, June 19, (2015)
7. Raurale, S.A.: Acquisition and processing real time EMG signals for prosthesis active hand movements (2014)
8. Shamim Kaiser, M., Chowdhury, Z.I., Hussain, A.: A neuro fuzzy control system based on feature extraction of sEMG signal for solar powered wheelchair **8**, 946–954 (2016)
9. Phinyomark, A., Thongpanja, S., Hu, H., Phukpattaranont, P., Limsakul, C.: The Usefulness of Mean and Median Frequencies in Electromyography Analysis (Chap. 8), pp. 195–212 (2012)

CPPG-Based Emotion Classification Based on Statistical and Eigen Space Features and Performance Analysis of Classifiers

Nivedita Daimiwal and Revati Shriram

1 Introduction

In optical sensor, the light source illuminates the tissue and the intensity variations of reflected light intensity show changes in blood volume at site of measurement by the photodetector. This optical sensor detects real-time changes in oxygenated blood and deoxygenated blood in prefrontal lobe [1–4].

Optical sensors are less in cost and require less maintenance as compared to imaging modalities like fMRI and PET [5]. Various constraints are there in these imaging modalities like person has to be steady during the scanning period so it is difficult to detect the activity in kids, infant, and the non-responding person [6, 7].

During the functional activity, blood flow and oxygen in the blood flow increase; this is detected by the optical sensor. This plethysmography effect is detected by CPPG sensor [8, 9].

2 Methods

CPPG is captured at central, left, and right side of prefrontal lobe for four different emotions. A Daubechies wavelet decomposition of CPPG signal is performed; the sixth level of the decomposed signal shows less noise signal. A_6 and D_6 component of the signal are used for feature extraction. Features are extracted using statistical and Eigen space analysis. Figure 1 shows the system block diagram [10–13].

N. Daimiwal (✉) · R. Shriram
Cummins College of Engineering for Women, Pune, India

S. K V and K. Guruprasad Rao (eds.), *Smart Sensors Measurements and Instrumentation*, Lecture Notes in Electrical Engineering 750,
https://doi.org/10.1007/978-981-16-0336-5_26

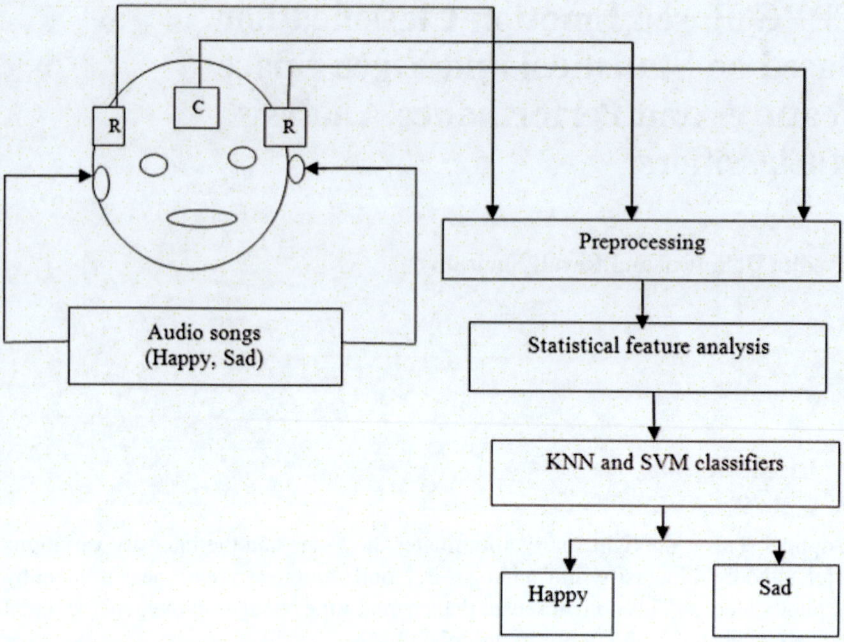

Fig. 1 System block schematic for CPPG acquisition

2.1 Data Collection for Various Emotions

Healthy volunteers are selected for data collection. CPPG signal is captured for two different emotions: Happy, sad, devotional, and random. Subject is supposed to listen Hindi songs for few minutes, while listening the Hindi songs corresponds to the emotions and the CPPG signal is captured for 4 s. The CPPG sensor were properly placed at three different positions (left, central, and right) prefrontal lobe of the subject. For each emotion, the CPPG is captured from these three positions. Subject is requested not to move during data collection so that noise-free signal is obtained. The signal is free from motion artifacts is recoded.

Happy Song: " Aaj Se Pehle Aaj Se Jyada".

Sad Song: "Tijhe Sab Pata Hai Na Maa".

Devotional Song: "Tu Pyaar Ka Sagar".

Random Song: "Mere Dholna Sun: Tu Pyaar.

Six-level decomposition is performed on the CPPG signal captured from both happy and sad emotions using the Daubechies 9 wavelet. The statistical parameters (mean, variance, energy, and entropy) were calculated for CPPG, A_6, and D_6 [14, 15].

3 CPPG Features Extraction

Statistical and Eigen features of happy and sad emotions are taken into consideration for classification. The programming is done to classify happy and sad emotions.

3.1 Statistical Features

Signal is decomposed to remove the noise signal and after decomposition, the statistical parameters of approximate and detailed component following statistical parameters are extracted [16].

Mean of the signal is calculated using Eq. (1):

$$\mu_i = \frac{1}{N}\sum_{j=1}^{N} D_{ij} \tag{1}$$

Standard deviation of the signal is calculated using Eq. (2):

$$\sigma_i = \left(\frac{1}{N-1}\sum_{j=1}^{N}(D_{ij} - \mu_i)^2\right)^{1/2} \tag{2}$$

Energy of the signal is calculated using Eqs. (3) and (4):

$$\mathrm{ED}_i = \sum_{j=1}^{N} |D_{ij}|^2 \tag{3}$$

$$\mathrm{EA}_i = \sum_{j=1}^{N} |A_{ij}|^2 \tag{4}$$

Entropy of the signal is calculated using Eq. (5):

$$\mathrm{ENT}_i = \sum_{j=1}^{N} D_{ij}^2 \log\left(D_{ij}^2\right) \tag{5}$$

Total statistical features are 36 as shown in Table 1. 12 features from central, left, and right prefrontal lobe each.

Table 1 Satistical features for happy, sad, devotional, and random emotions

	Original				A_6				D_6			
	Mean	Variance	Energy	Entropy	Mean	Variance	Energy	Entropy	Mean	Variance	Energy	Entropy
Central prefrontal												
No activity	119.981	138.843	5,799,060	−5.6E+07	996.284	4453.52	2.2E+07	−3E+08	2364.43	1.3E+07	52,017.5	−7E+09
D	120.006	164.984	5,811,864	−5.6E+07	887.388	8906.31	1.8E+07	−2.4E+08	2611.94	1.2E+07	57,462.6	−7.3E+09
S	118.34	198.707	5,666,785	−5.4E+07	990.536	3432.34	2.2E+07	−3E+08	4766.06	5E+07	104,853	−3E+10
H	123.18	147.146	6,112,760	−5.9E+07	901.067	5183.74	1.8E+07	−2.4E+08	2640.63	1E+07	58,093.8	−6.4E+09
R	121.276	81.2454	5,900,745	−5.7E+07	899.049	4349.62	1.8E+07	−2.4E+08	1077.2	2,424,424	23,698.4	−1.3E+09
Leftside												
No activity	122.457	221.915	6,071,652	−5.9E+07	1044.42	3577.73	2.4E+07	−3.3E+08	4834.04	4.6E+07	106,349	−2.9E+10
D	120.599	74.6591	5,832,818	−5.6E+07	929.812	2704.15	1.9E+07	−2.6E+08	780.261	1,313,575	17,165.7	−6.5E+08
S	120.642	111.56	5,851,605	−5.6E+07	959.675	533.849	2E+07	−2.8E+08	1881.14	6,907,367	41,385.1	−3.9E+09
H	122.901	97.5473	6,065,582	−5.8E+07	1011.64	803.79	2.3E+07	−3.1E+08	1651.15	5,173,964	36,325.4	−2.9E+09
R	120.466	98.5648	5,829,556	−5.6E+07	967.638	1149.72	2.1E+07	−2.8E+OS	1330.31	3,379,712	29,266.8	−1.8E+09
Right side												
No activity	122.885	59.704	6,048,923	−5.8E+07	911.882	2644.07	l.8E+07	−2.5E+08	729.2	1,484,321	16,042.4	−B.9E+0S
D	120.634	59.8947	5,830,319	−5.6E+07	943.452	790.193	2E + 07	−2.7E+08	774.8	1,231,816	17,045.6	−6.2E+08
S	121.649	62.2484	5,929,380	-5.7E+07	994.08	652.382	2.2E + 07	−3E+08	1038.55	2,670,306	22,848	−1.3E+09
H	121.157	56.8152	5,879,506	−5.6E+07	977.735	1168.39	2.1E + 07	−2.9E+08	791.114 1,173,932		17,404.5	−6E+08
R	122.871	88.7364	6,059,126	−5.8E+07	914.468	2399.29	l.SE + 07	−2.5E+08	1891.61	7,362,998	41,615.4	−4.1E+09

3.2 Eigen Space Features

If all CPPG from various prefrontal lobes are decomposed using wavelet and its levels, multivariate matrices are formed using similar subbands with an equal number of coefficients. Wavelet coefficients obtained from all CPPG with J-level wavelet decomposition are arranged in $J + 1$ subbands matrices. The columns in a subbands matrix correspond to CPPG at left, central, and right locations and the subband coefficients are represented by the rows. The approximation subband matrix is given by Eq. (6)

$$A_J = \left[cA^1_{J,k}, cA^2_{J,k}, \ldots, cA^m_{J,}\right] \tag{6}$$

Also, the detailed subband matrixes are given by Eq. (7),

$$D_j = \left[cD^1_{j,k}, cD^2_{j,k}, \ldots, cD^m_{j,k}\right] \tag{7}$$

where $m = 3$ is the CPPG at left, central, and right locations.

The subband matrices A_J and D_j obtained from the coefficients from each level of decomposition form 3 position are subjected to Eigen analysis. Eigen spaces are expected to give more precisely the segmented diagnostic information. [17, 22].The covariance matrices are evaluated from the subband matrices shown in Eq. (8) and Eq. (9) as follows:

$$C_{A_J} = \frac{1}{N_J - 1}\left([A_J][A_J]^T\right) \tag{8}$$

$$C_{D_j} = \frac{1}{N_j - 1}\left([D_j][D_j]^T\right) \tag{9}$$

where C_{A_J} is the covariance matrix at Jet approximation level and C_{D_j} is the covariance matrix at the details scale. The Eigen decomposition of covariance matrices are shown in Eqs. (10) and (11)

$$\mathrm{C_{A_J} V_{A_J} = V_{A_J} \Lambda_{A_J}} \tag{10}$$

$$C_{D_J} V_{D_J} = V_{D_J} \tag{11}$$

whereV_{A_J}, V_{D_J}. and Λ_{A_J}, Λ_{D_J} are the eigenvectors and eigenvalues for matrices of approximation and details subband matrices, respectively [20–22].

Steps for calculation of Eigen values:

- Subband matrices A_6 and D_6 are formed using the coefficients in the sixth subband formed after decomposition.
- The size of matrices A_6 and D_6 is 22×3

- The first column is the coefficients from the left. The second column contains the coefficients of the central lobe, and the third column contains the coefficients of the right lobe.
- The matrices are calculated for each specimen.
- The covariance matrices are then calculated from these two matrices. The size of the covariance matrix is 22 × 22.
- From the covariance matrix, the Eigen values are calculated.

The first six Eigen values obtained from the covariance matrices of both the approximate and detailed subband matrices are considered as features for classification. Therefore, a total of 12 Eigen values are used to form a feature vector for each patient.

The CPPG signals for happy and sad are decomposed using wavelet transform to different wavelet subbands based on their frequency content. Emotions alter the covariance structures and corresponding Eigen values. Eigen values of covariance structures at different scales are used as diagnostic features.

Thus, Eigen space analysis approach can be used for detection and localization of emotions in CPPG signals. Classification of emotions is done using three sets of features: Statistical features with 36 elements, using Eigen values with 12 elements, statistical and Eigen values, i.e., combined features with 48 elements this work 3 classifiers as K-nearest neighbor (KNN), support vector machine (SVM) linear and artificial feed forward (ANN) are used for classification. Classification of emotions are done using three sets of features: Statistical features with 36 elements, using Eigen values with 12 elements, statistical and Eigen values, i.e., combined features with 48 elements.

4 Classifier

A 12-element vector is used as a feature vector z_i for classification into two emotions happy and sad. The 12-elements of feature vector z_i include Eigen valued for happy and sad emotions For this purpose, CPPG signals from 20 healthy patients have been considered and the features are calculated based on the process described. Statistical features with 36 elements, Eigen values with 12 elements, statistical and Eigen values, i.e., combined features with 48 elements. The calculated feature vectors found for each patient of the two groups are stored in Microsoft EXCEL sheets used for further classification using either KNN or SVM for classification purposes.

4.1 Performance Evaluation Measures for Classifier

Performance of classifiers used for classification is evaluated based on the measures of accuracy, sensitivity, and specificity. These measures are calculated by comparing

Table 2 Performance of KNN, SVM, and ANN classifiers

Classifier	TP	TN	FP	FN	Accuracy (%)	Sensitivity (%)	Specificity (%)
Statistical features							
KNN	4	3	1	2	70	66.66	75
SVM Linear	2	3	3	2	50	50	50
ANN	3	3	2	2	60	60	60
Eigen values							
KNN	4	4	1	1	80	80	80
SVM Linear	3	2	3	2	50	60	40
ANN	2	4	1	3	60	40	80
Combined features							
KNN	4	3	1	2	70	66.66	75
SVM *Linear*	2	4	3	1	60	66.66	57.14
ANN	3	4	1	2	70	60	80

the actual input with the predicted output. A confusion matrix is determined based on the comparison. It visualizes the number of true positives (*TPs*), false positives (FPs), false negatives (FNs), and true negatives (TNs) for a classifier where true positives (TP) are cases in which the emotion is predicted to be happy and they do happy. Also, true negatives (TN) are those predicted as not happy, and they are sad (not happy). False positives (FP) are those patients predicted as having sad, but they are actually happy (not sad) (Also known as a "Type I error."), whereas false negatives (FN) are those predicted as sad, but they actually are happy (Also known as a "Type II error".). A confusion matrix is, therefore, a table used to describe the performance of a classification model (or "classifier") on a set of test data for which the true values are known. The definitions of the three performance measures based on the values of TP, TN, FP, and FN used for evaluation of the classifiers are as follows and shown in Table 2.

Sensitivity (SE): The sensitivity relates to the ability of a trained model either based on KNN or SVM to identify positive results of happy emotion is given by Eq. (12)

$$\mathrm{SE} = \frac{\mathrm{TP}}{\mathrm{TP} + \mathrm{FN}} \tag{12}$$

Specificity (SP): The specificity (SP) is related to the ability of identifying the negative outcomes. The specificity (SP) is evaluated using Eq. (13)

$$\mathrm{SP} = \frac{\mathrm{TN}}{\mathrm{TN} + \mathrm{FP}} \tag{13}$$

Accuracy (Acc): The classification accuracy (Acc) of a measurement system is the degree of closeness of measurements of a quantity to that of its actual (true) value, and it is calculated using Eq. (14) [22].

$$SP = \frac{TP + TN}{TP + TN + FP + FN} \tag{14}$$

5 Discussion

Brain neural activity is mostly related to change in blood flow and oxygen level in the blood. Functional activity can be detected by using optical sensor which can detects these parameters and shows variation in CPPG signal.

Cranial PPG of healthy persons is used for analysis using the wavelet transform. Statistical and Eigen space features can be used for emotion classification in prefrontal lobe. The first six Eigen values for happy and sad emotions obtained from the covariance matrices of both the approximate and detailed subband matrices are considered as features for classification.

Emotions alter the covariance structures of multivariate matrices at different scales and the corresponding Eigen values. Eigen values of covariance structures at different scales are used as diagnostic features. Thus, Eigen space analysis approach can be used for detection and localization of emotions in CPPG signals. In this study, three classifiers as k-nearest neighbor (KNN), support vector machine (SVM), linear and artificial feed forward (ANN) are used for classification. Scope of the present work is to classify the happy and sad emotions. But the variation in statistical parameters is observed for different emotions. As a future work, we plan to collect similar datasets and to classify the happy, sad, random, and devotional emotions simultaneously.

6 Conclusion

Classification of emotions are done using three sets of features: Statistical features with 36 elements, using Eigen values with 12 elements, statistical and Eigen values, i.e., combined features with 48 elements. CPPG is also captured for various emotions. The first six Eigen values for happy and sad emotions obtained from the covariance matrices of both the approximate and detailed subband matrices are considered as features for classification. After training the classifier with features calculated from CPPG signals of 10 patients, the KNN classifier was then tested with the features from 10 specimens that were not included for training the classifier. The accuracy of the KNN classifier is observed to be 80%, SVM classifier 60%, and ANN classifier 70%. Classification using statistical features shows an accuracy of 70% for KNN classifier and 60% for ANN classifier. Classification using Eigen features shows an accuracy of 80% for KNN classifier and 60% for ANN. Combined features shows an accuracy of 70% for KNN classifier, 60% for SVM, and 70% for ANN. Analysis with these features shows that these features are capable of discriminating functional activity and emotions.

References

1. Sun, Y., Thakor, N.: Photoplethysmography revisited: from contact to noncontact, from point to imaging. IEEE Trans. Biomed Engzion –resistance remote imaging phoplethysmography based on the optical properties of skin. IEEE Trans. Circ. Syst. Video Technol. 25(5):879–891 (2015)
2. Feng, L.: Lai-Man Po, Xuyuan Xu : Motion –resistance remote imaging phoplethysmography based on the optical properties of skin. IEEE Trans. Circ. Syst. Video Technol. **25**(5), 879–891 (2015)
3. Sun, Y, Hu, S, Azorin-Peris, V, Kalawsky R, Greenwald S.: Noncontact imaging photoplethysmography to effectively access pulse rate variability. J. Biomed. Opt. **18**(6), 061205-1–061205-9
4. Stricker, R., Müller, S., Gross, H.-M.: Non-contact video based pulse rate measurement on a mobile service robot. Conf. Proc. IEEE Int. Symp. Robot Hum .Interact. Commun. 1056–1062 (2014)
5. Sheth, S.A., Yanamadala, Eskandar, E.N.: Intraoperative human functional brain mapping using optical intrinsic signal imaging. Adv. Brain Imaging. ISBN 978-953-307-955-4 (2012)
6. Korhonen, V.O., Myllylä, T.S., Kirillin, M.Y., Popov, A.P., Bykov, A.V., Gorshkov, A.V., Sergeeva, E.A., Kinnunen, M., Kiviniemi, V.: Light propagation in NIR spectroscopy of the human brain. IEEE J. Sel. Top. Quant. Electron. **20**(2) (2014)
7. Hespos, S.J.: What is optical imaging? J. Cogn. Develop. **11**(1), 1–13 (2010)
8. Hou, X., Ding, H., Teng, Y.: NIRS study of cerebral oxygenation and hemoglobin in neonate at birth (2012)
9. Daimiwal, N: Non invasive FNIR and FMRI system for brain mapping. Int. J. Comput. Sci. Inform. Secur. (IJCSIS) **10**(11) (2012)
10. Cromwell, L., Weibell, F.J., Pfeiffer, E.A.: Biomedical and Measurements, 2nd edn. Eastern Economy Edition, New Delhi (2009)
11. OPT 101 Data sheet
12. Carr, J.J., Brown, J.M.: Introduction to biomedical equipment technology
13. Ahammad, N., Fathima, T., Joseph, P.: Detection of epileptic seizure event and onset using EEG. Hindawi Publishing Corporation, BioMed Res. Int. **2014**, Article ID 450573 (2014)
14. Cvetkovic, D., Ubeyli, E.D., Cosic, I.: Wavelet transform feature extraction from PPG, ECG and EEG signal responses to ELF PEMF exposures. Digital Signal Process. **18**(5), 861–874 (2008)
15. Mohd Tumari, S.Z., Sudirman, R., Ahmad, A.H.: Selection of a suitable wavelet for cognitive memory using electroencephalograph signal. Engineering **5**, 15–19 (2013). https://doi.org/10.4236/eng.2013.55B004. Published Online
16. Ting, W., Yan, G.-Z., Yang, B.-H., Hong, S.: EEG feature extraction based on wavelet packet decomposition for brain computer interface. Measurement **41**, 618–625 (2018)
17. Jayachandran, E.S., Joseph, K.P., Acharya, U.R.: Analysis of myocardial infarction using discrete wavelet transform. J. Med. Syst. **34**(6), 985–992 (2010)
18. Arif, M, Malagore, I.A., Afsar, F.A.: Detection and localization of myocardial infarction using k-nearest neighbor classifier. J. Med. Syst. **36**(1), 279–289 (2012)
19. Khandoker, A.H., Palaniswami, M., Karmakar, C.J.: Support vector machines for automated recognition of obstructive sleep apnea syndrome from ECG recordings. IEEE Trans. Inform. Technol. Biomed. **13**(1), 37–48 (2009).
20. Sun, L., Lu, Y., Yang, K.: ECG analysis using multiple instance learning for myocardial infarction detection. IEEE Trans. Biomed. Eng. **59**(12), 3348–3356(2012)
21. Sharma, L.N., Dandapat, S., Mahanta, A.: ECG signal denoising using higher order statistics in wavelet subbands. Biomed. Signal Process. Control **5**(3), 214–222 (2010)
22. Sharma, L.N., Tripathy, R.K., Dandapat, S.: Multiscale energy and eigenspace approach to detection and localization of myocardial infarction. IEEE Trans. Biomed. Eng. **62**(7), 1827–37 (2015)

Position-Sensorless Control System for Electric Vehicle with Brushless DC Motor

Akshaya Mhatre and Archana Thosar

1 Introduction

The advancement of electric vehicle technology has become necessity of the world due to dual pressure from environmental pollution and fossil fuel crisis. Efficiency and size are the crucial criteria for EVs and thus brushless DC (BLDC) motors are well liked for this application. However, BLDC motor requires additional mechanical position sensors like encoders, Hall sensors, and resolvers, which are responsible not only for increasing cost and size of the motor but also for restricting its application in harsh environments. Especially, Hall sensors, which make the BLDC motor drives less robust as well as their misalignment lead to significant torque/current ripple. Thus, position-sensorless operation of BLDC motor has become important. Numerous position-sensorless techniques for BLDC motor, namely (1) Estimation of back electromotive force (EMF); (2) Inductance variation method; and (3) Flux linkage variation method, have been discussed in literature [1].

In flux linkage variation method, measured terminal voltages and line currents are used to implement an algorithm for estimation of flux linkage and position of permanent magnet motor. At each time step, the two stages of the line current estimation are carried out for correcting the estimated position and estimated flux linkage [2–4]. A methodology for predicting the position of rotor and operating a BLDC motor fluently from zero speed is presented in [5]. Here, the responses of current to the variation in inductance with respect to position of rotor are tracked. The BLDC motor drive can be operated smoothly up to medium speed with this algorithm as compared to the traditional back EMF technique. However, these two methods are more appropriate for the permanent magnet (PM) motors with saliency. In case of

A. Mhatre (✉) · A. Thosar
College of Engineering Pune, Pune, India
e-mail: mhatreaa18.elec@coep.ac.in

A. Thosar
e-mail: agt.elec@coep.ac.in

S. K V and K. Guruprasad Rao (eds.), *Smart Sensors Measurements and Instrumentation*, Lecture Notes in Electrical Engineering 750,
https://doi.org/10.1007/978-981-16-0336-5_27

surface-mounted rotor where inherent saliency is absent, flux linkage and inductance variations with reference to rotor position arise only because of magnetic saturation.

The another method for operating BLDC motor drive in position-sensorless mode, using back EMF zero-crossing point (ZCP) estimation is proposed in [6–8], which is applicable to the machines with any type of the rotor. This method involves detection of ZCP of back EMF and then generation of appropriate commutation signal for the corresponding phase. In realistic implementation, the terminal voltages or phase voltages are used to estimate back EMF. A method to detect back EMF ZCPs using terminal voltages and without making use of neutral point voltage is presented in [9]. In [10], the ZCP detection signal of back EMF is predicted secondarily using terminal voltages calculated with reference to negative end of DC supply followed by delaying that signal by 30° to achieve appropriate commutation point. The same method is used for borewell application in [11], where low pass filters are used instead of sample and hold circuits as it is more advantageous.

Again with this method, the position estimation is possible only after attaining a threshold speed, when the magnitude of back EMF becomes considerable to be detected. Therefore, when the motor is to be operated from standstill, the starting acceleration is obtained under open-loop configuration by making use of a ramped frequency signal [12]. A simple start-up algorithm involving prepositioning of rotor is proposed in [13] where succeeding rotor rotation is achieved by triggering the switching devices sequentially, depending on ZCP of back EMF.

This paper adopts the ZCP detection of back EMF method for position-sensorless functioning of BLDC motor and initial acceleration of the rotor is obtained by energizing motor phases with ramped frequency signal. The mathematical modeling of BLDC motor and its specific position information requirement is explained in Sect. 2. The working principle of proposed technique is described in Sect. 3, whereas the simulation of proposed system is performed using MATLAB/Simulink and results of the same are discussed in Sect. 4.

2 Mathematical Modeling of BLDC Motor

PM synchronous machines with trapezoidal-shaped-induced EMF are identified as BLDC motor. A BLDC motor is considered to have a three-phase winding connected in star configuration and a surface-mounted magnet-type rotor structure. The commutation is achieved by sensing the rotor position followed by energizing appropriate phases using three-phase inverter as illustrated in Fig. 1.

The equations of phase voltages at stator winding terminals are [14],

$$V_a = i_a R_s + L_s \frac{\mathrm{d}i_a}{\mathrm{d}t} + e_a \tag{1}$$

$$V_b = i_b R_s + L_s \frac{\mathrm{d}i_b}{\mathrm{d}t} + e_b \tag{2}$$

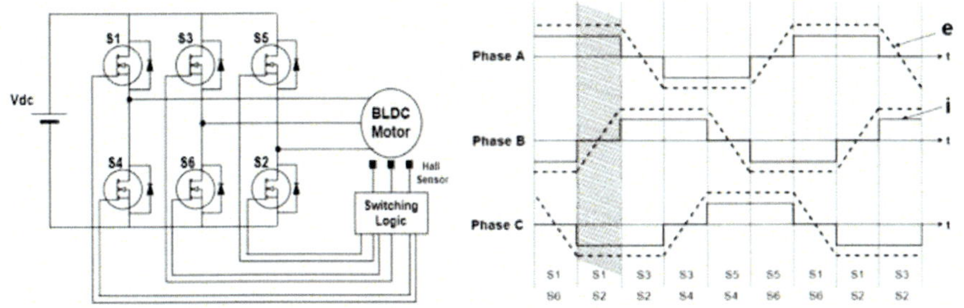

Fig. 1 Typical BLDC motor drive along with back EMF and phase current waveforms [10]

$$V_c = i_c R_s + L_s \frac{di_c}{dt} + e_c \tag{3}$$

where R_s is resistance (per phase) and L_s is inductance (per phase) of the stator windings, respectively. The electromagnetic torque is expressed by,

$$T_e = \frac{e_a i_a + e_b i_a + e_c i_c}{\omega_m} \tag{4}$$

where the instantaneous induced emfs are given by,

$$\begin{aligned} e_a &= f_a(\theta_r) k_b \omega_m \\ e_b &= f_b(\theta_r) k_b \omega_m \\ e_c &= f_c(\theta_r) k_b \omega_m \end{aligned} \tag{5}$$

where functions $f_a(\theta_r)$, $f_b(\theta_r)$, and $f_c(\theta_r)$ have shape as same as corresponding back EMF and their magnitude varies within ±1. k_b is back EMF constant. Additional mechanical position sensors are required for sensing correct rotor position, which not only raise the size and cost of the motor but also restrict its application. Therefore, position-sensorless operation of BLDC motor has become important.

3 Back EMF Estimation

The back EMF can be estimated using terminal phase voltages also [9]. However, it is not effectual during lower speed operations because of less magnitude of back EMF. Thus, in this work, difference of two-line voltages are used for back EMF calculation, indirectly. Figure 2 shows schematic of position-sensorless functioning of BLDC motor using proposed technique.

The line voltages can be calculated using (1), (2), and (3);

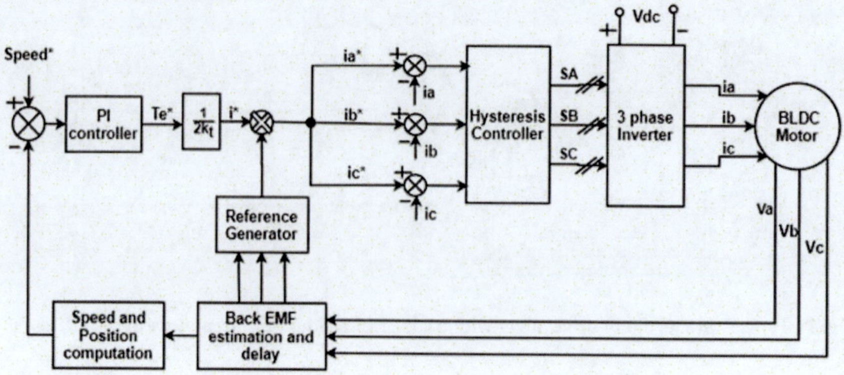

Fig. 2 Back EMF estimation-based closed-loop speed control of BLDC motor

$$V_{ab} = V_a - V_b = (i_a - i_b)R_s + L_s\frac{\mathrm{d}(i_a - i_b)}{\mathrm{d}t} + e_a - e_b \tag{6}$$

$$V_{bc} = V_b - V_c = (i_b - i_c)R_s + L_s\frac{\mathrm{d}(i_b - i_c)}{\mathrm{d}t} + e_b - e_c \tag{7}$$

$$V_{ca} = V_c - V_a = (i_c - i_a)R_s + L_s\frac{\mathrm{d}(i_c - i_a)}{\mathrm{d}t} + e_c - e_a \tag{8}$$

With reference to the negative DC bus, terminal voltages can be measured and their difference can be used to estimate these line voltages, eliminating requirement of star point. From Eqs. (6) and (7), the back EMF is calculated as,

$$\begin{aligned} V_{abbc} &= V_{ab} - V_{bc} \\ &= R_s(i_a - 2i_b + i_c) + L_s\frac{\mathrm{d}(i_a - 2i_b + i_c)}{\mathrm{d}t} + e_a - 2e_b + e_c \end{aligned} \tag{9}$$

Consider the shaded region in Fig. 1, when A and C phases are energized while B phase is open, $i_a = -i_c$, and $i_b = 0$. Also, $e_a = -e_c$ in the shaded region. Thus, for shaded interval, (9) can be simplified as,

$$V_{abbc} = -2e_b \tag{10}$$

Therefore, estimation of back EMF of phase B can be done using line voltages V_{ab} and V_{bc}. Likewise, $-2e_a$ and $-2e_c$ can be calculated from V_{caab} and V_{bcca}, respectively. The current flowing through the winding routes through the freewheeling diodes, because of the motor inductance. This causes the false identification of ZCPs of back EMF. Therefore, a low pass filter (LPF) is incorporated in this control technique which has several benefits over sample and hold circuit [11], where the required

Table 1 Commutation signals [11]

Difference in line voltages	Nature of zero crossing	Turn on switches
V_{abbc}	Positive	S5, S6
V_{abbc}	Negative	S3, S2
V_{bcca}	Positive	S1, S2
V_{bcca}	Negative	S5, S4
V_{caab}	Positive	S3, S4
V_{caab}	Negative	S1, S6

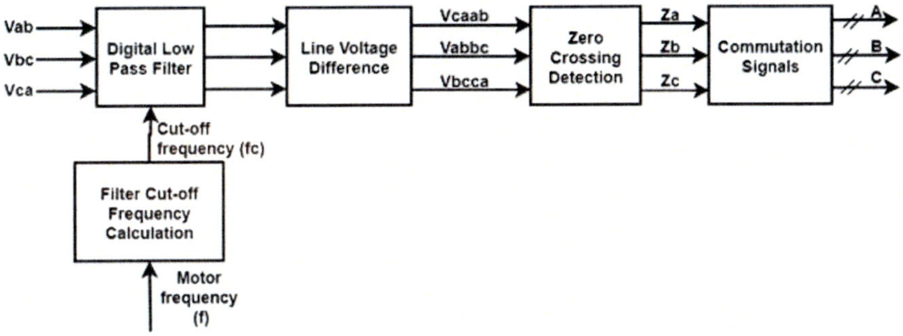

Fig. 3 Block diagram of position-sensorless operation of BLDC motor based on back EMF estimation

phase shift (θ_{LPF}) of 30° electrical is achieved by adjusting its cut-off frequency with respect to rotor speed.

Thus, the ZCP of back EMFs e_a, e_b, and e_c are acquired and appropriate commutation signal is generated. Table 1 shows the commutation signals with respect to the zero-crossing signals. BLDC motor speed is estimated with a speed observer [15], where the torque of a BLDC motor acts as an input variable and it is calculated using (4). Once the speed is estimated, the rotor position is predicted directly by its integration.

However, the back EMF is zero initially and the position of rotor is also unknown. Therefore, the motor is started using a starting algorithm [12]. The block diagram of the presented technique for position-sensorless control of BLDC motor is shown in Fig. 3.

4 Results and Discussion

The operation of the proposed method is observed by simulating it with MATLAB/Simulink. A 51 W, 28 V, 18,800 rpm BLDC motor is used for the simulation. Its detailed specifications are specified in Table 2. Initially for getting starting

Table 2 Specifications of BLDC motor

Parameters	Symbols	Values
Voltage	V	28 V
Per phase resistance	R_s	2 Ω
Per phase inductance	L_s	175 μH
Torque constant	k_t	10.3×10^{-3} Nm/A
Back EMF constant	k_b	10.31×10^{-3} V/rad/s
Number of poles	P	2
Rotor inertia	J	5.7×10^{-7} kg m^2

acceleration, the motor is operated in starting mode. As soon as the rotor speed reaches to 300 rpm, back EMF becomes sufficiently detectable and the control is latched to position-sensorless mode. Here, proportional-integral (PI) controller is used for closed-loop operation and hysteresis controller is used as current controller. Here, $K_p = 0.00006$ and $K_i = 0.000011$ are selected for PI controller using trial and error method.

The real position of rotor and estimated position of rotor, which are observed from simulation are shown in Fig. 4. It can be noticed from Fig. 4 that, the rotor position which is estimated is very much near to the actual rotor position. Therefore, speed observer [15] can be efficiently used for speed and position estimation. It also can be noticed from Fig. 5 that, the value of back EMF constant which is estimated is nearly equal to the actual value of the same that is 10.3×10^{-3} V/rad/s.

The back EMF and filtered line voltage difference (inverted) waveforms are presented in Fig. 6. It can be observed that the magnitude of estimated back EMF is twice that of actual one. Thus, this technique is efficient in low speed ranges also,

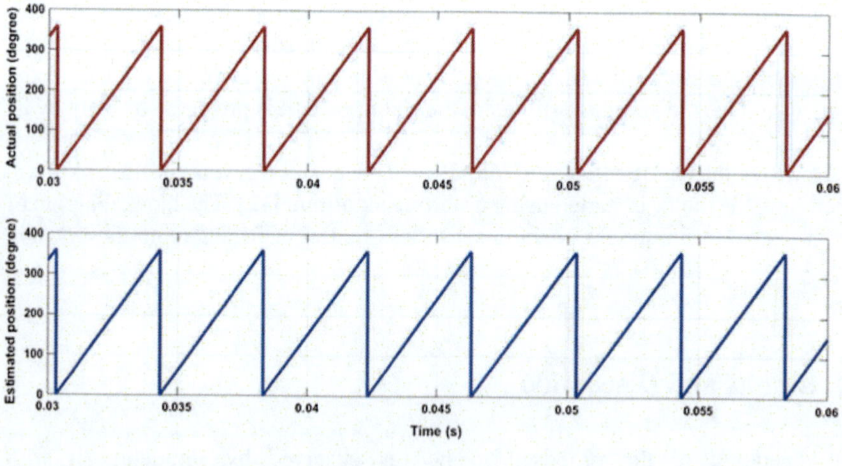

Fig. 4 Actual versus estimated angular rotor position

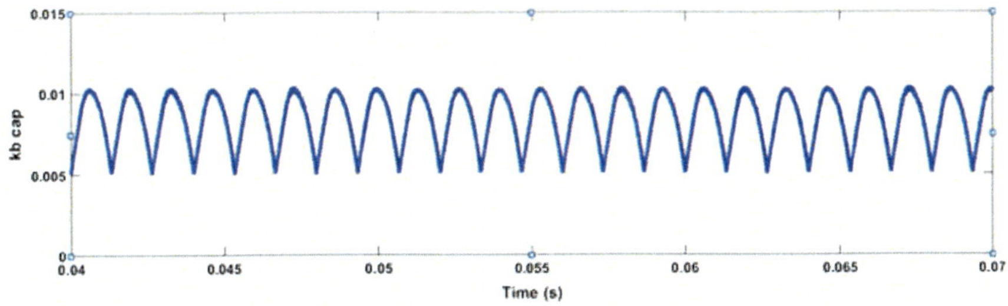

Fig. 5 Estimated back EMF constant

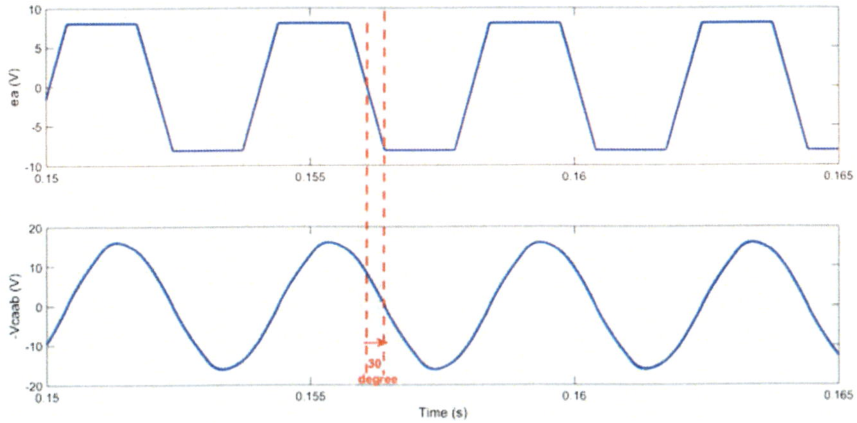

Fig. 6 Back EMF and inverted line voltage difference waveforms during position-sensorless running mode

where magnitude of actual back EMF is very less. It also can be seen that the estimated back EMF waveforms are delayed by 30° by setting cut-off frequency of LPF, to obtain appropriate commutation instant.

The effectiveness of the system for varying speed inputs at no-load condition is verified by applying reference speeds in five consecutive steps of magnitudes namely (1) 12,000 rpm at $t = 0$ s, (2) 7000 rpm at $t = 0.08$ s, (3) 15,000 rpm at $t = 0.13$ s, (4) 5000 rpm at $t = 0.2$ s, and (5) 17,000 rpm at 0.27 s. The speed waveform of the motor with respect to given reference speeds is shown in Fig. 7 and waveform of estimated back EMF for phase A is shown in Fig. 8. The current error in all the three phases is observed to be varying in the band of ± 0.3 A due to hysteresis controller as appeared in Fig. 9.

While operating the motor from standstill and in no-load condition, the speed increases constantly and comes to stable state where any deformation is absent, as noticed in Fig. 10. Then, the motor is subsequently loaded in running mode at 0.1, 0.2, and 0.3 s to check the reliability of the proposed technique in transient conditions. It can be noticed from Fig. 10 that, the motor runs continuously even after sudden

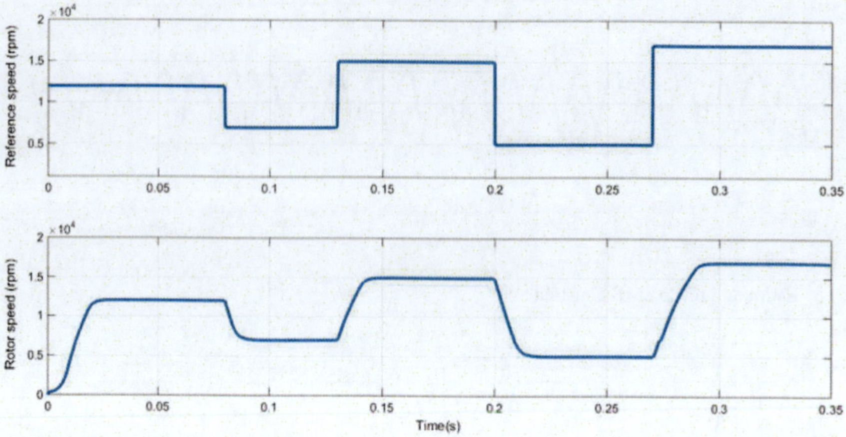

Fig. 7 Speed waveform on no-load

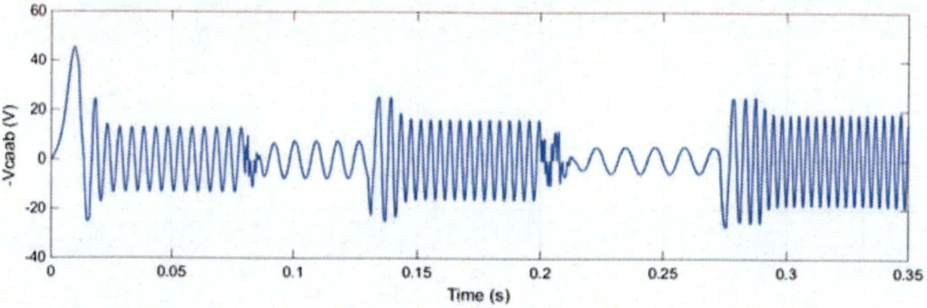

Fig. 8 Estimated back EMF waveform

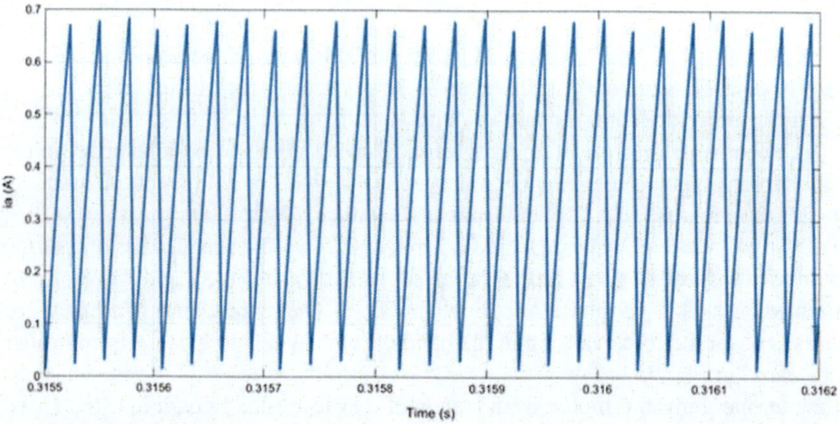

Fig. 9 Phase current waveform on load

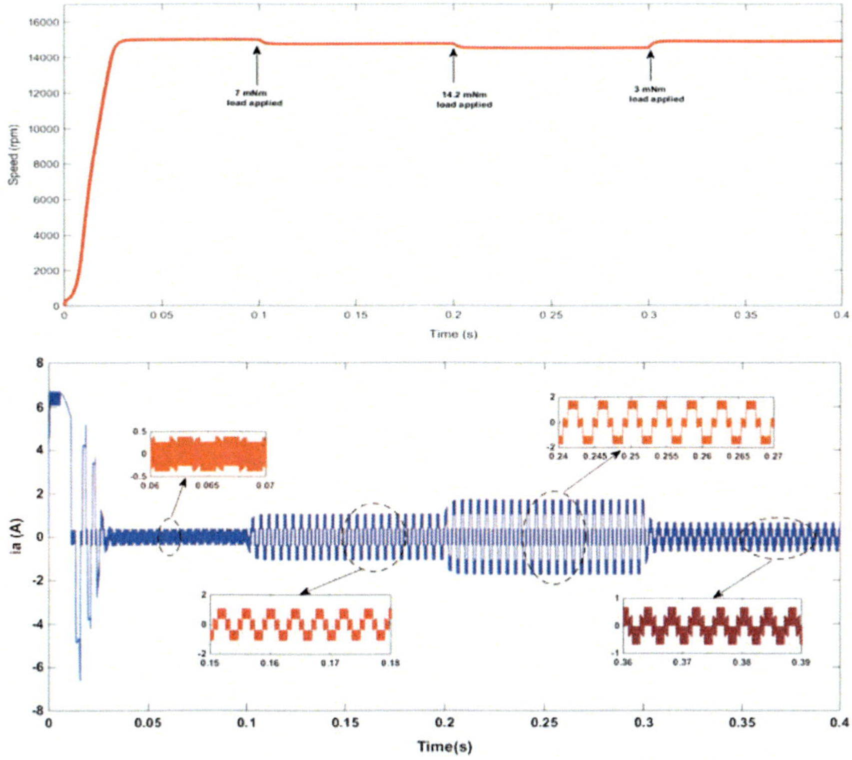

Fig. 10 Speed and phase current waveform during loading transient

application of load. This validates the performance of the proposed technique under transient load application.

5 Conclusion

Reliable and efficient BLDC motor drives are necessary for betterment of EV technology. A position-sensorless operation technique for BLDC motor is presented in this paper to eliminate the need of position sensors and thus to increase the reliability of BLDC motor drive in EV. This technique utilizes line–line voltage difference calculated at motor terminals for estimation of back EMF. Correct commutation signals are obtained directly from zero-crossing information of estimated back EMF by setting cut-off frequency of LPF. Thus, complexity of the control algorithm is reduced. The proposed method is simulated using a 51 W BLDC motor and tested for various reference speeds. It is observed that the BLDC motor can start and run smoothly without position sensors. Simulation results are shown which support the appropriateness of this work.

This paper has attempted to present an operation of BLDC motor in position-sensorless mode using PI controller. Another controller can be used to improve reliability of the system.

References

1. Acarnley, P.P., Watson, J.F.: Review of position-sensorless operation of brushless permanent-magnet machines. IEEE Trans. Ind. Electron. **53**(2), 352–362 (2006)
2. Kim, T.H., Ehsani, M.: Sensorless control of The BLDC motors from near-zero to high speeds. IEEE Trans. Power Electron. **19**(6), 1635–1645 (2004)
3. French, C., Acarnley, P.P.: Control of permanent magnet motor drives using a new position estimation technique. IEEE Trans. Ind. Appl. **32**(5), 1089–1097 (1996)
4. Ertugrul, N., Acarnley, P.P.: A new algorithm for sensorless operation of permanent magnet motors. IEEE Trans. Ind. Appl. **30**(1), 126–133 (1994)
5. Jang, G.H., Park, J.H., Chang, J.H.: Position detection and start-up algorithm of a rotor in a sensorless BLDC motor utilising inductance variation. IEE Proc. Electric Power Appl. **149**(2), 137–142 (2002)
6. Lai, Y., Lin, Y.: A Unified approach to zero-crossing point detection of back EMF for brushless DC motor drives without current and hall sensors. IEEE Trans. Power Electron. **26**(6), 1704–1713 (2011)
7. Shao, J.: An improved microcontroller-based sensorless brushless DC (BLDC) motor drive for automotive applications. IEEE Trans. Ind. Appl. **42**(5), 1216–1221 (2006)
8. Su, G.J., McKeever, J.W.: Low-cost sensorless control of brushless DC motors with improved speed range. IEEE Trans. Power Electron. **19**(2), 296–302 (2004)
9. Texas Instruments Inc.: Sensorless Trapezoidal Control of BLDC Motors. Application Report No. SPRABQ7A–July 2013–Revised Sept 2015
10. Damodharan, P., Vasudevan, K.: Sensorless brushless DC motor drive based on the zero-crossing detection of back electromotive force (EMF) from the line voltage difference. IEEE Trans. Energy Convers. **25**(3), 661–668 (2010)
11. Sashidhar, S., Guru Prasad Reddy, V., Fernandes, B.G.: A single-stage sensorless control of a PV-based bore-well submersible BLDC motor. IEEE J. Emerg. Sel Top. Power Electron. **7**(2), 1173–1180 (2019)
12. Krishnan, R., Ghosh, R.: Starting algorithm and performance of a PM DC brushless motor drive system with no position sensor. In: 20th Annual IEEE Power Electronics Specialists Conference, vol. 2, pp. 815–821 (1989)
13. Damodharan, P., Sandeep, R., Vasudevan, K.: Simple position sensorless starting method for brushless DC motor. IET Electr. Power Appl. **2**(1), 49–55 (2008)
14. Pillay, P., Krishnan, R.: Modeling, simulation, and analysis of permanent- magnet motor drives. II. The brushless DC motor drive. In: IEEE Trans. Ind. Appl. 274–279 (1989)
15. Park, J., Hwang, S., Kim, J.: Sensorless control of brushless DC motors with torque constant estimation for home appliances. IEEE Trans. Ind. Appl. **48**(2) (2012)

Performance Analysis of Sliding Mode-Controlled Bidirectional DC–DC Converter for Electric Vehicles

Mathew K. Kanthi and Abraham Dolly Mary

1 Introduction

One of the most prominent issues for the environment today is the pollution generated by hydrocarbon combustion, which is one of the primary origins of power for transportation. The popularity of electric vehicles is increasing due to their low carbon footprint and ease of integration with renewable energy [1, 2]. Three important barriers for electric vehicles to win wide acceptance are the high cost of batteries, complications of battery chargers and the lack of charging infrastructure. The current trend is to simplify battery charge and discharge management. A bidirectional DC–DC converter can accomplish to maintain a healthy battery and extend battery runtime [3].

Control of DC–DC converters plays a significant role in energy management of electric vehicles. The controllers for power converters play a vital role in efficient conversion and power transfer with guaranteed stability and performance [4]. The bidirectional DC–DC converters are time-variant systems and exhibit nonlinearity. Due to non-minimum phase nature of these converters, controller design is complicated. Robust performance of converters cannot be achieved with conventional proportional integral (PI) linear controller. Hence, SMC which is a nonlinear controller is necessary for electric vehicle applications [5].

Sliding mode control (SMC) is a powerful tool for obtaining robust performance for DC–DC power converters which is a variable structure model [6, 7]. The main drawback of SMC is the high-frequency switching operation called chattering effect which results in undesired oscillations. To overcome the high switching frequency

M. K. Kanthi (✉) · A. Dolly Mary
Department of Electrical Engineering, Rajiv Gandhi Institute of Technology, Kottayam, Kerala, India
e-mail: kanthimathew07@rit.ac.in

A. Dolly Mary
e-mail: dollymary@rit.ac.in

S. K V and K. Guruprasad Rao (eds.), *Smart Sensors Measurements and Instrumentation*, Lecture Notes in Electrical Engineering 750,
https://doi.org/10.1007/978-981-16-0336-5_28

limitation, a voltage-controlled sliding mode controller was designed incorporating the PWM technique for DC–DC converters operating in continuous conduction mode [8].

Constant switching frequency sliding mode controller can be obtained by employing hysteresis modulation [9] or pulse width modulation [10, 11]. The PWM-based SMC offers a simple control implementation. The sliding mode voltage controller for buck converter based on a parabolic modulation was proposed in [12]. The sliding mode control in [13] provides a PWM-based approach to avoid mismatched disturbance in a buck converter. In [14], sliding mode voltage control of the boost converter with good voltage regulation with unknown load demand was presented.

In this paper, a bidirectional DC–DC converter controlled by a PWM-based sliding mode control is discussed in detail for both buck and boost mode of operation. PWM-based SMC refers to a pulse width modulator that uses an equivalent control which is obtained by applying sliding mode control technique to generate a control signal. The equivalent control is then compared with the fixed frequency ramp signal [15]. Such a controller gives good stability for large load and line changes. The performance of the converter is validated using the Typhoon Virtual HIL emulator.

2 Bidirectional DC–DC Converter Model

The bidirectional DC–DC converter is shown in Fig. 1. The electric vehicle battery is represented by a low voltage v_{lv}, and DC bus is represented by a high voltage v_{hv}. The low-voltage side input current and capacitor current are represented by i_{lv} and i_{Clv}, while i_{hv} and i_{chv} represent high-voltage side input current and capacitor current, respectively. The inductor current is represented by i_{l}.

The sliding mode controller is designed with two basic steps: Initially a sliding surface is designed in state space, and then control law is derived. The design equation is formulated such that the system trajectory reaches the sliding surface from any

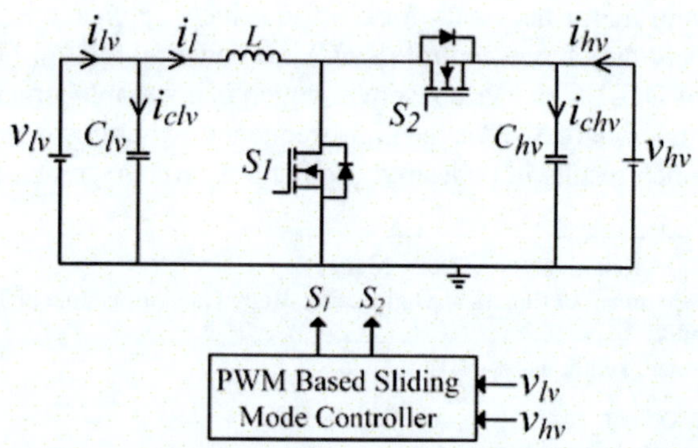

Fig. 1 Bidirectional DC–DC converter

arbitrary initial state in finite time to an equilibrium state, i.e., the origin point of the phase plane. Finally, the stability of the controller is based on three important factors, viz. existence condition, stability condition and hitting condition. The control variable of SMC is selected based on proportional integral derivative controller.

To analyze the performance of the bidirectional DC–DC converter operated in continuous conduction mode, two operating conditions are considered: the buck operation and the boost operation. First the state space average model for each operations is obtained and then SMC based on PWM was developed [16]. The control variables for state space modeling of each operation are voltage error x_1, the rate of change of voltage error x_2 and the integral of voltage error x_3.

2.1 Buck Operation

For a buck converter, v_{vh} and v_{lv} denote instantaneous input and output voltages, respectively, and the load is represented by R_{lv}. The circuit description for the buck operation is shown in Fig. 2.

The control variable for buck operation is given by

$$x_{\mathrm{bck}} = \begin{bmatrix} x_1 \\ x_2 \\ x_3 \end{bmatrix} = \begin{bmatrix} v_{\mathrm{lvref}} - \beta v_{\mathrm{lv}} \\ \frac{\mathrm{d}(v_{\mathrm{lvref}} - \beta v_{\mathrm{lv}})}{\mathrm{d}t} \\ \int (v_{\mathrm{lvref}} - \beta v_{\mathrm{lv}})\mathrm{d}t \end{bmatrix} = \begin{bmatrix} v_{\mathrm{lvref}} - \beta v_{\mathrm{lv}} \\ -\beta \frac{\mathrm{d}v_{\mathrm{lv}}}{\mathrm{d}t} \\ \int x_1 \mathrm{d}t \end{bmatrix} \tag{1}$$

where v_{lvref} is reference voltage and β the feedback ratio.

From the behavioral model of buck operation under continuous conduction mode [17], the following expression is derived:

$$-\beta \frac{\mathrm{d}v_{\mathrm{lv}}}{\mathrm{d}t} = \frac{\beta v_{\mathrm{lv}}}{R_{\mathrm{lv}} C_{\mathrm{lv}}} + \int \frac{\beta(v_{\mathrm{lv}} - v_{\mathrm{hv}} u)}{L C_{\mathrm{lv}}} \mathrm{d}t \tag{2}$$

where u is the switching state of switch. Substitution of Eq. (2) into Eq. (1) gives

Fig. 2 Buck operation

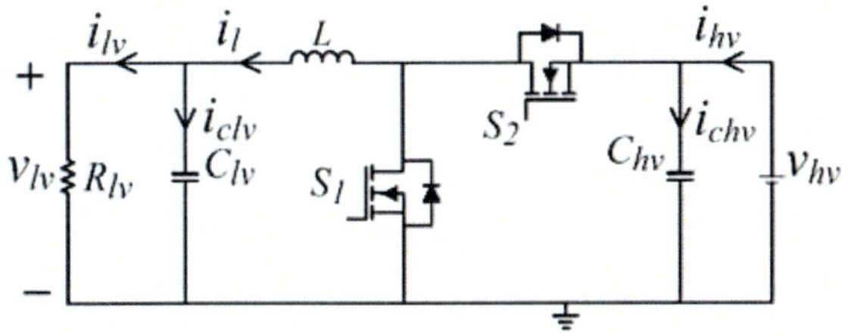

$$x_{\mathrm{bck}} = \begin{bmatrix} v_{\mathrm{lvref}} - \beta v_{\mathrm{lv}} \\ \frac{\beta v_{\mathrm{lv}}}{R_{\mathrm{lv}} C_{\mathrm{lv}}} + \int \frac{\beta (v_{\mathrm{lv}} - v_{\mathrm{hv}} u)}{L C_{\mathrm{lv}}} \\ \int x_1 \mathrm{d}t \end{bmatrix} \quad (3)$$

Differentiating Eq. (3) will give the state space description of the converter for the controller design as

$$\begin{bmatrix} \dot{x}_1 \\ \dot{x}_2 \\ \dot{x}_3 \end{bmatrix} = \begin{bmatrix} 0 & 1 & 0 \\ 0 & -\frac{1}{R_{\mathrm{lv}} C_{\mathrm{lv}}} & 0 \\ 1 & 0 & 0 \end{bmatrix} \begin{bmatrix} x_1 \\ x_2 \\ x_3 \end{bmatrix} + \begin{bmatrix} 0 \\ -\frac{\beta v_{\mathrm{hv}}}{L C_{\mathrm{hv}}} \\ 0 \end{bmatrix} + \begin{bmatrix} 0 \\ \frac{\beta v_{\mathrm{lv}}}{L C_{\mathrm{lv}}} \\ 0 \end{bmatrix} \quad (4)$$

2.2 Boost Operation

For a boost converter, v_{lv} and v_{hv} denote instantaneous input and output voltages, respectively. The circuit description of the boost operation is shown in Fig. 3. The control variable for this operation with v_{hvref} is the reference voltage given by

$$x_{\mathrm{bst}} = \begin{bmatrix} x_1 \\ x_2 \\ x_3 \end{bmatrix} = \begin{bmatrix} v_{\mathrm{hvref}} - \beta v_{\mathrm{hv}} \\ \frac{\mathrm{d}(v_{\mathrm{hvref}} - \beta v_{\mathrm{hv}})}{\mathrm{d}t} \\ \int (v_{\mathrm{hvref}} - \beta v_{\mathrm{hv}}) \mathrm{d}t \end{bmatrix} = \begin{bmatrix} v_{\mathrm{hvref}} - \beta v_{\mathrm{hv}} \\ -\beta \frac{\mathrm{d}v_{\mathrm{hv}}}{\mathrm{d}t} \\ \int x_1 \mathrm{d}t \end{bmatrix} \quad (5)$$

From the behavioral model of boost operation under continuous conduction mode, the following expression is derived:

$$-\beta \frac{\mathrm{d}v_{hv}}{\mathrm{d}t} = \frac{\beta v_{\mathrm{hv}}}{R_{\mathrm{hv}} C_{\mathrm{hv}}} + \int \frac{\beta (v_{\mathrm{hv}} - v_{\mathrm{lv}}) \overline{u}}{L C_{\mathrm{hv}}} \mathrm{d}t \quad (6)$$

where $\overline{u}$ is the inverse logic of u. Substitution of Eq. (6) into Eq. (5) gives

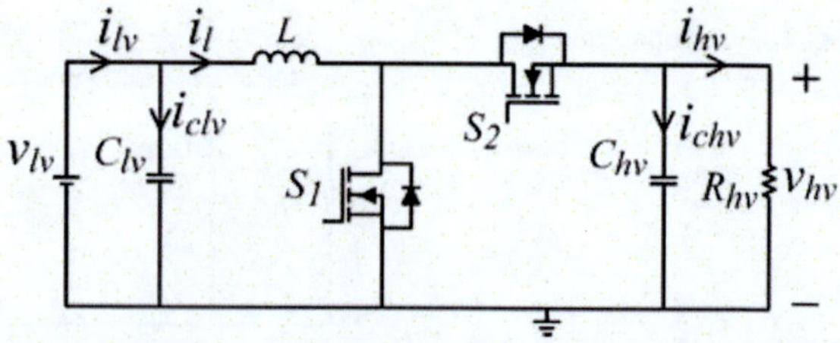

Fig. 3 Boost operation

$$x_{\text{bst}} = \begin{bmatrix} v_{\text{hvref}} - \beta v_{\text{hv}} \\ \frac{\beta v_{\text{hv}}}{R_{\text{Hv}} C_{\text{Hv}}} + \int \frac{\beta (v_{\text{lv}} - v_{\text{lv}}) \overline{u}}{L C_{\text{hv}}} \\ \int x_1 \text{d}t \end{bmatrix} \tag{7}$$

Differentiating Eq. (7) gives the state space description of the converter for the controller design.

$$\begin{bmatrix} \dot{x}_1 \\ \dot{x}_2 \\ \dot{x}_3 \end{bmatrix} = \begin{bmatrix} 0 & 1 & 0 \\ 0 & -\frac{1}{R_{\text{hv}} C_{\text{hv}}} & 0 \\ 1 & 0 & 0 \end{bmatrix} \begin{bmatrix} x_1 \\ x_2 \\ x_3 \end{bmatrix} + \begin{bmatrix} 0 \\ \frac{\beta (v_{\text{lv}} - v_{\text{lv}})}{L C_{hv}} \\ 0 \end{bmatrix} \overline{u} \tag{8}$$

3 Sliding Mode Controller Design

The sliding mode control is basically designed with a particular sliding surface with a control law that brings the state variable trajectory to the desired origin. The instantaneous state variable trajectory, S, is described as

$$S = \alpha_1 x_1 + \alpha_2 x_2 + \alpha_3 x_3 \tag{9}$$

where α_1, α_2 and α_3 are sliding coefficients. To define a control law, the switching function is described as

$$u = \begin{cases} 1 \text{ for } S > 0 \\ 0 \text{ for } S < 0 \end{cases} \tag{10}$$

3.1 Existence Conditions

In sliding mode control schemes, the existence condition of the controller is analyzed using the Lyapunov's method. The controller must satisfy the local reachability condition given by $\lim_{S \to 0} S \cdot \dot{S} < 0$ which can be expressed as

$$\begin{cases} \dot{S}_{S \to 0+} < 0 \\ \dot{S}_{S \to 0-} > 0 \end{cases} \tag{11}$$

For Buck Operation. The time differentiation of state variable trajectory, $\dot{S}$, is described as

$$\dot{S} = -\alpha_1 \frac{\beta i_{\text{clv}}}{C_{\text{lv}}} + \alpha_2 \frac{\beta i_{\text{clv}}}{R_{\text{lv}} C_{\text{lv}}^2} + \alpha_3 (v_{\text{lvref}} - \beta v_{\text{lv}}) + \alpha_2 \frac{\beta v_{\text{lv}}}{L C_{\text{lv}}} - \alpha_2 \frac{\beta v_{\text{hv}}}{L C_{\text{lv}}} u \tag{12}$$

Substituting $u = 1$ and $u = 0$ into Eq. (12), the reachability conditions of buck converter are derived as

$$\begin{cases} -\alpha_1 \frac{\beta i_{\text{clv}}}{C_{\text{lv}}} + \alpha_2 \frac{\beta i_{\text{clv}}}{R_{\text{lv}} C_{\text{lv}}^2} + \alpha_3 (v_{\text{lvref}} - \beta v_{\text{lv}}) + \alpha_2 \frac{\beta v_{\text{lv}}}{L C_{\text{lv}}} - \alpha_2 \frac{\beta v_{\text{hv}}}{L C_{\text{lv}}} < 0 \\ -\alpha_1 \frac{\beta i_{\text{clv}}}{C_{\text{lv}}} + \alpha_2 \frac{\beta i_{\text{clv}}}{R_{\text{lv}} C_{\text{lv}}^2} + \alpha_3 (v_{\text{lvref}} - \beta v_{\text{lv}}) + \alpha_2 \frac{\beta v_{\text{lv}}}{L C_{\text{lv}}} > 0 \end{cases} \tag{13}$$

Combining the two inequalities yields

$$0 < -\alpha_1 \frac{\beta i_{\text{clv}}}{C_{\text{lv}}} + \alpha_2 \frac{\beta i_{\text{clv}}}{R_{\text{lv}} C_{\text{lv}}^2} + \alpha_3 (v_{\text{lvref}} - \beta v_{\text{lv}}) + \alpha_2 \frac{\beta v_{\text{lv}}}{L C_{\text{lv}}} < \alpha_2 \frac{\beta v_{\text{hv}}}{L C_{\text{lv}}} \tag{14}$$

Multiplying the above equation by $\frac{LC_{\text{lv}}}{\alpha_2}$ gives the existence condition for buck operation.

$$0 < -\beta L \left(\frac{\alpha_1}{\alpha_2} - \frac{1}{R_{\text{lv}} C_{\text{lv}}} \right) i_{\text{clv}} + L C_{\text{lv}} \frac{\alpha_3}{\alpha_2} (v_{\text{lvref}} - \beta v_{\text{lv}}) + \beta v_{\text{lv}} < \beta v_{\text{hv}} \tag{15}$$

For Boost Operation. The time differentiation of state variable trajectory, $\dot{S}$, is described as

$$\dot{S} = -\alpha_1 \frac{\beta i_{\text{chv}}}{C_{\text{hv}}} + \alpha_2 \frac{\beta i_{\text{chv}}}{R_{\text{hv}} C_{\text{hv}}^2} + \alpha_3 (v_{\text{hvref}} - \beta v_{\text{hv}}) + \alpha_2 \frac{\beta (v_{\text{hv}} - v_{\text{lv}})}{L C_{\text{hv}}} \overline{u} \tag{16}$$

Substituting $\overline{u} = 0$ and $\overline{u} = 1$ into Eq. (16), the reachability condition for boost converter is derived as

$$\begin{cases} -\alpha_1 \frac{\beta i_{\text{chv}}}{C_{\text{lv}}} + \alpha_2 \frac{\beta i_{\text{chv}}}{R_{lv} C_{\text{lv}}^2} + \alpha_3 (v_{\text{hvref}} - \beta v_{\text{hv}}) < 0 \\ -\alpha_1 \frac{\beta i_{\text{chv}}}{C_{\text{hv}}} + \alpha_2 \frac{\beta i_{\text{chv}}}{R_{\text{hv}} C_{\text{hv}}^2} + \alpha_3 (v_{\text{hvref}} - \beta v_{\text{hv}}) + \alpha_2 \frac{\beta (v_{\text{hv}} - v_{\text{lv}})}{L C_{\text{hv}}} > 0 \end{cases} \tag{17}$$

Combining the two inequalities yields

$$0 < \alpha_1 \frac{\beta i_{\text{chv}}}{C_{\text{hv}}} - \alpha_2 \frac{\beta i_{\text{chv}}}{R_{\text{hv}} C_{\text{hv}}^2} - \alpha_3 (v_{\text{hvref}} - \beta v_{\text{hv}}) < \alpha_2 \frac{\beta (v_{\text{hv}} - v_{\text{lv}})}{L C_{\text{hv}}} \tag{18}$$

Multiplying the above equation by $\frac{LC_{\text{hv}}}{\alpha_2}$ gives the existence condition for boost operation.

$$0 < -\beta L \left(\frac{\alpha_1}{\alpha_2} - \frac{1}{R_{\text{hv}} C_{\text{hv}}} \right) i_{\text{chv}} + L C_{\text{hv}} \frac{\alpha_3}{\alpha_2} (v_{\text{hvref}} - \beta v_{\text{hv}}) < \beta (v_{\text{hv}} - v_{\text{lv}}) \tag{19}$$

3.2 Control Parameter Selection

The control parameters α_1, α_2 and α_3 are selected based on the desired dynamic properties using Ackermann's formula [18]. Substituting $S = 0$ in Eq. (9) gives a linear second-order equation:

$$\alpha_1 x_1 + \alpha_2 \frac{\mathrm{d}x_1}{\mathrm{d}t} + \alpha_3 \int x_1 \mathrm{d}t = 0 \tag{20}$$

Differentiating the above equation and comparing with the second-order system give

$$\begin{cases} \frac{\alpha_1}{\alpha_2} = 2\zeta\omega_n \\ \frac{\alpha_3}{\alpha_2} = \zeta\omega_n^2 \end{cases} \tag{21}$$

where ζ is the undamped natural frequency, and ω_n is the damping ratio. Sliding coefficients can be selected using Eq. (21) based on the desired dynamic property.

3.3 Derivation of PWM Control Law

To derive the control equations, initially the equivalent control signal u_{eq} is derived by setting $\dot{S} = 0$ and then the equivalent control function,u_{eq}, is compared with the instantaneous duty cycle function d of the PWM method. The sliding mode control structure is implemented based on the derived control law with control signal, v_c and ramp signal,v_{ramp}. Comparing these two signals, we get a pulse having constant frequency. Fixed frequency operation improves the line and load regulations of the converter and makes the process of input and output filter design easier.

For Buck Operation. The control law for buck conversion is developed by setting $\dot{S} = 0$ in Eq. (12) given by

$$-\alpha_1 \frac{\beta i_{\mathrm{clv}}}{C_{\mathrm{lv}}} + \alpha_2 \frac{\beta i_{\mathrm{clv}}}{R_{\mathrm{lv}} C_{\mathrm{lv}}^2} + \alpha_3 (v_{\mathrm{lvref}} - \beta v_{\mathrm{lv}}) + \alpha_2 \frac{\beta v_{\mathrm{lv}}}{LC_{\mathrm{lv}}} - \alpha_2 \frac{\beta v_{\mathrm{hv}}}{LC_{\mathrm{lv}}} u_{\mathrm{eq}} = 0 \tag{22}$$

From the above equation, the equivalent control function is obtained as

$$u_{\mathrm{eq}} = -\frac{\beta L}{\beta v_{\mathrm{hv}}} \left(\frac{\alpha_1}{\alpha_2} - \frac{1}{R_{\mathrm{lv}} C_{\mathrm{lv}}} \right) i_{\mathrm{clv}} + \frac{\alpha_3}{\alpha_2} \frac{LC_{\mathrm{lv}}}{\beta v_{\mathrm{hv}}} (v_{\mathrm{lvref}} - \beta v_{\mathrm{lv}}) + \frac{v_{\mathrm{lv}}}{v_{\mathrm{hv}}} \tag{23}$$

Substituting the above equation into the inequality $0 < u_{eq} < 1$ gives

$$0 < -\frac{\beta L}{\beta v_{\mathrm{hv}}} \left(\frac{\alpha_1}{\alpha_2} - \frac{1}{R_{\mathrm{lv}} C_{\mathrm{lv}}} \right) i_{\mathrm{clv}} + \frac{\alpha_3}{\alpha_2} \frac{LC_{\mathrm{lv}}}{\beta v_{\mathrm{hv}}} (v_{\mathrm{lvref}} - \beta v_{\mathrm{lv}}) + \frac{v_{\mathrm{lv}}}{v_{\mathrm{hv}}} < 1 \tag{24}$$

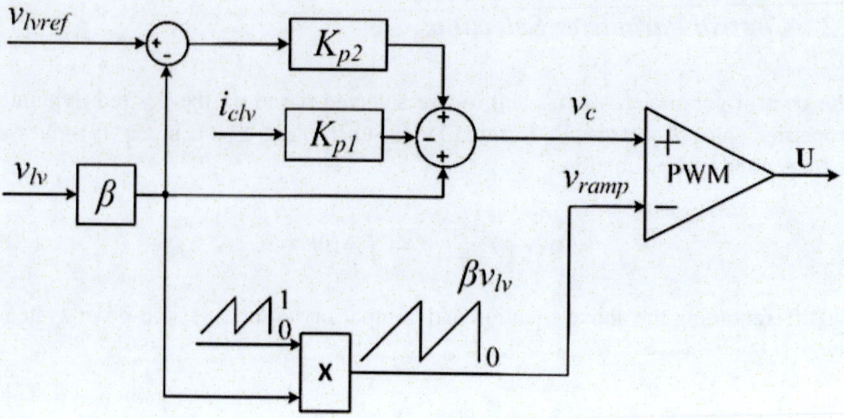

Fig. 4 Control structure for buck operation

Multiplying the above equation by βv_{hv} gives

$$0 < -K_{p1}i_{\mathrm{clv}} + K_{p2}(v_{\mathrm{lvref}} - \beta v_{\mathrm{lv}}) + \beta v_{\mathrm{lv}} < \beta v_{\mathrm{hv}} \tag{25}$$

where $K_{p1} = \frac{\beta L}{\beta v_{\mathrm{hv}}}\left(\frac{\alpha_1}{\alpha_2} - \frac{1}{R_{\mathrm{lv}}C_{\mathrm{lv}}}\right)$ and $K_{p2} = LC_{\mathrm{lv}}\frac{\alpha_3}{\alpha_2}$.

The duty ratio control of conventional converters is $0 < d = \frac{v_c}{v_{\mathrm{ramp}}} < 1$

$$0 < v_c < v_{\mathrm{ramp}} \tag{26}$$

For mapping the equivalent control function onto the duty ratio control, Eq. (25) and Eq. (26) are compared, and the following relationship is obtained (Fig. 4)

$$\begin{cases} v_c = -K_{p1}i_{\mathrm{clv}} + K_{p2}(v_{\mathrm{lvref}} - \beta v_{\mathrm{lv}}) + \beta v_{\mathrm{lv}} \\ v_{\mathrm{ramp}} = \beta v_{\mathrm{hv}} \end{cases} \tag{27}$$

For Boost Operation. By setting $\dot{S} = 0$ in Eq. (16) gives

$$-\alpha_1\frac{\beta i_{\mathrm{chv}}}{C_{\mathrm{hv}}} + \alpha_2\frac{\beta i_{\mathrm{chv}}}{R_{\mathrm{hv}}C_{\mathrm{hv}}^2} + \alpha_3(v_{\mathrm{hvref}} - \beta v_{\mathrm{hv}}) + \frac{\beta(v_{\mathrm{hvref}} - \beta v_{\mathrm{lv}})}{LC_{\mathrm{hv}}}\overline{u}_{\mathrm{eq}} = 0 \tag{28}$$

From the above equation, the equivalent control function is obtained as

$$\overline{u}_{\mathrm{eq}} = \frac{\beta L}{\beta(v_{\mathrm{hv}} - v_{\mathrm{lv}})}\left(\frac{\alpha_1}{\alpha_2} - \frac{1}{R_{\mathrm{hv}}C_{\mathrm{hv}}}\right)i_{\mathrm{chv}} - \frac{\alpha_3}{\alpha_2}\frac{LC_{\mathrm{hv}}}{\beta(v_{\mathrm{hv}} - v_{\mathrm{lv}})}(v_{\mathrm{hvref}} - \beta v_{\mathrm{hv}}) \tag{29}$$

Since $\overline{u}_{\mathrm{eq}}$ is the inverse logic of u_{eq},

$$u_{\mathrm{eq}} = -\frac{\beta L}{\beta(v_{\mathrm{hv}} - v_{\mathrm{lv}})}\left(\frac{\alpha_1}{\alpha_2} - \frac{1}{R_{\mathrm{hv}}C_{\mathrm{hv}}}\right)i_{\mathrm{chv}} + \frac{\alpha_3}{\alpha_2}\frac{LC_{\mathrm{hv}}}{\beta(v_{\mathrm{hv}} - v_{\mathrm{lv}})}(v_{\mathrm{hvref}} - \beta v_{\mathrm{hv}}) + 1 \tag{30}$$

Substituting the above equation into the inequality $0 < u_{eq} < 1$ gives

$$0 < -\frac{\beta L}{\beta(v_{\mathrm{hv}} - v_{\mathrm{lv}})}\left(\frac{\alpha_1}{\alpha_2} - \frac{1}{R_{\mathrm{hv}}C_{\mathrm{hv}}}\right)i_{\mathrm{chv}} + \frac{\alpha_3}{\alpha_2}\frac{LC_{\mathrm{hv}}}{\beta(v_{\mathrm{hv}} - v_{\mathrm{lv}})}(v_{\mathrm{hvref}} - \beta v_{\mathrm{hv}}) + 1 < 1 \tag{31}$$

Multiplying the above equation by $\beta(v_{\mathrm{hv}} - v_{\mathrm{lv}})$ gives

$$0 < -K'_{p1}i_{\mathrm{chv}} + K'_{p2}(v_{\mathrm{hvref}} - \beta v_{\mathrm{hv}}) + \beta(v_{\mathrm{hv}} - v_{\mathrm{lv}}) < \beta(v_{\mathrm{hv}} - v_{\mathrm{lv}}) \tag{32}$$

where $K'_{p1} = \beta L\left(\frac{\alpha_1}{\alpha_2} - \frac{1}{R_{\mathrm{hv}}C_{\mathrm{hv}}}\right)$ and $K'_{p2} = LC_{hv}\frac{\alpha_3}{\alpha_2}$.

For mapping the equivalent control function onto the duty ratio control, Eq. (26) and (32) are compared, and the following relationship is obtained (Fig. 5)

$$\begin{cases} v_c =< -K'_{p1}i_{\mathrm{chv}} + K'_{p2}(v_{\mathrm{hvref}} - \beta v_{\mathrm{hv}}) + \beta(v_{\mathrm{hv}} - v_{\mathrm{lv}}) \\ v_{\mathrm{ramp}} = \beta(v_{\mathrm{hv}} - v_{\mathrm{lv}}) \end{cases} \tag{33}$$

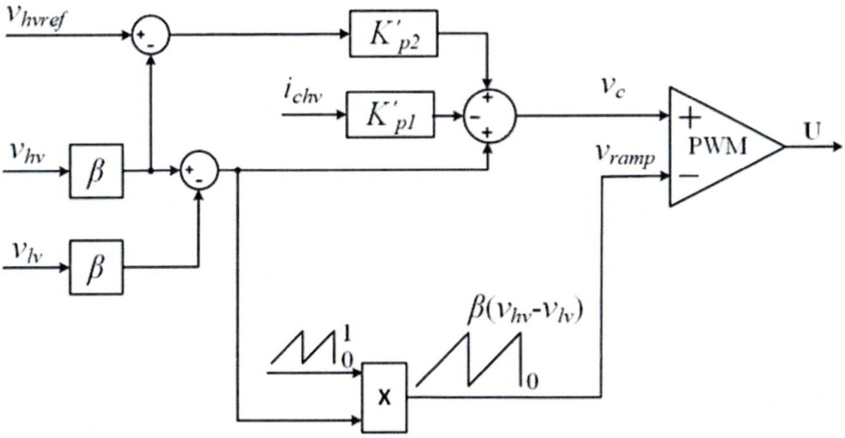

Fig. 5 Control structure for boost operation

4 Simulation and Experimental Results

The dynamic analysis of bidirectional DC–DC converter was performed using PWM-based SMC and conventional PI controller. The real-time simulations were done with the Typhoon Hardware-in-the-loop (HIL) + HIL 402 device. The parameters of bidirectional DC–DC converter are described in Table 1.

The evaluations were carried out under variable load and variable input voltage. The PWM-based sliding mode controller is designed for an undamped natural frequency of $\omega_n = 1.57$ krad/s, and the damping ratio is selected as $\zeta_n = 0.2$. The sliding coefficients (α_1/α_2) and (α_3/α_2) are determined using (21).

The control signal v_c, input ramp v_{ramp} and gate pulse during buck and boost operation at full load are shown in Fig. 6a, b respectively. It is clear that the frequency is maintained constant at 20 kHz, which reduces the chattering phenomenon of the sliding mode controller.

Table 1 Bidirectional DC–DC converter parameters

Description	Parameter	Nominal value
Low voltage	V_{lv}	144 V
High voltage	V_{hv}	370 V
Maximum power	P_{max}	3.3 kW
Switching frequency	f	20 kHz
Capacitance	C_{lv} and C_{hv}	5 mF
Capacitor ESR	R_{clv} and R_{chv}	10 m Ω
Inductance	L	1.5 mH
Inductor resistance	R_l	150 mΩ

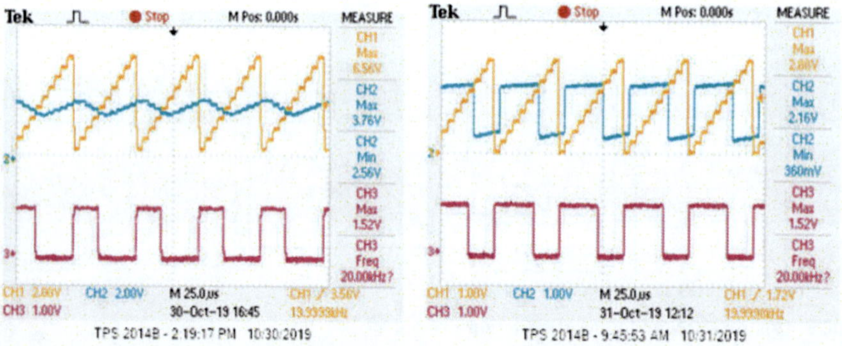

(a) (b)

Fig. 6 Experimental waveforms of control signal v_c, input ramp v_{ramp} and gate pulse of sliding mode controller with **a** buck operation at full load current of $i_{lv} = 23.0$ A and **b** boost operation at full load current of $i_{hv} = 9.0$ A

To evaluate the transient behavior of bidirectional converter during buck operation, the load current is shifted from full load ($i_{\mathrm{lv}} = 23.0$ A) and half load ($i_{\mathrm{lv}} = 11.5$ A) conditions, while the input voltage is kept at nominal value ($v_{\mathrm{hv}} = 370$ V). The transient behavior of bidirectional converter during boost operation is evaluated by shifting the load current from full load ($i_{\mathrm{hv}} = 9.0$ A) and half load ($i_{\mathrm{hv}} = 4.5$ A) conditions, while the input voltage is kept at nominal value ($v_{\mathrm{lv}} = 144$ V). The real-time simulated waveforms for buck and boost operation are shown in Figs. 7 and Fig. 8, respectively.

The comparison result for both operations during load changes from full load and half load is illustrated in Table 2 from which it is clear that voltage-controlled sliding mode controller has better performance.

To study the steady-state performance of buck converter, the input voltages were changed from minimum ($v_{hv} = 340$ V), nominal ($v_{hv} = 340$ V) and maximum ($v_{\mathrm{hv}} = 400$ V), and the results are plotted in Fig. 9. The steady-state performance of boost converter was studied by changing input voltage from minimum ($v_{\mathrm{lv}} = 120$ V), nominal ($v_{\mathrm{lv}} = 144$ V) and maximum ($v_{\mathrm{lv}} = 168$ V), and the results are plotted in Fig. 10.

5 Conclusion

The bidirectional DC–DC converter controlled by a PWM-based sliding mode controller is presented along with the detailed design steps. The effectiveness of the sliding mode controller for bidirectional DC–DC converter is verified using the results obtained from real-time simulations using the Typhoon Hardware-in-the-loop environment. The sliding mode controller is operated at fixed frequency operation in all operating conditions for both buck and boost mode of operation. The dynamic behavior of output voltage ripple in both direction is similar for all operating input and load conditions revealing the robustness of SMC which makes it suitable for electric vehicle applications.

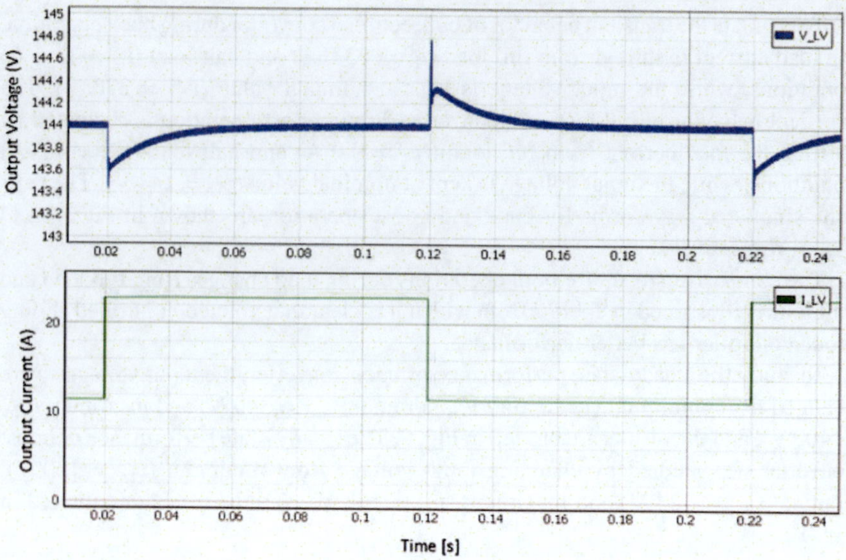

(a)

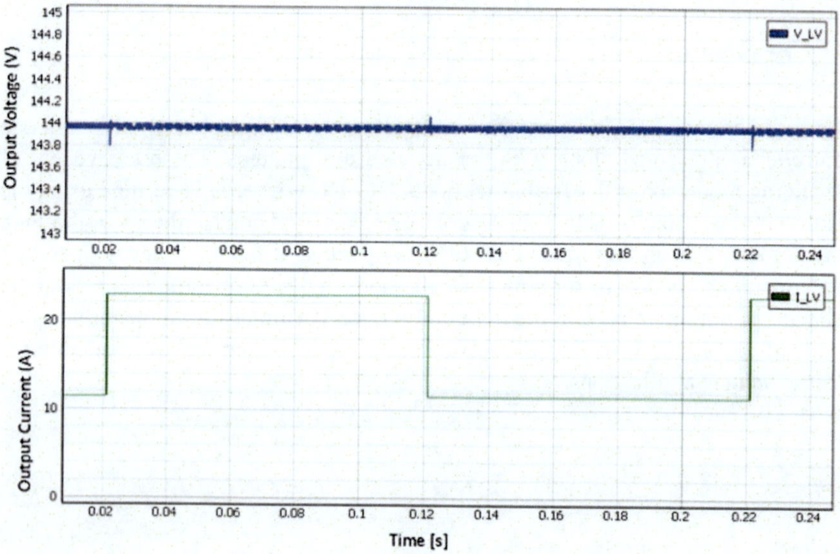

(b)

Fig. 7 Real-time simulated waveforms of output voltage v_{lv} and output current i_{lv} during buck operation operating at input voltage, $v_{hv} = 370$ V alternating between full load and half load conditions with **a** PI controller and **b** sliding mode controller

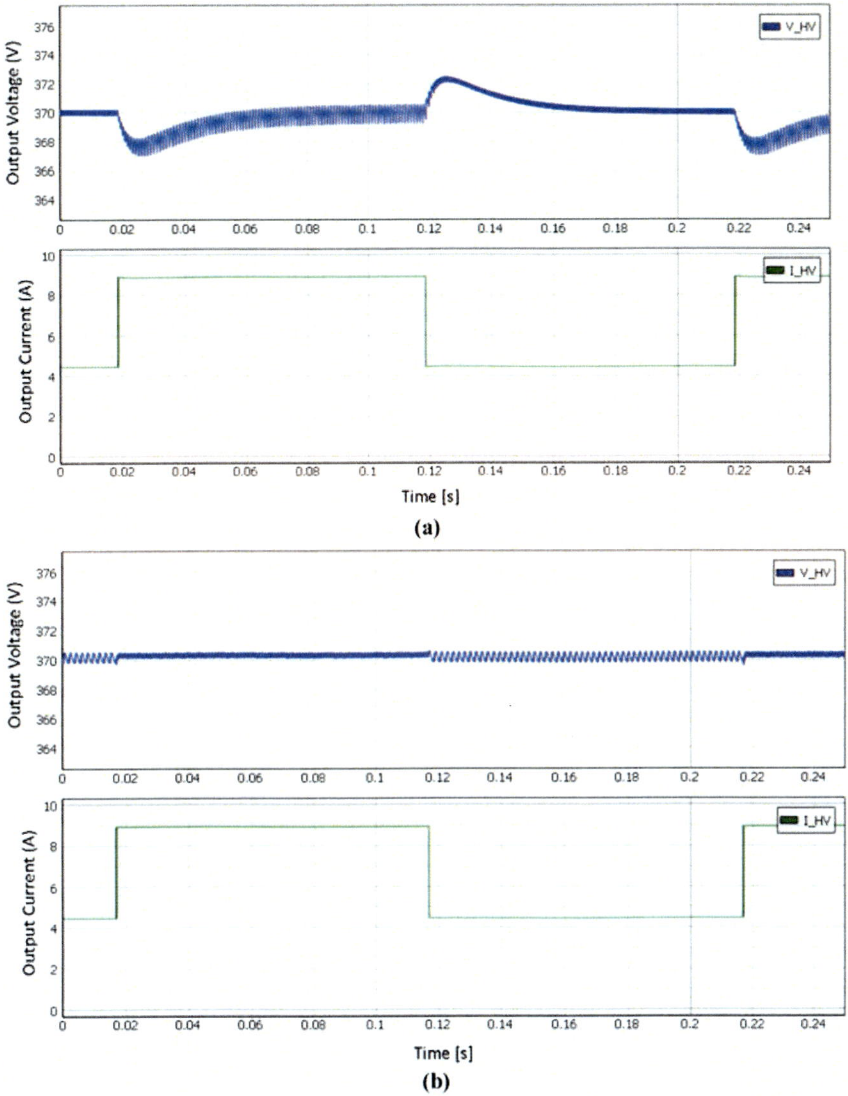

Fig. 8 Real-time simulated waveforms of output voltage v_{hv} and output current i_{hv} during boost operation operating at input voltage, $v_{lv} = 144$ V alternating between full load and half load conditions with **a** PI controller and **b** sliding mode controller

Table 2 Analysis for change in load from full load and half load

Mode of operation	Transient voltage ripple (% of nominal output voltage)		Settling time	
	PI	SMC	PI	SMC
Buck	0.5% of v_{lv}	0.1% of v_{lv}	High	Low
Boost	0.8% of v_{hv}	0.2% of v_{hv}	High	Low

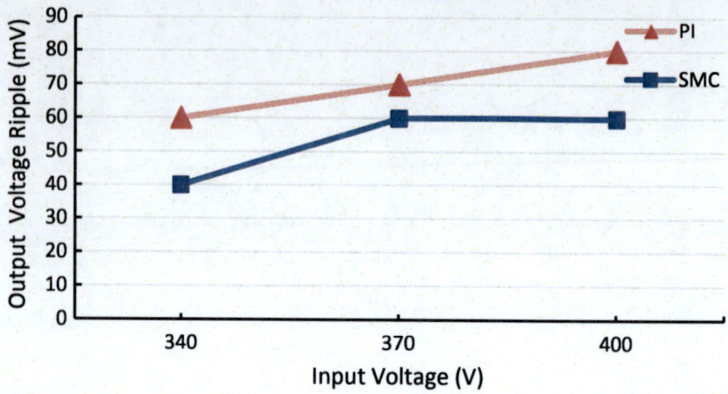

Fig. 9 Plot of output voltage ripple against input voltage v_{hv} during buck operation at minimum, nominal and maximum input voltages

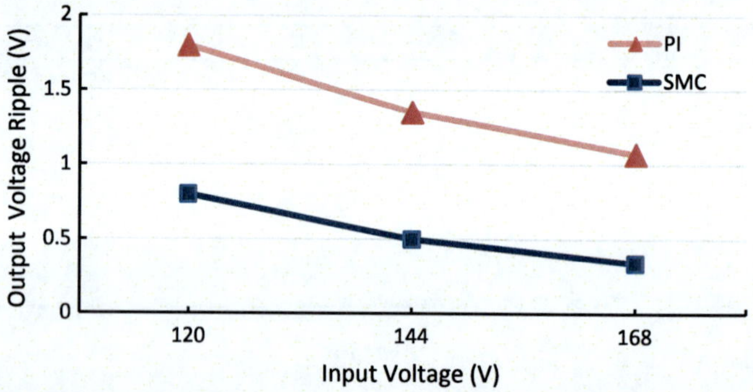

Fig. 10 Plot of output voltage ripple against input voltage v_{lv} during boost operation at minimum, nominal and maximum input voltages

References

1. Singh, K.V., Bansal, H.O., Singh, D.: A comprehensive review on hybrid electric vehicles: architectures and components. J. Mod. Transport. **27**(2), 77–107 (2019)
2. Pavić, I., Capuder, T., Kuzle, I.: Low carbon technologies as providers of operational flexibility in future power systems. Appl. Energy **168**, 724–738 (2016)

3. Minh, V.T., Pumwa, J.: Simulation and control of hybrid electric vehicles. Int. J. Control Autom. Syst. **10**(2), 308–316 (2012)
4. Han, S., Aki, H., Han, S., Kwon, B., Park, J.B.: Optimal charging strategy for a residential PEV battery considering bidirectional trade and frequency regulation. Int. J. Control Autom. Syst. **14**(2), 587–597 (2016)
5. Gorji, S.A., Sahebi, H.G., Ektesabi, M., Rad, A.B.: Topologies and control schemes of bidirectional DC–DC power converters: an overview. IEEE Access **7**, 117997–118019 (2019)
6. Nagarale, R.M., Patre, B.M.: Exponential function based fuzzy sliding mode control of uncertain nonlinear systems. Int. J. Dynamics Control **4**(1), 67–75 (2016)
7. Martinez-Treviño, B.A., El Aroudi, A., Vidal-Idiarte, E., Cid-Pastor, A., Martinez-Salamero, L.: Sliding-mode control of a boost converter under constant power loading conditions. IET Power Electron. **12**(3), 521–529 (2018)
8. Salimi, M., Soltani, J., Zakipour, A., Abjadi, N.R.: Hyper-plane sliding mode control of the DC–DC buck/boost converter in continuous and discontinuous conduction modes of operation. IET Power Electron. **8**(8), 1473–1482 (2015)
9. Wang, L., Meng, F., Sun, Z., Yang, S., Yang, W.: An adaptive hysteresis sliding mode control method for double-switch buck-boost converter. In: 2017 IEEE Transportation Electrification Conference and Expo, Asia-Pacific (ITEC Asia-Pacific), pp. 1–6 (2017)
10. Yasin, A.R., Ashraf, M., Bhatti, A.I.: A novel filter extracted equivalent control based fixed frequency sliding mode approach for power electronic converters. Energies **12**(5), 853 (2019)
11. Tan, S.C., Lai, Y.M., Tse, C.K., Cheung, M.K.: A fixed-frequency pulsewidth modulation based quasi-sliding-mode controller for buck converters. IEEE Trans. Power Electron. **20**(6), 1379–1392 (2005)
12. Qi, W., Li, S., Tan, S.C., Hui, S.R.: Parabolic-modulated sliding-mode voltage control of a buck converter. IEEE Trans. Ind. Electron. **65**(1), 844–854 (2017)
13. Pandey, S.K., Patil, S.L., Ginoya, D., Chaskar, U.M., Phadke, S.B.: Robust control of mismatched buck dc–dc converters by pwm-based sliding mode control schemes. Control Eng. Pract. **84** (2019)
14. Cucuzzella, M., Lazzari, R., Trip, S., Rosti, S., Sandroni, C., Ferrara, A.: Sliding mode voltage control of boost converters in DC microgrids. Control Eng. Practice **73**, 161–170 (2018)
15. Tan, S.C., Lai, Y.M., Tse, C.K.: Sliding mode control of switching power converters: techniques and implementation. CRC press (2018)
16. Chincholkar, S.H., Jiang, W., Chan, C.Y.: An improved PWM-based sliding-mode controller for a DC–DC cascade boost converter. IEEE Trans. Circ. Syst. II: Express Briefs **65**(11) (2017)
17. Mohan, N., Undeland, T.M., Robbins, W.P.: Power Electronics: Converters, Applications, and Design. Wiley, New York (2003)
18. Utkin, V.: Sliding mode control of DC/DC converters. J. Franklin Institute **350**(8), 2146–2165 (2013)

Comparative Study of Strain Using 2D Digital Image Correlation and Extensometer on Glass Fiber

Abhilash Salian and T. P. Ashwini

1 Introduction

In experimental stress analysis its emphasized on measurement of stress and strain of the materials. As the complexity of structures or component have increased over the years, new development has taken place in creating new composite materials. Using these newly made composite materials may results in catastrophic. Keeping in mind the safety of structures and human life, the testing of the materials are done. It also becomes important to make analysis of these materials for very long term. As technology has advanced, the measuring method under the hood of experimental analysis has improved tremendously. The conventional method are getting replaced by the non-conventional method. These advancements are beneficial for both the engineering aspect as well as financial aspect of the experiments. Gone are the days were the material are destroyed or deformed to measure the strain or stress. These were called contact method of measuring. In past decades, non-contact method of measuring strain and stress are picking pace in the filed of experimental analysis. They could be as complex as holography or simple as ultrasonic method. Optical method and sound application has become prominent in non contact method of measuring the stress and strain in structures. In particular DIC technique has taken centre stage because of its easy to use approach and also being inexpensive. In this experiment, the strain obtained from extensometer is compared with strain values found in the DIC technique. Glass fiber with other epoxy resin makes a good composite material. Even though the application of composite material are a lot, they are brittle in nature. Therefore the accessing their stress and strain is also important. It also gives an insight

A. Salian (✉) · T. P. Ashwini
Mangalore Institute of Technology and Engineering, Mangalore, India
e-mail: k.abhisalian@gmail.com

T. P. Ashwini
e-mail: ashwinitp@mite.ac.in

S. K V and K. Guruprasad Rao (eds.), *Smart Sensors Measurements and Instrumentation*, Lecture Notes in Electrical Engineering 750,
https://doi.org/10.1007/978-981-16-0336-5_29

how the this composite material behaves under tensile load. Under DIC technique we could measure the surface deformation of the composite material. This technique helps in understanding of stress localisation.

In recent days there has been tremendous growth in the field of DIC. DIC is one of the foremost techniques in digital image processing to measure strain. In 2016, Supriya and Navthar [1] presented a comapative study of strain of aluminium plate under UTM and used DIC method. The DIC results were close to the values found in UTM. In 1980s the dawn of Digital Image correlation happened and after that it simply grew to twenty-first century. At University of South Carolina it all started and the application of DIC was applied in experiment of Experimental Analysis. In spite of limited application of DIC in those days, DIC outperformed relatively other optical methods by constant improvement in algorithm and the technique. In the year 1982, Peters and Ranson [2], first used in Digital Image Correlation in the experimental stress analysis. The approach in the experiment for surface displacement was to correlate two speckle images mathematically. In this the, system cross relates the reference image stored and the deformed image. In the paper they concluded that subsets of the image simplify the numerical calculation. It also suggested that the DIC method is effective in finding surface deformation. In 1983, Sutton et al. [3] using the same approach developed the numerical algorithms for computing the deformation of a cantilever beam with an end load and apply theory of digital image correlation. In the experiment the intensity spreading of light reflected by the used specimen is stored in memory as set of gray levels. The conversion of the light intensity of the reflected light to a digital number is maintained by the digitizer. It was noted that, the improved algorithm can successfully calculate displacements larger than 0.10 pixels by using bi linear interpolation procedure.

Anderson et al. [4], used the same algorithm for application of 2D digital image correlation method to rigid body mechanics. In this experiment the it involved the rigid body rotation of a shaft and one-dimensional vibratory motion of rod. Using the theory of digital image correlation using the least square method with bi linear interpolation the experiment was conducted. The result presented gave enough evidence suggesting that DIC can be effectively employed to analyze the rigid body deformation of a body undergoing two-dimensional motion. In 1985, Chu et al. [5] worked on basic theory of deformation and digital image correlation and how they are combined for the measurement technique. Four experiment was performed to validate the viability of the method for actual measurements. The experimental results pointed that the correlation method was very accurate in determining of rigid body translation and rotations. It was concluded that the errors can be reduced by increasing the image resolution. In 1986, to reduce the computation time Sutton et al. [6] employed an advanced correlation method. The modified iterative algorithm included Newton–Raphson technique with differential corrections to increase the speed of search time. In the paper it was evident that the time taken was reduced and giving much better accuracy than the previous experiments. It was also noted that if the physical characteristics of the image is improved or in both the algorithm to obtain consistently accurate values for the local deformation. Tian and Huhns [7] explained the application of subpixel accuracy. It was inscribed that the registration accuracy can be further improved by reducing noise in the imaging system, increasing the number of bits per pixel, or pre smoothing the images to remove some of the high-frequency details. McNeill et al. [8] in 1987, found a method to demonstrate the stress intensity factor using digital image correlation. It was concluded that stress intensity can be founded through the use of digital image correlation and least squares fitting seems to be a viable proposition. Sutton et al. [9], performed 1-D numerical simulations to the results of initial studies to determine the key parameters affecting the performance of a computer-based measurement system.

In 1989, Bruck et al. [10] concluded after the experiment that the Newton–Raphson technique alternative to the coarse-fine search technique for determining deformations between digital images. The Newton–Raphson technique used less CPU time than the coarse-fine search method and determined not only displacements, but displacement gradients as well. The Newton–Raphson technique used less CPU time than the coarse-fine search method and determined not only displacements, but displacement gradient as well. In 1989, Russel et al. [11] studied strain fields surface of reinforced composite specimen. The X ray radiograph was applied to get results on image correlation. It was observed that the damage due to low energy impact of a glass fiber reinforced composite induced a shift in the strain field, which was detected as localized, high compression and tensile strains by the X-ray correlation technique. In 1993, Chen et al. [12] suggested alternative approach to fully automated speckle displacement measurement. The technique provides a whole-field survey of 2-D displacements by analyzing two speckle patterns, one taken before and one after specimen deformation. By using the technique of peak finding approach the displacements were successfully determined.

Choi et al. [13], in 1991 applied image correlation technique to measure strain in wood and also paper. It was found that image correlation technique was suitable because any other method would have altered the physical properties of wood and paper. In the concluding note it was said that image correlation technique was quick and accurate in assessing the strain of both wood and paper. Dawike and Sutton [14] measured crack opening displacement using 2D-DIC. In 1998, Vendroux and Knauss [15–17] set up an experiment to study submicron deformations in micromechanical system. A special digital scanning tunneling microscope (STM) was used for that very purpose which is to be attached to a mechanically deforming specimen. In this work it was formulated that the implementation of a least square's coefficient together with Hessian matrix computation greatly simplifies the Newton–Raphson scheme, leading improvement not only in speed but also in the convergence robustness of the code. It was further stated that with both STM and image correlation method, the probe could be of microscopy for scales ranging from tens of microns to atomistic dimensions. Schreier et al. [18] addressed the errors occurred in iteration of spatial cross correlation algorithm used for gray scale interpolation.

In 2003, Hung et al. [19] presented simple detection algorithm based on digital image correlation. This proved to be much faster that the Newton–Raphson method and results very in tangent with theoretical results. There was tremendous improvement in the algorithm used for digital image correlation. Pan et al. [20] proposed universally acceptable DIC method with zero-mean normalized cross correlation

(ZNCC) coefficient used to identify the dependability of the point calculated. It was concluded that the new method is applicable to experimental images with shadows, discontinuous areas, and deformation discontinuity. Anton et al. [21] the method of calculating digital image correlation with Fast Fourier transform algorithm was proposed. This method provided much improvement in speed compared to spatial domain correlation when large displacements are analyzed. The proposed method provides the same correlation function as the direct spatial domain correlation. As the application of DIC increased over the years Silva et al. [22] combined thermoelastic stress analysis and DIC to find the strain in the aluminium alloy plates. Enfedaque et al. [23] extended the application of DIC to find strain in glass fiber reinforced cement. DIC method was found to be an appropriate tool in analyzing the fracture mechanisms that appeared in the fracture tests performed. The analysis of fracture test results was equivalent to the DIC analysis and also justified the behavior of the materials used in the experiment.

Rejab et al. [24] investigated tensile properties of aluminium A1100, glass fiber-reinforced plastics, and pure resin plastics. In experiment three methods was used, (i) Strain gauge, (ii) experimental extensometer and (iii) DIC. It was stated in the result, from the method perspective, the DIC method offers reasonable results compared to the other two methods. Moreover, it shows practically closer values with experimental results, which is suggested to predict the tensile properties as a simple evaluation method. Saidi and Gabor [25] applied the 2D DIC for the understanding the properties of cementitious matrix composites. The results were more reliable with respect to DIC when compared to the other methods. In the year 2018, Yang and Bhattacharya [26] presented augmented-Lagrangian digital image correlation (ALDIC), that has application of both the local (fast) and global (compatible) methods. It was concluded this method is faster and effective method of combining global point and local grid in the image in finding out the deformation. Khoo et al. [27] presented a review of development of DIC in terms of algorithm and also method of capturing the images. It was concluded that the DIC setup can be easily set up in the laboratories. It was mentioned that 2D-DIC technique will keep expanding and the DIC method will become a famous tool for the surface deformation and strain measurements.

2 Experimental Analysis

The context of this experiment is to provide a comparison between contact method of measuring the strain with non-contact technique of measuring. Since the material is composite the results depends upon the set up and also effectiveness of algorithm used for the computation.

2.1 Methods of Strain Measurement

In this experiment we are employing two methods to find the strain. In the conventional method, the specimen is deformed and the data are computed. In the non conventional method, images are captured at different instances for further computation. The two methods for finding strain

1. Tensile test using UTM
2. By using DIC method

Tensile test using UTM The use of UTM or extensometer is conventional method of determining the strain in the specimen. This set up gives the deformation measurement of the specimen with respect to the loading done to the specimen. In this set up load is applied gradually increased with passing of the time. The specimen is fixed in UTM machine. The load is applied in kN starting from zero. The setup consist of two units. One is to load the specimen linearly and the other is to compute the variation in load as well as deformation. These days UTM comes in digital connection and they are connected to the computer. In this experiment the specimen undergoes tensile test at increasing load and they are measured against the time. The strain is plotted to the corresponding stress value for respective time. The measurement of the strain is done in conventional way by applying the engineering formulas. The stress and strain graph is plotted for the values received from this experiment (Fig. 1).

DIC Technique The other method in this experiment is DIC technique. The application of DIC is much popular because of easy set up under natural light. The camera is positioned directly in parallel to the specimen placed in the extensometer. The

Fig. 1 UTM for tensile testing

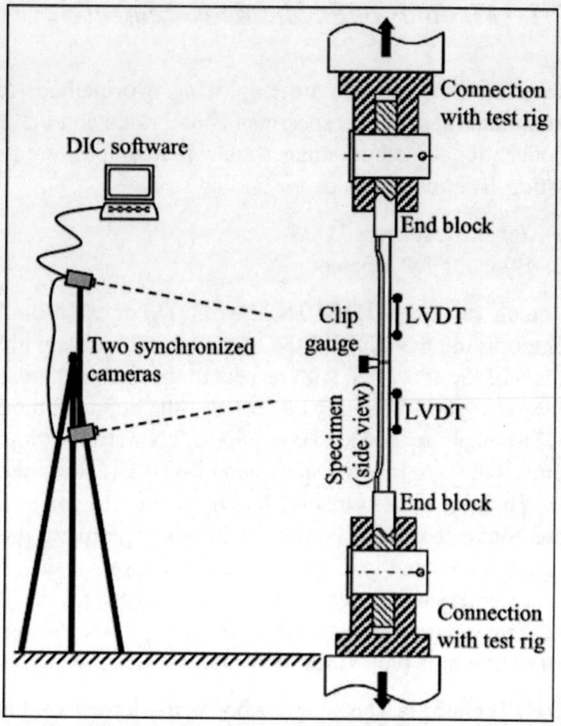

Fig. 2 DIC setup (*Courtesy* Matthias Verstraete)

measure of distance in between the specimen and the camera is defined according to the presence of natural light. In fact, sometimes artificial lighting are provided for the set up so as to extract much information about pixel deformation occurring in the specimen during the experiment. Since the specimen had natural speckle pattern, no additional pattern was applied on to the specimens surface. The image is captured as the loading is done on the specimen. The time interval between the two images is kept as 5 s. The time interval is subjective. The pixel movement happens in 3 dimensional in spatial plane, hence its difficult to predict the accuracy dependency on the time interval. These images are kept in the computer's memory in digital form. Later these images are processed in MATLAB for the analysis of the strain (Fig. 2).

3 Lagrangian Formulations and DIC Algorithm

The study of deformation is pivotal to the study of all solid mechanics engineering problems. Solid mechanics equations are expressed by tracking a volume of material as it translates, rotates, and deforms. This is called a Lagrangian formulation.

3.1 Coordinate Systems and Displacements

Let us consider coordinates 'X' original location of a specimen particle. It can be considered that point 'X' is that point which is attached on the surface till the deformation is completed. The coordinate system is named as the material coordinate system. At a particular time, t, the particle has travels to a new point $x = x(X, t)$. It is assumed that both the sets of coordinates have the same orientation and origin. The coordinates 'x' is present in the spatial coordinate system. The spatial coordinate system is static in space, and the material coordinate system is fixed on to the body. The change in vector pointing from original point to the point it is displaced is given as displacement vector $u(X, t)$. The Lagrangian formulation is done because of the original coordinates are also independent coordinates. This provides the changes from material frame to the spatial frame.

The displacement field usually do not represent a rigid body motion, there will be some changes in dimensions of the material, defined as strains. The changes occurring in the specimen could be change in area, volume or local small do-main. The strain causes internal localization of stress and sometimes even the failure of the specimen. There are different methods to define the material behavior and also deformation occurring in the material. Sometimes they are described as independent of rigid body deformation.

Continuum Mechanics Formulation for Deformation The deformation in the planar is best explained by using the continuum mechanics [28]. Figure 5 depicts the basic concept of deformation the spatial planes.

In the image the point P where rigid rotation may occur. We are incorporating the continuum mechanics to explain the deformation of pixel. Since pixel has the freedom to get deformed in all three planes. Describing the deformation with respect to point P in three principal direction could be $x = (X_1, X_2, X_3)$, then the deformation can be expressed as,

$$\begin{aligned} x &= X + d \\ \mathrm{d}x &= f(X)\mathrm{d}X \\ F(X) &= x\nabla \cdot \mathrm{d}x \end{aligned} \tag{1}$$

$$x\nabla_{i,j} = \begin{bmatrix} \frac{\partial x_1}{\partial X_1}(X_1, X_2, X_3) & \frac{\partial x_1}{\partial X_2}(X_1, X_2, X_3) & \frac{\partial x_1}{\partial X_3}(X_1, X_2, X_3) \\ \frac{\partial x_2}{\partial X_1}(X_1, X_2, X_3) & \frac{\partial x_2}{\partial X_2}(X_1, X_2, X_3) & \frac{\partial x_2}{\partial X_3}(X_1, X_2, X_3) \\ \frac{\partial x_3}{\partial X_1}(X_1, X_2, X_3) & \frac{\partial x_3}{\partial X_2}(X_1, X_2, X_3) & \frac{\partial x_3}{\partial X_3}(X_1, X_2, X_3) \end{bmatrix}$$

In this F is the primary deformation gradient of each component and $x\nabla_{i,j}$is the vector. Defining x$(X_1, X_2, X_3) = (X_1, X_2, X_3) +$ d(X_1, X_2, X_3), where component of each component and 'd' is given as displacement vector of the point P. The deformation gradient can be given as

$$F(X) = I + d\nabla$$

$$I_j = \begin{bmatrix} 1\,0\,0 \\ 0\,1\,0 \\ 0\,0\,1 \end{bmatrix}$$

$$x\nabla_{i,j} = \begin{bmatrix} \frac{\partial d_1}{\partial X_1}(X_1, X_2, X_3) & \frac{\partial d_1}{\partial X_2}(X_1, X_2, X_3) & \frac{\partial d_1}{\partial X_3}(X_1, X_2, X_3) \\ \frac{\partial d_2}{\partial X_1}(X_1, X_2, X_3) & \frac{\partial d_2}{\partial X_2}(X_1, X_2, X_3) & \frac{\partial d_2}{\partial X_3}(X_1, X_2, X_3) \\ \frac{\partial d_3}{\partial X_1}(X_1, X_2, X_3) & \frac{\partial d_3}{\partial X_2}(X_1, X_2, X_3) & \frac{\partial d_3}{\partial X_3}(X_1, X_2, X_3) \end{bmatrix} \tag{2}$$

The undeformed state from the deformed state is given by

$$\mathrm{d}X = F^{-1}(x) \cdot \mathrm{d}x$$
$$F^{-1}(x) = X\nabla$$

$$x\nabla_{i,j} = \begin{bmatrix} \frac{\partial X_1}{\partial x_1}(x_1, x_2, x_3) & \frac{\partial X_1}{\partial x_2}(x_1, x_2, x_3) & \frac{\partial X_1}{\partial x_3}(x_1, x_2, x_3) \\ \frac{\partial X_2}{\partial x_1}(x_1, x_2, x_3) & \frac{\partial X_2}{\partial x_2}(x_1, x_2, x_3) & \frac{\partial X_2}{\partial x_3}(x_1, x_2, x_3) \\ \frac{\partial X_3}{\partial x_1}(x_1, x_2, x_3) & \frac{\partial X_3}{\partial x_2}(x_1, x_2, x_3) & \frac{\partial X_3}{\partial x_3}(x_1, x_2, x_3) \end{bmatrix} \tag{3}$$

Strain Tensor This strain tensor is defined as the deformation of the material before the rotation, and all the components are equivalent to zero. i.e. undeformed state. The component for can be written as

$$E\nabla_{i,j} = \frac{1}{2}\left(\frac{\partial d_i}{\partial X_j} + \frac{\partial d_j}{\partial X_i} + \frac{\partial d_k}{\partial X_i} \cdot \frac{\partial d_k}{\partial X_j}\right) \tag{4}$$

where summation over repeated indices is assumed. The Green-Lagrange strain tensor is ignored when the rigid body rotation is small. This means the displacement of the point is negligible. This leads to the well-known engineering strain tensor,

$$\varepsilon = \frac{1}{2}\left(\frac{\partial d_i}{\partial X_j} + \frac{\partial d_j}{\partial X_i}\right) \tag{5}$$

having components such as,

$$\varepsilon_{XX} = \frac{\partial d}{\partial X} \tag{6}$$

The direct strain or normal strain is nothing but diagonal strain. The displacement obtained by the algorithm, which is in the form of Lagrangian–Green strain tensor is equated to the Eq. (7), [28] to get the direct comparison of to the extensometer results.

$$\varepsilon_{\text{engr}}^{i} = \sqrt{1 + 2(E_{i,j})} - 1$$

$$\varepsilon_{\text{engr}}^{\text{avg}} = \frac{\sum_{i=1}^{N} \varepsilon_{engr}^{i}}{N} \tag{7}$$

3.2 DIC Algorithm

Digital image correlation is a technique for measuring full field surface displacement and engineering strain. Comparison of the images are done in this process. The images before deformation are taken reference images and the image after deformation is compared with the defined reference image. There are many algorithms used for the purpose of finding displacement in deformed image. These included Newton-Raphson method, least square method, these methods are incorporated in the algorithm to reduce the computation speed. The algorithm used in this project is application of most of these methods. In the algorithm processing of local subset is done. It also computes global subset in the process of anlaysis of the specimen. Disadvantages of Local subset is DIC is relatively noisy because of smaller size of the subset. Global DIC is computationally expensive. The (ALDIC), combines the advantages of both the local and global methods.

Algorithm of Proposed Digital Image Correlation Algorithm
Input: Reference Image before deformation at time 0 (f), Subsequent deformed images. (g)
Output: Displacement tensor (u), Deformation gradient tensor (Fi),
Steps:

1. Initialize using FFT integer pixel search method.
2. Preliminary computation of finite difference operator D.
3. Identify the initial parameters for real scalars.
4. While displacement of 2 pixels is greater that epsilon
   - Update local gradient for local subset and global displacement.
   - Update global gradient for global displacement.
   - Update the initial parameters for scalars.

End

The FFT is used to decompose an image into Fourier or frequency domain, while the input image is the spatial domain equivalent. In the Fourier domain image, each point represents a particular frequency contained in the spatial domain image. where D is the discrete gradient operator that depends on the discretization. While local update, holding displacement and deformation tensor fixed, minimize L over to obtain displacement and deformation of next iteration. While global update, holding deformation gradient and displacement fixed, we minimize L over displacement to obtain displacement of next iteration.

Algorithm for Lagrangian Strain Tensor
Input: Pixel displacement tensor from ALDIC
Output: Lagrangian strain tensor
Steps:

1. While for each value in displacement tensor
   - Calculate the parameter du/dx, dv/dx, dv/dy, du/dy
   - Apply LR strain calculation formula Green-Lagrangian strain: E = (C-I)/2.
   - Store the value in LR tensor

   end
2. Return the LR tensor

Flow Chart of Digital Image Correlation Software Figure 3 explains the algorithm used in the project. In this the image is fed into the DIC software. i.e. MATLAB. The Image tool box process the image in the algorithm. Then we define the Region of Interest (ROI), to speed up computational process. We also define the subset in the step. After this step we initiate the Fast Fourier Transformation to select the pixel in the ROI defined. Later this image is computed by different methods, such Gauss method, Least square method. It then calculates the image displacement in the defined ROI. It later presents the value of strain in 3D graph. The last step in this process is that calculating Lagrangian strain tensor.

Fig. 3 Glass fiber specimen

4 Results and Verification

Figures 4, 5 and 6 depicts the graph plotted between the stress and strain. The top first figure is the graph obtained in the experiment setup using UTM. The graph is the values plotted between the stress and the strain values obtained by the DIC. In the graph you can observe that, the pixel deformation is deviated a lot in the beginning of the graph and at the higher range the DIC graph tends to move towards the value obtained by the UTM. This is because, the material used here is Glass Fiber, which is brittle material and also composite material. The molecules are tightly packed

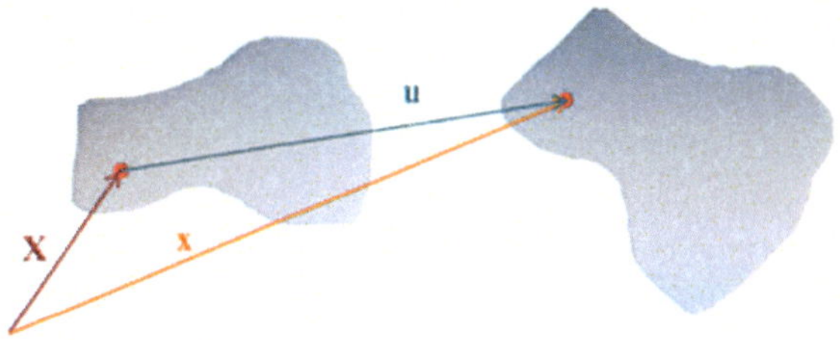

Fig. 4 Relation between the material and spatial coordinates of a point (*Courtesy* https://www.comsol.co.in)

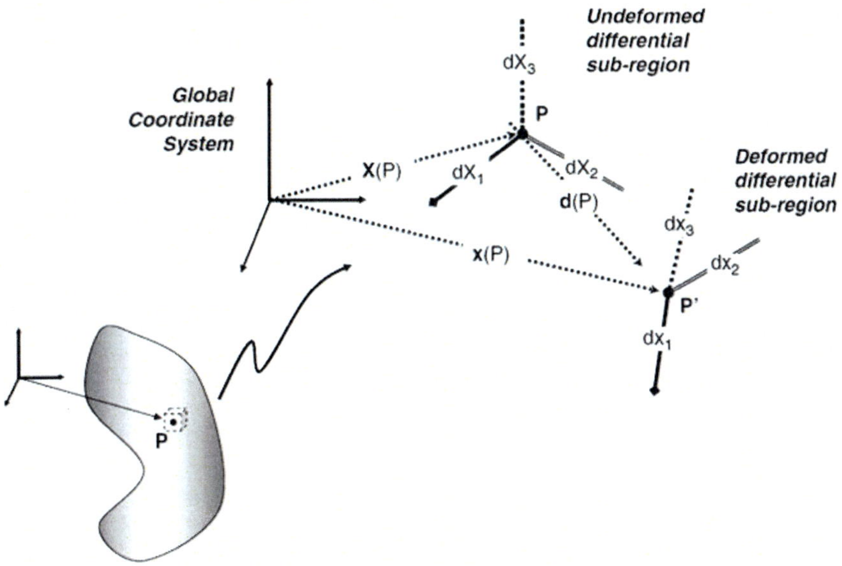

Fig. 5 The movement of the point in its principal direction

Fig. 6 Flow chart of DIC Algorithm

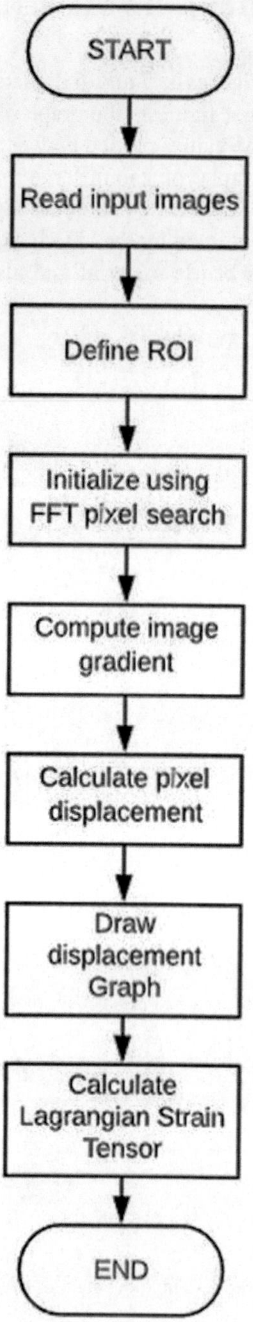

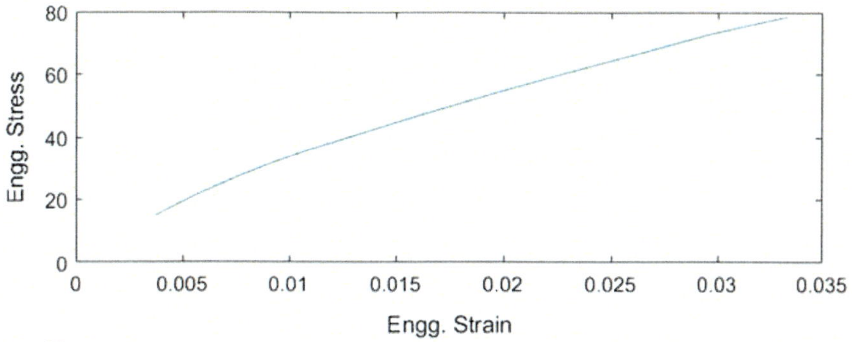

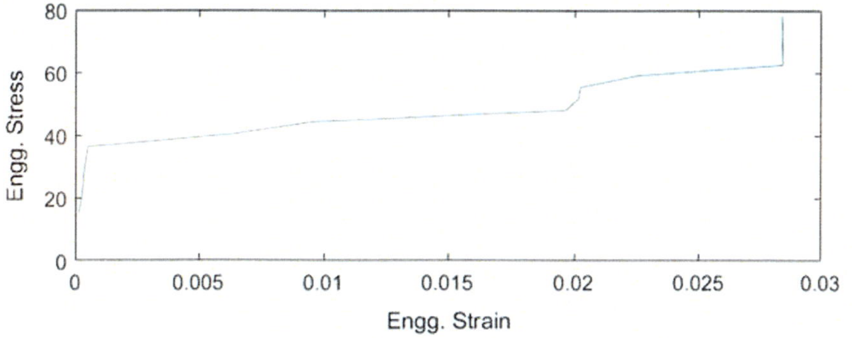

Fig. 7 Stress versus strain graph (i) Results from the UTM. (ii) Results from DIC method

and deformation is only observed after certain value loading on the specimen. The parameter considered here for different trials are load and distance of the lens. Figure 4 depicts the graph when the lens is 10 cm away from the specimen and loading is done at higher rate. The set up for trial 2 and 3 are same, where the lens is placed 15 cm away yet the subset selected are different for both the specimen, therefore the difference in the graph Figs. 5 and 6.

There is also difference in the values obtained in UTM and DIC is because 2D-DIC is averaging the strain data over the region containing the localized Region of Interest, whereas UTM is the elongation of the specimen in one direction. In 2D-DIC it's also has to observe that, the values are taken surface of the material. The composition of the specimen also determines the values obtained by UTM and DIC, because the strain obtained in image processing is the surface values of the specimen (Figs. 7, 8 and 9).

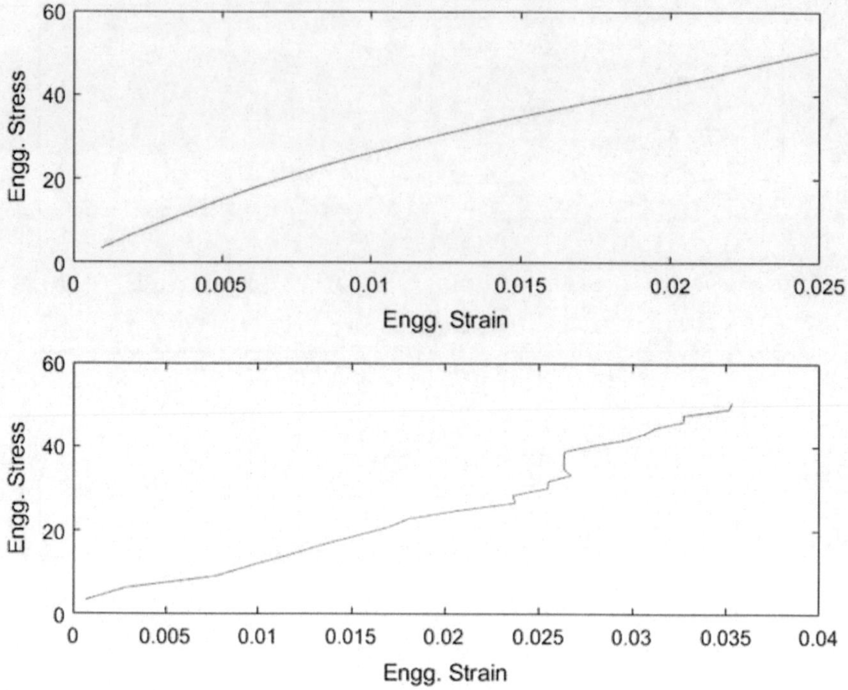

Fig. 8 Stress versus strain graph (i) Results from the UTM. (ii) Results from DIC method

5 Conclusion

The Digital Image Correlation method is one of the easiest ways of finding strain in non-contacting method. The development in the experimental analysis, especially in DIC has been tremendous over the years. In the project, the use of glass fiber which is brittle material was used. The specimen being brittle material, it's difficult to capture the pixel displacement. For that very reason, through the MATLAB platform, FFT approach was used. The algorithm used in this project used global point and also local point to detect the pixels and also register the pixel values in its neighbors. The result was 72% matching with the results found in the UTM. The results can be improved by performing the experiment under the right light condition and also applying various loads and record the results with respect to the loads applied. There are many parameters impacting the overall performance of DIC results. Therefore it's very inconclusive to predict the results without proper assessment of the parameters.

Though, this project has opened up the opportunity to explore the application of DIC in brittle materials. There has to be lots results awaited all over the world on real time application of the DIC. With improved algorithm and also proper setup, the error values can be reduced. The improved algorithm can measure full body deformation. This might also reduce the computation time. There has been tremendous work going

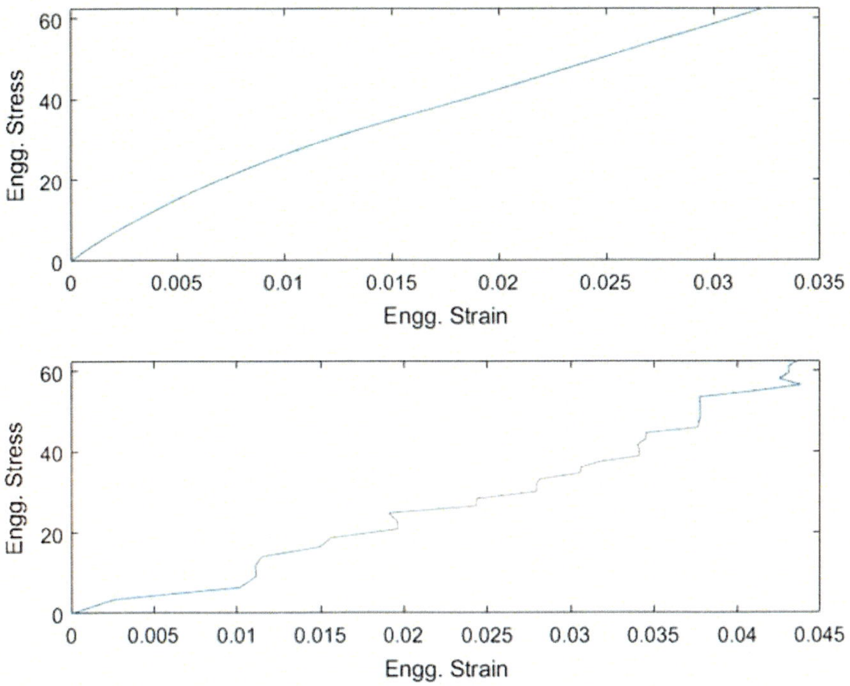

Fig. 9 Stress versus strain graph (i) Results from the UTM. (ii) Results from DIC method

on in the field of Artificial Intelligence and Machine learning. The improved speed in computing the pixels and also the accuracy of prediction in the movement of the pixel point on the image will give us better results. There is a lot of impact if the DIC method is used extensively.

References

1. Navthar, R.R., Gadhe, S.S.: Digital image correlation technique for strain measurement of aluminium plate. Int. J. Eng. Trends Technol. (IJETT) **39**(6), 306–311 (2016)
2. Peters, W.H., Ranson, W.F.: Digital imaging techniques in experimental stress analysis. Opt. Eng. **21**(3), 427–431 (1982)
3. Sutton, M.A., Walters, W.J., Peters, W.H., Ranson, W.F., McNeill, S.R.: Determination of displacements using an improved digital correlation method. Image Vision Comput. **1**(3), 133–139 (1983)
4. Peters, W.H., Ranson, W.F., Sutton, M.A., Chu, T.C., Anderson, J.: Application of digital correlation methods to rigid body mechanics. Opt. Eng. **22**(6), 738–742 (1983)
5. Chu, T.C., Ranson, W.F., Sutton, M.A., Peters, W.H.: Applications of digital image correlation techniques to experimental mechanics. Experim. Mech. **25**(3), 232–244 (1985)

6. Sutton, M.A., Cheng, M., Peters, W.H., Chao, Y.J., McNeill, S.R.: Application of an optimized digital correlation method to planar deformation analysis. Image Vision Comput. **4**(3), 143–150 (1986)
7. Tian, Q.I., Huhns, M.N.: Algorithms for subpixel registrations. Computer Vision Graph. Image Process. **35**(2), 220–233 (1986)
8. McNeill, S.R., Peters, W.H., Sutton, M.A.: Estimation of stress intensity factor by digital image correlation. Eng. Fract. Mech. **28**(1), 101–112 (1987)
9. Bruck, H.A., McNeill, S.R., Sutton, M.A., Peters III, W.H.: Digital image correlation using Newton-Raphson method of partial differential correction. Experim. Mech. **29**(3), 261–267 (1989)
10. Russell, S.S., Sutton, M.A.: Strain-field analysis acquired through correlation of X-ray radiographs of a fiber reinforced composite laminate. Experim. Mech. **29**(2), 237–240 (1989)
11. Sutton, M.A., McNeill, S.R., Jang, J., Babel, M.: Effects of subpixel image restoration on digital correlation error estimates. Opt. Eng. **27**(10), 870–877 (1988)
12. Chen, D.J., Chiang, F.P., Tan, Y.S., Don, H.S.: Digital speckle-displacement measurement using a complex spectrum method. Appl. Opt. **32**(11), 1839–1849 (1993)
13. Choi, D., Thorpe, J.L., Hanna, R.B: Image analysis to measure strain in wood and paper. Wood Sci. Technol. **25**(4), 251–262 (1991)
14. Dawicke, D.S., Sutton, M.A.: CTOA and Crack-tunneling measurements in thin sheet 2024–T3 aluminum alloy. Experim. Mech. **34**(4), 357–368 (1994)
15. Vendroux, G., Knauss, W.G.: Submicron deformation field measurements: part 1. Developing a digital scanning tunneling microscope. Experim. Mech. **38**, 18–23 (1998)
16. Vendroux, G., Knauss, W.G.: Submicron deformation field measurements: part 2. Developing a digital scanning tunneling microscope. Experim. Mech. **38**, 86–92 (1998)
17. Vendroux, G., Knauss, W.G.: Submicron deformation field measurements: part 3. Developing a digital scanning tunneling microscope. Experim. Mech. **38**, 154–160(1998)
18. Schreier, H.W., Braasch, J.R., Sutton, M.A.: Systematic errors in digital image correlation caused by intensity interpolation. Opt. Eng. **39**(11), 2915–2921 (2000)
19. Hung, P.-C., Voloshin, A.S.: In-plane strain measurement by digital image correlation. J. Braz. Soc. Mech. Sci. Eng. **25**(3), 215–221 (2003)
20. Pan, B.: Reliability-guided digital image correlation for image deformation measurement. Appl. Opt. **48**(8), 1535–1542 (2009)
21. Grebenyuk, A.A., Ryabukho, V.P.: Digital image correlation with fast Fourier transform for large displacement measurement In: Saratov Fall meeting , pp. 1–5. Optical Technologies in Biophysics and Medicine XII 79990B, Saratov, Russian Federation (2010)
22. Silva, M.L., Ravichandran, G.: Combined thermoelastic stress analysis and digital image correlation with a single infrared camera. J. Strain Anal. Eng. Design **46**(8), 783–793 (2011)
23. Enfedaque, A., Gálvez, J.C., Suárez, F.: Analysis of fracturetests of glass fibre reinforced cement (GRC) using digital imagecorrelation. Construct. Build. Mater. **75**, 472–487 (2015)
24. Ma, Q., Rejab, M.R.M., Halim, Q., Merzuki, M.N.M., Darus, M.A.H: Experimental investigation of the tensile test using digital image correlation (DIC) method. Mater. Today Proc. Mater. **27**(2), 757–763(2020)
25. Saidi, M., Gabor, A.: Adaptation of the strain measurement in textile reinforced cementitious matrix composites by distributed optical fibre and 2D digital image correlation. Strain **1–18**, (2019)
26. Yang, J., Bhattacharya, K.: Augmented Lagrangian digital image correlation, experimental mechanics. Experim. Mech. **59**(9), 187–205 (2019)
27. Yang, J., Bhattacharya, K.: A Review of surface deformation and strain measurement using two-dimensional digital image correlation, experimental mechanics. Metrol. Measure. Syst. **23**(3), 461–480 (2016)
28. Schreier, H., Orteu, J.-J., Sutton, M.A.: ImageCorrelation for Shape. Motion and Deformation Measurements. Springer, New York (2009)

Parts of Speech Tagging and Extractive Summarization Techniques for Kannada Documents

Alaka Ananth, Sachin Bhat, Rushali Naik, Suraksha, U. Pooja Nair, and Vinutha

1 Introduction

Summaries of documents were initially done manually, but today, huge data is available on the Internet and everywhere, where manual summarization is failing miserably. A massive amount of data is being put on the Internet every single day. The Internet is being inflated, since the last few years. The increasing availability of information online has created the problem of information overload which has also increased the demand for automatic text summarization. Instead of reading a huge document with lot of theory and examples, the reader would always prefer to read the summarized document which has all the important details. Our proposed work of Kannada text summarization is meant for the same purpose. The reader is provided with most important and relevant information from the original input text document.

Business vanguard, specialists, students and pedagogic advisers have to go through a massive number of documents every day which consumes a large amount of their time. Extracting the most significant part of the documents and creating a summary will help them all to save a lot of time as it will be worth reading only the most important part of the document. Text summarization can be used in various fields like getting news headlines, creating an abstract summary of a bulky technical paper and can also be helpful to create reviews of the books.

Despite the fact that several approaches have been preferred for summarization of major languages like English, Swedish and many other European languages, text summarization is still a challenging problem for other languages in the world. India

A. Ananth
Department of Information Science and Engineering, NMAM Institute of Technology, Nitte, Karnataka 574110, India

S. Bhat (✉) · R. Naik · Suraksha · U. Pooja Nair · Vinutha
Department of Electronics and Communication Engineering, Shri Madhwa Vadiraja Institute of Technology and Management,, Bantakal, Udupi, Karnataka 574115, India
e-mail: sachinbhat88@gmail.com

S. K V and K. Guruprasad Rao (eds.), *Smart Sensors Measurements and Instrumentation*, Lecture Notes in Electrical Engineering 750,
https://doi.org/10.1007/978-981-16-0336-5_30

has around 1500 different languages, out of which, 120 major languages are used all over the country. Kannada, which is the official language of Karnataka state, is spoken by around 70 million people across the globe.

While performing an affiliated search, it is noticed that from recent years several researchers have been working towards text summarization in many languages. However, very less work has been done on Indian languages. There are several built-in libraries available for the preprocessing and processing part of the document for English, but did not come across any such built-in libraries for Kannada. There is a lack of proper text summarization system, although it is being used by millions of people in India. Hence, we have designed a system which can be used to summarize Kannada text documents. Parts of speech(POS) tagging is also carried out where each word of a sentence is tagged with appropriate syntactic label such as noun, pronoun, verb and adjective. POS tagging is one of the most basic natural language processing applications widely used in text summarization and entity recognition of natural languages. Kannada is morphologically rich and agglutinative language with a very ancient history of 2500 years. Its vast dictionary and grammar pose many challenges in natural language processing. An illustration of the POS tags for Kannada language is shown below.

ಕನ್ನಡವು ಭಾರತದ ಅತಿ ಪುರಾತನ ಭಾಷೆಗಳಲ್ಲಿ ಒಂದಾಗಿದೆ.
Noun Noun Determiner Adjective Noun Quantifier

2 Literature Survey

Many of the parts of speech taggers have been developed predominantly for English and other European languages. Relatively fewer attempts have been made to develop POS taggers for Indian languages. Nisheeth et al. [11] proposed a POS tagger for Hindi based on hidden Markov model(HMM) that achieves an accuracy of 92% using Indian language (IL) POS tagset. A Tamil POS tagger developed by Jaybal et al. [8] which achieves an accuracy of 87.74% is one of the first attempts for a Dravidian language.

Few literatures have accomplished notable performance in POS tagging for Kannada. A POS tagger for Kannada using support vector machine(SVM) was proposed by Antony et al. [5]. Their proposal achieves an accuracy of precision of 86% using a hierarchical tagset. Siva Reddy et al. [17] developed a POS tagger for Kannada using Telugu resources with a HMM. They argue that morphological similarity between these two languages allows for cross-language POS (CLP) tagger. Although their tagger achieves 77.7% precision, they rely heavily on the quality of lemmatization for which very few accurate resources exist. Work by Shambhavi et al. [19] uses EMILLE corpus to compare the performance between maximum entropy (Maxent), HMM and conditional random fields (CRF). Their CRF model outperformed Maxent and HMM models with an accuracy of 84.6%. Most recent and notable of these works is the Kannpos POS tagger for Kannada developed by Pallavi et al. [15]. They utilized a large annotated corpus of 80,000 words from TDIL and achieved an accuracy of 92.94% using n-gram CRF model. It can also be noted that, apart from the CLP tagger from Siva Reddy et al. [17], implementation details of none of the other POS taggers for Kannada are publicly available. Table 1 shows the brief outline of different extractive summarization techniques used in various Indian languages.

3 Methodology

The proposed methodology has been split into two major parts. First part deals with the POS tagging to the collected large row corpora, and second part is the summary extraction from a Kannada text document.

3.1 Kannada POS Tagging

POS tagging is labelling parts of speech to individual tokens depending on their meaning and context. Labels like noun, verb and adjective, etc., will be assigned to each word in the text. To build a tagset, we have analysed the POS tagging techniques used in various other Indian languages including existing works in Kannada as well. They have used 12–40 tags depending on the algorithm. Using multiple sets, create difficulty in analysis. We have chosen 14 important tags which are most repeatedly used in Kannada.

Corpus is collected from various Internet sources related to news, literature and essays containing around 2000 sentences with more than 15,500 words. They are manually annotated with labels from the defined set. Table 2 illustrates the use of individual tags with description and example.

Corpus is divided into training and testing sets with a ratio of 80:20, i.e. 12,400 words are taken for training and remaining words for testing. Corpus is trained with a hybrid bidirectional long short-term memory(BLSTM)-chain conditional random fields(CRF)-based model proposed in [4]. In this model, word embedding, i.e word and character level distributions, is built specifically to understand Kannada. Word embedding will enable to capture the position, and character-level embedding will capture prefix and suffix information of words. These word and character representations are combined and passed through hybrid BLSTM-CRF network as shown in Fig. 1.

Unlike RNN which can store only recent information, LSTM can also retrieve long-term dependencies in the sequence. LSTMs use gate mechanism to overcome the drawback of RNNs. Single LSTM cell has input, output and forget gates to control the information coming in and going out in a cell. In text analysis, both earlier and later information are important for the prediction of current output. Input will be passed through two LSTMs to compute forward and backward sequence computations.

Table 1 Literature review outline for extractive method

References	Language	Method of preprocessing	Type of summarization algorithm	Result/ accuracy	Evaluation metric
[16]	Bengali	TF-IDF, sentence length	Fuzzy C means	40% summary	Precision, Recall and F-score
[1]	Bengali	Stop words, text split	Word2vec model	–	Skip gram, CBOW
[6]	Bengali	Corpus, inter annotation	Theme clustering	72.15%	ROUGE
[3]	Bengali	Labelling	RNN	–	ROUGE
[21]	Bengali	Noise removal, Tokenization, Stop word removal	K-means clustering algorithm	40% summary	Not Mentioned
[2]	Bengali	Tokenization, Stop word and noise removal, Stemming	K-means clustering algorithm	30% summary	–
[20]	Hindi	Same as above	Genetic algorithm	–	Manual evaluation
[9]	Hindi	Same as above	Fuzzy logic.	73%	Precision, Recall and F-score
[7]	Kannada	Tagging and parsing of text, using a lexicon	AutoSum	Nil	Comparative analysis
[10]	Kannada	Tokenization, stop word removal	Artificial neural networks	Nil	ROUGE
[12]	Malayalam	Tokenization, filtration, POS tagging	NLP	Nil	Intrinsic method
[13]	Malayalam	Sentence Tokenization, Word tokenizer	Sentence Ranking.	Nil	ROUGE
[18]	Marathi	Punctuation marks removal, Stop words elimination, Stemming and lemmatization	TextRank algorithm	Nil	ROUGE (1-4)
[14]	Tamil	Stop word removal	PageRank and its variants	–	Precision

Table 2 Description of tagset

No.	Tag	Description	Example
1	NN	Noun	ರಾಮು, ಶ್ರೀ, ಕನ್ನಡ
2	PRO	Pronoun	ಅವನು, ನನ್ನ , ಅವಳ
3	VM	Verb	ಹೋಗುತ್ತಾನೆ, ತಿಂದೆನು, ಇದ್ದೇನೆ
4	RB	Adverb	ಚೆನ್ನಾಗಿ, ಇತ್ತೀಚೆಗೆ, ಸಹ
5	JJ	Adjective	ಪ್ರಸಿದ್ಧ , ಉತ್ತಮ, ದೂರ
6	DEM	Demonstrative	ಈಗ, ಇನ್ನು , ಈ
7	DET	Determiner	ತುಂಬ, ಬೇಗ, ಹೆಚ್ಚೆಚ್ಚು
8	CC	Conjunction	ಹಾಗೂ, ಮತ್ತು
9	NUM	Numeric	೧, ೨೦೨೦, ೧೫
10	PUNC	Punctuation	“,?,!
11	UT	Quote	ಎಂದು, ಎಂದೇ, ಎನ್ನುವಂತೆ
12	PSP	Postposition	ಮುಂದಿನ, ಅದುವರೆಗೆ, ಜೊತೆ
13	QF	Quantifier	ತುಂಬಾ, ಒಂದು, ದೊಡ್ಡದು
14	WQ	Question	ಏನು, ಯಾರು, ಹೇಗೆ

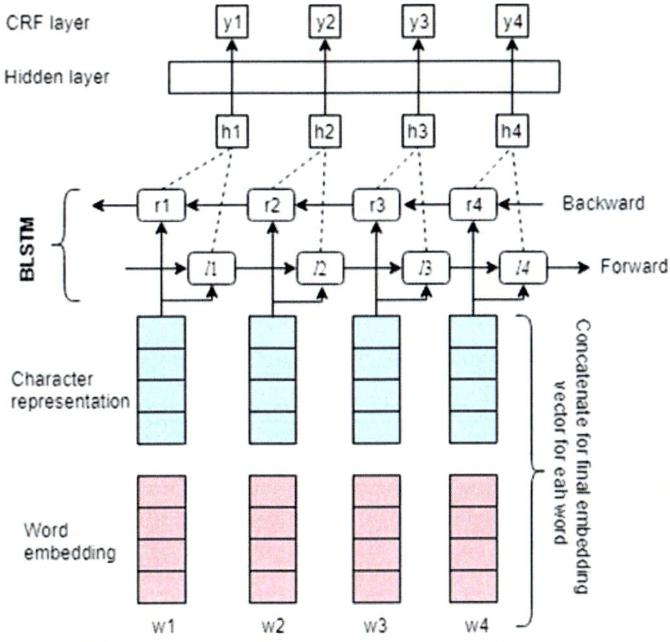

Fig. 1 BLSTM-CRF architecture for the POS tagging system

Words are represented as a continuous vector space where semantically similar words are mapped to nearby points which is the basic idea of Word2vec model. We have used gensim implementation for the purpose of word embedding. Feature extraction has been done using continuous bag-of-words(BOW) algorithm. Context window of five words is defined with size of feature vector being 150. Feature set of any word 'w' includes the features of its next two(w+1, w+2) and earlier two(w-1, w-2) words. In character-level representation, a random vector is passed to the forward and backward LSTMs to learn the prefix and suffix information. Their final states were concatenated to form a vector for the word. This vector was then concatenated with word embeddings from a pretrained model as shown in Fig. 1. The feature vectors obtained by BLTSM are passed through a CRF layer which predicts the POS tags of the words in the corpus. Probability of the POS tag of the current word(y) with feature set (h) is given by

$$P(y \vee h) = \exp \tag{1}$$

Where T is sequence length, K is number of features, f_k is feature function of current word and λ_k is weight for the given feature function.

3.2 *Extractive Summarization*

The model proposed in this paper used text ranking algorithm for the extractive summarization of Kannada documents. Kannada document is fetched to the model which first undergoes preprocessing. In this stage, document is split into paragraphs, sentences and words. Unwanted parts of the text are removed, and root words are identified. Five types of features are extracted during feature extraction part, and summary is generated according to these features (Fig. 2).

Preprocessing: This is a first step where text should be made ready for summarization. In this stage, text will be undergo immense cleansing. Unwanted components are removed through regular expressions. Contractions (ಸಂಕೋಚನ) like ಡಾ, ಉದಾ will be replaced by their full forms. Following four phases are done in the remaining part of the preprocessing.

i Sentence separation: In this phase, text is broken into separate paragraphs and sentences which have independent meaning.
ii Tokenization: Sentence is broken into tokens or words remove punctuations, space, special characters and commas.
iii Stop word removal: It has been observed that any Kannada text contains 20–30% redundant words which carry less important meaning. Such words will be identified and removed. Following category of words are identified as stop words.

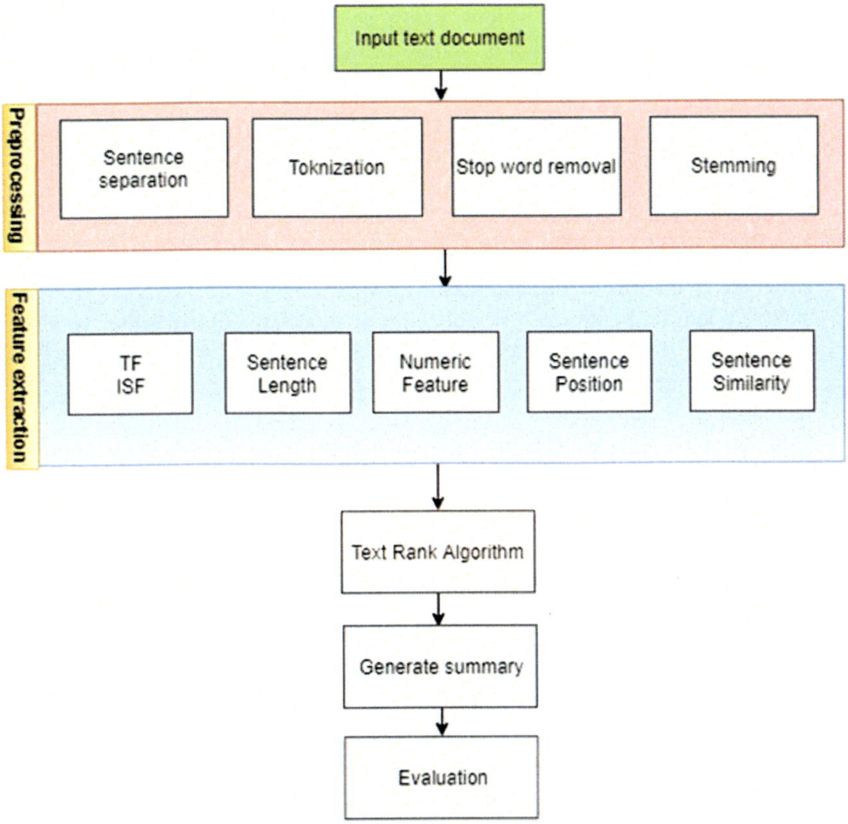

Fig. 2 Workflow of the system

(a) ವಿಶೇಷಣ: ಪರಿಮಾಣವಾಚಕ (ಅಷ್ಟು , ಇಷ್ಟು), ಪ್ರಕಾರವಾಚಕ (ಅಂತಹ, ಇಂತಹ)
(b) ಅವ್ಯಯ: ಸಾಮಾನ್ಯಾವ್ಯಯ (ಸುಮ್ಮನೆ, ನೆಟ್ಟಗೆ), ಅನುಕರಣಾವ್ಯಯ (ದರದರನೆ, ಬೌಬೌ), ಭಾವಸೂಚಕಾವ್ಯಯ (ಅಯ್ಯೋ, ಆಹಾ), ಸಂಬಂಧಾರ್ಥಕ (ಅಥವಾ, ಆದ್ದರಿಂದ, ಹಾಗೂ), ಸಂಬೋಧಕಾವ್ಯಯ (ಎಲಾ, ಎಲವೋ), ಅವಧಾರಣಾರ್ಥಕಾವ್ಯಯ (ಅದೇ, ಅವಳೇ)
(c) ಪ್ರತ್ಯಯ: ಇಕೆ, ಇಗೆ, ಗೆ
(d) ಕ್ರಿಯಾವಿಶೇಷಣ: ವೇಗವಾಗಿ, ಜೋರಾಗಿ

iv Stemming: Words are converted to their root forms. Words from same root but with different suffixes will be considered as a same word. For example, ಅವನು, ಅವರು, ಅವನನ್ನು , ಅವರನ್ನು , ಅವನಿಂದ, ಅವರಿಂದ, ಅವನಿಗೆ, ಅವರಿಗೆ, ಅವನ, ಅವರ, ಅವನಲ್ಲಿ, ಅವರಲ್ಲಿ, ಅವನಿಗೆ, ಅವರಿಗೆ, ಅವಳನ್ನು , ಅವಳಿಂದ will all be converted to ಅವ.

Feature Extraction: In feature extraction, relevance of the sentences is marked by the appropriate features. Every sentence is represented by the features vector ranging from 0 to 1 and assigned with weights using weight ranking algorithm. Sentences have scores based on the weights of feature terms. This score is used for sentence ranking. Following important features are employed on the text.

i TF-ISF: Term frequency (TF) is a ratio of number of times a word appears in sentence to total number of words in the text. Few words occur repeatedly in the text with very little information. Such words will be punished by giving less weights, and rare words will be given more preference. Inverse sentence frequency(ISF) is used for this purpose which is the log of the ratio of all the sentences in the text to the sentences with a particular word.

ii Sentence length: Sentences with too short or long size contain either very less or redundant information. Such sentences will be omitted while generating the summary. Sentence length computes least and highest thresholds to exclude such sentences as shown in Fig. 3.
Calculation of SL is given by
SL=0, if minL < L < maxL
else SL = sin θ

iii Numeric feature: Numeric values will have important information in the text and should be retained. Such values will be assigned with higher weights and are retained.

iv Sentence position (SL): Beginning sentences of a text describe theme, and ending sentences describe summary. Hence, position of a sentence in text is an important criteria. Threshold value(TH) will be the percentage of sentences retained in the summary at the beginning and end with weight SP=1 as given in Fig. 4. For remaining sentences with minV and maxV being the minimum and maximum value of the sentences, SP is calculated as

$$\text{SP} = \cos\left[\frac{\max\,\theta - \min\,\theta}{\max\,V - \min\,V} \times (CP - \min\,V)\right] \tag{2}$$

v Sentence similarity (SS): Weighted matrix will be constructed by comparing individual sentence with every other sentence depending on the number of words overlap between these two. SS is given by

$$\text{SS} = \text{similarity}(x, y) \quad \text{where}\, x \neq y \tag{3}$$

4 Results and Analysis

4.1 Result of POS Tagger

The trained BLSTM-CRF model is run for 15 epochs which showed the training accuracy of 90.26% and test accuracy of 81.02%. Graph of the training accuracy is

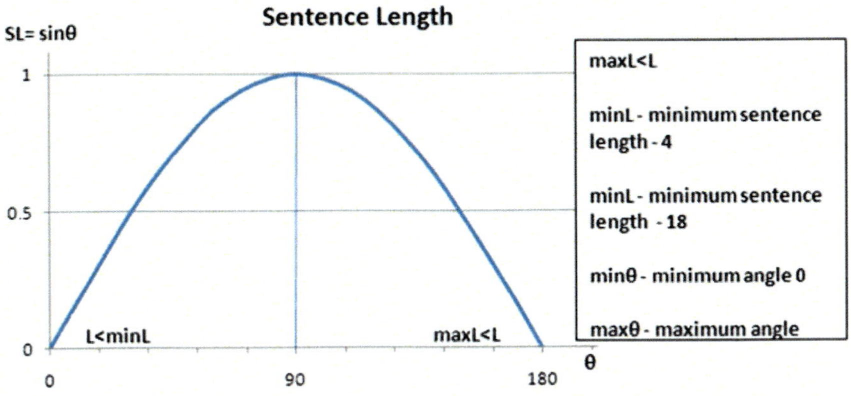

Fig. 3 Visualization sentence length

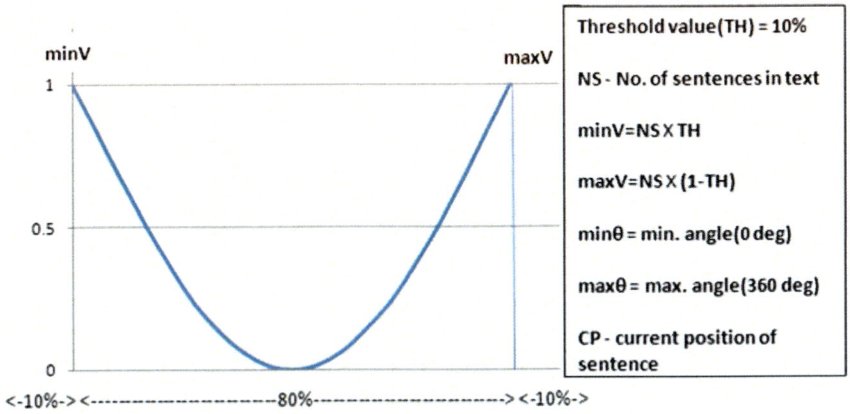

Fig. 4 Visualization of sentence position

as in Fig. 5. Backpropogation is used to update the parameters on every epoch. ADAM with a learning rate of 0.001 is used. Dimension of character representation is fixed to 128, and word embedding is to 150. For regularization, dropout mask with dropout ratio of 0.5 is utilized. It is tested on a dataset of 3100 words, and the following classification report is obtained as shown in Fig. 6. To evaluate the performance of the model, we calculated precision, recall and F1-score for individual tagsets.

We have also presented overall accuracy, precision, recall and F1-scores on the test set in Table 3.

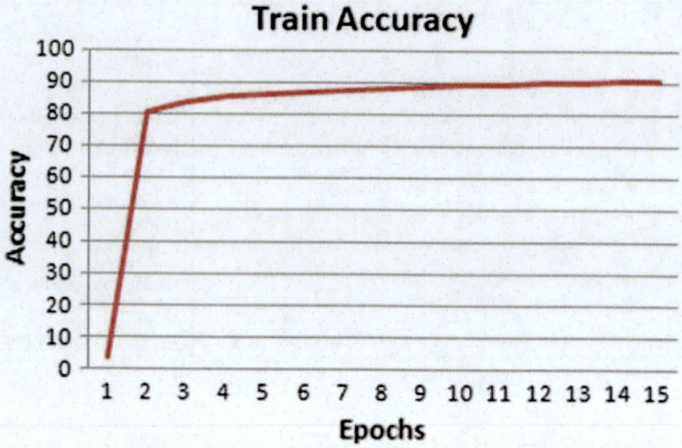

Fig. 5 Train accuracy of BLSTM-CRF model

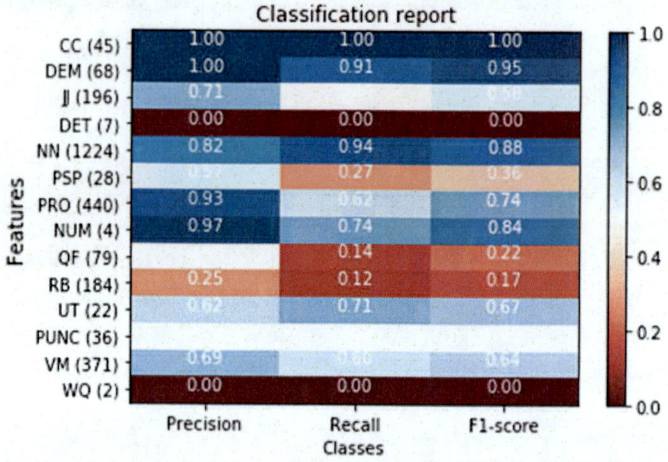

Fig. 6 Classification report

4.2 Result of Extractive Summarization

A part of Kannada Wikipedia article is given as input to the proposed model as in Fig. 7. Aggregated feature values calculated using sentence ranking are shown in Fig. 8. Summaries will be displayed according to the decreasing order of sentence ranks with relevant amount of content. Figure 9 shows the 50% of the summary generated for the input document. For every article chosen, we have considered one reference summary for evaluation. ROUGE metric is used for evaluation which compares the machine generated summary against this reference summary. Three cri-

Table 3 Test set results

Exp	Precision	Recall	F1-score	Support	Test accuracy
Micro avg	0.82	0.82	0.82	3100	0.82
Macro avg	0.61	0.5	0.54	3100	
Weighted avg	0.81	0.82	0.82	3100	

Fig. 7 Input Kannada text

teria of ROUGE metric based on the overlap between system generated and reference summaries are computed as follows.

$$\text{ROUGE-based Recall Score} = \frac{|R| \bigcap |S|}{|R|} = 0.8045 \tag{4}$$

$$\text{ROUGE-based Precision Score} = \frac{|R| \bigcap |S|}{|S|} = 0.7846 \tag{5}$$

$$\text{F Measure Score} = 2 \times \frac{|R| \, * \, |S|}{|R| \, + \, |S|} = 0.7944 \tag{6}$$

where $|R|$ being the length of the reference summary, $|S \cap R|$is the total number of overlapping sentences and $|S|$ is the length of machine generated summary.

```
' ಕನ್ನಡ ಲಿಪಿ ಸುಮಾರು - ವರ್ಷಗಳಿಗಿಂತಲೂ ಹಿಂದಿನದು ': 2.6470735658131286,
' ಅದಲ್ಲದೆ ಸಾವಿರ ವರುಷಗಳ ಸಾಹಿತ್ಯ ಪರಂಪರೆ ಕನ್ನಡಕ್ಕಿದೆ ': 3.3350623701563906,
' ಇವರಲ್ಲಿ  ಕೋಟಿ ಜನಗಳ ಮಾತೃಭಾಷೆ ಕನ್ನಡವಾಗಿದೆ ': 2.4457082133611645,
' ಐದನೆಯ ಶತಮಾನದ ಹಲ್ಮಿಡಿ ಶಾಸನದ ಸಮಯಕ್ಕಾಗಲೇ ಕನ್ನಡವು ಸಾಕಷ್ಟು ಅಭಿವೃದ್ಧಿ
ಹೊಂದಿತ್ತು ': 3.0799400578488183,
' ಕನ್ನಡ ಕರ್ನಾಟಕ ರಾಜ್ಯದ ಆಡಳಿತ ಭಾಷೆ ಜಗತ್ತಿನಲ್ಲಿ ಅತ್ಯಂತ ಹೆಚ್ಚು ಮಂದಿ ಮಾತನಾಡು
ವ ಭಾಷೆಯೆಂಬ ನೆಲೆಯಲ್ಲಿ ಇಪ್ಪತ್ತೊಂಬತ್ತನೆಯ ಸ್ಥಾನ ಕನ್ನಡಕ್ಕಿದೆ ': 2.901121985536191,
' ಕನ್ನಡ ಬರಹದ ಮಾದರಿಗಳಿಗೆ ಸಾವಿರದ ಐನೂರು ವರುಷಗಳ ಚರಿತ್ರೆಯಿದೆ ': 2.9053282
785164236,
' ಕ್ರಿಶ ಆರನೆಯ ಶತಮಾನದ ಪಶ್ಚಿಮ ಗಂಗ ಸಾಮ್ರಾಜ್ಯದ ಕಾಲದಲ್ಲಿ ಮತ್ತು ಒಂಬತ್ತನೆಯ ಶತ
ಮಾನದ ರಾಷ್ಟ್ರಕೂಟ ಸಾಮ್ರಾಜ್ಯದ ಕಾಲದಲ್ಲಿ ಹಳಗನ್ನಡ ಸಾಹಿತ್ಯ ಅತ್ಯಂತ ಹೆಚ್ಚಿನ ರಾಜಾಶ್ರ
ಯ ಪಡೆಯಿತು ': 2.7960740500278294,
' ದ್ರಾವಿಡ ಭಾಷಾತಜ್ಞರ ಅಭಿಪ್ರಾಯದಂತೆ ಕನ್ನಡದ ಭಾಷಿಕ ಚರಿತ್ರೆಯನ್ನು ಹಳಗನ್ನಡ ನಡುಗ
ನ್ನಡ ಹಾಗೂ ಹೊಸಗನ್ನಡವೆಂಬ ಮೂರು ವಿಧವಾಗಿ ವಿಂಗಡಿಸಬಹುದು ': 2.2431356426470
836,
' ಬ್ರಾಹ್ಮಿ ಲಿಪಿಯಿಂದ ರೂಪುಗೊಂಡ ಕನ್ನಡ ಲಿಪಿಯನ್ನು ಉಪಯೋಗಿಸಿ ಕನ್ನಡ ಭಾಷೆಯನ್ನು
ಬರೆಯಲಾಗುತ್ತದೆ ': 3.4758822176471735,
' ರ ಜನಗಣತಿಯ ಪ್ರಕಾರ ಜಗತ್ತಿನಲ್ಲಿ  ಕೋಟಿ ಜನಗಳು ಕನ್ನಡ ಮಾತನಾಡುತ್ತಾರೆ ಎಂದು ತಿಳಿ
ದುಬಂದಿದೆ ': 3.486680639562052,
' ವಿನೋಬಾ ಭಾವೆ ಕನ್ನಡ ಲಿಪಿಯನ್ನು ಲಿಪಿಗಳ ರಾಣಿಯೆಂದು ಹೊಗಳಿದ್ದಾರೆ ': 2.874350
618892056,
'\ufeffದ್ರಾವಿಡ ಭಾಷೆಗಳಲ್ಲಿ ಪ್ರಾಮುಖ್ಯವುಳ್ಳ ಭಾಷೆಯೂ ಭಾರತದ ಪುರಾತನವಾದ ಭಾಷೆಗಳ
ಲ್ಲಿ ಒಂದೂ ಆಗಿರುವ ಕನ್ನಡ ಭಾಷೆಯನ್ನು ಅದರ ವಿವಿಧ ರೂಪಗಳಲ್ಲಿ ಸುಮಾರು  ದಶಲಕ್ಷ ಜ
ನರು ಆಡು ನುಡಿಯಾಗಿ ಬಳಸುತ್ತಲಿದ್ದಾರೆ ': 2.6666666666666665}
```

Fig. 8 Aggregated feature values

```
Enter the percentage of summary you want of original paragraph 50

 ಕನ್ನಡ ಕರ್ನಾಟಕ ರಾಜ್ಯದ ಆಡಳಿತ ಭಾಷೆ ಜಗತ್ತಿನಲ್ಲಿ ಅತ್ಯಂತ ಹೆಚ್ಚು ಮಂದಿ ಮಾತನಾಡುವ
 ಭಾಷೆಯೆಂಬ ನೆಲೆಯಲ್ಲಿ ಇಪ್ಪತ್ತೊಂಬತ್ತನೆಯ ಸ್ಥಾನ ಕನ್ನಡಕ್ಕಿದೆ
 ಜನಗಣತಿಯ ಪ್ರಕಾರ ಜಗತ್ತಿನಲ್ಲಿ  ಕೋಟಿ ಜನಗಳು ಕನ್ನಡ ಮಾತನಾಡುತ್ತಾರೆ ಎಂದು ತಿಳಿದು
ಬಂದಿದೆ
 ಬ್ರಾಹ್ಮಿ ಲಿಪಿಯಿಂದ ರೂಪುಗೊಂಡ ಕನ್ನಡ ಲಿಪಿಯನ್ನು ಉಪಯೋಗಿಸಿ ಕನ್ನಡ ಭಾಷೆಯನ್ನು ಬ
ರೆಯಲಾಗುತ್ತದೆ
 ಕನ್ನಡ ಬರಹದ ಮಾದರಿಗಳಿಗೆ ಸಾವಿರದ ಐನೂರು ವರುಷಗಳ ಚರಿತ್ರೆಯಿದೆ
 ಅದಲ್ಲದೆ ಸಾವಿರ ವರುಷಗಳ ಸಾಹಿತ್ಯ ಪರಂಪರೆ ಕನ್ನಡಕ್ಕಿದೆ
 ಐದನೆಯ ಶತಮಾನದ ಹಲ್ಮಿಡಿ ಶಾಸನದ ಸಮಯಕ್ಕಾಗಲೇ ಕನ್ನಡವು ಸಾಕಷ್ಟು ಅಭಿವೃದ್ಧಿ
ಹೊಂದಿತ್ತು
```

Fig. 9 Compressed summary to 50%

5 Conclusion

This paper discusses the POS tagging and extractive summarization of Kannada documents. Many models are available for the summarization of English text and for some of the Indian languages like Bengali. But, only few researchers have attempted to create a POS tag for Kannada. A hybrid LSTM-CRF-based architecture is designed for a customized corpus of 15000 words and has obtained promising results. A larger corpus would definitely increase the performance of both word embedding and classification models. This paper also deals with the extractive summarization where the input document is preprocessed by removing unwanted details, five types of features are calculated, and summary is generated according to these features.

Acknowledgements This work was supported by VTU TEQIP competitive research grant.

References

1. Abujar, S., Masum, A.K.M., Mohibullah, M., Hossain, S.A., et al.: An approach for Bengali text summarization using word2vector. In: 2019 10th International Conference on Computing, Communication and Networking Technologies (ICCCNT), pp. 1–5. IEEE (2019)
2. Akter, S., Asa, A.S., Uddin, M.P., Hossain, M.D., Roy, S.K., Afjal, M.I.: An extractive text summarization technique for Bengali document (s) using k-means clustering algorithm. In: 2017 IEEE International Conference on Imaging, Vision & Pattern Recognition (icIVPR), pp. 1–6. IEEE (2017)
3. Al Munzir, A., Rahman, M.L., Abujar, S., Hossain, S.A., et al.: Text analysis for Bengali text summarization using deep learning. In: 2019 10th International Conference on Computing, Communication and Networking Technologies (ICCCNT), pp. 1–6. IEEE (2019)
4. Alam, F., Chowdhury, S.A., Noori, S.R.H.: Bidirectional lstms–crfs networks for bangla pos tagging. In: 2016 19th International Conference on Computer and Information Technology (ICCIT), pp. 377–382. IEEE (2016)
5. Antony, P., Soman, K.: Kernel based part of speech tagger for kannada. In: 2010 International Conference on Machine Learning and Cybernetics, vol. 4, pp. 2139–2144. IEEE (2010)
6. Das, A., Bandyopadhyay, S.: Opinion summarization in Bengali: a theme network model. In: 2010 IEEE Second International Conference on Social Computing, pp. 675–682. IEEE (2010)
7. Embar, V.R., Deshpande, S.R., Vaishnavi, A., Jain, V., Kallimani, J.S.: saramsha—a kannada abstractive summarizer. In: 2013 International Conference on Advances in Computing, Communications and Informatics (ICACCI), pp. 540–544. IEEE (2013)
8. Ganesh, J., Parthasarathi, R., Geetha, T., Balaji, J.: Pattern based bootstrapping technique for tamil pos tagging. In: Mining Intelligence and Knowledge Exploration, pp. 256–267. Springer, Berlin (2014)
9. Gupta, M., Garg, N.K.: Text summarization of hindi documents using rule based approach. In: 2016 international conference on micro-electronics and telecommunication engineering (ICMETE), pp. 366–370. IEEE (2016)
10. Jayashree, R., Murthy, K.S., Anami, B.S.: An artificial neural network approach to text document summarization in the kannada language. In: 13th International Conference on Hybrid Intelligent Systems (HIS 2013), pp. 45–48. IEEE (2013)
11. Joshi, N., Darbari, H., Mathur, I.: Hmm based pos tagger for hindi. In: Proceeding of 2013 International Conference on Artificial Intelligence, Soft Computing (AISC-2013), pp. 341–349 (2013)
12. Kishore, K., Gopal, G.N., Neethu, P.: Document summarization in malayalam with sentence framing. In: 2016 International Conference on Information Science (ICIS), pp. 194–200. IEEE (2016)
13. Krishnaprasad, P., Sooryanarayanan, A., Ramanujan, A.: Malayalam text summarization: an extractive approach. In: 2016 International Conference on Next Generation Intelligent Systems (ICNGIS), pp. 1–4. IEEE (2016)
14. Mohamed, S.S., Hariharan, S.: An investigation on graphical approach for tamil text summary generation. In: 2017 International Conference on Intelligent Computing and Control (I2C2), pp. 1–5. IEEE (2017)

15. Pallavi, A.S.P., Pillai, A.: Parts of speech (pos) tagger for kannada using conditional random fields (crfs). In: Proceedings of the National Conference on Indian Language Computing, NCILC (2014)
16. Rahman, A., Rafiq, F.M., Saha, R., Rafian, R.: Bengali text summarization using TextRank, Fuzzy C-means and aggregated scoring techniques. Ph.D. thesis, BRAC University (2018)
17. Reddy, S., Sharoff, S.: Cross language pos taggers (and other tools) for indian languages: an experiment with kannada using telugu resources. In: Proceedings of the Fifth International Workshop on Cross Lingual Information Access, pp. 11–19 (2011)
18. Sarwadnya, V.V., Sonawane, S.S.: Marathi extractive text summarizer using graph based model. In: 2018 Fourth International Conference on Computing Communication Control and Automation (ICCUBEA), pp. 1–6. IEEE (2018)
19. Shambhavi, B., Ramakanth, K.P., Revanth, G.: A maximum entropy approach to kannada part of speech tagging. IJCA **41**, 9–12 (2012)
20. Thaokar, C., Malik, L.: Test model for summarizing hindi text using extraction method. In: 2013 IEEE Conference on Information & Communication Technologies, pp. 1138–1143. IEEE (2013)
21. Tumpa, P., Yeasmin, S., Nitu, A., Uddin, M., Afjal, M., Mamun, M.: An improved extractive summarization technique for Bengali text (s). In: 2018 International Conference on Computer, Communication, Chemical, Material and Electronic Engineering (IC4ME2), pp. 1–4. IEEE (2018)

Swarm-Based Optimizers for PID Tuning of Bioreactor

Mrinalini P. Sirsat and Rajashree V. Sarwadnya

1 Introduction

Proportional-integral-derivative (PID) controller is still widely used feedback control strategy in control engineering even though there are significant developments in control schemes like model predictive control (MPC), internal model control (IMC), and sliding mode control (SMC). This is because of its simple architecture and satisfactory control performance. The basic function of the controller is to maintain the output at a level, so that there is no difference between the process variable and setpoint [1]. While designing a PID controller, the key methodology is to determine the finest set of three gains viz. proportional gain (K_p), integral gain (K_i) and derivative gain (K_d) to achieve optimal control performance. Zeigler–Nichols (Z–N) and Cohen–Coon (C–C) are frequently used tuning methods for single-input, single-output (SISO) control systems. These are conventional tuning techniques. Neural network, fuzzy-based approach, and metaheuristic algorithm approach are the modern techniques. In control literature, PID tuning attempts using various metaheuristic techniques like genetic algorithm (GA), differential evolution (DE), ant colony optimization (ACO) etc. for SISO system have been reported by several researchers [2]. These techniques are optimization-based stochastic algorithms. Unlike deterministic algorithms, they are gradient-free algorithm and hence, derivative terms are not involved in this approach. Particle swarm optimization (PSO) and bacterial foraging algorithms (BFO) are swarm-based stochastic optimization techniques. The best combination of elements that maximize or minimize some objective function can be obtained by using these techniques. It involves four steps namely initialization, selection of fitness function, introduction of genetic operators, and termination.

M. P. Sirsat (✉) · R. V. Sarwadnya
Instrumentation Engineering Department, S.G.G.S. Institute of Engineering and Technology, Nanded, Maharashtra, India

S. K V and K. Guruprasad Rao (eds.), *Smart Sensors Measurements and Instrumentation*, Lecture Notes in Electrical Engineering 750,
https://doi.org/10.1007/978-981-16-0336-5_31

- **Initialization =** Initial population is created randomly.
- **Selection of fitness function =** Each particle is evaluated according to fitness function. Thus, it is important to select a fitness function that represents process/system data. It is problem specific.
- **Introduction of genetic operators =** New generation of particles is introduced into the current generation, so that the algorithm will not get stuck into local minima or maxima.
- **Termination =** Algorithm is terminated after satisfying one of the criteria given below, and final solution is selected at termination point.

(1) It has reached maximum number of iterations.
(2) It has reached some fitness threshold value.

In this paper, swarm-based optimization techniques are proposed to find the control gains of PID controller for bioreactor process. Bioreactor plays vital role in chemical process industries [3]. Simple PID loops can be used to promote the cell growth in reactor by controlling the substrate flow rate [4]. In [3], mathematical model for different stages of microbial growth in bioreactor has been developed. Intelligent controller tuning using PSO algorithm is proposed in [5]. Bioreactor control using BFO algorithm is attempted in [6]. The main contribution of the work involves the correct choice of initial values of acceleration coefficients in PSO. The paper is organized as follows. Section 2 describes the bioreactor process. In Sect. 3, PSO algorithm is described. Section 4 explains the BFO algorithm. In Sect. 5, PSO and BFO strategies are applied to bioreactor process for finding the PID parameters. Tuning parameters are compared, and finally, conclusion is drawn in Sect. 6.

2 Bioreactor Process Description

In chemical process industries, bioreactor is used essentially to produce biochemical marketable products from microbes which are either intermediate or final products. In bioreactor, biological reactions occurred in a liquid medium. This process is called as biosynthesis. Equation (1) shows the biosynthesis operation [3].

$$(\text{Substrate} + \text{Microbes}) \rightarrow (\text{More Microbes} + \text{Products}) \quad (1)$$

Microbes consume the nutrients from substrate in order to produce more microbes and products like medicines, beverages, industrial solvents, etc. Biosynthesis is a slow process having three stages viz. initial stage, growth stage and stationary stage. In initial stage, microbes are placed in a reactor and substrate is gradually fed. This is the lag stage as microbes take some time to adjust to the new environment. In growth stage, microbes consume nutrients and grow in size. They undergo reproduction by using binary fission process. The growth of microbes is directly proportional to the mass of microbes and nutrient concentration. In this stage, toxic materials and organic acids are produced due to which the growth of microbes is limited. This stage is called as stationary stage. During these stages, bioreactor is kept at controlled environmental conditions. Bioproducts are formed in growth and stationary stages. Thus, it is necessary to place a controller in these stages to increase the productivity of microbes concentration (X) and substrate concentration (S). Equations (2) and (3) are mass balance equations used for the mathematical modeling of bioreactor [3].

$$\frac{\mathrm{d}X}{\mathrm{d}t} = (\mu - D)X \tag{2}$$

$$\frac{\mathrm{d}S}{\mathrm{d}t} = D\left(S_f - S\right) - \frac{\mu X}{Y} \tag{3}$$

where
$X =$ microbes concentration.
$S =$ substrate concentration.
$\mu =$ growth rate $= \frac{\mu_{\max} S}{K_m + S + K_1 S^2}$
$D =$ dilution rate.
$S_f =$ substrate feed concentration.
$Y =$ cell mass yield.
$\mu_{\max} =$ maximum growth rate.
$K_m =$ substrate saturation concentration.
$K_I =$ substrate inhibition concentration.
Values of parameters considered in equations are given in Table 1 [3].

Equations (4) and (5) represent mathematical model for growth and stationary stage for the given conditions [3].

$$G(s)_{\text{growth}} = \frac{-0.9951s - 0.2985}{s^2 + 0.1302s - 0.0509} \mathrm{e}^{-0.1s} \tag{4}$$

$$G(s)_{\text{stationary}} = \frac{-1.53s - 0.4588}{s^2 + 2.5645s - 0.6792} \mathrm{e}^{-0.1s} \tag{5}$$

It can be observed that the growth and stationary phases of bioreactor are second-order, time-delayed and unstable processes. In this work, simulation study for growth phase of bioreactor is conducted using PSO and BFO algorithms.

Table 1 Parameters of bioreactor

Stage				Parameter values				
	X	*S*	*D*	$\mu_{\max}$	*Km*	*K1*	*Sf*	*Y*
Growth stage	0.9951 g/lit	1.5122 g/lit	0.3 /hr	0.53/hr	0.12 g/lit	0.4545 g/lit	4.0 g/lit	0.4
Stationary stage	1.5302 g/lit	0.1746 g/lit	0.3 /hr	0.53/hr	0.12 g/lit	0.4545 g/lit	4.0 g/lit	0.4

3 Particle Swarm Optimization

PSO is metaheuristic computational technique developed by Kennedy and Eberhart in 1995 [7]. It is inspired by bird flocking or fish schooling. PSO model consists of swarm of particles which randomly searches food in the search space. There is only one piece of food in the area being searched. All the particles in the group do not know where the food is. But they know how far is the food. Best strategy to find the food is to follow the particle which is nearest to the food. Each single solution is particle in the search space. Each particle has its own velocity and it keeps track of its location by means of coordinates of search space. Also, they have fitness values viz. also called as best solutions. Best solution achieved so far by each particle is p_{best}. The best value, obtained so far by any particle in the swarm is the global best value and is called as g_{best}. Along with p_{best} and g_{best} values, each particle remembers its own best position attained within local neighborhood called as local best l_{best}. Velocity, position, l_{best} and g_{best} value of each particle in a population are represented by vectors v^{id}, x^{id}, pl_{best}, and pg_{best}.

PSO is initialized by particles with random positions and velocities. They move though the D-dimensional search space iteratively. In each iteration, particle updates its position and velocity to achieve gbest value, according to following equations:

$$v_{k+1}^{\text{id}} = v_k^{\text{id}} + C_1 R_1 \left(p_{\text{lbest}}^{\text{id}} - x_k^{\text{id}}\right) + C_2 R_2 \left(p_{\text{gbest}}^{\text{id}} - x_k^{\text{id}}\right) \quad (6)$$

$$x_{k+1}^{\text{id}} = x_k^{\text{id}} + v_{k+1}^{\text{id}} \quad (7)$$

where R_1 and R_2 are random numbers between 0 and 1. C_1 and C_2 are learning factors ranges from 0 to 4 which pulls each particle toward p_{best} and g_{best} values. Inappropriate parameter values of cognitive factors C_1 and C_2 may result in undesirable effect on PSO performance. Each particle is evaluated by fitness function.

3.1 Particle Swarm Optimization Algorithm

Step 1: Initialization of parameters.

- Initialize swarm size N, step size, cognitive and social learning factors C_1 and C_2, dimension of search space D, momentum or inertia weight W.
- Initialize random numbers R_1 and R_2 and current fitness as

$$R_1 = \text{rand}(D, N)$$
$$R_1 = \text{rand}(D, N)$$
$$\text{Current fitness} = 0 \times \text{ones}(N, 1)$$

Step 2: Initialize swarm velocity and position as

$$\text{Current position} = 10 \times (\text{rand}(D, N) - 0.5)$$
$$\text{Velocity} = 0.3 \times \text{rand}(D, N)$$
$$\text{Local best position} = \text{Current position}$$

Step 3: Evaluate local and global best fitness for each particle in swarm on objective function.

Step 4: Compare the current fitness values of each particle with its optimal best fitness values viz. p_{best} and g_{best}. If the current fitness values are good, then take it as new p_{best} and g_{best} values.

Step 5: Update position and velocity of particles according to Equations (1) and (2) for achieving better fitness value in D-dimensional search space.

Step 6: Repeat Step 2–Step 5 if stopping criteria has not been reached. Stopping criteria can be predefined value of objective function or number of iterations. When stopping criteria is met, swarm stops the search and we will get result as optimized output.

4 Bacterial Foraging Optimization

Bacterial foraging optimization (BFO) belongs to the family of nature-inspired optimization algorithms. Foraging behavior of E. Coli bacteria is the key idea behind this algorithm [8]. Bacteria search for nutrients to maximize energy E obtained per unit time T spent for foraging. Four principle mechanisms are observed in foraging theory namely chemotaxis, swarming, reproduction and elimination dispersal.

Chemotaxis

Motion pattern generated by bacteria in chemical environment is called as chemotaxis. It has a tendency to move toward or away from the certain chemical stimuli. Bacteria perceive the direction of nutrient rich environment and try to avoid the noxious environment via swimming and tumbling mechanism. In noxious or neutral place, it tumbles and changes its orientation. During swimming, it will move in its current direction. This step continues until the bacteria are in the positive gradient of nutrient rich substances.

Swarming

Several species of bacteria exhibit the property of forming intricate stable spatiotemporal patterns. Bacteria communicate with each other with the help of attractant and repellent signals. Attractant signal is used for swarming, while repellent signal is used to signal others to be at minimum distance from it. So that they can arrange themselves to form a riding ring in positive nutrient gradient.

Reproduction
After the swarming process, bacteria are sorted in descending order according to the health status. The best half of the bacteria will undergo reproduction and the rest of the bacteria will be eliminated. Thus, population of bacteria is maintained constant.

Elimination Dispersal
Due to undesirable environmental conditions such as change in temperature, noxious surface, it may possible that the group of bacteria is trapped around local optima position. So in order to escape from local optima, either all the bacteria in the region are eliminated or the group of bacteria will dispersed into new food location.

4.1 Bacterial Foraging Optimization Algorithm

Step 1: Initialize $D, N, N_c, N_s, N_{\text{re}}, N_{\text{ed}}, P_{\text{ed}}, C(i)$, and position of each bacterium in *D*-dimensional search space.
where
$D =$ Dimension of search space.
$N =$ Number of bacteria.
$N_c =$ Number of chemotactic steps.
$N_s =$ Length of swim.
$N_{\text{re}} =$ Number of reproduction steps.
$N_{\text{ed}} =$ Number of elimination–dispersal steps.
$N_r = N/2$ Number of bacteria reproductions (splits) per generation.
$P_{\text{ed}} =$ The probability that each bacteria will be eliminated/dispersed.
$C(i) =$ Size of the step taken by *i*th bacterium in each swim or tumble.
$\theta_i =$ Initial position of each bacterium in *D*-dimensional search space.
Step 2: Start elimination-dispersal loop with index *l*.
Step 3: Start reproduction loop with index *k*.

Step 4: Start chemotaxis loop with index *j*.

- Take chemotactic step for each bacterium *i* where $i = 1, 2, \ldots.S$.
- Compute $J(i, j, k, l)$.
- Let $J_{last} = J(i, j, k, l)$ to save this value since we may find better cost value via a run.
- Tumbling or swimming decision is taken for each bacterium.
- The bacterium is moved in the direction of tumble or swim having step size $C(i)$. This movement is represented by following equation:

$$\theta^i(j+1,k,l) = \theta^i(j,k,l) + C(i)\varphi(j) \tag{8}$$

where $\varphi(j) = \frac{\Delta(i)}{\sqrt{\Delta^T(i)\Delta(i)}}$.

$\varphi(j)$ defines the direction of movement of bacterium. During swim, direction is same with the last chemotactic step. When the bacteria tumble, it moves in a random direction.

A. Tumbling: The bacteria will update its location using Equation (8).

- Compute $J(i, j+1, k, 1)$.

B. Swimming:

- Let $m = 0$
- While $m < N_s$, $m = m + 1$
  if $J(i, j+1, k, l) <$ Jlast; Jlast $= J(i, j+1, k, l)$
  and use Equation (8) to compute new $J(i, j+1, k, l)$
  Else, $m = N_s$.
- If $i \neq S$, go to next bacterium.
- Continue chemotaxis till $j < N_c$

Step 5: Reproduction

- Let, for given k and l, health of bacteria is

$$J_{\text{health}} = \sum J(i, j, k, l) \tag{9}$$

- Sort bacteria in ascending order of J_{health}.
- Half of the population having lower health will die and rest of them will be splitted.
- Continue next generation in chemotactic loop till $k < N_{\text{re}}$.

Step 6: Elimination Dispersal

- Eliminate a bacterium and disperse one to a random location to keep the number of bacteria in the population constant.
- Continue elimination-dispersal events till $l < N_{\text{ed}}$.

5 Design of PID Controller Using Swarm Optimizers

For closed-loop system, PID controllers' objectives are to minimize the time domain characteristics such as steady-state error, peak overshoot, and settling time and the frequency domain characteristics such as gain margin, phase margin and resonant peak. But frequency domain design techniques require a reduced order model in case of higher-order processes. While time domain design techniques do not necessitate the reduced order model. Multiple objectives can be combined with appropriate weights to form a single objective function. This type of function is called as aggregate objective function (AOF). In this work, following AOF is considered for minimization.

$$J(\theta) = \min \sum_{i=0}^{N} (W_1 \cdot \mathrm{ISE} + W_2 \cdot M_p) \tag{10}$$

where $J(\theta)$ is objective function consisting of integral square error (ISE) and % overshoot (M_p). The weighing coefficients W_1 and W_2 are selected as 10.

5.1 Experiment 1

In this experiment, PID controller is designed using PSO algorithm. For this approach, dimension of search space $D = 3$, number of particles $N = 30$ and weight of inertia $w = 0.9$ are chosen. To find the effect of learning coefficients on PSO performance, different values of C_1 and C_2 are considered. Simulation is terminated when it has reached the maximum number of iteration. For all the cases, maximum number of iteration is set as 40. The learning coefficients are chosen by assuming $C_1 + C_2 = 4$. C_1 is a cognitive parameter which forces the particles to move toward the local optima, while C_2 is the social parameter which drives the algorithm to reach to the global extrema.

Case 1: Learning coefficients $C_1 = C_2$.

In this case, cognitive and social parameter values are same. Due to this, all the particles will be attracted toward average of personal best and global best values.

Case 2: Learning coefficients $C_1 > 0$, $C_2 = 0$.

In this case, due to absence of social parameter, the particles will be moved with the velocity

$$v_{k+1}^{i}\mathrm{d} = v_k^i\mathrm{d} + C_1 R_1 \left(p_l\mathrm{best}^i\mathrm{d} - x_k^{id}\right) \tag{11}$$

Equation (11) suggests that particles move independently and search for their personal best values.

Case 3: Learning coefficients $C_1 = 0$, $C_2 > 0$.

In this case, the particles will be moved with the velocity

$$v_{k+1}^{i}\mathrm{d} = v_k^i\mathrm{d} + C_2 R_2 \left(p_{\mathrm{gbest}}^{i}\mathrm{d} - x_k^i\mathrm{d}\right) \tag{12}$$

Equation (12) suggests that all the particles will be attracted toward the single point.

Case 4: Learning coefficients $C_1 >> C_2$.

In this case, the cognitive parameter has more impact on the searching strategy of particles. Each particle is more strongly influenced by its personal best position.

Case 5: Learning coefficients $C_1 = C_2 = 0$.

Table 2 PSO PID Parameters for Case 1

Run	Optimum PID values				
	K_p	K_i	K_d	ISE	CPU Time (Seconds)
Run 1	−0.7895	−1.7889	−3.8623	2.043	115.49
Run 2	−1.3168	−0.5765	−3.3810	0.6156	115.77
Run 3	−1.5950	−1.1893	−2.3845	3.618	100.87
Run 4	−2.8206	0.2773	−5.1256	$8.253 * 10^{10}$	102.73
Run 5	−0.0278	−2.4483	−7.3292	18.77	121.20
Run 6	−1.9185	−0.2877	−5.600	18.57	120.85
Run 7	−0.3430	1.5137	−1.8166	$1.179 * 10^{17}$	119.34
Run 8*	−1.2386	−1.1119	−4.9423	0.6412	116.48
Run 9	1.0031	−0.0015	−0.0032	$2.001 * 10^{17}$	115.35
Run 10	−1.0843	−0.1450	−2.1066	0.6754	123.54

* Best PID parameters for Case 1

In this case, the particles will be moved with the velocity

$$v^i_{k+1}\mathrm{d} = v^i_k\mathrm{d} \tag{13}$$

All particles will be continued to fly at their current speed. They have no impact of social as well as cognitive parameter. There is no communication between the particles. They will not find their personal best position which is a local optima and global best position which is a global optima, which results in excessive wandering.

Case 6: Learning coefficients $C_1 << C_2$.

In this case, the social parameter has more impact on the foraging strategy of particles. Each particle is more strongly influenced by personal best position of best particle.

PSO PID parameters for all the cases with ten trials are given in Tables 2, 3, 4, 5, 6 and 7. Best PID parameters for each case are highlighted.

Reference-tracking performance of PID controller obtained for different values of learning coefficients is shown in Fig. 1.

5.2 *Experiment 2*

In this experiment, BFO based PID controller is designed. BFO is initialized with dimension of search space $D = 3$, number of particles $N = 10$, number of chemotactic steps $N_c = 3$, number of reproduction steps $N_{re} = 3$, number of elimination-dispersal loop $N_{ell} = 2$, number of bacteria that undergo reproduction per generation $N_r = 5$ and probability of elimination-dispersal event $P_{ed} = 5$. Considering these values,

Table 3 PSO PID parameters for Case 2

Run	Optimum PID values				
	K_p	K_i	K_d	ISE	CPU time (Seconds)
Run 1	0.7401	573.7468	−3.1112	$1.938 * 10^{17}$	118.13
Run 2	−2.096	−1.0277	4.1614	$1.732 * 10^{14}$	121.17
Run 3	−2.4664	−2.8646	0.9125	$5.315 * 10^{14}$	119.95
Run 4	−0.5201	1.1811	1.0480	$7.43 * 10^{16}$	123.38
Run 5	−3.0828	−4.3070	2.2167	$6.977 * 10^{14}$	116.20
Run 6	4.4758	2.6203	−2.5608	$3.129 * 10^{-14}$	115.82
Run 7	−0.2531	2.4676	4.1826	$1.607 * 10^{17}$	123.77
Run 8	−1.2703	3.0480	−0.9983	$4.721 * 10^{16}$	120.38
Run 9*	−4.4337	−0.3706	−5.5698	20.64	117.50
Run 10	−1.2688	2.4058	−1.6566	$1.934 * 10^{16}$	108.22

* Best PID parameters for Case 2

Table 4 PSO PID parameters for case 3

Run	Optimum PID values				
	K_p	K_i	K_d	ISE	CPU time (Seconds)
Run 1	2.5577	1.8076	0.1549	$2.27 * 10^{17}$	183.97
Run 2	−2.2213	−1.2336	−6.7985	37.24	108.25
Run 3	−0.1562	−3.4811	−5.8201	9.25	120.95
Run 4	−55.6325	−1.8120	−0.1218	807.1	117.76
Run 5	−3.6072	−3.8891	−4.8726	$1.23 * 10^{-14}$	115.78
Run 6	0.3499	−0.0180	0.6037	$1.723 * 10^{17}$	119.95
Run 7*	−0.3697	−2.1294	−7.0875	7.388	121.61
Run 8	−1.9126	0.3480	−2.6486	$1.35 * 10^{14}$	115.72
Run 9	−12.3777	2.4141	−4.7562	$1.296 * 10^{15}$	117.18
Run 10	−121.8264	460.4042	−71.7118	$1.718 * 10^{17}$	115.29

* Best PID parameters for Case 3

BFO algorithm is terminated after executing for loops having 180 total number of iterations.

BFO PID parameters with ten trials are given in Table 8.

Reference-tracking performance of PID controller obtained for different values of learning coefficients is shown in Fig. 2.

Table 5 PSO PID parameters for Case 4

Run	Optimum PID values				
	K_p	K_i	K_d	ISE	CPU time (Seconds)
Run 1	−1.4482	−0.9254	−3.0123	1.256	119.69
Run 2	−0.9611	−0.7510	−4.060	1.152	115.51
Run 3	−0.8831	−3.1796	−4.6416	1.772	263.40
Run 4	−1.7764	3.1962	−3.9724	$1.068 * 10^{17}$	130.34
Run 5	−7.1296	−0.2991	−2.8380	$8.57 * 10^{-14}$	118.85
Run 6	−1.6446	−1.7476	−15.8621	9.81	125.62
Run 7	−2.2722	−0.2022	−4.5680	22.57	128.02
Run 8	1.2182	0.1112	1.0687	$2.17 * 10^{17}$	93.80
Run 9	−0.5893	−1.3456	−3.7917	3.425	116.91
Run 10*	−1.2816	−0.2446	−3.5258	0.5679	122.71

* Best PID parameters for Case 4

Table 6 PSO PID parameters for Case 5

Run	Optimum PID values				
	K_p	K_i	K_d	ISE	CPU time (Seconds)
Run 1	−0.5132	−1.3711	0.1657	$4.178 * 10^{14}$	98.63
Run 2	−1.2501	−2.8221	4.6124	$2.260 * 10^{14}$	118.65
Run 3	−1.2827	−2.2296	6.1794	$2.92 * 10^{15}$	85.40
Run 4	3.0181	−0.4763	5.7316	$2.27 * 10^{17}$	116.34
Run 5	−2.966	6.1909	2.992	$2.268 * 10^{17}$	120.77
Run 6	7.0487	1.5575	−5.0787	$2.251 * 10^{17}$	120.45
Run 7	2.2119	−1.3191	0.5218	$1.14 * 10^{16}$	99.72
Run 8	5.2606	−4.5455	1.9589	$2.036 * 10^{15}$	121.52
Run 9	−0.0343	−3.0998	1.6212	$2.511 * 10^{13}$	119.81
Run 10	−0.4441	−4.6478	−1.3904	$1.342 * 10^{14}$	121.77

5.3 Experiment 3

In this experiment, PID controller is tuned using BFO assisted PSO algorithm. PSO algorithm is hybridized with BFO algorithm to improve the quality of solution obtained by the individual algorithm.

BFO assisted PSO PID parameters with ten trials are given in Table 9.

Reference-tracking performance of PID controller obtained for different values of learning coefficients is shown in Fig. 3.

Table 7 PSO PID parameters for Case 6

Run	Optimum PID values				
	K_p	K_i	K_d	ISE	CPU time (Seconds)
Run 1	−3.5845	0.2068	−3.9277	$1.366 * 10^7$	115.30
Run 2	−0.9654	−2.8509	22.6369	$4.04 * 10^{16}$	118.18
Run 3	−2.0092	−0.9320	−4.7528	33.35	104.41
Run 4	1.3300	−1.6079	8.1721	$3.294 * 10^{16}$	88.82
Run 5	−3.877	−0.3073	−14.2773	26.03	97.43
Run 6	−159.72	−0.2566	−139.4513	18.49	108.93
Run 7	−1.9874	−1.1346	−4.3747	36.54	112.68
Run 8	1.4917	−20.0815	23.994	$7.153 * 10^{14}$	119.75
Run 9	−3.5322	−0.6837	−1.2363	$4.51 * 10^{14}$	110.56
Run 10*	−0.4275	−2.5464	−6.3489	5.724	112.33

* Best PID parameters for Case 6

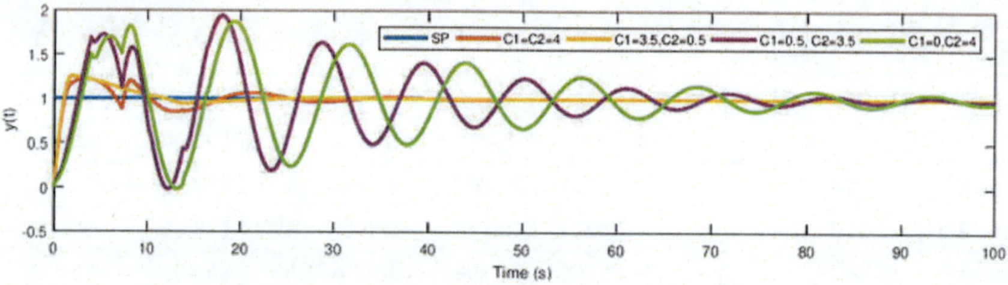

Fig. 1 PSO PID controller performance: process output

Table 8 BFO PID parameters

Run	Optimum PID values				
	K_p	K_i	K_d	ISE	CPU time (Seconds)
Run 1	−0.6645	−0.1609	−0.6646	1.6981	111.78
Run 2	−1.2682	−0.4524	−0.0933	0.4853	165.99
Run 3	23.5757	−0.0619	0.1433	$2.23*10^{17}$	166.77
Run 4	−0.8480	−0.3182	0.0672	0.7614	152.63
Run 5	−0.9216	−0.6645	0.1058	0.6347	166.86
Run 6	−0.7992	−0.2172	−0.7521	1.2512	160.93
Run 7	30.0913	0.0529	0.0987	$2.23*10^{17}$	185.67
Run 8	−8.098	−0.1639	−0.0208	0.9406	164.79
Run 9	−0.7805	−0.5330	−0.0722	0.9238	111.46
Run 10	−0.8735	−0.6818	0.0056	0.7412	118.69

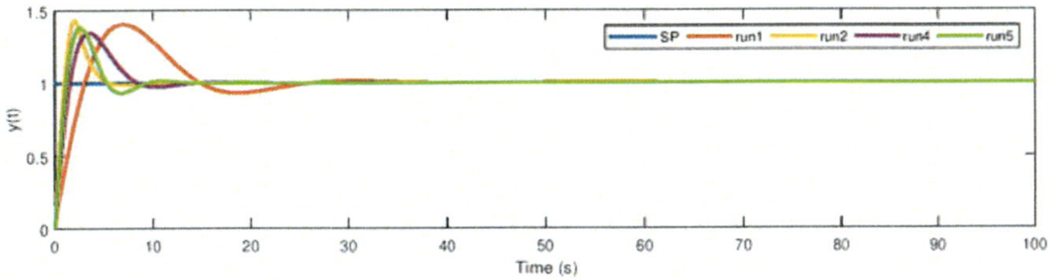

Fig. 2 BFO PID controller performance for $J(\theta)$

Table 9 BFO assisted PSO PID parameters

Run	Optimum PID values				
	K_p	K_i	K_d	ISE	CPU time (Seconds)
Run 1	−1.1294	−3.2217	−1.9390	0.9925	240.08
Run 2	−1.1226	−1.4813	−4.0432	0.7211	298.08
Run 3	−1.1661	−1.5742	−1.1597	0.6347	300.90
Run 4	−1.1629	−1.4545	−2.2712	0.6338	282.20
Run 5	−1.1111	−1.8663	−31.4389	2.263	273.56
Run 6	−1.1103	−1.5304	−1.2088	0.6912	306.12
Run 7	−1.1355	−1.2525	−77.3468	478.2	273.18
Run 8	−1.0688	−4.0478	−19.8120	$1.192*10^{13}$	276.36
Run 9	−1.1760	−1.8918	−1.4224	0.6712	307.33
Run 10	−1.3795	−2.1848	−2.3252	0.714	300.61

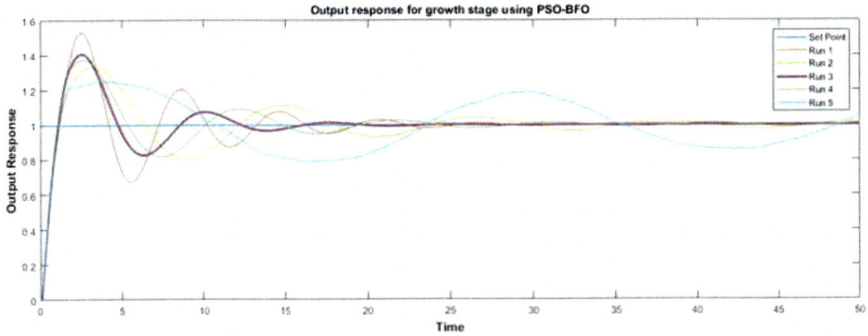

Fig. 3 PSO + BFO PID controller performance for $J(\theta)$

6 Conclusion

In this work, PID controller is designed for growth stage of bioreactor process using swarm-based evolutionary algorithms. Gain parameters of PID controller are estimated by minimizing the objective function using PSO, BFO and hybridized PSO–BFO. Each algorithm is simulated ten times. Simulations are carried out in MATLAB R2019b Simulink software on Core i3-7020U CPU, 2.30 GHz processor and 4 GB RAM laptop. Proportional, integral and derivative gains of controller K_p, K_i and K_d as well as integral square error (ISE) and CPU time are reported for each run. It is observed that acceleration coefficients C_1 and C_2 in PSO affect the convergent behavior of particles in the swarm. Global searching ability is improved in BFO algorithm due to the reproduction and elimination-dispersal mechanisms. Bacteria go under reproduction and 50% of them with less optimized objective function gets eliminated. In this study, research efforts are made to incorporate these BFO algorithm mechanisms in PSO to obtain better performance in terms of convergent behavior of objective function. From the tabulated data, it is observed that the objective function is converged to global optima for more number of runs in BFO assisted PSO algorithm as compared to that in individual PSO and BFO algorithms. In terms of real CPU time, optimization with individual PSO and BFO algorithm is faster than that of hybridized PSO–BFO.

References

1. Ghosal, S., Darbar, R., Neogi, B., Das, A., Tibarewala, D.N.: Application of swarm intelligence computation techniques in PID controller tuning: a review. In: Proceedings of the International Conference on Information Systems Design and Intelligent Applications 2012 (India 2012) held in Visakhapatnam, India, January 2012, pp. 195–208. Springer, Berlin, Heidelberg (2012)
2. Banga, J.R., Balsa-Canto, E., Moles, C.G., Alonso, A.A.: Dynamic optimization of bioreactors: a review. Proc.-Indian National Sci. Acad. Part A **69**(3/4), 257–266 (2003)
3. Rajinikanth, V., Latha K.: Modeling, analysis, and intelligent controller tuning for a bioreactor: a simulation study. International Scholarly Research Notices (2012)
4. Henson, M.A.: Biochemical reactor modeling and control. IEEE Control Syst. Mag. **26**(4), 54–62 (2006)
5. Rajinikanth, V., Latha K.: Tuning and retuning of PID controller for unstable systems using evolutionary algorithm. ISRN Chemical Engineering (2012)
6. Rajinikanth, V., Latha, K.: Bacterial foraging optimization algorithm based PID controller tuning for time delayed unstable system. Mediterranean J. Measurement Control **7**(1), 197–203 (2011)
7. Russell, E., Kennedy, J.: Particle swarm optimization. In: Proceedings of the IEEE International Conference on Neural Networks, vol. 4, pp. 1942–1948. Citeseer (1995)
8. Passino, Kevin M.: Bacterial Foraging Optimization. In Innovations and Developments of Swarm Intelligence Applications, pp. 219–234. IGI Global (2012)

A Vision-Based GPS-Aided Quadcopter for Emergency Vehicle Assistance

Rahul B. Patil, Ron Chaudhuri, VTNS Pavan Kumar, and S. Ramya

1 Introduction

Drones have become much more popular in recent times, and their applications have increased rapidly in various fields. A drone is an unmanned aircraft or ship that is guided remotely. These UAVs are now finding more practical and innovative uses and applications. Studies show that 20% of patients needing emergency treatment have died on their way to the hospital because of delays due to traffic jams and uncooperative motorists. There are a few systems in place to ensure that emergency vehicles encounter the least amount of traffic, but they are only used for very critical and time sensitive cases, such as creating green corridors for ambulances carrying live organs like a heart. These systems require a lot of planning and coordination on the part of officials and regular citizens.

This project proposes an idea that requires less planning and can be used frequently. The project tackles a couple of problems that cause these delays, such as vehicles getting confused as to which route will be taken by the emergency vehicle and navigation problems on the side of the driver. The visualization of the project is shown in Fig. 1.

2 Methodology

Figure 2 is a block diagram representation of all the basic components that are essential to the project. The quadcopter is connected with camera and GPS module. The recorded video from the quadcopter camera is transmitted to road detector module

R. B. Patil · R. Chaudhuri · V. P. Kumar · S. Ramya (✉)
Electronics and Communication Department, Manipal Institute of Technology, MAHE, Manipal, India

S. K V and K. Guruprasad Rao (eds.), *Smart Sensors Measurements and Instrumentation*, Lecture Notes in Electrical Engineering 750,
https://doi.org/10.1007/978-981-16-0336-5_32

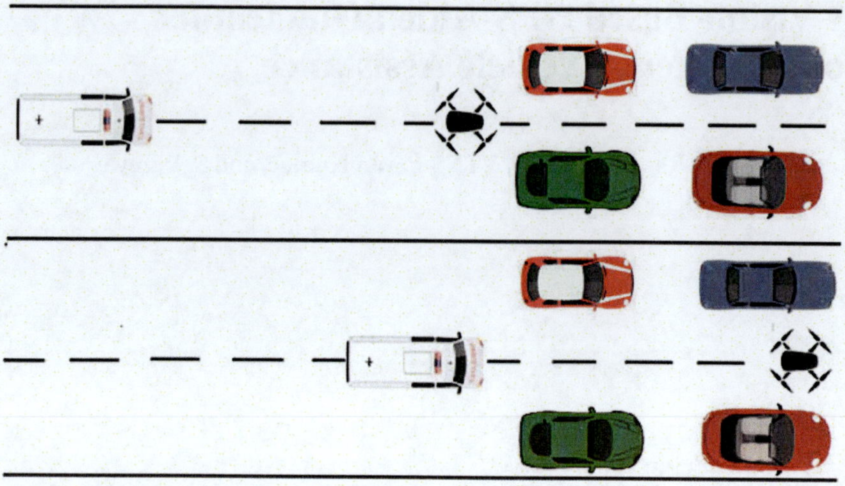

Fig. 1 Visualization of a quadcopter assisting an ambulance

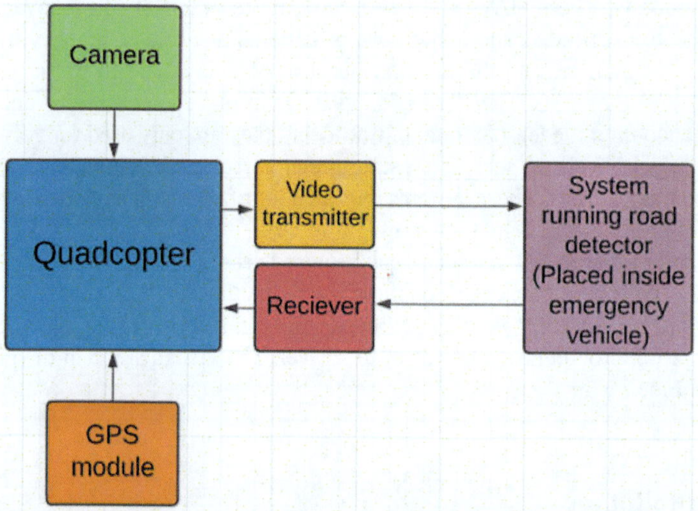

Fig. 2 Block diagram of the basic components of the project

placed in the emergency vehicle through video transmitter. Whenever there is a correction required, these corrections sent to quadcopter.

The project is divided into three modules as shown in Fig. 3, namely designing the quadcopter, obtaining waypoints to trace GPS route. The third module is the road detector program which receives a video feed from a camera mounted on the

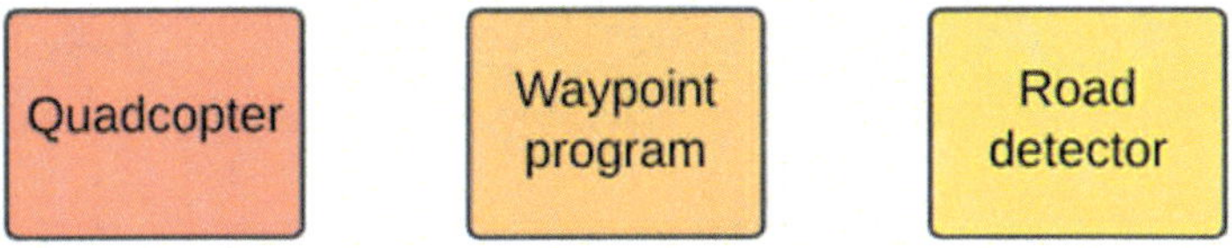

Fig. 3 Schematic block diagram showing the division of the project

quadcopter. The road detector ensures that the quadcopter is flying in the middle of the road by sending back correction signals.

Following section describes each modules in detail. Initially how waypoints are obtained and road is detected is explained, and then how quadcopter makes use of these modules to navigate the path is explained.

2.1 Using Waypoints to Trace GPS Route

Waypoints are the points uploaded to the quadcopter to follow in a sequence from the source to the destination. The Directions Application Peripheral Interface (API) is a service that calculates directions between locations [1]. By providing a source and destination, extract a JavaScript Object Notation (JSON) file containing the information regarding the route. The JSON file contains four different types of response elements, but the project is concerned with "routes." "Routes" contains an array of routes from the origin to the destination. It consists of nested "legs" and "steps." Each element in the "steps" array defines a single step of the calculated directions. The waypoints are to be placed at every change in direction of the route. The field "end_location" in "steps" contains the coordinates of the end of a straight road. To access the Directions API, an API key has to be obtained. This is done on the Google Maps platform.

A graphical user interface (GUI) is created to help the user enter the source and destination. The Tkinter package of Python is used to create this GUI. After entering the source and destination, the user clicks the "Enter" button which saves the locations in a text file.

2.2 Road Detection and Following

The whole process of the road detection program is represented in the flowchart shown in Fig. 4. The input image from the quadcopter camera captured and sent to the fully connected network (FCN). Output of FCN is sent to canny edge detector and Hough transform to find the line. Following subsection discusses the function of these modules in detail.

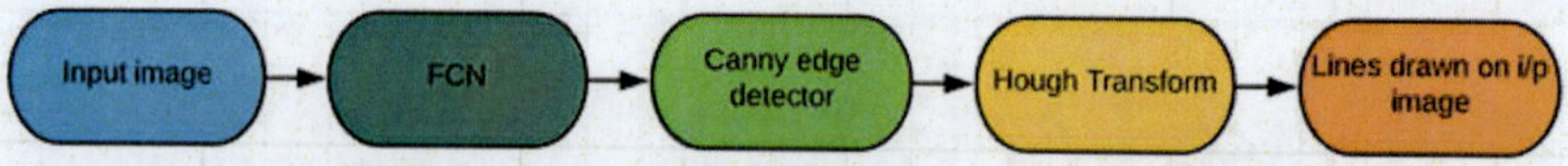

Fig. 4 Flowchart of the road detection program

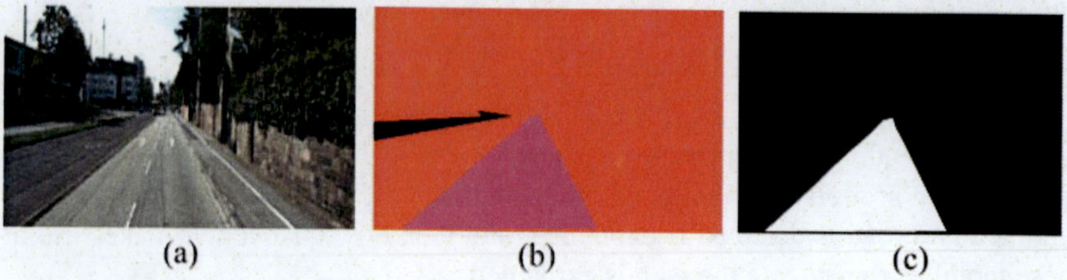

Fig. 5 **a** Sample image from the KITTI dataset **b** Original ground truth of the image **c** Processed ground truth

Pre-processing the training data

The KITTI road/lane dataset was used to train the FCN [2]. It includes 289 images for training in different environments and 290 images for testing. Figure 5a shows sample image from the KITTI road/lane dataset. The ground truth images, as shown in Fig. 5b, the road or lane pixels are pink, the opposite road/lane pixels are black, and all other pixels are red.

There is no necessity in the project to detect the opposite road/lane; therefore, the ground truth images were converted to greyscale images using thresholding method, as shown in Fig. 5c. This, in turn, reduced the training time.

All the training images and processed ground truth images are then read and resized to 640×320. This is done to reduce training time while retaining important information. The preprocessed data is input to fully convolutional network (FCN). The UNET architecture is used for the project.

The UNET architecture

The UNET architecture was originally created for bio-medical image segmentation [3]. It gets its name from the shape of its flowchart. The UNET architecture consists of a contraction path and an expansion path as shown in Fig. 6.

The first path involves contraction, hence called the contraction path (or encoder). This path captures the context of the image. The contraction path includes a stack of convolutional and max-pooling layers in each step. The contraction path has five steps. Each step has two convolutional layers with a dropout layer between them, then a max-pooling layer at the end. The fifth step does not have a max-pooling layer. The contraction path learns the different features of the image.

The second path involves expansion symmetrically, hence called the expansion path (or decoder). This path makes use of transposed convolutions to acquire precise

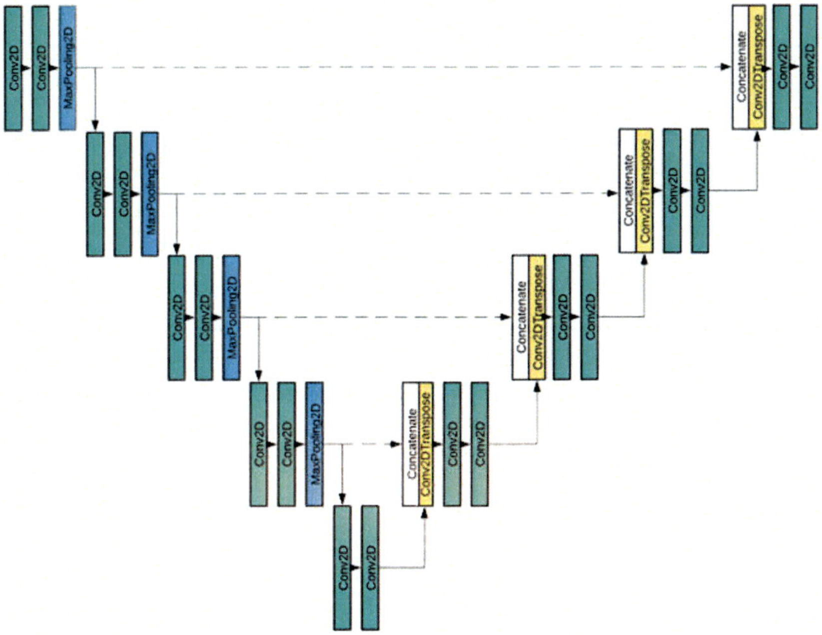

Fig. 6 UNET architecture

localization. The size of the input and the output images are the same because the network only contains convolutional networks and no dense layers; therefore, it is a fully convolutional network. The expansion path has four steps. The first layer is the transpose convolutional layer. The contraction path recovers the lost location information of the features. This is done by adding skip connections by concatenating the output of the transposed convolutional layers with the feature maps of the same level. This layer is then followed by two regular convolutional layers with a dropout layer between them.

After training the model, the weights of the model are saved in a JSON file so that it can be loaded in any program for use. The prediction images contain pixels that range between 0 and 255. This is because the prediction of each pixel is a probability based on how sure the FCN is that a certain pixel is road-pixel or not.

Canny edge detector

The output of the FCN is a black and white image, in which the white pixels represent the road. The canny edge detector is used to obtain the edges of the road detected. The canny edge detector has a multi-stage process. First, the image is smoothed by Gaussian convolution. A simple 2D first derivative operator smooths image to highlight regions of the image with high first spatial derivatives. Ridges in the gradient magnitude image are caused by edges. The algorithm then tracks along the top of these ridges and sets zero as all pixels that are not actually on the ridge top to give a thin line in the output, a process known as non-maximal suppression.

Hough Transform

The canny edge detector's output is an image that highlights edges of the output from the FCN. These edges are 1 pixel wide white pixels, but there is no relation between these pixels. The association required is that a majority of them lie on a straight line. The Hough transform is used to fit a straight line to these points.

Highlighting road edges

The quadcopter requires correction signals, which are to be determined by its position relative to the road. After acquiring the lines for both edges of the road, two thresholds are defined to decide whether the UAV has to turn left or right or keep going straight. The thresholds are lines to the left and right of the center of the camera, assuming the camera is placed in the center of the drone. The position of the drone is evaluated by finding the midpoint of the road edges and checking which side of the threshold it lies.

3 Simulating the Drone

The project demonstration is done on a drone simulator using a Python package "dronekit-sitl." "sitl" stands for "Simulation in the loop." This package allows users to simulate various UAVs including quadcopter. A ground control station (GCS) can be used to visualize the simulation on a digital map. The project uses "QGroundControl" as its GCS. Before starting the simulation, the first point is obtained from the text file containing all the waypoints. This point will be the home location of the simulated quadcopter. The simulation is started by passing various arguments, such as vehicle type, home location, etc. The simulation starts by opening three transfer control protocols (TCP) ports on the local host. The waypoints in the file are again accessed and are sent to the drone in sequence.

When the quadcopter receives a waypoint, it travels the displacement from the present point to the waypoint. This means the quadcopter always travels in straight lines. This is a problem when a curved road arrives as the quadcopter does receive any signals to correct its direction until it reaches the next waypoint. To solve this issue, the computer vision program writes corrections in a text file for each frame, which the drone reads immediately.

As the project makes use of a simulated quadcopter, this is demonstrated by writing corrections signals randomly in a text file. The corrections are either "STRAIGHT," "RIGHT," or "LEFT." To move the drone in a particular direction, the velocity variables are changed accordingly.

There are three velocity variables attributed to a drone corresponding to three axes, i.e., x, y, and z [4].

- Positive x variable means forward
- Positive y variable means right
- Positive z variable means downward

Depending on the correction read, the appropriate encoded message is sent to the drone for every frame. After setting a request, the obtained JSON file is decoded to "uint8" format, and then, the data are loaded into a readable format. The loaded object is a dictionary containing more nested lists and dictionaries. The extracted points are then written in a file, which will later be accessed by another program that uploads the points to the quadcopter. The last line of the program launches the Python script that starts the simulation of the drone.

4 Results and Discussions

Following sections provide the accuracy and loss for the training of FCN. The test results depict the road detection using FCN, canny edge detector, and Hough transform.

4.1 Training Results of the FCN

The FCN was trained for 30 epochs, and a callback was defined to only save the best model on the basis of "validation loss." Epoch no. 28 had the least validation loss. Therefore, the model saved finally had the following parameters:

1. Accuracy: 97.89%
2. Loss: 5.25%
3. Validation accuracy: 96.10%
4. Validation loss: 10.72%

The accuracy and loss graphs were plotted for all 30 epochs. They are shown in Figs. 7 and 8.

4.2 Testing the FCN

Random images of roads under different lighting conditions were used to test the trained FCN, including images at night and images with cars present. Some predictions of the FCN.

Figure 9a is one frame of the input video. The red lines indicate the road edges. The green lines indicate the center of the camera and hence the center of the drone. The intersection of the two blue lines indicates the midpoint of the two road edges

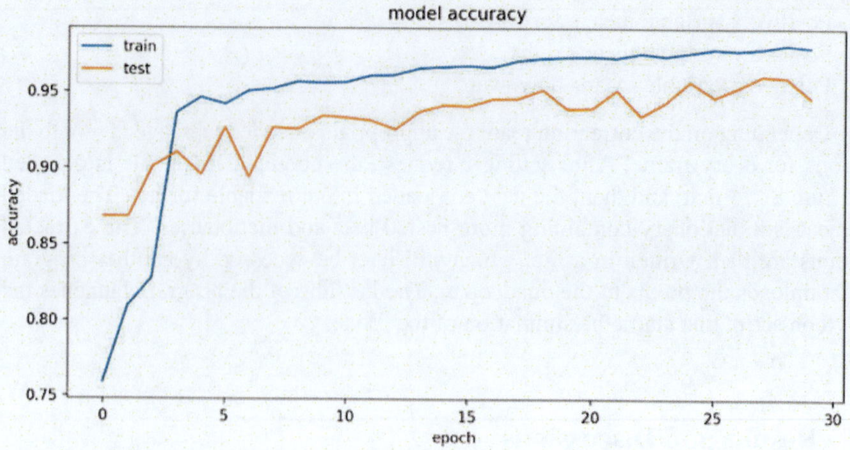

Fig. 7 Plot of accuracy of the neural network

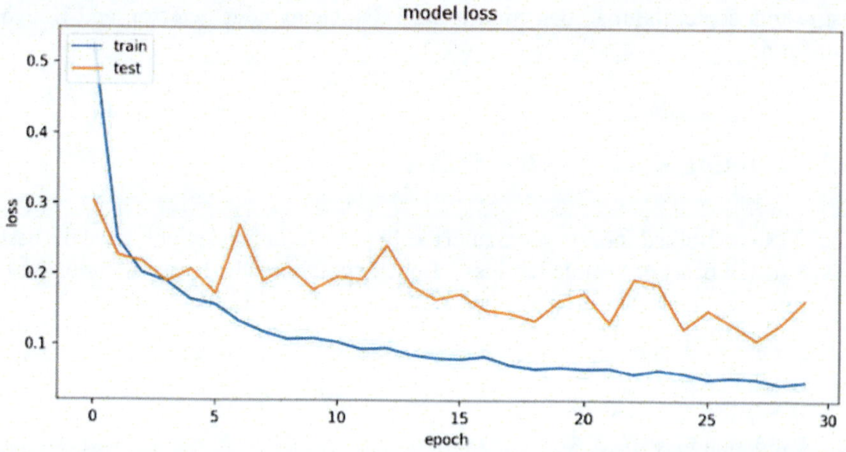

Fig. 8 Plot of loss of the neural network

Fig. 9 **a** Sample frame of the input video with road edges highlighted **b** Prediction of the frame

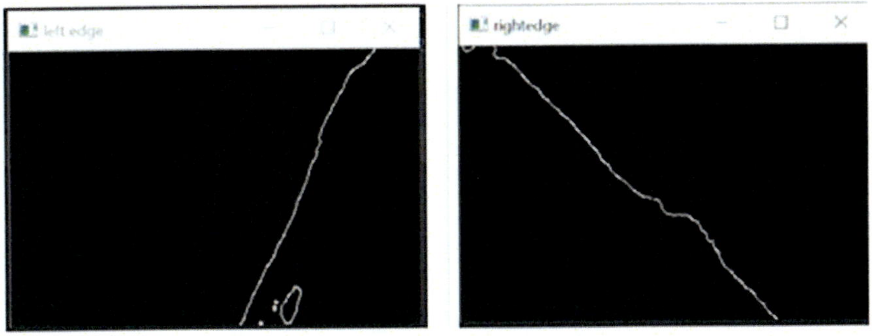

Fig. 10 Left and right edges of the road from the frame in Fig. 9**b**

detected. Figure 9b is the prediction of the FCN for the respective frame of the video shown. Left and right edges of the road from the frame are as in Fig. 10.

5 Conclusion

The project aimed at creating a vision-based GPS-aided quadcopter to assist emergency vehicles such as ambulances and fire trucks, through a city's traffic. The quadcopter will always fly in front of the vehicle, warning other vehicles of the route that it will take. The user only has to enter the source and destination, the same way as using Google Maps normally. The waypoints are then uploaded to the flight controller on the quadcopter before starting the mission. The computer vision-based road detection module is implemented so that the quadcopter always follows the road in between two waypoints. This was done by creating a fully convolutional network based on the UNet architecture and trained by the KITTI road/lane detection dataset. The corrections from the computer vision code are written in a text file which is simultaneously read by the simulation to change the course of the simulated quadcopter. A limitation of using a simulated drone is that it does not consider the various obstacles on the road such as low hanging traffic lights, tree branches, sign boards or tall vehicles. Going forward, the project can be implemented on a real quadcopter. For this, the drone quadcopter receives instructions from the computer vision program in real time over a Wi-Fi network.

Acknowledgements We would like to thank our internal guide, Dr. Ramya S. for assisting us with our project. It would have not been possible to do this without her guidance and help.

References

1. University of Michigan, Using Python to Access Web Data.: https://www.coursera.org/learn/python-network-data/home
2. Fritsch, J., Kuehnl, T., Geiger, A.: A new performance measure and evaluation benchmark for road detection algorithms. In: 16th International IEEE Conference on Intelligent Transportation Systems (ITSC 2013), pp. 1693–1700. IEEE Press (2013)
3. Ronneberger O., Fischer P., Brox T.: U-Net: Convolutional networks for biomedical ımage segmentation. In: Medical Image Computing and Computer-Assisted Intervention MICCAI 2015, Lecture Notes in Computer Science, vol. 9351, pp. 234–241. Springer Cham (2015). https://doi.org/10.1007/978-3-319-24574-4_28
4. 3DR Dronekit Python Manual: https://dronekitpython.readthedocs.io/en/latest/examples/simple_goto.html

Object Detection and Mapping in 3D Space for Quad-Copter Control Application

Mohamed Hannan Sohail, Pawan Kumar Dhakal, and Gurunayk Nayak

1 Introduction

The main challenge in the paper resides in estimating and localising the object in a 2D frame (say, wall frame) and then converting the estimated position into the thrusts required to reach the estimated position in the other 2D frame [1] (say, ground plane frame) as shown in Fig. 1. Object detection or object recognition and localisation in computer vision refers to detection or recognition of an object in a digital image with a fixed frame for indoor application. It is based on the principle that every object around us has a specific feature [2]. In other words, every object has a feature associated with it which distinguishes itself from the other object. Hence for our work, the first step involves feature extraction form the image. The main advantage of feature extraction is that it reduces the size of the data to be processed. Hence, the amount of resources or processor power required for the application is significantly saved through this. To apply the feature extraction, we have used two filters. The first filter is used for preprocessing of the image, and the other one is for segmentation of the image.

1.1 Median Filter

The first one is the median filter. The median filter is a nonlinear digital filtering technique, often used to remove noise from an image or signal. Median filtering is widely

M. H. Sohail (✉)
Department of Electrical Engineering, National Institute of Technology, Patna, India

P. K. Dhakal
Department Electrical and Computer Engineering, University of Oviedo, Ovideo, Spain

G. Nayak
Department of Electrical and Electronics, M. S. Ramaiah Institute of Technology, Bangalore, India

S. K V and K. Guruprasad Rao (eds.), *Smart Sensors Measurements and Instrumentation*, Lecture Notes in Electrical Engineering 750,
https://doi.org/10.1007/978-981-16-0336-5_33

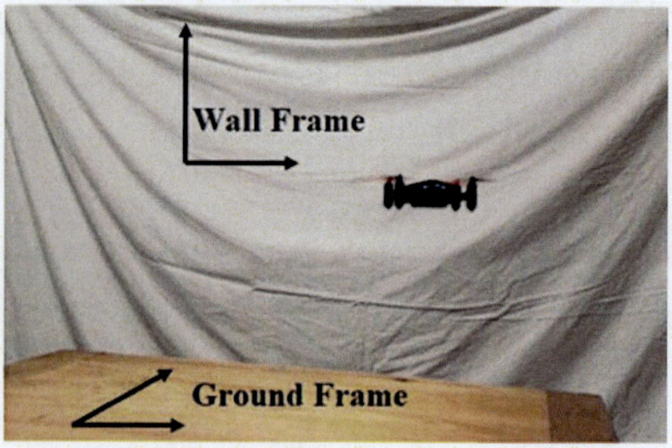

Fig. 1 Indoor navigation of quad-copter

used in digital image processing because, under certain conditions, it preserves edges while removing noise. The other image processing technique which we have used for our work is colour-based thresholding which helps in detection of the object. Image thresholding is the process of partition a digital image into multiple set of pixels.

The 1D nonlinear function of the median filter can be expressed as shown in Eq. 1.

$$y(n) = \mathrm{med}[x(n-k), x(n-k+1), \ldots, x(n), \ldots x(n+k-1, x(n+k)]. \quad (1)$$

where $y(n)$ is the output and $x(n)$ the input signals. The filter "collects" a window containing $N = 2\,k + 1$ samples of the input signal and then performs the median operation on this set of samples. The median filter itself is simple, and in the standard form, there is only one design parameter, namely the filter length $N = 2\,k + 1$.

1.2 Colour-Based Thresholding

There are many feature extraction algorithm available such as edge detection [3], corner detection, blob detection and thresholding. The method which we have used in our work is thresholding, which is a simplest technique of segmentation for computer vision application. The main reason for using this technique is that it is fast, and it provides very accurate result.

The simplest example of thresholding is converting a greyscale image to a binary or black and white image. The thresholding method which we have used in our work is colour-based thresholding, i.e. the region of the object to be detected in the image is converted into two segments based on the colour bound. One segment is formed by the pixels having colour within the colour bound, while the other segment is formed by the rest of the other pixels of the image. This technique helps in detecting a specific colour object in the entire region.

For our work, we have used HSV colour space instead of RGB colour space for the image processing. In case of RGB colour space, the each pixel is represented in terms of component of red value, component of green value and component of blue value. The main reason of using HSV colour space is that it helps in reducing the effect of luminescence.

1.3 Kalman Optimal Estimator and Extrapolation

After detection of object, the next step required to control movement of the quad-copter is the future coordinates. In our work, we are using Kalman optimal estimator to predict the coordinates of the object for the next time step, and then, we are using the Euler's extrapolation technique for further prediction of coordinates after multiple time steps by using the previous predicted coordinates. The reason for using the Euler extrapolation after the Kalman optimal estimator is that the processing time is higher for Kalman optimal estimator as compared to the Euler's extrapolation [4].

Kalman estimator is also known as linear quadratic estimation. It is an algorithm which requires very high measurement over the period of time. Kalman optimal estimator algorithm works in two steps, namely measurement step and prediction step. Kalman gain is a factor which determines how these measurements will be used for determining the new estimated value. In prediction step, this filter produces the estimations of the current state variables along with certain uncertainties. The block diagram representation of the same is shown in Fig. 2.

The state vector (X), transition matrix (A) and measurement matrix (H) need to be defined for the Kalman filter. Since we are considering the object to move in a 2D plane in our work, we define a state vector consisting of six state variables which includes position, velocity and acceleration of object in both horizontal and vertical direction. This is shown in Eq. 2

$$X = [x, y, v\_x, v\_y, a\_x, a\_y] \tag{2}$$

where x and y are position of the object in horizontal and vertical direction, respectively. v_x and v_y are the velocities in the horizontal and vertical direction, respectively. a_x and a_y are the accelerations in horizontal and vertical directions, respectively. The transition matrix (A) is designed based on kinematic equations of motion as shown in Eq. 3.

$$A = \begin{bmatrix} 1 & 0 & 1 & 0 & 0.5 & 0 \\ 0 & 1 & 0 & 1 & 0 & 0.5 \\ 0 & 0 & 1 & 0 & 1 & 0 \\ 0 & 0 & 0 & 1 & 0 & 1 \\ 0 & 0 & 0 & 0 & 1 & 0 \\ 0 & 0 & 0 & 0 & 0 & 1 \end{bmatrix}. \tag{3}$$

In the Kalman filter, your next state, $X(t+1)$, is equal to the previous state $X(t)$ multiplied by the transition matrix A, so with the transition matrix you posted, you

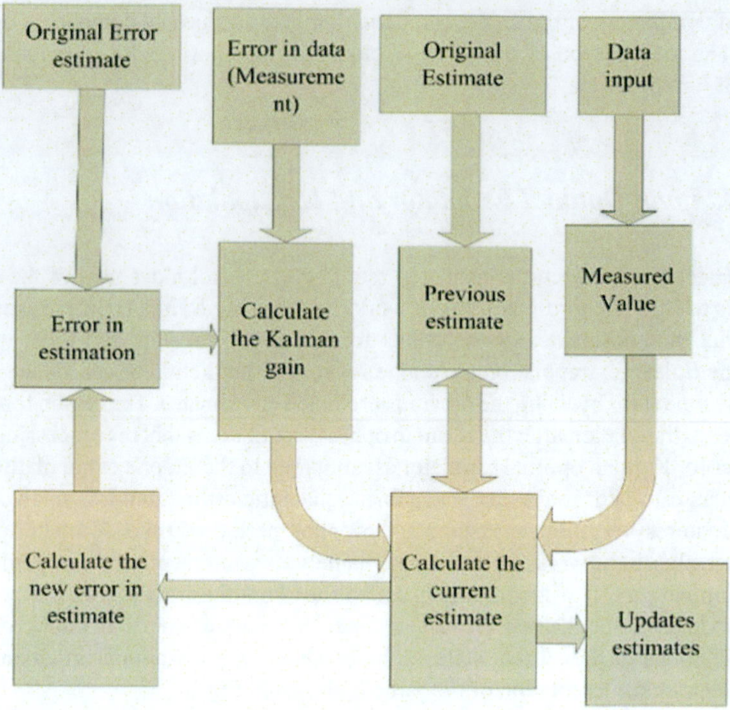

Fig. 2 Functional block diagram of Kalman filter/estimator

would have:

$$\begin{aligned}
x(t+1) &= x(t) + v_x(t) + 0.5a_x(t) \\
y(t+1) &= y(t) + v_y(t) + 0.5a_y(t) \\
v_x(t+1) &= v_x(t) + a_x(t) \\
v_y(t+1) &= v_y(t) + a_y(t) \\
a_x(t+1) &= a_x(t) \\
a_y(t+1) &= a_y(t).
\end{aligned} \tag{4}$$

In our work, the current position of the object is obtained from the video source after object detection using image processing techniques. The velocity and acceleration of the object cannot be measured directly from the input source but can be estimated using previous and present object position values. Therefore, the only measurable state is the position of the object. The measurement matrix (H) is shown in Eq. 5.

$$H = \begin{bmatrix} 1\,0\,0\,0\,0\,0 \\ 0\,1\,0\,0\,0\,0 \end{bmatrix}. \tag{5}$$

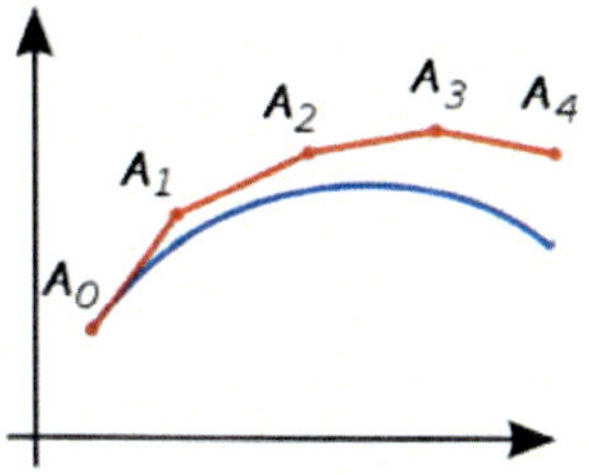

Fig. 3 Visual illustration of Euler's method, blue curve is the unknown curve, and the red curve is the approximation curve using Euler's method

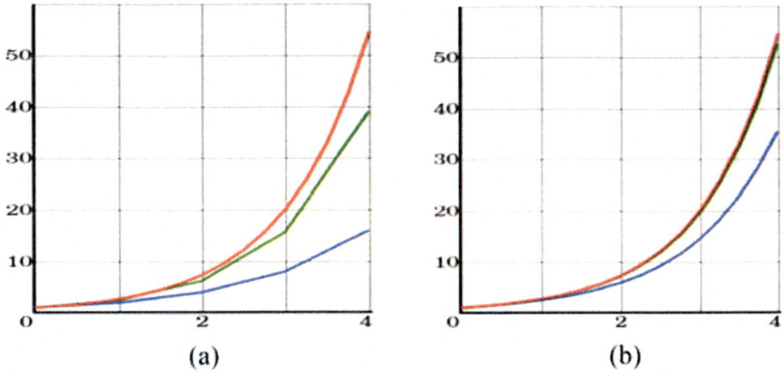

Fig. 4 Numerical integration of e^x with time (**a**) step = 1.0 and (**b**) step = 0.25

Euler's extrapolation method or forward Euler method is a first-order numerical-based approach to solve ordinary differential equation (ODE) with some initial value. It is the simplest Range–Kutta method, and it is the most basic explicit method of numerical integration of an ordinary differential equation.

The local error, i.e. the error per step, is proportional to the square of the step size, and the global error, i.e. error at a given time, is proportional to the step size because Euler method is a first-order method. Euler method serves as a basic for constructing more complex method such as predictor–corrector method [5].

Figures 3 and 4 demonstrate the effect of the time step in approximation. The green curve in Fig 4(a) is result of midpoint method. From the above example, it is evident that as the time step decreases the approximation curve which is more closely related to the actual curve.

2 Implementation

We have used an external webcam and a laptop for acquiring the data and processing it. It is done through a Python program. This involves first step as acquisition of data through the webcam and sending it to for thresholding. The image obtained after thresholding is further processed to detect the object in the image frame. To do so, through the program, we first find the largest contour in the processed frame. Then, we find the minimum circle required to enclose the contour. Then, the circle coordinates in the frame including its radius is replicated in the RGB colour space of the original image to display the detected object in the image frame. Once the tracking of the ball is done, then the processed coordinates are passed through Kalman optimal estimator for finding an intermediate predicted value. The intermediate predicted value is used as the initial value for the Euler's extrapolation. The time step for the Euler method for our case is 0.003, and number of iteration is 30.

2.1 *Algorithm for Path Prediction*

- Initialise all the camera and video parameters and the variables.
- Start capturing the videos through both the cameras
- Apply median filter to the captured video and convert it to HSV representation
- Detect the object in HSV representation using the thresholds to create a mask
- Use erosion and dilation on the mask to remove any undesirable points
- Crete the contour of the detected object on the mask.
- Obtain the coordinates of the contour circle centre and create the same contour circle on the video with RGB representation
- Draw the tracking path.
- Use the same coordinates as input to the Kalman optimal estimator.
- Extrapolate the output of the Kalman optimal estimator to get the final predicted coordinate.
- Display the predicted 2D coordinate in wall frame on the RGB colour space.
- Send the predicted coordinates to the companion system for further processing.
- Combine both the camera video and display as final output with detected object, its trajectory and its predicted position.

2.2 *Discussion of Tuning the Algorithm*

We have used a yellow colour ball as the object to be detected. The image of the ball in HSV colour space after thresholding is as follows. In Fig. 5(a), the white colour patch is the regions of the pixels of the yellow ball in the image but in the HSV colour space. Once the ball is detected and the future position of the ball is predicted, it is displayed on the same image but in RGB colour space as shown in Fig. 5.

In Fig. 5(b), the yellow circle is the actual position of the ball, and the red circle is future predicted location of the ball. Then, the predicted value is converted into suitable format for the input to the quad-copter and is sent to it through the telemetry. The pixels are used to determine the distance between quad-copter and the ball, and the pixel per length is given by Eq. 6.

(a) (b)

Fig. 5 **a** Result of thresholding in HSV colour space **b** Tracked ball and its predicted position

$$L = \frac{720\,\text{px}}{300\,\text{cm}}. \tag{6}$$

where L is the length per pixel. The value of the 300 cm comes from the fact that the camera was positioned in such a way that it covers 300 cm by 300 cm of visual area.

References

1. Li, X., Wang, K., Wang, W., Li, Y.: A multiple object tracking method using Kalman filter. In: Proceedings of the 2010 IEEEInternational Conference on Information and Automation, June 20–23, Harbin, China (2010)
2. Engel, J., Sturm, J., Cremers, D.: Camera-based navigation of a low-cost quadcopter. In: IEEE/RSJ International Conference on Intelligent Robots and Systems(IROS), October 2012, pp. 2815–2821. (2012) ISBN 2153–0858
3. OpenCV Documentation: https://docs.opencv.org
4. Chui, C.K., Chen, G.: Kalman filtering: with real-time applications. In: Springer Publication, 4th edn (2009)
5. Euler method, Wikipedia: https://en.wikipedia.org/wiki/Euler_method

Response Control of Wind-Excited Outrigger Structural System

Kavyahsree, Shantharam Patil, and Vidya S. Rao

1 Introduction

One of the basic needs for physical well-being is shelter, to withstand environmental vulnerabilities. Construction of the permanent structure evolved throughout civilization, where stability is the main concern regarding it. The stability and serviceability of the structure are taken into consideration to provide a tall and slender structure [1]. Tall and slender structure is preferred, so the space occupied is less to accommodate more people in densely populated areas. As the tall structure is preferred as the economic eminence of the nation, the tall building is in the contest of construction to reach skywards. As the building becomes slender, the stiffness and damping of the building decreases and becomes more susceptible to wind lateral load [2]. The wind load is considered in the design of a structure with more than ten floors [3]. As the height of the building increases, the wind load will pose discomfort like the perception of motion by the occupants. The perception of motion in the building causes sickness and dilemma in the occupants if the acceleration of the building crosses the limit and all the furniture's and fixture in the building tend to move [4]. There are different building code limit for the acceleration of the building for human comfort [5]. The wind load also poses a problem for the pedestrian because of the collapse of some building fixtures in the perimeter of the structure. The shielding effect from tall structures can create low-pressure regions in the vicinity of those buildings [6]. Also, the tunneling effect created by the presence of two or more nearby tall buildings can ultimately result in creating high-velocity regions. To withstand these effects from the environment, the different forms of structures are proposed as internal and external structural forms [7]. One of the internal structural forms is the outrigger structure

Kavyahsree (✉) · S. Patil
Manipal School of Architecture and Planning, Manipal Academy of Higher Education, Manipal, India

V. S. Rao
Manipal Institute of Technology, Manipal Academy of Higher Education, Manipal, India

S. K V and K. Guruprasad Rao (eds.), *Smart Sensors Measurements and Instrumentation*, Lecture Notes in Electrical Engineering 750,
https://doi.org/10.1007/978-981-16-0336-5_34

which has proven to be promising against the lateral load. Historically, the outrigger arms were used in the canoes as a single outrigger and two outriggers on both sides to improve rolling stability [8]. Outrigger is a beam that is deep and has large stiffness; it connects the inner core of the structure to the perimeter column which helps in keeping the columns in position and also sway of the building [9]. Outrigger helps in reducing the movement of the core when compared to the system without outriggers, reducing the lateral drift at the top, increasing the stiffness by 25–30% of the original structure, providing additional stiffness to resist the overturning forces and making the entire structure to act as one unit in resisting the lateral loads [10]. The parameter such as core moment, column forces, outrigger moment, optimum positioning of single and multiple outriggers, structural drift, and flexural rigidity of the outrigger has been already studied [11]. These outrigger structures with the incorporation of the external structural protective system will increase the performance of the system against the wind load, and this concept is called damped outrigger [12, 13]. In damped outrigger, the dampers will be placed at the end of the outrigger and connected to the perimeter column which will break the rigid connection between them, and this structural system has been analyzed to study its behavior [14, 15]. These dampers are considered as a structural protective system such as a passive protective system, active protective system, semi-active protective system, and hybrid protective system [16, 17]. Passive protective systems work without using an external power supply, and the maintenance of the system is easy compared to other protective systems. Some of the passive protective system include base isolation systems, energy dissipaters, and energy absorbers [18]. Viscous dampers are one of the passive dampers; their arrangement is made up of orifices, pistons, and seals that produce the resistive force with the action of fluid passage through it. The force by the damper which is generated will be the velocity function between the ends of the device [19, 20]. There have been an ample number of studies on the use of a protective system for different structural systems and behavior of outrigger structural systems in the past and present. This paper focuses on simulating outrigger structural system behavior with the viscous damper in MATLAB, where the structure is influenced by the wind load.

2 Methodology

The structural dimension considered for static analysis is a 40-story rectangular building with 4 m story height, encounters story height of 160 m, and the base dimension is 30 × 30 m. The aspect ratio will be 5.33 with the floor slab of 0.15 m. The core dimension is 14 × 14 m made of concrete with a thickness of 0.4 m. The outrigger is the stiff arm that connects the core to the perimeter column which spans 8 m, and there will be two outriggers in each direction summing up to eight outriggers in four directions. There are four columns in each direction, two supporting the outrigger and the other two at the ends made up of steel which will be participating in resisting the lateral load. Figure 1 shows the plan and elevation of the structure

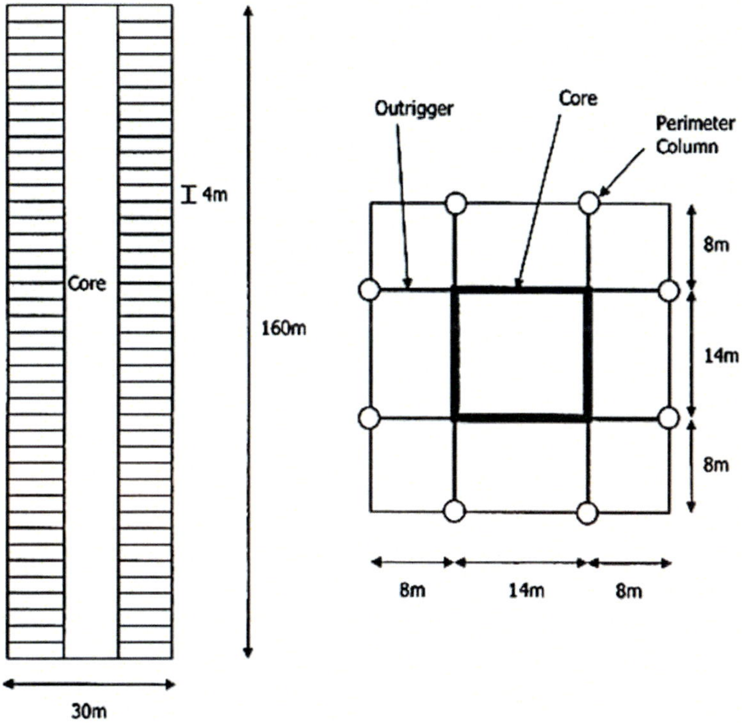

Fig. 1 Dimension of a building **a** elevation **b** top plan

considered in this study [10]. The building core is considered as a cantilever beam with constant cross-sectional area throughout, and stiffness is considered as uniform throughout. Outrigger is considered to have a rigid connection to the core, and thus, the core and outrigger rotate the same amount. The columns are considered to have a pinned connection to the ground. For static analysis, the load is considered as a quasi-static load uniform throughout the height of the building [15].

The modulus of elasticity of concrete = 24.82 GPa, the modulus of elasticity of steel = 200 GPa, the base length of the building, $x = 30$ m, base length of the core, $B = 14$ m, thickness of the core, $t = 0.4$ m, floor height, $c = 4$ m, floor thickness, $f = 0.15$ m, concrete density, $\rho_c = 2400$ kg/m3, area of the core, $A = 21.76$ m2, moment of inertia of core, $I = 671.37$ m4, M = mass of the floor + mass of the core, mass of the floor, mf = 253,400 kg, mass of the core, mc = 208,896 kg, $M = mf + mc =$ 462,336 kg, $J = 6{,}723{,}666$ kg m2 [10].

The dead load is the permanent load due to the weight of the structure. It is calculated as the product of the unit weight of the member to the volume of the member. The deadweight will be of the beam, column, wall, slab or structural member, enduring member, partition wall and material weight of floor finish, etc. It is calculated based on the dimension and the materials used in the building. The structure is considered to have the identical mass throughout the height as $M_1 = M_2 = \ldots = M_{40} = m$. The mass is obtained from the slab, mass of concrete core of one floor. The rotational inertia is also equal through the full height of the building. $J_1 = J_2 = \ldots = J_{40} = J$. Equation of motion of the structure is given in (1), and overall mass matrix M is given in (2),

$$M\dddot{U} + C\dot{U} + KU = P \tag{1}$$

In the above equation M represents the mass matrix, C represents the damping matrix, K represents the stiffness matrix of the structure, and P represents the lateral force. U represents the floor displacement, $\dot{U}$ represents the floor velocity, and $\ddot{U}$ represents the floor acceleration, respectively. The dimension of the mass matrix of the outrigger structure M is 80×80 as shown in (2) as surveyed in the literature [5].

$$M = \begin{bmatrix} M1 & & & & & \\ & J1 & & & 0 & \\ & & \ddots & & & \\ & & & \ddots & & \\ & 0 & & & M40 & \\ & & & & & J40 \end{bmatrix} \tag{2}$$

There are 80 degrees of freedom for this structure representing forty floors obtained by the direct stiffness matrix. The dimension of the stiffness matrix of the outrigger structure is K is 80×80, where $u_1..u_{40}$ and $\theta_1..\theta_{40}$ are deflection and rotation of the beam element, as surveyed in the literature [6].

$$K = \begin{bmatrix} k_{1BB} + k_{1AA} & k_{1AB} & & & & & \\ k_{1BA} & k_{1BB} + k_{2AA} & k_{2AB} & & & 0 & \\ & k_{2BA} & k_{2BB} + k_{3AA} & & & & \\ & & & \ddots & k_{38AB} & & \\ & 0 & & k_{38BA} & k_{38BB} + k_{39AA} & k_{39AB} \\ & & & & k_{39BA} & k_{39BB} + k_{40AA} \end{bmatrix}$$

$$\begin{bmatrix} u_1 \\ \theta_1 \\ \vdots \\ \vdots \\ u_{40} \\ \theta_{40} \end{bmatrix} \tag{3}$$

The stiffness matrix of the member which is considered as beam element for core and bar element for a column is given by,

$$k_{AA} = \begin{bmatrix} \left(\frac{AE}{L}\sin^2 y + \frac{12EI}{L^3}\cos^2 y\right) & \frac{6EI}{L^2}\cos y \\ \frac{6EI}{L^2}\cos y & \frac{4EI}{L} \end{bmatrix} \tag{4}$$

$$k_{AB} = \begin{bmatrix} \left(-\frac{AE}{L}\sin^2 y + \frac{12EI}{L^3}\cos^2 y\right) & \frac{6EI}{L^2}\cos y \\ -\frac{6EI}{L^2}\cos y & \frac{2EI}{L} \end{bmatrix} \tag{5}$$

$$k_{BA} = \begin{bmatrix} \left(-\frac{AE}{L}\sin^2 y + \frac{12EI}{L^3}\cos^2 y\right) & -\frac{6EI}{L^2}\cos y \\ \frac{6EI}{L^2}\cos y & \frac{2EI}{L} \end{bmatrix} \tag{6}$$

$$k_{BB} = \begin{bmatrix} \left(\frac{AE}{L}\sin^2 y + \frac{12EI}{L^3}\cos^2 y\right) & -\frac{6EI}{L^2}\cos y \\ -\frac{6EI}{L^2}\cos y & \frac{4EI}{L} \end{bmatrix} \tag{7}$$

A = area of the core in (m^2).
E = elastic modulus of the core in (G Pa).
I = moment of inertia of the core concerning the bending axis in (m^4).
L = floor height in (m).
y = angle of reference concerning the global coordinate in degree.

When the structural mass and stiffness are known, the mode shape and Eigenvalues are found by Eigenvalue problem solution, where frequency and natural period of the building are calculated as shown in Eqs. (8) and (9)

$$K\varphi = \omega^2 M\varphi \tag{8}$$

φ—Eigenvectors represent the mode shape.
ω^2—Eigenvalues.
ω—Angular frequency (rad/sec).

From the frequency of the structure, the natural period can be obtained by (9),

$$T = \frac{1}{\text{frequency}} = \frac{2\pi}{\omega} \tag{9}$$

The outrigger structural deflection is found by the combination of the deflection profile of a cantilever beam and the rotational spring, where the cantilever beam deflection is given in (10)

$$U(x) = \frac{wx^2}{8EI}\left(6H^2 - 4Hx + x^2\right) \tag{10}$$

Deflection of the cantilever beam at the top is as given in (11)

$$U(H) = \frac{wH^4}{8EI} \tag{11}$$

The deflection of the outrigger structure is due to the deflection of the cantilever and deflection of the rotational spring. Thus, the formula for the deflection is given by superimposing the deflection of the cantilever beam with the rotating spring as stated in the literature [10],

$$U(H) = \frac{wH^4}{8EI} - \left[\int_0^{aH} \frac{M_S x}{EI} \mathrm{d}x + \int_{aH}^{H} \frac{M_S aH}{EI} \mathrm{d}x \right] \tag{12}$$

$$U(H) = \frac{wH^4}{8EI} - \frac{M_S}{2EI}(aH)^2 - \frac{M_S aH}{EI}(H - aH) \tag{13}$$

$$U(H) = \frac{wH^4}{8EI}\{1 - f[4a(1 - a) + 2a^2]\} \tag{14}$$

The design of the structural components is to be stable, and it should able to resist wind load specified by the code. For the lateral load resisting system, calculation of the wind load is important according to the code specification, where it includes structural elements and their components, claddings, etc. [6]. Lateral wind load is calculated manually according to IS-875-Part3 (1987) and applied at every floor level. The design wind speed (V_b) is selected according to the location of the structure.

The structure is considered to be located in the Tiruchirapalli region which has a basic wind speed (V_b) of 47 m/s (clause 5.2, IS: 875 (Part 3)-1987) [21].

3 Results and Conclusion

A forty-story structure is considered in this study to find a response to the wind load. When the structural mass and stiffness are known, the mode shape and Eigenvalues are found by Eigenvalue problem solution, where frequency and natural period of the building are calculated and given in Table 1. Figure 2 shows the structural deformation under the vibrating natural frequency representing its behavior under the loads, where the structure shows the first, second, and third mode depending on the modal analysis.

The wind load is considered for static analysis. The wind load has its intensity uniform through the height of the structure. The response of the structure, i.e., displacement profile of the structure, under the wind load is shown in Fig. 3. The displacement of the outrigger structure with the viscous damper is 0.23 m, and the undamped outrigger displacement is 0.28 m. This proves that when the outrigger structure installed with the viscous damper as an external damper it reduces the displacement of the structure.

Table 1 Frequency and period of the structure

Modes	Angular frequency ω, (rad/s)	Period T, (second)
1	1.757	3.576
2	10.003	0.628
3	27.846	0.226

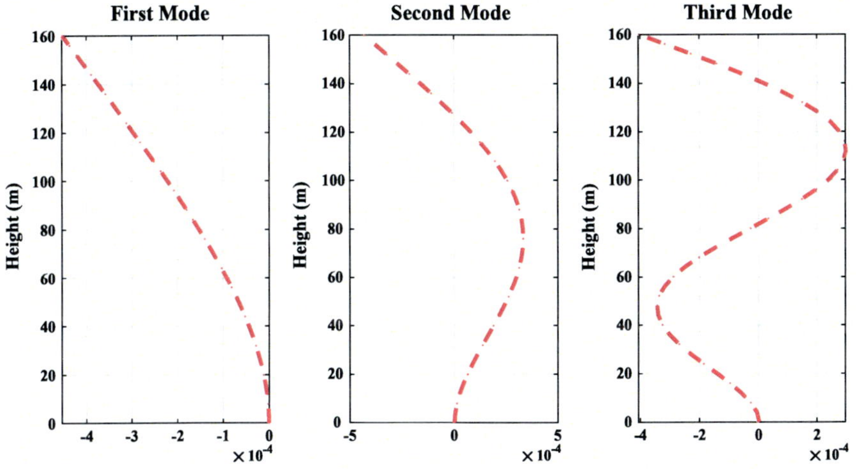

Fig. 2 Mode shape of the structure

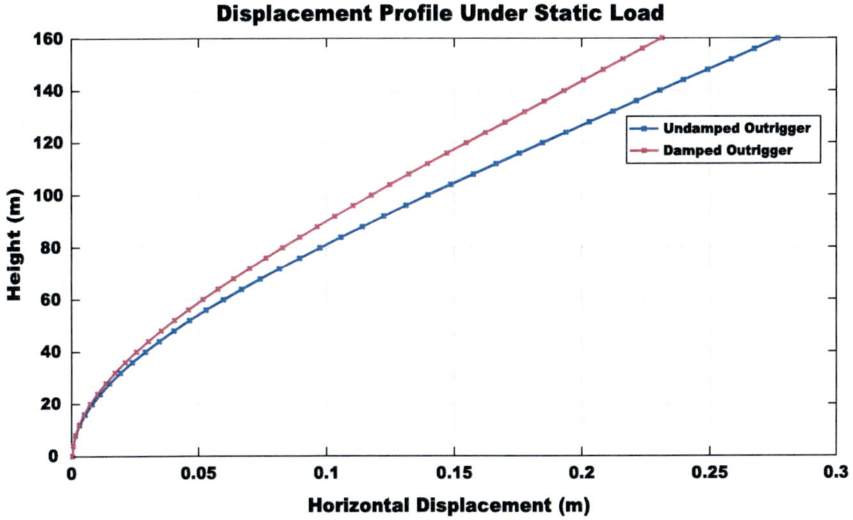

Fig. 3 Displacement of the outrigger structure for wind load

4 Conclusions

A forty-story structure is considered in this work to find the response to wind load. The results are shown for a damped outrigger structure model installed with a viscous damper. The damping matrix is collaborated with the stiffness matrix of the structure to form a complex stiffness matrix. The static analysis is done by taking wind load to be uniform throughout the height in one direction, which shows the difference in the deflection of the structure with the presence of viscous damper in the outrigger structure from that of only outrigger structure. The analysis shows the comparison of deflection of the outrigger structure with and without viscous damper. Thus, results prove that outrigger structure performance is enhanced with viscous damper installation by exhibiting the reduced structural deflection.

References

1. Smith, B., Coull, A.: Tall Building Structures: Analysis and Design. Wiley, New York, USA (1991)
2. Taranath, B.S.: Reinforced Concrete Design of Tall Buildings. CRC Press, Boca Raton, USA (2010)
3. Taranath, B.S.: Steel Concrete and Composite Design of Tall Buildings. McGraw-Hill, New York, USA (1998)
4. Mendis, P., Mohotti, D., Ngo, T.: Wind design of tall buildings, problems, mistakes and solutions. In: 1st International Conference on Infrastructure Failures and Consequences, July (2014)
5. Chopra, A.K.: Dynamics of Structures–Theory and Applications to Earthquake Engineering. Prentice-Hall, New Jersey (2001)
6. Taranath, B.S.: Wind and Earthquake Resistant Buildings. Marcel Dekker, Cimarron Road, Monticello, New York (2005)
7. Ali, M.M., Moon, K.S.: Structural developments in tall buildings: current trends and future prospects. Architect. Sci. Rev. **50**, 205–223 (2007)
8. Abramovitch, D.: The outrigger: a prehistoric feedback mechanism. IEEE Control System. **2**, 57–72 (2005)
9. Choi, H., Ho, G., Joseph, L., Mathias, N.: Outrigger Design for High Rise Buildings An output of the CTBUH Outrigger Working Group. Council on Tall Buildings and Urban Habitat, Chicago (2012)
10. Gamaliel, R.: Frequency-Based Response of Damped Outrigger Systems for Tall Buildings. M. Eng. Dissertation, Massachusetts Institute of Technology, USA (2007)
11. Smith, B.S., Salim, I.: Parameter study of outrigger-braced tall building structures. J. Struct. Div. **107**, 2010–2014 (1981)
12. Smith, R.J., Willford, M.R.: The damped outrigger concept for tall buildings. Struct. Design Tall Spec. Build. **16**, 501–517 (2007)
13. Smith, R.: The damped outrigger–design and implementation. Int. J. High-Rise Build. **5**, 63–70 (2016)
14. Chen, Y., McFarland, D.M., Wang, Z., Spencer, B.F., Bergman, L.A.: Analysis of tall buildings with damped outriggers. J. Struct. Eng. **136**, 1435–1443 (2010)
15. Asai, T., Chang, C.M., Phillips, B.M., Spencer, B.F.: Real-time hybrid simulation of a smart outrigger damping system for high-rise buildings. Eng. Struct. **57**, 177–188 (2013)
16. Anvar, N., Aung, T.H., Najam, F.: Smart systems for smart structures. Technology 33–39 (2016)
17. Housner, G.W., Bergman, L.A., Caughey, T.K., Chassiakos, A.G., Claus, R.O., Masri, S.F., Skelton, R.E., Soong, T.T., Spencer, B.F., Jr., Yao, T.P.: Structural control: past, present, and future. J. Eng. Mech. **123**, 897–971 (1997)
18. Spencer, B.F., Nagarajaiah, S.: State of the art of structural control. J. Struct. Eng. **129**, 845–856 (2003)
19. Willford, M., Smith, R., Scott, D., Jackson, M.: Viscous Dampers Come of Age, Structure Magazine, pp. 15–18 (2008)
20. Silvestri, S., Trombetti, T.: Physical and numerical approaches for the optimal insertion of seismic viscous dampers in shear-type structures. J. Earthquake Eng. **11**, 787–828 (2007)
21. IS 875 (Part 3):1987 Code of Practice for Design Loads for Buildings and Structures. Bureau of Indian Standards, New Delhi (1987)

Performance Analysis of Harmonic Suppression Techniques

K. U. Vinayaka, N. Komal Raj, and C. G. Parvathi

1 Introduction

Nonlinear loads are the major contributors for power quality-related problems. According to some industry experts, around 75% to 90% of the loads used are nonlinear. Inductive loads, variable speed drives, switched mode power supplies, and power electronic devices such as converters and inverters are some of the best examples of nonlinear loads [1]. Randomly increasing usage of power electronic devices in industries, hospitals, households and for commercial purposes affects the quality of power [2, 3].

A power system network in which voltage and current waveforms deviate from their sinusoidal periodic waveforms and having a frequency which is an integral multiple of the fundamental frequency is called as system harmonics. These are commonly characterized by waveform distortion and have an undesirable effect on power system components and their operations [4, 5].

Some of the major sources of harmonics include power converters, arc furnaces, power transformers, cycloconverters, static VAR compensators, ballast inductors, etc. These nonlinear loads are known to draw currents from the source or the utility grid in an abrupt way in the form of short pulses. These pulses cause a distortion in the current waveforms and results in the harmonic components of the fundamental current to be injected back into the grid, thus affecting the grid as well as the other consumers connected to it [3, 6, 7].

Harmonics can be broadly classified into varies types like odd and even harmonics, characteristic and non-characteristic harmonics, positive, negative and zero sequence harmonics, etc. The positive sequence harmonics produce excessive heating effects. The negative sequence harmonics produce motor torque-related problems, while the zero sequence harmonics add voltages or currents in the neutral wire causing heating effects. In view of the negative impacts that occur on the power systems due

K. U. Vinayaka (✉) · N. K. Raj · C. G. Parvathi
Department of E&EE, Siddaganga Institute of Technology, Tumakuru, Karnataka, India

S. K V and K. Guruprasad Rao (eds.), *Smart Sensors Measurements and Instrumentation*,
Lecture Notes in Electrical Engineering 750,
https://doi.org/10.1007/978-981-16-0336-5_35

Table 1 Voltage distortion limit

Bus voltages at point of common coupling	Individual harmonic (%)	THD
$V \leq 1.0$ KV	5.0	8.0
1 KV $< V \leq$ 69KV	3.0	5.0
69 KV $< V \leq$ 161 KV	1.5	2.5
161KV $< V$	1.0	1.5

to harmonics injected into the grid, it is necessary to suppress and mitigate these harmonics by various control and filtering techniques.

This paper briefs about the voltage and current harmonic limits as specified by IEEE-519 standards. The following section briefs about the various available harmonic suppression techniques along with their advantages and drawbacks. Active filtering method incorporated with a three-phase grid feeding a nonlinear load is simulated in MATLAB Simulink software for its FFT analysis. The results of the Fourier analysis of all the circuits are discussed.

It has to be noted here that in the present day world, the commonly used electrical devices like motors, electric fans, chokes, coil based electrical components, etc., are majorly based on the theory of magnetic fields and not on electrostatic theory. Thus, most of the loads are considered to be inductive rather than being capacitive in nature. Thus, for all the simulations discussed in the paper, a six-pulse diode-based rectifier with RL load is used as a nonlinear load. Also, the load is kept constant across all the simulations for better comparison of results.

2 Harmonic Limits Specified by IEEE Standards

Harmonics cause abnormal operations of circuit breakers, fuses, and other protection equipment's, additional heating and failure of fuses present in capacitor banks. Harmonics of higher amplitudes weaken the insulations of windings and capacitors. Torque pulsation of motors can cause overheating, additional vibrations, and increased noise level [4, 8]. Thus, reducing harmonics in any power system to meet with the acceptable limits as specified by IEEE-519 is necessary. The standards mentioned in 2014 for TDD and THD are given in Tables 1 and 2.

3 Harmonic Suppression

Harmonic suppression can be done through the following approaches:

- By reducing the harmonic currents produced by the load.

Table 2 TDD limits for systems rated 120 v through 69 kv

Isc/IL	$3 \leq h < 11$	$11 \leq h < 17$	$17 \leq h < 23$	$23 \leq h < 35$	$35 \leq h < 50$	TDD
< 20	4.0	2.0	1.5	0.6	0.3	5.0
20–50	7.0	3.5	2.5	1.0	0.5	8.0
50–100	10.0	4.5	4.0	1.5	0.7	12.0
100–1000	12.0	5.5	5.0	2.0	1.0	15.0
> 1000	15.0	7.0	6.0	2.5	1.4	20.0

- By limiting the harmonics generated by the load from being injected into the source by the use of harmonic filters.
- By modifying the system frequency response by using inductors, capacitors, or other filters.

While increasing the pulse numbers and adding line reactors in series with the load falls under the first category, use of filters such as passive and active filters along with their various configurations belong to the second way of harmonic suppression. Increasing the stiffness of supply mode, zig-zag transformers, high power factor pre-regulators, and network topology reconfigurations sums up the third way of suppressing harmonics [2, 9].

This paper deals with the different types of filters that are used for harmonic suppressions. The various filtering approaches and their hybridizations are discussed in the following section.

3.1 Passive Filters

Passive filtering technique is the simplest conventional method for reducing harmonics and is also least expensive. It uses passive elements such as inductor, capacitor, and resistor. They are tuned to a particular frequency in order to suppress the harmonic current of that particular frequency [1].

Passive filters are broadly classified into series passive filters and shunt passive filters. Some of the key features of passive power filters are discussed below.

Series passive filter: The series passive filter is characterized by its high inductance and low capacitance. Due to this, the filter effectively blocks the harmonic current and hence induces spikes in the source current waveform. This is an undesirable character. Although series passive filters are cheaper, series passive filters cannot be used in power systems where technical precision is more important that cost saving.

Shunt passive filter: The spikes in the source current waveforms are reduced since the shunt passive filter is characterized by higher capacitance value and a lower inductance. One of the major advantages of shunt passive filter is that it carries only a fraction of current to the whole system, i.e. only the harmonic components of current and not the current corresponding to the fundamental frequency component. Hence, the AC power losses are also reduced compared to the series passive filter.

Hybrid passive filter: The series passive filter acts as a band pass filter. On the other hand, the shunt passive filter behaves as a band stop filter. Also, since the series passive filter offers low impedance at fundamental frequency, it absorbs the voltage harmonics, thus reducing it. The shunt passive filter is found to suppress the current fed type of harmonics. The main objective of the hybrid configuration is to extract the functionalities of both the filters in order to find a suitable solution for the reduction of harmonics [1].

The disadvantages of passive filters are the filtering method is largely affected by the source impedance and the number of single passive filters used is equal to the number of filtered harmonics. In order to overcome these drawbacks, active power filters are used.

3.2 Active Power Filtering

There are two types of active filters, namely series active power filter and shunt active power filter. The major drawback of the series active power filter is that, it has to be designed for very high ratings. Also, components such as the coupling transformers become bulky with increased power ratings. Since we are only concerned about the current harmonics in the system, the SAPF is taken into consideration (Fig. 1).

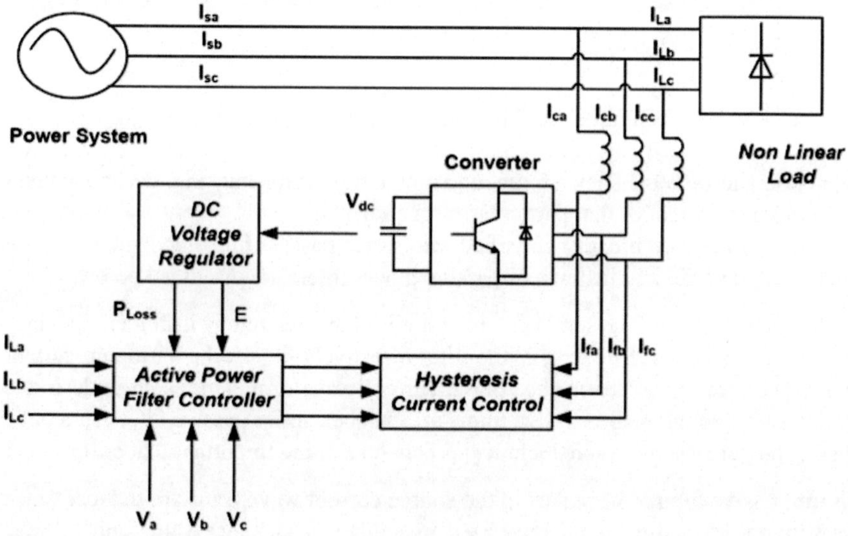

Fig. 1 Block diagram of circuit with SAPF based on PQ theory with hysteresis current control (HCC)

A shunt active power filter (SAPF) functions as current source that injects components of harmonics that are generated by the load, but are phase shifted by 180° [10]. In this paper, instantaneous PQ theory-based SAPF has been discussed and is used for simulations.

The PQ block is used here to separate the reactive power in order to compensate it. The instantaneous active and reactive power theory implements Clarke Transformation, which transforms the stationary reference system co-ordinates as a-b-c into $\alpha - \beta - 0$ co-ordinates.

Main intension of this transformation is to separate the zero sequence component, active power, and reactive power components [11]. The mathematical expressions for the Clarke's transformations involved in PQ theory are as given below.

$$\begin{bmatrix} V_\alpha \\ V_\beta \end{bmatrix} = \sqrt{2/3}\begin{bmatrix} 1 & -1/2 & -1/2 \\ 0 & \sqrt{3}/2 & -\sqrt{3}/2 \end{bmatrix}\begin{bmatrix} V_a \\ V_b \\ V_C \end{bmatrix} \tag{1}$$

$$\begin{bmatrix} i_\alpha \\ i_\beta \end{bmatrix} = \sqrt{2/3}\begin{bmatrix} 1 & -1/2 & -1/2 \\ 0 & \sqrt{3}/2 & -\sqrt{3}/2 \end{bmatrix}\begin{bmatrix} i_a \\ i_b \\ i_C \end{bmatrix} \tag{2}$$

$$\begin{bmatrix} p = p^- + p'' \\ q = q^- + q'' \end{bmatrix} = \begin{bmatrix} V_\alpha & V_\beta \\ -V_\beta & V_\alpha \end{bmatrix}\begin{bmatrix} i_\alpha \\ i_\beta \end{bmatrix} \tag{3}$$

$$\begin{bmatrix} i_\alpha \\ i_\beta \end{bmatrix} = \frac{1}{V_{\alpha^2} + V_{b^2}}\begin{bmatrix} V_\alpha & V_\alpha \\ V_\beta & -V_\beta \end{bmatrix}\begin{bmatrix} p^\sim \\ q^\sim \end{bmatrix} \tag{4}$$

$$\begin{bmatrix} i^*_{ca} \\ i^*_{cb} \\ i^*_{CC} \end{bmatrix} = \frac{1}{\sqrt{2/3}}\begin{bmatrix} 1 & 0 \\ -1/2 & \sqrt{3}/2 \\ -1/2 & -\sqrt{3}/2 \end{bmatrix}\begin{bmatrix} i_\alpha \\ i_\beta \end{bmatrix} \tag{5}$$

In a hysteresis current controller (HCC), the compensating current and the reference currents are compared in each of the three phases and six pulses are generated in order to drive the six switches of the inverter circuit in a shunt active power filter. The actual output waveform of the inverter circuit is compared with the reference current that is generated using the instantaneous *p–q* theory [12].

4 Results and Discussions

A system consisting of a utility grid feeding a nonlinear load is considered [13–15]. A three-phase diode-based bridge rectifier with R–L load is used as a nonlinear load. The three-phase source or the utility grid has a phase-to-phase RMS voltage of 13,800 V.

The source current waveform of the grid feeding a nonlinear load without any compensation is shown in (Fig. 2).

The FFT analysis at the source end of this system gives a total harmonic distortion of 13.73%. Since the nonlinear load used is a six-pulse rectifier, triplen harmonics are absent. The order of existing harmonics here are fifth, seventh, eleventh, thirteenth, so on. It is found that the fifth and the seventh harmonics are dominant and constitute 12.69 and 4.56%, respectively, while the eleventh and the higher order harmonics are of the order 2.11% and less (Fig. 3).

When network topology reconfiguration is employed, a linear load, which is a resistive load of suitable magnitude is connected in parallel to the nonlinear industrial

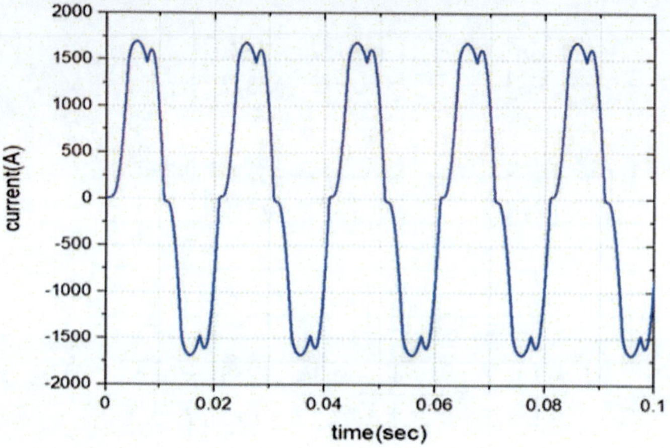

Fig. 2 Source current waveform of a circuit without compensation

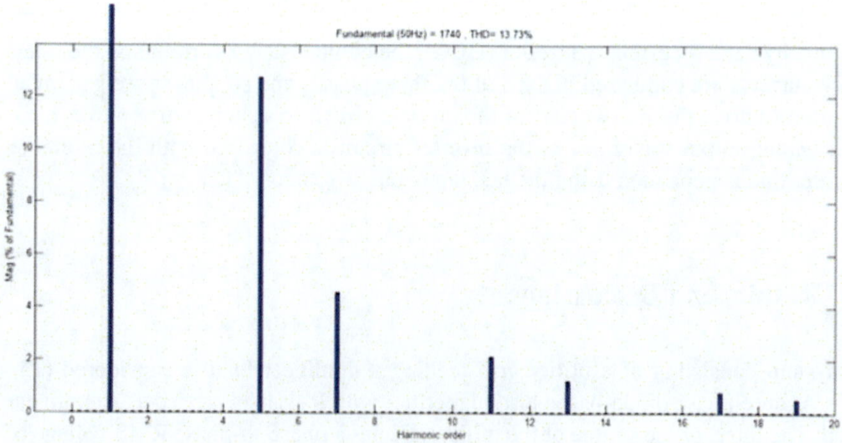

Fig. 3 THD window of a circuit without compensation

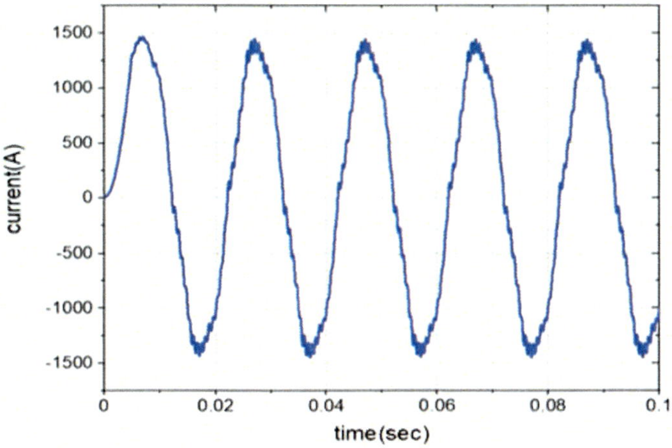

Fig. 4 Source current waveform of a circuit with series passive filter

load. This resistive load connected in parallel to the nonlinear load helps in reducing the harmonic distortion that is injected to the source by the inductive nonlinear load, as the circuit is more linear compared to the previous case. When an FFT analysis of the reconfigured circuit is carried out, it is observed that the %THD is reduced from 13.73 to 13.25%. But this is not seem to be within the 5% limit specified by IEEE 519 standards.

Therefore, when a series passive filter is employed, the filter is tuned to eliminate the dominating fifth- and seventh-harmonic components. The source current waveform when a series passive filter is used is shown in Fig. 4. The waveforms is seen to have spikes due to the nature of the filter. The FFT analysis results show that the dominating fifth and seventh harmonics are reduced to 5.27 and 1.94% from 12.24 and 4.43%, respectively. Also, the %THD is reduced to 5.79%.

When a shunt passive filter is employed, the filter is again designed to suppress the dominating fifth and seventh harmonics. It is seen that the shunt passive filter offers better compensation as compared to the series passive filter. The %THD further comes down to 2.19%, while the other harmonic components have also reduced significantly. The source current waveform is shown in Fig. 5. The spikes which were earlier observed in case of series passive filter are seen to have reduced when a shunt passive filter is employed.

A combination of series and shunt passive filter further reduces the fifth and seventh harmonic components to 0.16% and 0.02%, respectively. Also, the %THD is reduced to 1.22% [16].

When a shunt active power filter with hysteresis current controller is employed, the compensating current waveform and the source current waveform after compensation using SAPF are shown in Figs. 6 and 7, respectively, [12, 17, 18]. The waveform of current at the source end after compensation by using SAPF can be observed to be

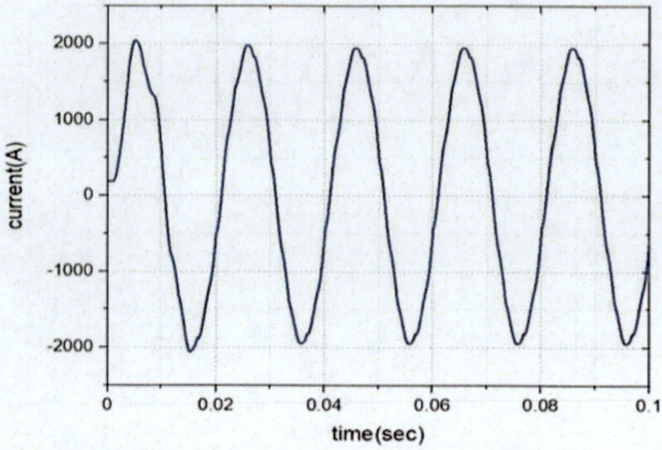

Fig. 5 Source current waveform of a circuit with shunt passive filter

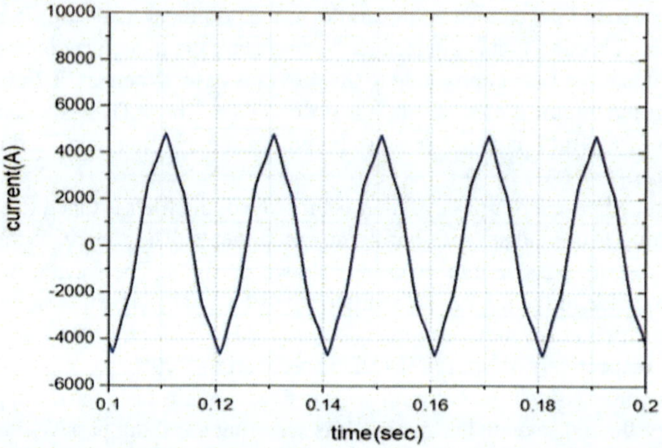

Fig. 6 Source current waveform with hybrid passive filter

almost sinusoidal in nature, unlike the waveform without compensation as shown in Fig. 5.

Also, the %THD of the circuit with SAPF and HCC is 0.82%, which is well within the limits of current harmonics as specified by the IEEE recommended limits as shown in Fig. 6 and Fig. 7 respectively and the comparison amongst the differing filtering approaches for current harmonic suppression is as represented in Fig. 8.

From Table 3, it can be inferred that SAPF proves to be the most-efficient harmonic suppression method among all the other configurations discussed above,

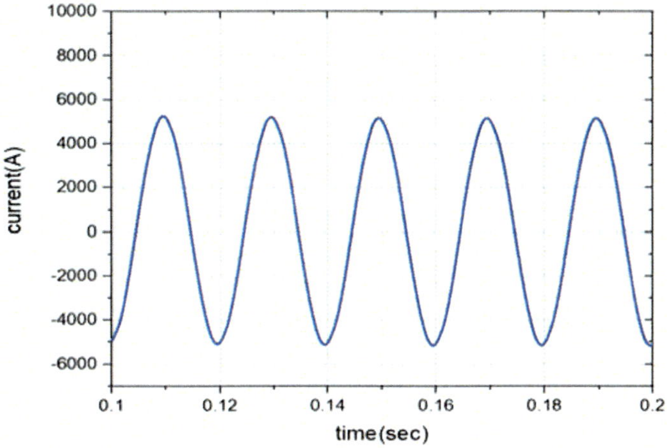

Fig. 7 Source current waveform of a circuit with SAPF

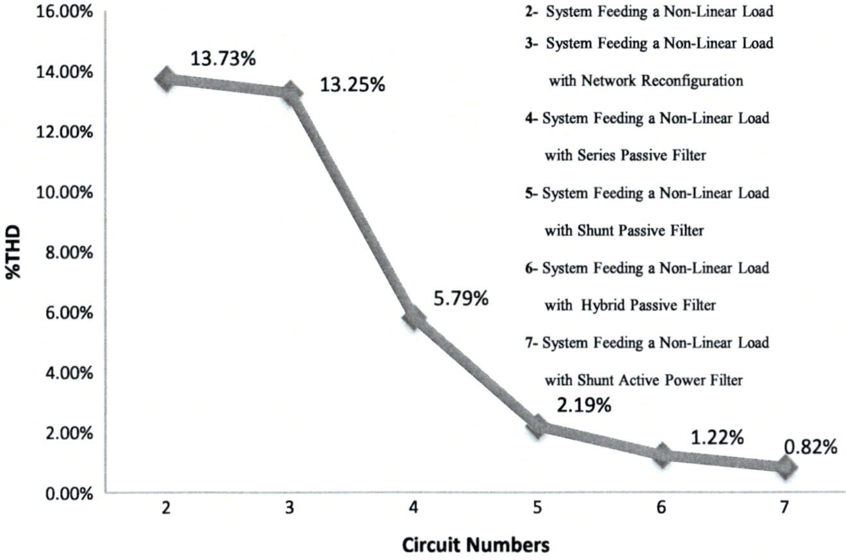

Fig. 8 Current THD comparison of system with various harmonic filtering techniques

by presenting a %THD of 0.82% at the source end. Also, the dominating fifth-harmonic component is reduced to 0.68% from 12.69% in case of a circuit with no compensation. These values are found to be well within the IEEE limits.

Table 3 Comparison of %THD in all circuits

CktNo	Circuit description	Fifth (%)	Seventh (%)	Eleventh (%)	%THD
1	Grid with linear load	0.00]	0.00	0.00	0.00%
2	Grid with nonlinear load	12.69	4.56	2.11	13.73%
3	Network reconfiguration	12.24	4.43	2.00	13.25%
4	Series passive filter	5.27	1.94	1.07	5.79%
5	Shunt passive filter	0.42	0.06	1.82	2.19%
6	Series passive filter + shunt passive filter	0.16	0.02	0.96	1.22%
7	SAPF with HCC	0.68	0.24	0.11	0.82%

5 Conclusion

The injection of harmonics due to the usage of nonlinear loads in industrial and domestic applications imposes various consequences on the power system equipments in an integrated system. Although passive filters provide the benefit of harmonic suppression and reactive power compensation, they are unreliable due to inadaptability based on load conditions and system changes. Active filters provide the required compensation and proves to be effective in limiting the harmonics as per the IEEE standards. The application of SAPF proves to be the best harmonic reduction methodology among all the different topologies in view of mitigation, economic considerations, and performance.

References

1. Singh, B., Chandra, A., Al-Haddad, K.: In: Power Quality Problems and Mitigation Techniques. Wiley publications (2015)
2. Dugan, R.C., McGranaghan, M.F., Santoso, S., Wayne Beaty, H.: In: Electrical Power System Quality. Mc Graw-Hill Publications (2002)
3. Bam, L., Jewell, W.: Review: power system analysis software tools. IEEE PES General Meeting **1**, 139–144 (2005)
4. JawadGhorbani, M., Mokhtari, H.: Impact of harmonics on power quality and losses in power distribution systems. Int. J. Electri. Comput. Eng. (IJECE) (2015)
5. Rosa, F.C.: Harmonics and Power Systems. CRC Press, Taylor & Francis Group, Boca Raton, London, New York (2006)
6. Harrison, A.: The Effects of Harmonics on Power Quality and Energy Efficiency. Technological University, Dublin (2010)
7. Mikkili, S., Panda, A.K.: Power Quality Issues-Current Harmonics. CRC Press, Boca Raton, London, New York (2016)
8. McGranaghan, M., Roettger, B.: Economic evaluation of power quality. IEEE Power Eng. Rev. (2012)
9. McGranaghan, M., Roettger, B.: Challenges and trends in analysis of electric power quality measurement data. EURASIP J. Adv. Signal Process. (2010)
10. Shah, A., Vaghela, N.: Shunt active power filter for power quality improvement in distribution systems. Int. J. Eng. Develop. Res. (IJEDR) (2005)

11. AkagiI, H., Kanazawa, Y., Nabae, A.: Generalized theory of instantaneous reactive power in three-phase circuits. In: International Power Electronics Conference (IPEC'83). Tokyo, Japan (1983)
12. Dessaint, L.A., Al-Haddad, K., Hoang, L.H., Sybille, G., Brunelle, P.: A power system simulation tool based on simulink. IEEE Trans. Industr. Electron. **46**, 1252–1254 (2013)
13. McEachern, A.: A free simulator program for teaching power quality concepts. In: 9th International Conference Electrical Power Quality and Utilisation (2014)
14. Hadi, S.: In: Power System Analysis. 2nd edn. McGraw Hill (2014)
15. Rashmi, M.N., Namratha, M., Nayana, S.V., Archana, K.: Power quality ımprovement using hybrid filters. In: 3rd IEEE International Conference on Recent Trends in Electronics, Information and Communication Technology (RTEICT), Bengaluru, India (2018). https://ieeexplore.ieee.org/author/37088216743
16. Milano, F.: An open source power system analysis toolbox. IEEE Trans. Power Syst. **20**, 1199–1206 (2005)
17. Singh, B., Al-Haddad, K., Chandra, A.: A review of active filters for power quality improvements. IEEE Trans. Ind. Electron. **46**(5), 960–971 (1999)
18. Patel, M.: Twelve pulse converter with differential delta connected transformer arrangement with reduced KVA capacities for line side harmonic reduction. IJEDR **2**(2) (2014). ISSN: 2321–9939

Charge/Discharge Control Design Models of Li-Ion Battery in Electric Vehicles Using MATLAB/Simulink

K. U. Vinayaka and J. Shwetha Katari

1 Introduction

The major issues faced by the global automotive industry are emission of carbons from conventional transport. To overcome the issues faced by conventional vehicles, electric vehicles (EVs) transport takes over the place of internal combustion engine (ICE)-based vehicles. EVs are eco-friendly transport as they are more energy efficient, i.e. less energy is needed to travel the same distance, reduces air pollution and prevents climate change due to emissions [1]. Electric vehicle runs using electricity as their fundamental power source. Figure 1 shows the overview of the electric vehicle (EV).

The electricity which serves as fuel for the EVs is stored using energy storage devices such as batteries, super capacitors and fuel cells. In EVs, batteries are used as the fundamental storage device. Traditionally, lead-acid and nickel metal hydride batteries have been used for this purpose, On the other hand, lithium-ion batteries (LIBs) are advancing popularity due to their several advantages such as high energy density, high columbic efficiency, longer cycle life, low self-discharge, fast charging capacity, less polluting electrode material and light weight [2]. Hence, LIBs are used in electric vehicle applications. For lithium-ion battery, the charging is done through constant current (CC) and constant voltage (CV) mode to maintain the higher efficiency and accuracy [3]. Due to safety concern for lithium-ion battery, it is necessary for reliable system to limit the charging and discharging of battery.

In this paper, different control strategies for controlled charging and discharging of battery are realized by the aid of MATLAB/Simulink software, and the simulated results of the charging and discharging of battery with SOC, current and voltage waveforms are plotted and analysed.

K. U. Vinayaka (✉) · J. S. Katari
Department of E&EE, Siddaganga Institute of Technology, Tumakuru, Karnataka, India

S. K V and K. Guruprasad Rao (eds.), *Smart Sensors Measurements and Instrumentation*, Lecture Notes in Electrical Engineering 750,
https://doi.org/10.1007/978-981-16-0336-5_36

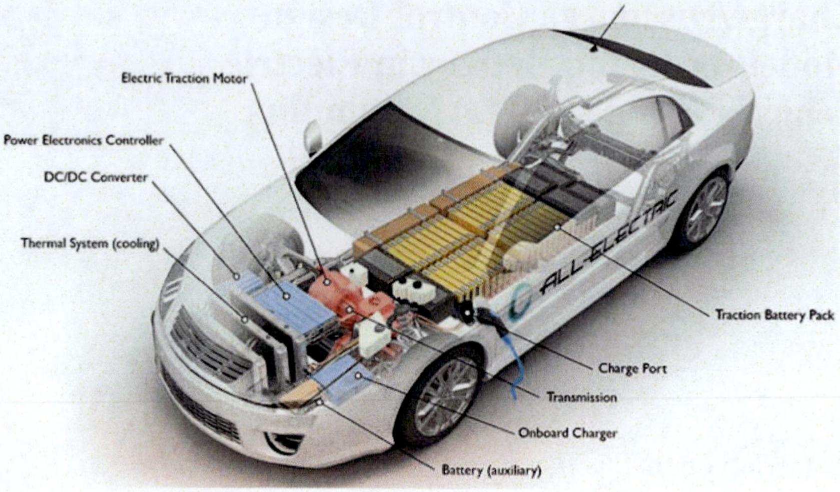

Fig. 1 Electric vehicle (EV)

2 Bidirectional DC/DC Converter

The bidirectional half bridge buck–boost DC/DC converter is shown in Fig. 2. The role of this topology is to charge and discharge of the battery. Charging of the battery can be achieved when the converter is made to operate in the buck mode. Discharging of the battery can be achieved when the converter is made to operate in the boost mode [4]. The charging and discharging can be controlled directly from the duty cycle.

$$\mathrm{d} = \left(\frac{V_b}{V_{dc}}\right) \tag{1}$$

Fig. 2 Bidirectional buck–boost topology

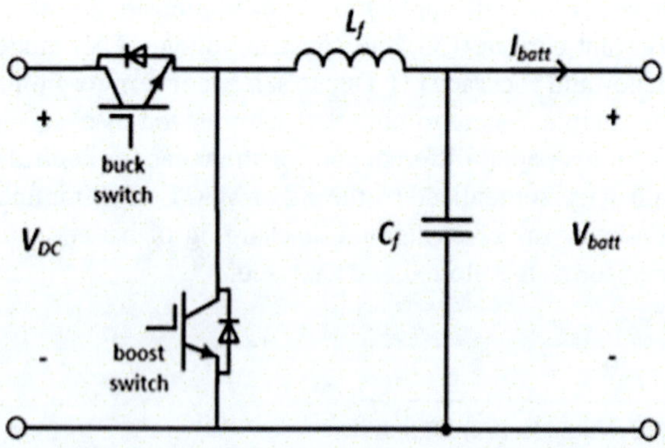

During battery charging, its terminal voltage increases, and during battery discharging, its terminal voltage decreases due to the series resistance of the battery. The open-circuit duty cycle d_0 is defined as the no charge and no discharge duty cycle as follows:

$$\mathrm{d} = \left(\frac{V_{oc}}{V_{dc}}\right) \tag{2}$$

If the DC/DC converters were to operate at $d = d_0$, no current would flow in or out of the battery. If $d < d_{0,}$ then $V_{\text{batt}} < V_{oc,}$ and the battery is discharging current. If $d > d_{0,}$ then $V_{\text{batt}} > V_{oc}$ and the battery is being charged. Bidirectional DC/DC converter is responsible for charging and discharging [5]. So as to achieve control battery charge/discharge through different control strategies are modelled with the help of MATLAB/Simulink are described in below.

3 Proposed Models

In this paper, Li-ion battery specifications are taken for the simulation of battery with 50Ah capacity, and its nominal voltage is 24 V. In MATLAB/Simulink software during battery charging, the current will be negative while during discharging of battery, the current will be positive. Control algorithm can be added for control of charging and discharging modes. Different control strategies are discussed below.

Duty Cycle Control Strategy

For 10khz frequency with $1e^{-4}$ period by varying the pulse generator duty cycle, the battery is to charge and discharge, respectively, in addition to that charge current changes with respect to duty cycle (d_0) [6]. In the proposed model for $0 < d_0 \leq 50$, battery starts to discharge, whereas for $51 \leq d_0 < 100$, the battery starts to charge. The proposed bidirectional DC/DC converter with duty cycle control-based model is simulated in MATLAB/Simulink as shown in Figs. 3 and 5.

Figure 3 is simulated for the $d_0 = 50$, and the simulated model of the battery works in discharging mode. The discharging mode waveforms of the battery SOC, current and voltage are shown in Fig. 4.

Figure 5 is simulated for the $d_0 = 60$, and the simulated model of the battery works in charging mode. The charging mode waveforms of the battery SOC, current and voltage are shown in Fig. 6.

Current Control Strategy

The proposed bidirectional DC/DC converter with current control charging-based model is simulated in MATLAB/Simulink as displayed in Fig. 7. Similarly for building the current control circuit, PI controller is taken with maximum duty cycle of 0.95, and PWM generator (DC/DC) is used with 10khz switching frequency. After building the current control PI, respective negative value reference current (I_{ref}) is

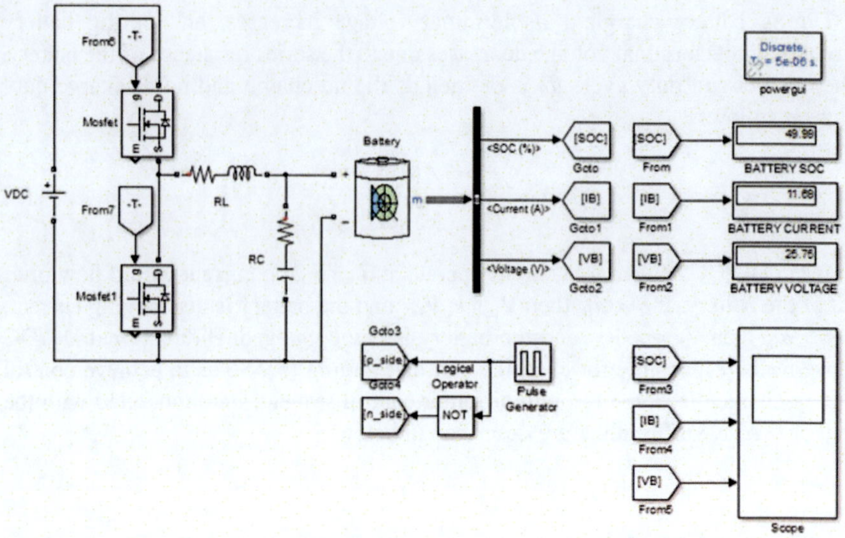

Fig. 3 Duty cycle control discharging model

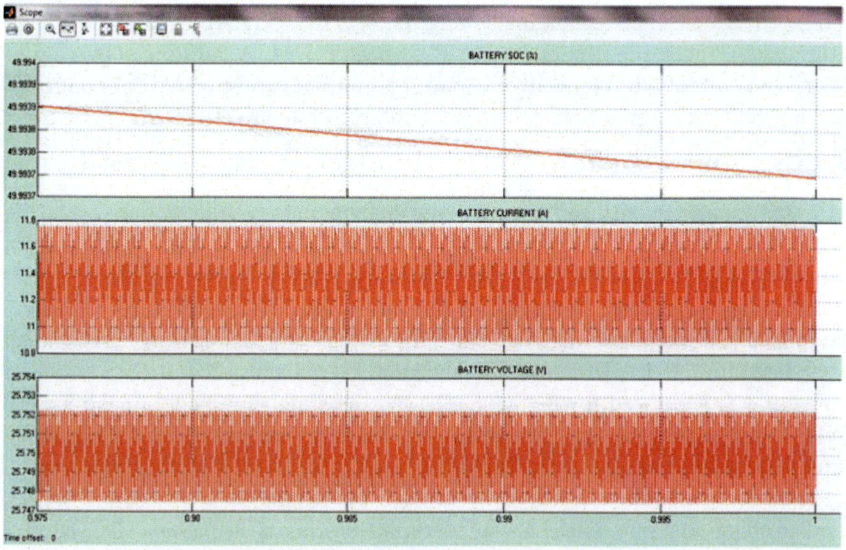

Fig. 4 Duty cycle control discharging waveforms

entered for charging operation through this battery which starts to charge and battery current is maintained correspondingly with the value of reference current [5]. And by increasing the I_{ref} value, the battery used to charge faster as compared to lower I_{ref} value.

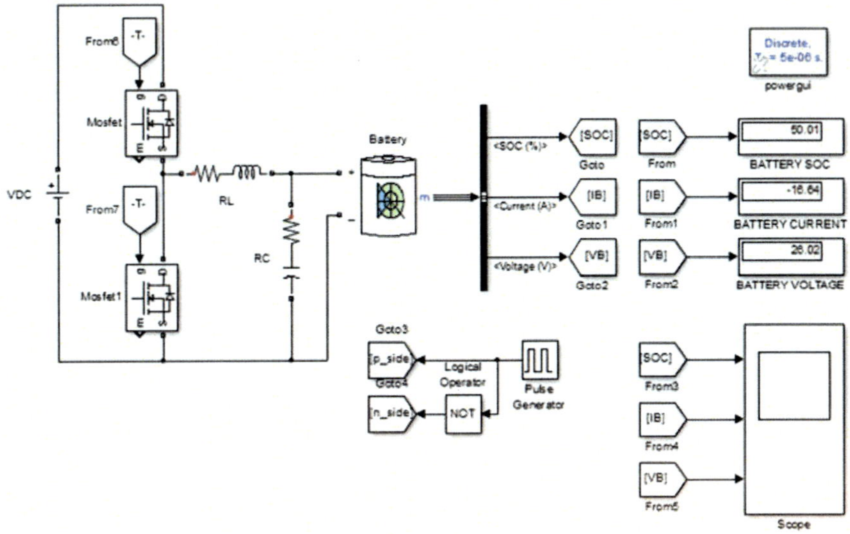

Fig. 5 Duty cycle control charging model

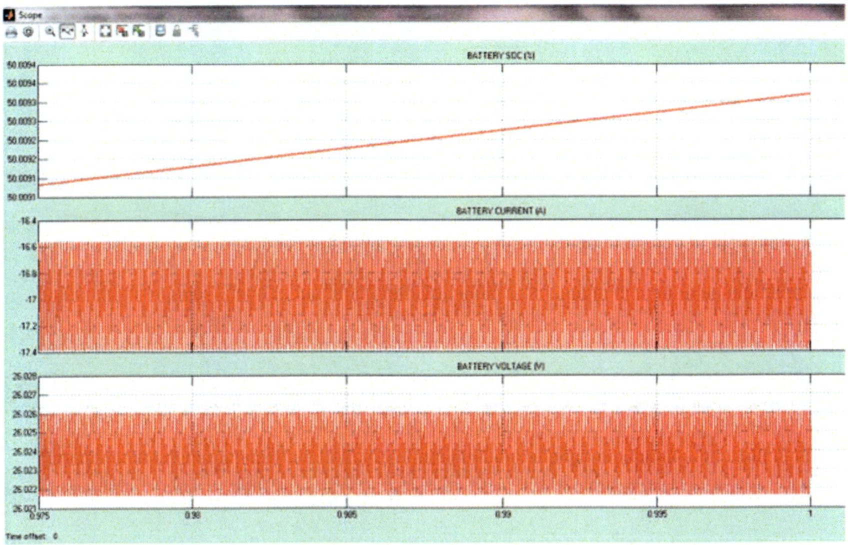

Fig. 6 Duty cycle control charging waveforms

Figure 7 is simulated for the $I_{\text{ref}} = -15$, and the simulated model of the battery works in charging mode. The current control charging waveforms of the battery SOC, current and voltage are shown in Fig. 8.

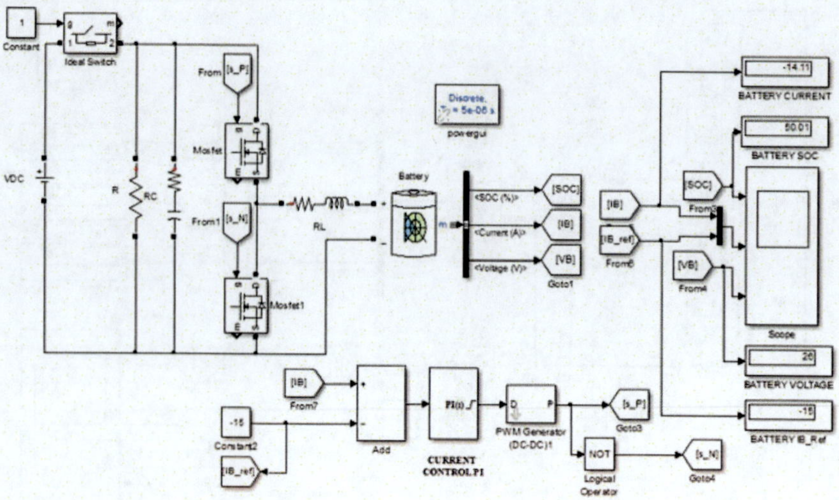

Fig. 7 Current control charging model

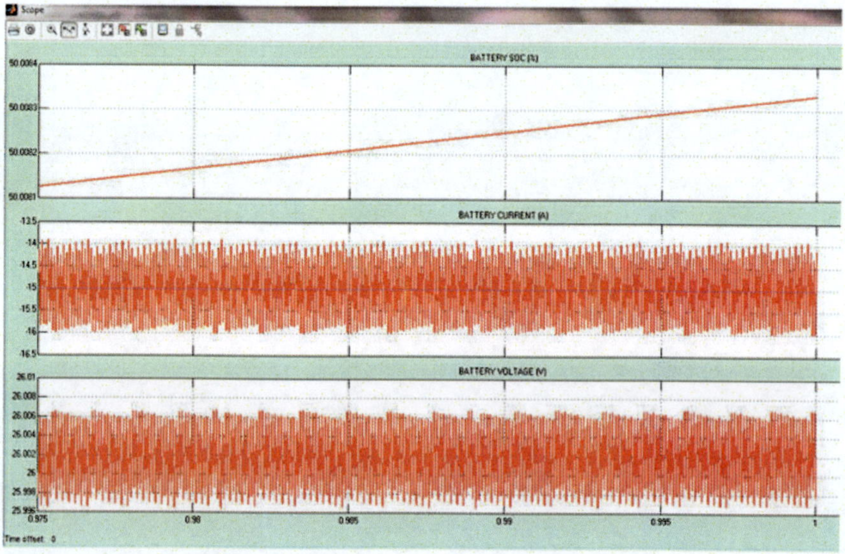

Fig. 8 Current control charging waveforms

Figure 9 is simulated for the $I_{\text{ref}} = 15$, and the simulated model of the battery works in discharging mode. The current control discharging waveforms of the battery SOC, current and voltage are shown in Fig. 10

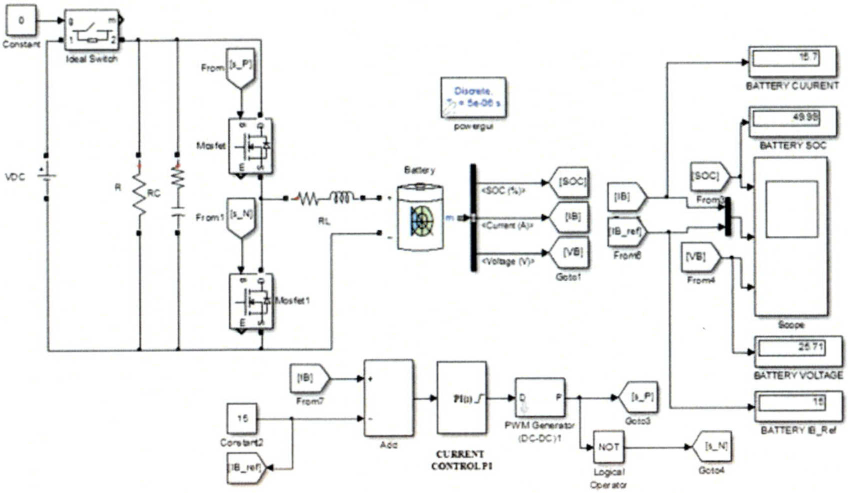

Fig. 9 Current control discharging model

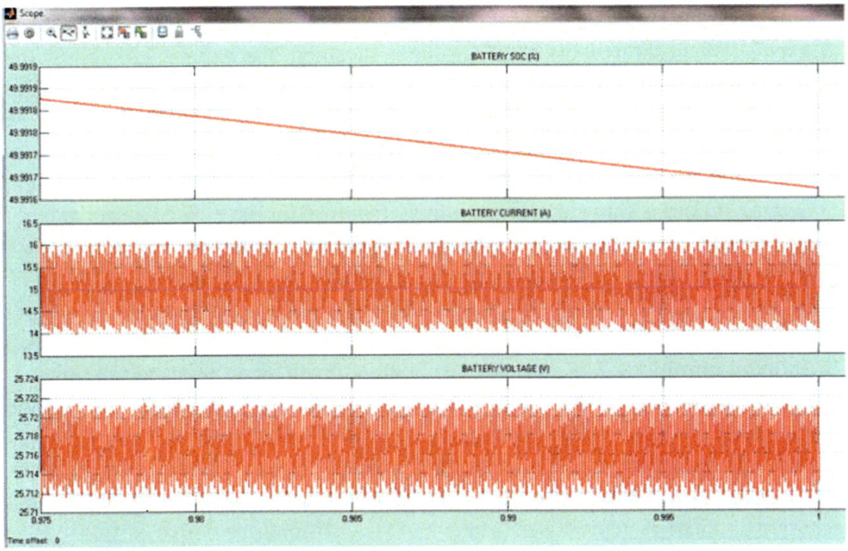

Fig. 10 Current control discharging waveforms

Voltage Control Strategy

The bidirectional DC/DC converter with voltage-based controlled charging model is simulated with the help of MATLAB/Simulink software as shown in Fig. 11. The working of below model has a two case of operations:

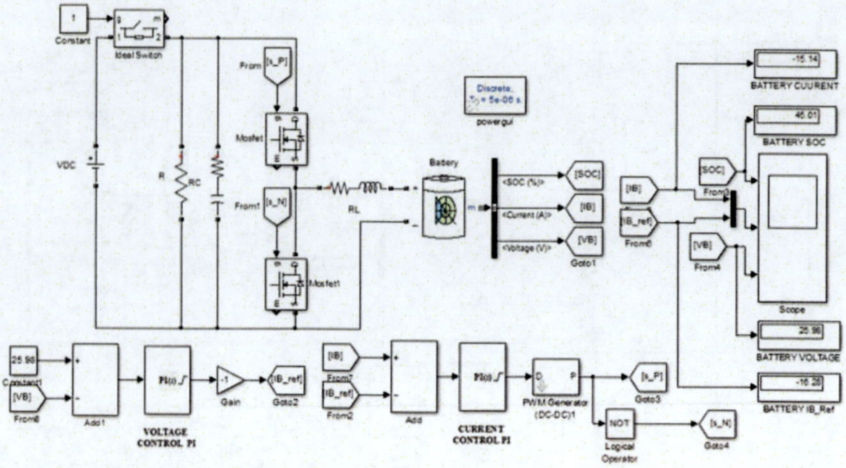

Fig. 11 Voltage control charging model

Case (1) When the voltage source value is enabled, it will charge and voltage source will supply the load.

Case (2) When the voltage source value is disabled, the battery will supply the load by discharging.

These two cases are modelled separately in this section. Since the batteries are charging in two modes CC and CV. For the desired battery with specific capacity has a maximum current of 22A, this indicates it is the maximum allowed current for charging process. Many batteries are manufactured for the 80% SOC but another separate control algorithm is needed for charging the battery above 80% level to 100%. By observing open-circuit voltage (OCV) value at 80%, SOC is 25.98 V [7]. In this test, 80% SOC is taken for consideration with voltage-based control. In order to determine the reference battery current, another PI controller is needed, upper limit of maximum current of 22A. Voltage of 25.98 is given as an input to the voltage-controlled PI, and the battery starts charging. The voltage control charging of the battery waveforms is shown in Fig. 12.

Similar to the voltage control charging model, another PI controller is introduced to have the discharging process, and the proposed for voltage-based controlled discharging model is reproduced using MATLAB/Simulink which is laid out in Fig. 13, and respective waveforms are shown in Fig. 14. The load is supplied with 48 V, discharging of battery takes place with decrease in SOC, and current will be in positive values. In Figs. 11 and 13, models are built to work separately for charging and discharging. In order to combine these two models, switch control action is built which is described below.

Switch Control Strategy

In this section, two cases are mentioned, and the proposed switch control-based model will combine both the cases of charging and discharging in single model by adding

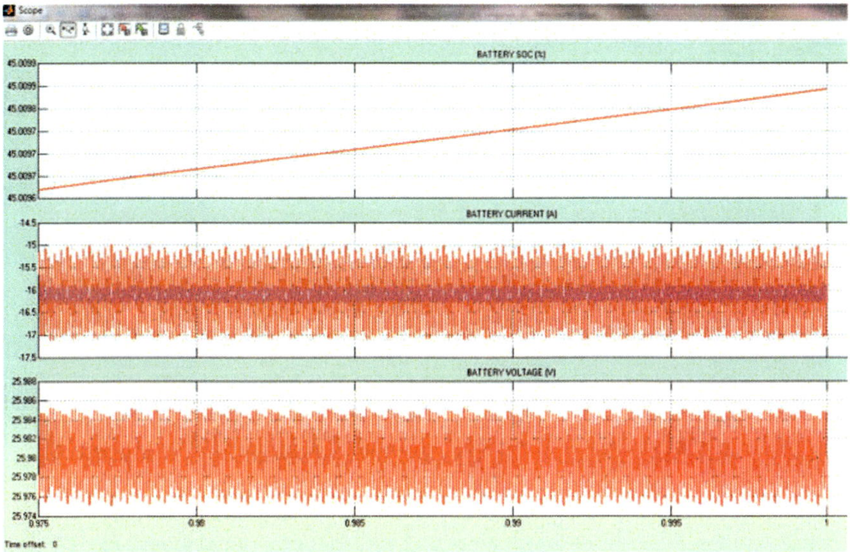

Fig. 12 Voltage control charging waveforms

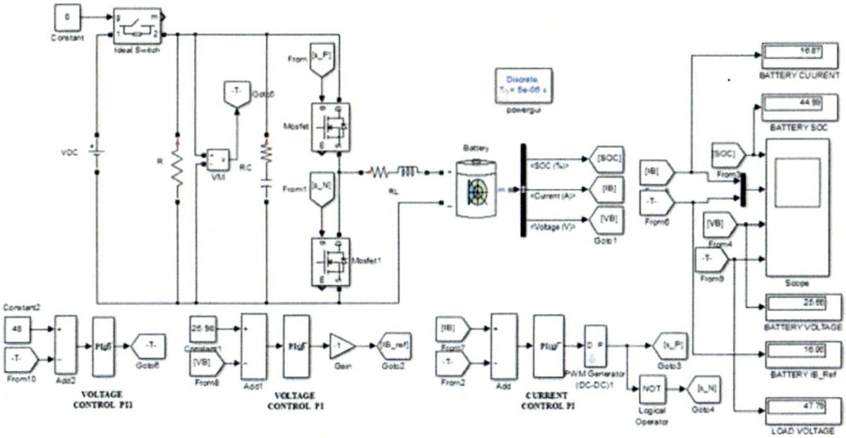

Fig. 13 Voltage control discharging model

switch control circuit to it. Figure 15 works in case 2 that is simulated by disabling the voltage source by giving source = 0, battery works in discharging mode, and load is supplied from the battery. The switch control discharging of battery waveforms is shown in Fig. 16.

Similarly, the proposed model for charging the battery is shown in Fig. 17 works in case 1, that is simulated by enabling the voltage source by giving source = 1, battery works in charging mode, and load is supplied from the source. The switch

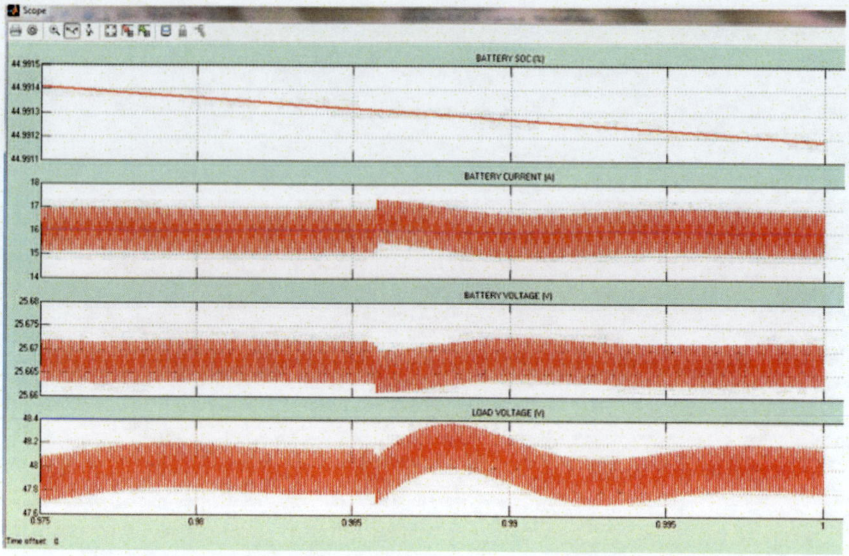

Fig. 14 Voltage control discharging waveforms

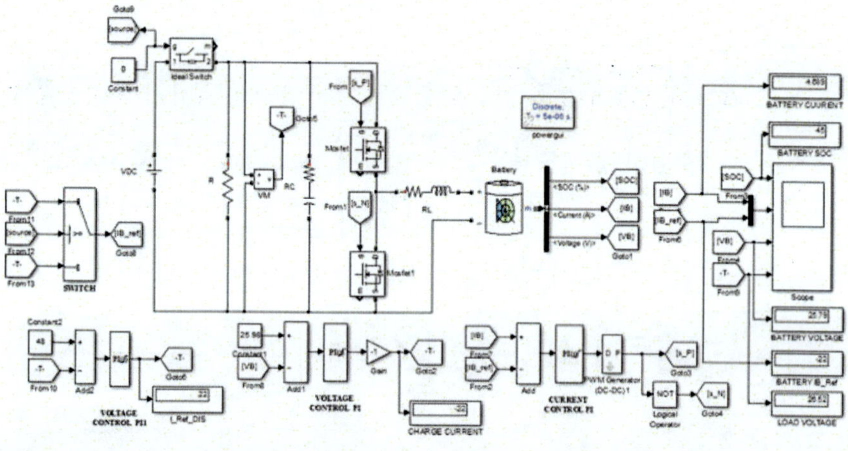

Fig. 15 Switch control discharging model

control charging of battery SOC, current, voltage and load voltage waveforms is shown in Fig. 18.

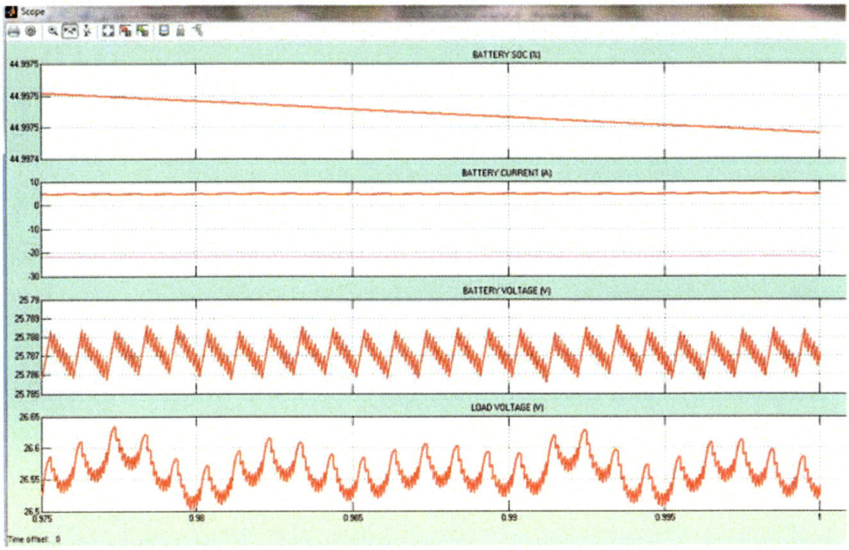

Fig. 16 Switch control discharging waveforms

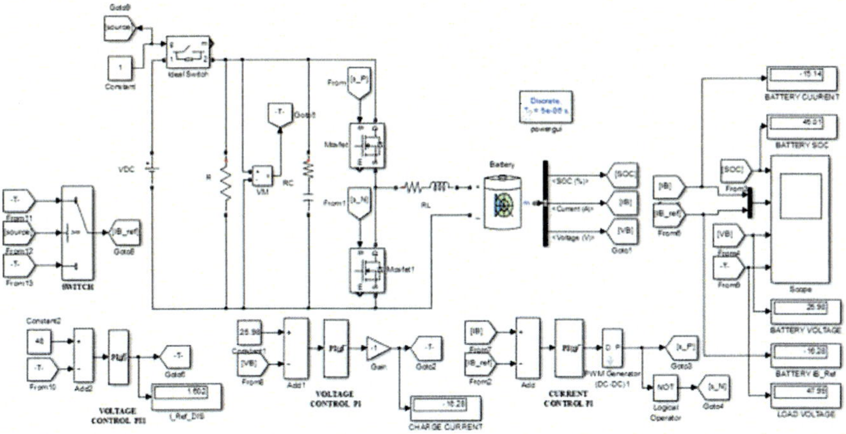

Fig. 17 Switch control charging model

4 Conclusion

The proposed model is specially designed and simulated for Li-ion battery using MATLAB/Simulink software, with desired capacity of 50Ah with its nominal voltage of 24 V. The proposed model provides the control in charging and discharging for Li-ion battery.To achieve the control over charging and discharging, duty cycle control, current control, voltage control and switch-based control are the different methods

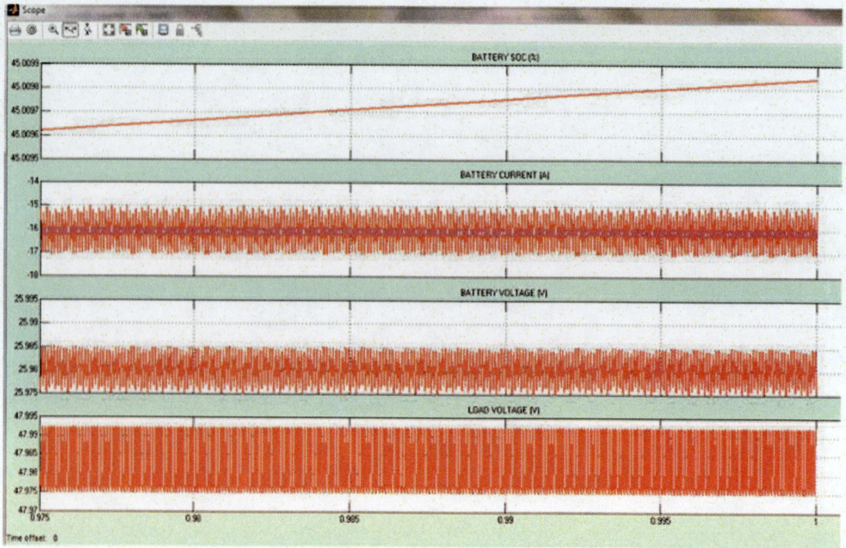

Fig. 18 Switch control charging waveforms

which are exhibited to have control in charging and discharging. Hence, it shows that the proposed models can control the charging and discharging of Li-ion battery with different methods of control strategies. All the results of these strategies show that the proposed model can charge/discharge the battery, and hence, the battery will maintain the higher efficiency and accuracy.

References

1. Pollet, B.G., Staffell, I., Lei Shangc, J.: Current status of hybrid, battery and fuel cell electric vehicles: from electrochemistry to market prospects. ElectrochimicaActa **84**, 235-249 (2012)
2. Miao, Y., Yokochi, A., von Jouanne, A., Hynan, P.: Current Li- Ion battery technologies in electric vehicles and opportunities for advancements. Energies **12**, 1074 (2019)
3. Dung, L., Yuan, H., Chen, C.: A robust, intelligent CC-CV fast charger for aging lithium batteries. In: 2016 IEEE 25th International Symposium on Industrial Electronics (ISIE), Santa Clara, CA, pp. 268-273. (2016)
4. Wang, J., Shan, X., Zhang, M., Xiang, S., Jia, S., Wang, C.: Charge and discharge control and equalization control strategy of charger rear stage. In: 2019 14th IEEE Conference on Industrial Electronics and Applications (ICIEA), pp. 1338–1341. Xi'an, China (2019)
5. Chaurasiya, S., Singh, B.: A G2V/V2G off-board fast charger for charging of lithium-ion based electric vehicles. In: 2019 IEEE International Conference on Environment and Electrical Engineering and 2019 IEEE Industrial and Commercial Power Systems Europe (EEEIC / I&CPS Europe), Genova, Italy, pp. 1–6. (2019)
6. Yong, S.O., Rahims, N.A.: Development of on-off duty cycle control with zero computational algorithm for CC-CV Li ion battery charger. In: 2013 IEEE Conference on Clean Energy and Technology (CEAT), Lankgkawi, pp. 422–426. (2013)

7. Lam, L., Kelder, E., Bauer, P.: A practical circuit-based model for Li-ion battery cells in electric vehicle applications. In: 2011 IEEE 33rd International Telecommunications Energy Conference (INTELEC), Amsterdam, pp. 1–9. (2011)

An Accelerating Power-Based Approach for Transient Stablity Assessment

P. K. Chandrashekhar and S. G. Srivani

1 Introduction

Transient stability in power system generally refers to ability of synchronous machines to be in synchronism after being exposed to a large disturbance [1]. Power system transient stability assessment following a large disturbance such as large generating unit shutdown, loss of a heavily loaded line, occurrence of a three-phase fault, etc., is an important requirement of an interconnected power system [2]. Due to economic reasons, nowadays, most of the power systems are operated very close to their stability limits [3]. After a large disturbance, evaluation of the power system operating stability limit or status in terms of transient stability margin plays an important role in modern interconnected power system [4]. In this paper, an accelerating power-based transient stability margin index (TSMI) derived after transforming a multimachine power system into its single machine equivalent (SIME) after clearing the fault. The proposed index not only gives the information about stability status (i.e., stable or unstable) but also provides information about proximity to loosing stability. Contingency ranking and critical clearing time (CCT) evaluation are the bi-products of the proposed method. Main focus of this paper is on first swing stability.

The methods used for assessment of transient stability are largely classified as time domain simulation (TDS), direct, and AI-based methods [5]. In TDS method, the stability of a system is assessed by simultaneously solving nonlinear equations [6]. In direct or energy function-based methods, the assessment is done by determination and comparison of energy stored during pre- and post-fault conditions [7, 8]. This requires more and accurate computations. The AI-based methods demand for more

P. K. Chandrashekhar (✉)
VTU Research Centre, CPRI, Bengaluru, India
e-mail: pkcsmce@gmail.com

S. G. Srivani
Department of E & EE, RVCE, Bengaluru, India
e-mail: srivanisg@rvce.edu.in

S. K V and K. Guruprasad Rao (eds.), *Smart Sensors Measurements and Instrumentation*, Lecture Notes in Electrical Engineering 750,
https://doi.org/10.1007/978-981-16-0336-5_37

time consuming off-line database generation and training process. The proposed approach is carried out using SIME method which is a mixture first two classes, i.e., TDS and equal area criterion (EAC) methods.

2 Single Machine Equivalent (SIME) Method

2.1 Fundamentals

SIME method [3] is a hybrid method which is obtained by combining results of time domain simulation and equal area criterion concept. Using SIME method, transient stability of any power system can be assessed irrespective of complexity of the models used for simulation. It also assesses the system in terms of CCT, stability margin, and contingency ranking. SIME is obtained by reducing the multimachine system into OMIB by observing rotor angles obtained by TDS. Based on post-fault rotor angles, generators are classified into two groups critical machines group and non-critical machines group. Critical machines (CM) are the one which swing together and are likely to lead the post-fault power system into unstable condition. Non-critical machines (NM) are the one which swing together and remain in stable condition even after the power system becomes unstable.

2.2 OMIB Formulation

After each time step of the TDS, the post-fault rotor angles are observed and the multimachine power system is decomposed into two clusters, 'critical' and 'non-critical'. Then these two are replaced by equivalent OMIB using following procedure.

1. At each time step, the rotor angles and omega obtained by TDS are converted from synchronous reference frame to center-of-inertia (COI) or center-of-angle (COA) frame of reference.

$$\left.\begin{aligned} \delta_{i\mathrm{COI}} &= \delta_i - \delta_0 \\ \omega_{i\mathrm{COI}} &= \omega_i - \omega_0 \end{aligned}\right\} \tag{1}$$

$$\delta_0 = \frac{1}{M_T}\sum_{i=1}^{n} M_i\delta_i \quad \omega_0 = \frac{1}{M_T}\sum_{i=1}^{n} M_i\omega_i \quad M_i = \frac{2H_i}{2\pi f} \quad M_T = \sum_{i=1}^{n} M_i$$

where

δ_i Angular displacement of machine i
ω_i Angular velocity of machine i

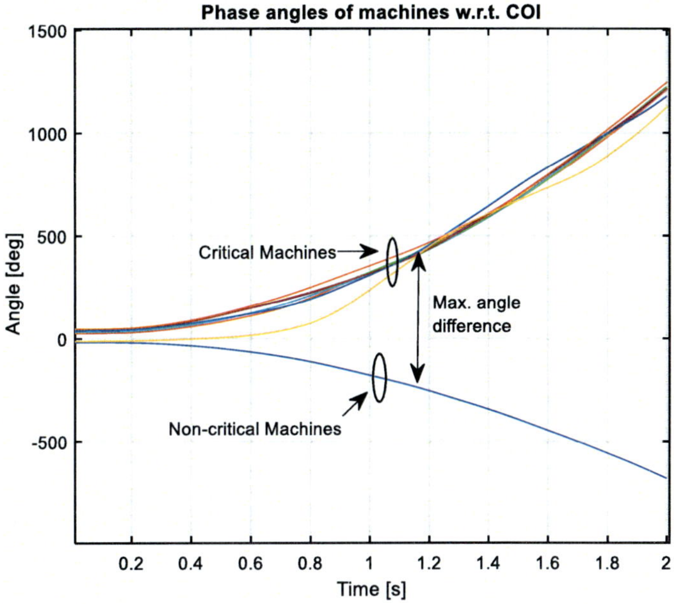

Fig. 1 Grouping of machines

$\delta_{i\text{COI}}$ Angular displacement w.r.t. COI
$\omega_{i\text{COI}}$ Angular velocity w.r.t. COI
H Inertia constant in MW-s/MVA
M inertia coefficient.

2. Thus obtained rotor angles are sorted and two groups are identified based on maximum difference between rotor angles of two adjacent generators. The group with more deviation is termed critical group and the other one as non-critical (Fig. 1).

3. Corresponding rotor angles and speed are calculated.

For critical machines, C:

$$\delta_C(t) = \frac{1}{M_C} \sum_{k \in C} M_k \delta_k(t) \tag{2}$$

$$\omega_C(t) = \frac{1}{M_C} \sum_{k \in C} M_k \omega_k(t) \tag{3}$$

For non-critical machines, N:

$$\delta_N(t) = \frac{1}{M_N} \sum_{j \in N} M_j \delta_j(t) \tag{4}$$

$$\omega_N(t) = \frac{1}{M_N}\sum_{j\in N} M_j\omega_j(t) \quad (5)$$

4. The equivalent OMIB parameters are calculated.

Rotor angle:

$$\delta_{\text{OMIB}} = \delta_C(t) - \delta_N(t) \quad (6)$$

Rotor speed:

$$\omega_{\text{OMIB}} = \omega_C(t) - \omega_N(t) \quad (7)$$

Mechanical power:

$$P_{\text{m}}(t) = M\left(\frac{1}{M_C}\sum_{k\in C} P_{\text{m}k}(t) - \frac{1}{M_N}\sum_{j\in N} P_{\text{m}j}(t)\right) \quad (8)$$

Electrical power:

$$P_{\text{e}}(t) = M\left(\frac{1}{M_C}\sum_{k\in C} P_{\text{e}k}(t) - \frac{1}{M_N}\sum_{j\in N} P_{\text{e}j}(t)\right) \quad (9)$$

Accelerating power:

$$P_{\text{a}}(t) = P_{\text{m}}(t) - P_{\text{e}}(t) \quad (10)$$

$$M_C = \sum_{k\in C} M_k; \quad M_N = \sum_{j\in N} M_j; \quad M = \frac{M_C M_N}{M_C + M_N}$$

3 Proposed Approach

From the OMIB trajectory, the proposed accelerating power-based transient stability margin index (TSMI) is determined as given below (Fig. 2).

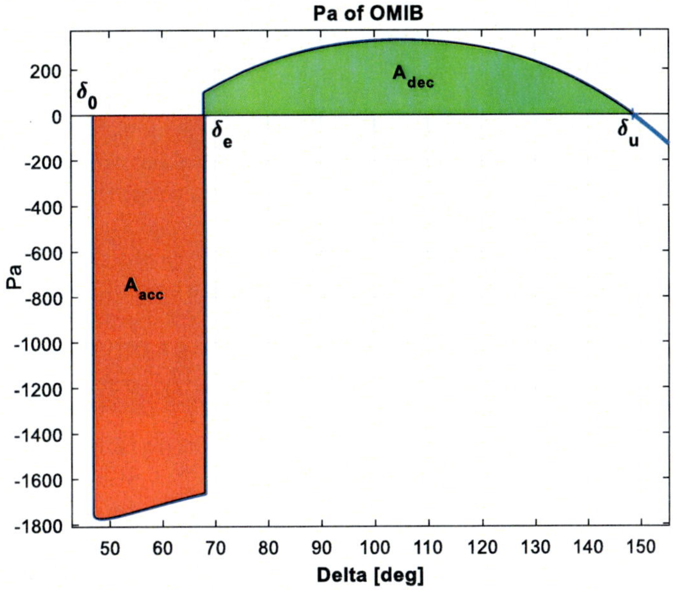

Fig. 2 P_a-δ OMIB trajectory

$$A_{\text{acc}} = \int_{\delta_0}^{\delta_e} P_a \mathrm{d}\delta \tag{11}$$

$$A_{\text{dec}} = \int_{\delta_e}^{\delta_u} P_a \mathrm{d}\delta \tag{12}$$

where,

δ_0 fault starting instant.
δ_e fault clearing instant.
δ_u end of observation.

Normalized accelerating power-based transient stability margin (TSM),

$$\eta = \begin{cases} \frac{A_{\text{dec}} - A_{\text{acc}}}{A_{\text{dec}}} \text{If } A_{\text{dec}} > A_{\text{acc}} \text{ (Stable)} \\ \frac{A_{\text{dec}} - A_{\text{acc}}}{A_{\text{acc}}} \text{If } A_{\text{acc}} > A_{dec} \text{ (Unstable)} \end{cases} \tag{13}$$

The above-defined stability margin is positive and lies between 0 to 1 for stable cases where the accelerating area is less than decelerating area. It is negative and lies between -1 to 0 for unstable cases. Thus, the margin lies between -1 to 1.

4 Simulations and Results

The proposed technique is verified using New England 39-bus test system. It contains ten generating units, 29 loads, and 46 lines. The load flow simulation is carried out using MATPOWER [9]. The dynamic simulation is carried out using MatDyn [10]. Consider a three-phase fault on bus 4 for duration of five cycles. The fault is cleared by removal of connected line between buses 4 and 14. This results in first swing stable case. The value of stability margin determined by the proposed method is 0.3357. The corresponding OMIB accelerating power trajectory is as shown in Fig. 3. If the same fault is cleared in ten cycles, then the resulting system becomes unstable and the stability margin obtained by the proposed method is −0.4352. The negative sign indicates the unstable case. The corresponding accelerating power trajectory is as shown in Fig. 4.

The results obtained with proposed method for three-phase fault at several buses and the corresponding results of time domain simulation are given in Table 1. The results obtained by proposed method are compared with results obtained by time domain simulation for first swing stability. From the results, it is observed that the proposed approach is able to clearly indicate the system status after exposed to a three-phase fault. Compared to the regular time domain simulations used for assessment of transient stability, the proposed method is faster. Unlike TDS, for the proposed method, it is not required to run the simulation for a long duration for stability assessment. After the post-fault period, as soon as the accelerating power reaches zero at instance δ_i as shown in Fig. 2, the simulation is stopped. This reduces the computation time required for the first swing transient stability assessment.

The proposed method not only classifies the post-fault system as stable or unstable, it also gives the stability margin giving information about severity of the contingency.

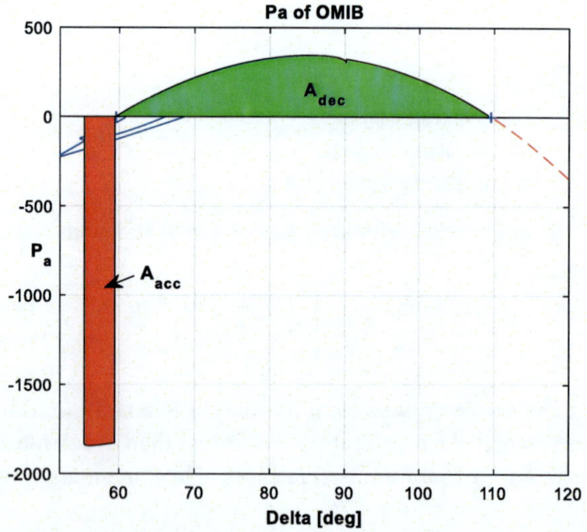

Fig. 3 Stable case corresponding to fault of 5 cycles

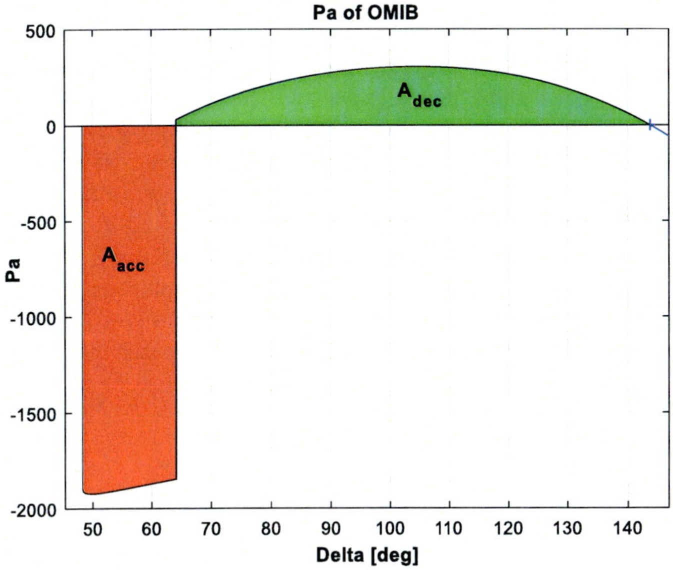

Fig. 4 Unstable case corresponding to fault of 10 cycles

Table 1 Comparison of results obtained by proposed method and TDS method

Faulted bus	Removed line	Clearing time (cycles)	TDS method	Proposed method	
				TSMI (η)	Assessment
1	1–2	5	Stable	0.9600	Stable
3	3–4	10	Unstable	−0.5360	Unstable
5	4–5	8	Stable	0.0634	Stable
8	8–9	6	Stable	0.3697	Stable
10	10–13	9	Unstable	−0.1668	Unstable
13	13–14	11	Unstable	−0.6995	Unstable
17	16–17	4	Stable	0.3587	Stable
27	17–27	12	Unstable	−0.9831	Unstable
39	1–39	7	Stable	0.8789	Stable

If the stability margin is positive and nearer to 1 that means the system is more stable. If the obtained stability margin is nearer to -1 and negative that indicates that the corresponding contingency is more unstable. Similarly, if the stability margin is positive and nearer to 0 that indicated that the corresponding contingency is nearer to losing stability or the system is more likely to become unstable.

5 Conclusions

In this paper, an accelerating power-based transient stability assessment method using normalized transient stability margin index (TSMI) is proposed. The proposed method is fast and less time consuming for first swing transient stability assessment. This method can be applied to any multimachine power system and is independent of the machine model used for simulation. This makes it a model-free approach. This method not only classifies a contingency into stable or unstable but also provides severity of the contingency in terms of stability margin. Simulations are carried out using New England 39-bus test system for various fault locations of different clearing time. From the results obtained, it can be seen that the proposed method is reliable and accurate in prediction of first swing transient stability.

References

1. Kundur, P., Paserba, J., Ajjarapu, V., Andersson, G., Bose, A., Canizares, C., Hatziargyriou, N., Hill, D., Stankovic, A., Taylor, C. and Van Cutsem, T.: Definition and classification of power system stability IEEE/CIGRE joint task force on stability terms and definitions. IEEE Trans. Power Syst. **19**(3), 1387–1401 (2004)
2. Kundur, P., Balu, N.J., Lauby, M.G.: Power system stability and control, vol. 7. McGraw-hill, New York (1994)
3. Pavella, M., Ernst, D., Ruiz-Vega, D.: Transient stability of power systems: a unified approach to assessment and control. Springer Science & Business Media (2012)
4. Savulescu, S.C.: Real-Time Stability Assessment in Modern Power System Control Centers, vol. 42. Wiley, New York (2009)
5. Wadduwage, D.P., Annakkage, U.D., Wu, C.Q.: Hybrid algorithm for rotor angle security assessment in power systems. J. Eng. 241–251 (2015)
6. Sauer, P.W., Pai, M.A.: Power System Dynamics and Stability. Prentice Hall, Upper Saddle River (1998)
7. Pai, M.A.: Energy Function Analysis for Power System Stability. Kluwer Academic (1989)
8. Fouad A.A., Vittal, V.: Power System Transient Stability Analysis Using the Transient Energy Function Method. Prentice-Hall (1992)
9. Zimmerman, R.D., Murillo-Sanchez, C.E.: MATPOWER [Software] (2015). Available https://matpower.org
10. MatDyn website. [Online]. Available: https://www.esat.kuleuven.be/electa/teaching/matdyn/

Level Control of Quadruple Tank System with Feedback Linearization

Ashwini Pachare and Archana Thosar

1 Introduction

The quadruple tank process is a four tank system. It has various applications in pharmaceutical industries and industrial chemical processing plants. The quadruple tank process consists of four interconnected tanks, two pumps and two control valves (three-way control valves). QTP is a multiple input multiple output (MIMO) system which has multiple loops. Due to its complexity and coupled nature, it is used to test various controllers. The objective of the control scheme is to set the liquid level according to the desired height of Tank 1 and Tank 2.

Modeling of a system is important to determine the behavior of the process. Mathematical representation shows system behavior. Here, due to interconnection, coupling is present which makes the system nonlinear. The modeling and design of a suitable controller as well as the effect of multivariable zero for QTP are discussed in [1, 2]. In modeling, nonlinearity is present in the form of square root term which can be overcome by linearization of the nonlinear model using Taylor series and Jacobian conversion is given in [3]. To control the liquid levels of lower two tanks, a controller needs to be designed. The design of the feedback linearization scheme for QTP to control liquid levels of lower two tanks is described. Also, an adaptive feedback linearization(AFL) and model predictive controller(MPC) are compared in [4]. Another technique, neural model predictive controller, for multivariable process is designed to get the desired output liquid level in QTP is given in [5]. To control liquid level, PI controller is designed. In addition to designing a PI controller for a stable system, N. Z. Abidin, S. Sahlan and N. A. Wahab designed PI controllers and investigated for various conditions like minimum phase and non-minimum phase [6]. Now comming to intelligent control system, R. J. A. Raj and S. N. Deepa modified the PID controller with a non-conventional controller with FUZZY and designed

A. Pachare (✉)
College of Engineering Pune, SPP University, Pune, India
e-mail: ashwini.pachare1994@gmail.com

A. Thosar
College of Engineering Pune, Pune, India
e-mail: agt.elec@coep.ac.in

S. K V and K. Guruprasad Rao (eds.), *Smart Sensors Measurements and Instrumentation*, Lecture Notes in Electrical Engineering 750,
https://doi.org/10.1007/978-981-16-0336-5_38

FUZZY PID controller for QTP [7]. Responses for different control strategies like the IMC controller and optimal PID controller have been compared as well as different tunning methods are discussed in [8] .

The remainder of the paper is arranged as follows. Sect. 2 contains plant description, and a nonlinear model of the quadruple tank process is derived based on physical data. Linearization of the model is carried out by Taylor series and Jacobian conversion and conditions for minimum and non-minimum phase are described in Sect. 3. To get the desired output from the plant, controllers are designed. Section 4 discusses the design of the PI controller and feedback linearization. Further by considering required values for QTP, simulation is carried out in MATLAB, and results are presented as well as the comparison discussed in Sect. 5. The advantages of using feedback linearization and conclusion are discussed in Sect. 6.

2 Quadruple Tank Process

2.1 Plant Description

Quadruple tank process (QTP) has four interconnected tanks, two pumps and two control valves (three-way control valves). Input voltage to the pump 1 and pump 2 be V_1 and V_2 resp. Tanks and pumps arrangement is as shown in Fig. 1. This shows interconnection between four tanks. Output of each pump shared between two tanks, one lower tank and other upper diagonal tank. Liquid received by lower and upper diagonal tank depends on valve position called as valve ratio which is denoted as γ. Position of control valve determines wheather the system is in the minimum phase or in the non-minimum phase.

Discharge valve is present at the bottom of each tank. Tank 1 receives the discharged liquid from Tank 3 by the action of gravity, when manual valve kept open. While Tank 2 receives the discharge liquid from Tank 4. And the reservoir collects liquid from Tank 1 and Tank 2. Due to the presence of interaction, a strong coupling is created between tanks which makes the system nonlinear in behavior.

2.2 Modeling of QTP

Modeling of a quadruple tank process is mandatory to determine the behavior of the process and variations with time due to the presence of external disturbances [3]. Schematic diagram of quadruple tank process (QTP) is shown in Fig. 1. For each tank $i = 1, 2, 3, 4$, total mass accumulation is determined by mass balance equation, and input flow rate defined by Bernoulli's law for mathematical modeling.

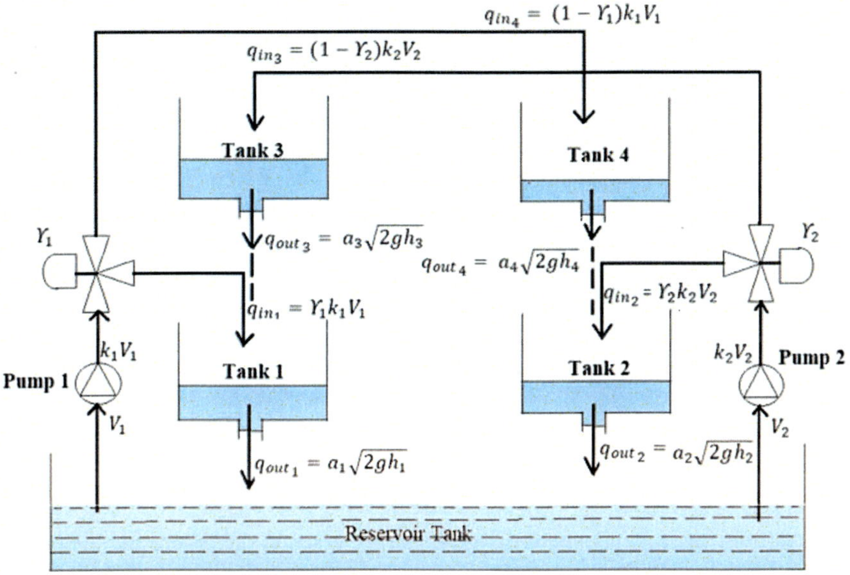

Fig. 1 Schematic diagram of quadruple tank process (QTP)

Assumptions:

1. The valve ratio is $\gamma_1, \gamma_2[0, 1]$.
2. Disturbances like leakage from tank and valve are neglected.

The applied input voltage is directly proportional to the flow rate through the pump. The flow rate to the Tank 1 and Tank 2 from the valve 1 and valve 2 is $\gamma_1 k_1 V_1$ and $\gamma_2 k_2 V_2$, respectively. The flow rate to the Tank 4 and Tank 3 from the valve 1 and valve 2 is $(1 - \gamma_1)k_1 V_1$ and $(1 - \gamma_2)k_2 V_2$, respectively. According to mass balance equation is given as

$$\begin{aligned}\text{[Rate of mass accumulation]} &= \text{[flow of mass into the system]}\\ &\quad - \text{[Flow of mass out of the system]}\end{aligned}$$

Using the law of conservation of mass in each tank,

$$\frac{dm_T}{dt} = m_{\text{in}} - m_{\text{out}} \tag{1}$$

where
m_T = total mass accumulation
m_{in} = mass entering in the tank

m_{out} = mass going out from the tank
m_T= Volume of tank(v)*Density of liquid (ρ)
m_{in}= Volumetric input flow rate (q_{in}) * Inlet stream liquid density (ρ_1)
m_{out}= Volumetric output flow rate (q_{out})* Output stream liquid density (ρ_2)

The liquid used is the same. Hence, density of inlet stream liquid and output stream liquid is same. Considering above equations, modified equation (1) is given as

$$A_i \frac{dh_i}{dt} = q_{\text{in}_i} - q_{\text{out}_i} \tag{2}$$

where
A_i= Cross-sectional area of each tank
h_i= Liquid level
q_{in_i}= Mass in—flow rate of the tank $i = 1, 2, 3, 4$ and
q_{out_i}= Mass out—flow rate of the tank $i = 1, 2, 3, 4$

The in-flow rate (q_{in_i}) depends on input voltage given to the pump, and the mass out-flow rate (q_{out_i}) depends on acceleration due to gravity and height of liquid present in the tank. Flow rate of the liquid and Bernoulli's equation is used to determine (q_{out_i}).
Therefore,

$$\begin{aligned} q_{\text{in}_1} &= \gamma_1 k_1 V_1 \\ q_{\text{in}_2} &= \gamma_2 k_2 V_2 \\ q_{\text{in}_3} &= (1 - \gamma_2) k_2 V_2 \\ q_{\text{in}_4} &= (1 - \gamma_1) k_1 V_1 \end{aligned} \tag{3}$$

where K_1 and K_2 are the pump constants, γ_1 and γ_2 are the valve ratios.

$$(q_{\text{out}_i}) = a_i \sqrt{2gh_i} \tag{4}$$

where
a_i= cross-sectional area of the outlet pipe present at the bottom of each tank,
g = acceleration due to gravity,
h_i= height of liquid in each tank $i = 1, 2, 3, 4$.

According to the law of conservation of mass, change of liquid level with respect to time using Eqs. (2)–(4) is given as for tank 1, 2, 3 and 4 as follows,

$$\begin{aligned} [\text{Rate of mass accumulation}] &= [\text{flow of mass into the system}] \\ &\quad - [\text{Flow of mass out of the system}] \end{aligned}$$

$$
\begin{aligned}
A_1\frac{dh_1}{dt} &= q_{in_1} + q_{out_3} - q_{out_1}\\
A_2\frac{dh_2}{dt} &= q_{in_2} + q_{out_4} - q_{out_2}\\
A_3\frac{dh_3}{dt} &= q_{in_3} - q_{out_3}\\
A_4\frac{dh_4}{dt} &= q_{in_4} - q_{out_4}
\end{aligned} \tag{5}
$$

Rearranging Eq. (5), final equations are

$$
\begin{aligned}
\frac{dh_1}{dt} &= \frac{1}{A_1}(\gamma_1 k_1 V_1 + a_3\sqrt{2gh_3} - a_1\sqrt{2gh_1})\\
\frac{dh_2}{dt} &= \frac{1}{A_2}(\gamma_2 k_2 V_2 + a_4\sqrt{2gh_4} - a_2\sqrt{2gh_2})\\
\frac{dh_3}{dt} &= \frac{1}{A_3}((1-\gamma_2)k_2 V_2 - a_3\sqrt{2gh_3})\\
\frac{dh_4}{dd} &= \frac{1}{A_4}((1-\gamma_1)k_1 V_1 - a_4\sqrt{2gh_4})
\end{aligned} \tag{6}
$$

Mathematical model of quadruple tank process is represented by above nonlinear differential equation (6). Above equations show coupling between interconnected system. Due to this, system becomes nonlinear. This mathematical model is validated by running the simulation in open loop.

The open loop response is simulated for different initial conditions shown below. Open loop response of QTP is shown in Fig. 2. For $h_1 = 0.125$ cm, $h_2 = 0.125$ cm $h_3 = 0.15$ cm, $h_4 = 0.15$ cm

When these results are validated with practical phenomenon, it is found that lower tanks level will be certainly greater than upper one. Because in lower tanks, additional liquid is received from upper tanks due to gravitational force. As time goes, level of all tanks will be increased. It is observed that open loop respone is validated.

QTP shows strong nonlinearity in the system due to presence of square root term in differential equations. But due to nonlinearity and uncertainty present in the system, development of controller is difficult. To overcome this drawback, linearization of Eq. (6) is necessary which is carried out by Taylor series and Jacobian matrix transformation.

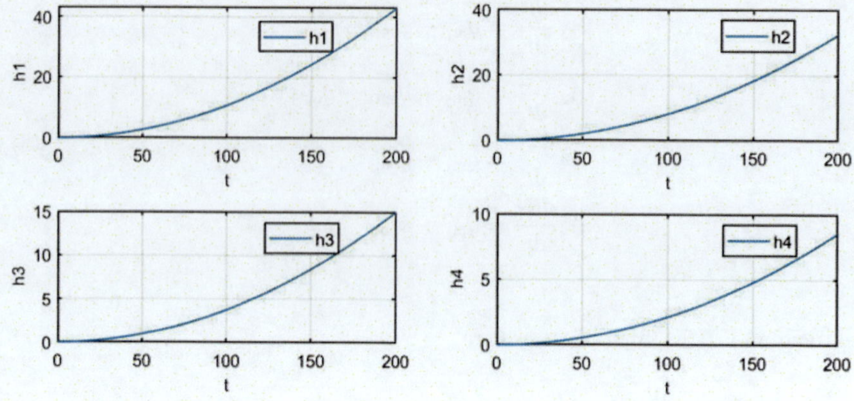

Fig. 2 Open loop simulation results

3 Linearization of QTP

The nonlinearity in the tank system equation is due to the presence of square root term in differential equations. To overcome this nonlinearity, linearization is required. These equations solved by Taylor series and Jacobian matrix transformation to obtain a state space model form of the QTP. Linearization of the differential equations is done by Taylor series. At an equilibrium point, differential equations are solved by Jacobian. The state space equation of linearized quadruple tank process (QTP) is given as

$$\dot{X} = \begin{bmatrix} \frac{-1}{T_1} & 0 & \frac{A_3}{A_1 T_3} & 0 \\ 0 & \frac{-1}{T_2} & 0 & \frac{A_4}{A_2 T_4} \\ 0 & 0 & \frac{-1}{T_3} & 0 \\ 0 & 0 & 0 & \frac{-1}{T_4} \end{bmatrix} X + \begin{bmatrix} \frac{\gamma_1 k_1}{A_1} & 0 \\ 0 & \frac{\gamma_2 k_2}{A_2} \\ 0 & \frac{(1-\gamma_2)k_2}{A_3} \\ \frac{(1-\gamma_1)k_1}{A_4} & 0 \end{bmatrix} u \tag{7}$$

$$y = \begin{bmatrix} K_c & 0 & 0 & 0 \\ 0 & K_c & 0 & 0 \end{bmatrix} X$$

where

$$T_i = \frac{A_i}{a_i}\sqrt{\frac{2h_i}{g}}, \quad X = \begin{bmatrix} h_1 \\ h_2 \\ h_3 \\ h_4 \end{bmatrix}, \quad u = \begin{bmatrix} V_1 \\ V_2 \end{bmatrix} \tag{8}$$

State space model form of the QTP is used for frequency-domain analysis and stability analysis. Transfer function (TF) matrix is

$$G(s) = \begin{bmatrix} \frac{B_1}{(1+sT_1)} & \frac{B_1(1-\gamma_2)}{(1+sT_1)(1+sT_3)} \\ \frac{B_2(1-\gamma_1)}{(1+sT_2)(1+sT_4)} & \frac{B_2}{(1+sT_2)} \end{bmatrix} \tag{9}$$

where

$$B_1 = \frac{T_1 K_c k_1}{A_1} \quad \text{and} \quad B_2 = \frac{T_2 K_c k_2}{A_2} \tag{10}$$

The TF matrix G has two finite zeros $\gamma_1,\gamma_2 \in (0,1)$. Location of the multivariable zeros of G(s) states that whether the system model is in the minimum phase or non-minimum phase.
The system is in non-minimum phase if
$0 < (\gamma_1 + \gamma_2) < 1$
The system is in minimum phase if
$1 < (\gamma_1 + \gamma_2) < 2$
And the multivariable zero is at the origin if
$\gamma_1 + \gamma_2 = 1$

Physical interpretation based on multivariable zero being in the left or right half plane shows system stability. As q_i denotes liquid flow through pump i, let $q_1 = q_2$, i.e., flow through each pump is same. Then, consider sum of flow to upper tanks and sum of flow to lower tanks. Sum of flow to the upper tanks is $[2 - (\gamma_1 + \gamma_2)]q_1$, and sum of flow to the lower tanks is $(\gamma_1 + \gamma_2)q_1$. Considering this flow rate if $1 < (\gamma_1 + \gamma_2)$, this shows that lower tanks receive greater flow than upper tanks. Then, the system is in minimum phase. If $(\gamma_1 + \gamma_2) < 1$ then flow to the lower tank is less than upper tanks, then system is in non-minimum phase.

4 Design of the Controller

The main aim of controller design is to control the liquid level of lower two tanks, i.e., of Tank 1 and Tank 2. Conventional controller and feedback linearization controller with PI are used to set the liquid level according to desired height.

4.1 Conventional Controller

Conventional controller like proportional integral controller (PI) is widely used as a feedback controller. The PI controller output which is used as a system input is given by

$$u(t) = k_p e(t) + k_i \int_0^t e(\tau)\,\mathrm{d}\tau \tag{11}$$

Controller configuration with two PI controllers is shown in Fig. 3,

The performance is improved by careful tuning of controller gains to get desired results. However, the output may become unstable by poor tunning of controller gains.

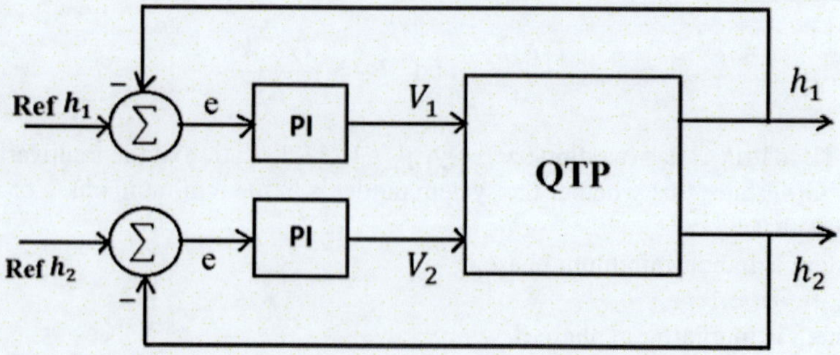

Fig. 3 Control configuration with two PI controllers

The model with PI controller has been tested for following cases and further campared with feedback linearization scheme.

Case I: When all tanks are empty
For $h_1 = 0$ cm, $h_2 = 0$ cm $h_3 = 0$ cm, $h_4 = 0$ cm
When all tanks are empty, simulation result of QTP with PI controller is shown in Fig. 4.

Case II: When all tanks are half filled
For $h_1 = 10$ cm, $h_2 = 10$ cm $h_3 = 10$ cm, $h_4 = 10$ cm

Simulation result of QTP with PI controller is shown in Fig. 5. Similar to first case, it is settling to desired reference value.

4.2 *Feedback Linearization Scheme*

In recent years, nonlinear control design can be converted into the linear design using a feedback linearization approach that has attracted lots of research. It is one of the nonlinear control methods that can linearize the system in all working points. The goal is to design a feedback control law that perfectly cancels the nonlinear term in the system dynamics.

In this paticular application, the main advantage of this method is the elimination of the coupling existed in the considered multivariable systems and that feedback can completely linearize the plant if an accurate model of the plant is used in the designing process [4]. But the feedback linearization scheme requires exact knowledge of the system. Consider the system in the following form to apply feedback linearization as shown in Eq. (12).

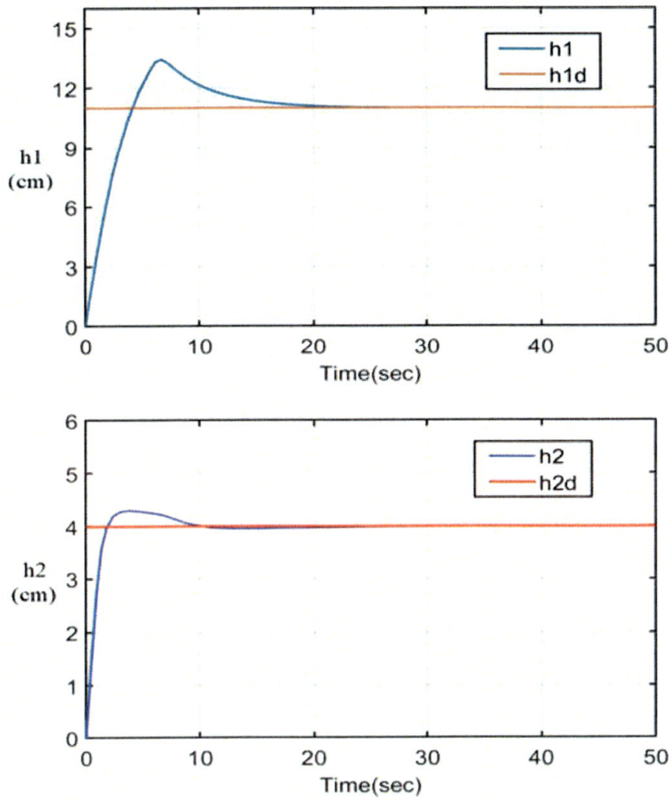

Fig. 4 Variation of level with respect to time of QTP with PI Controller when all tanks are empty

$$\begin{bmatrix} \dot{x}_1 \\ \dot{x}_2 \\ \dot{x}_3 \\ \dot{x}_4 \end{bmatrix} = \underbrace{\begin{bmatrix} \frac{-a_1}{A_1}\sqrt{2gx_1} + \frac{a_3}{A_1}\sqrt{2gx_3} \\ \frac{-a_2}{A_2}\sqrt{2gx_2} + \frac{a_4}{A_2}\sqrt{2gx_4} \\ \frac{-a_3}{A_3}\sqrt{2gx_3} \\ \frac{-a_4}{A_4}\sqrt{2gx_4} \end{bmatrix}}_{f(x)} + \underbrace{\begin{bmatrix} \frac{\gamma_1 k_1}{A_1} & 0 \\ 0 & \frac{\gamma_2 k_2}{A_2} \\ 0 & G_{32} \\ G_{41} & 0 \end{bmatrix}}_{g(x)} u$$
$$\begin{bmatrix} y_1 \\ y_2 \end{bmatrix} = \underbrace{\begin{bmatrix} K_c x_1 \\ K_c x_2 \end{bmatrix}}_{h(x)}, \tag{12}$$

where

$$G_{41} = \frac{(1-\gamma_1)k_1}{A_4}, \quad G_{32} = \frac{(1-\gamma_2)k_2}{A_3} \tag{13}$$

The matrices of $f(x)$, $G(x)$ and $h(x)$ are obtained. For designing the feedback, it is necessary to calculate the relative degree of the outputs. To derive the relationship

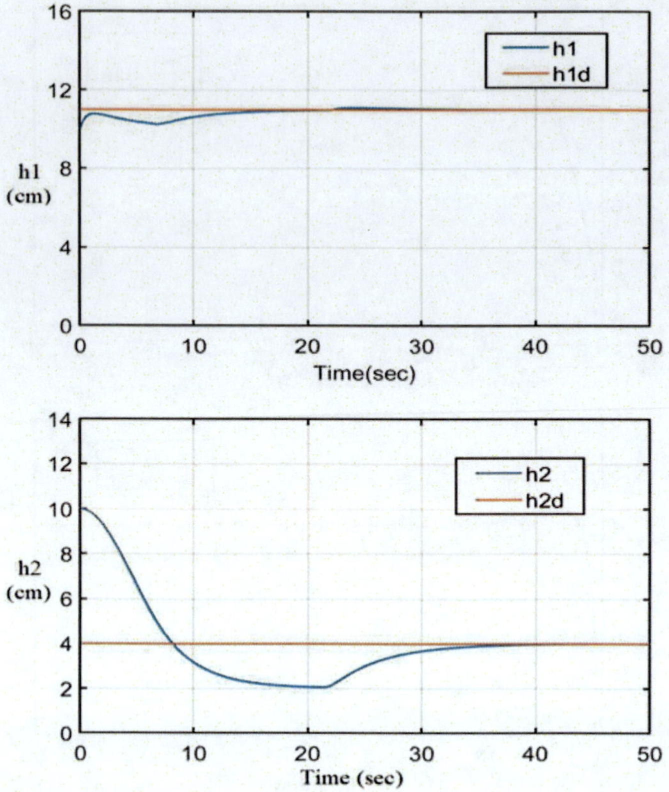

Fig. 5 Variation of level with respect to time of QTP with PI controller when all tanks are half filled

between the output y and input u, differentiate the output of a system r times. Then, the system has a relative degree r. We also need to differentiate output until input occurs in that equation. Differentiate first output of the system just one time for this particular system. To understand the target system, we use Lie derivative.

$$\dot{y} = \frac{\mathrm{d}h(x)}{\mathrm{d}x} f(x) + \frac{\mathrm{d}h(x)}{\mathrm{d}x} g(x)u \tag{14}$$

Now, the Lie derivative of $h(x)$ along $f(x)$ is defined as

$$L_f h(x) = \frac{\mathrm{d}h(x)}{\mathrm{d}x} f(x) \tag{15}$$

And similarly, the Lie derivative of $h(x)$ along $g(x)$ is defined as

$$L_g h(x) = \frac{\mathrm{d}h(x)}{\mathrm{d}x} g(x) \tag{16}$$

$\dot{y}$ is expressed as

$$\begin{aligned} \dot{y_1} &= L_f h_1(x) + L_{g_1} h_1(x) u_1 + L_{g_2} h_1(x) u_2 \\ \dot{y_1} &= K_c \left(\frac{-a_1}{A_1} \sqrt{2gx_1} + \frac{a_3}{A_1} \sqrt{2gx_3} + \frac{\gamma_1 k_1}{A_1} u_1 \right) \end{aligned} \tag{17}$$

where the operator L is Lie derivative (refer to [4]). The first derivative of the y_1 is directly proportional to u_1, so the relative degree of y_1 is unit. By defining a new input variable v_1 such that:

$$\dot{y_1} = v_1 \tag{18}$$

$$\frac{v_1}{K_c} = \frac{-a_1}{A_1} \sqrt{2gx_1} + \frac{a_3}{A_1} \sqrt{2gx_3} + \frac{\gamma_1 k_1}{A_1} u_1 \tag{19}$$

By rearranging, we have

$$u_1 = \frac{A_1}{\gamma_1 k_1} \left(\frac{v_1}{K_c} + \frac{a_1}{A_1} \sqrt{2gx_1} - \frac{a_3}{A_1} \sqrt{2gx_3} \right) \tag{20}$$

Therefore, the transfer function from input v_1 to output y_1 is given by

$$\frac{y_1}{v_1} = \frac{1}{s} \tag{21}$$

Similarly, for the second output, we have

$$\begin{aligned} \dot{y_2} &= L_f h_2(x) + L_{g_1} h_2(x) u_1 + L_{g_2} h_2(x) u_2 \\ \dot{y_2} &= K_c \left(\frac{-a_2}{A_2} \sqrt{2gx_2} + \frac{a_4}{A_2} \sqrt{2gx_4} + \frac{\gamma_2 k_2}{A_2} u_2 \right) \end{aligned} \tag{22}$$

i.e., the relative degree of y_2 is unit too. By defining a new input v_2 such that:

$$\dot{y_2} = v_2 \tag{23}$$

The relation between u_2 and v_2 is then given by

$$\frac{v_2}{K_c} = \frac{-a_2}{A_2} \sqrt{2gx_2} + \frac{a_4}{A_4} \sqrt{2gx_4} + \frac{\gamma_2 k_2}{A_2} u_2 \tag{24}$$

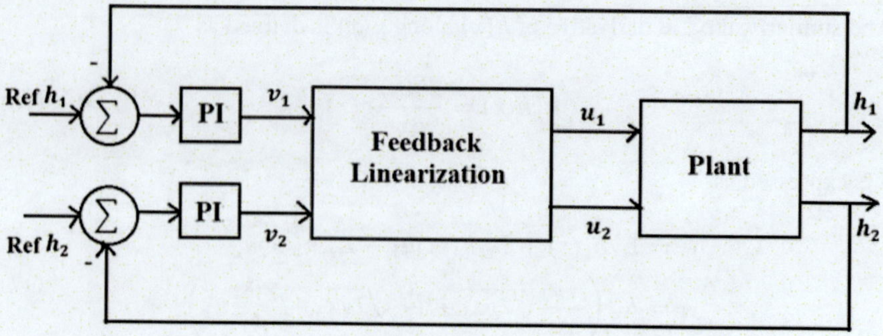

Fig. 6 Controller configuration of the feedback linearization unit with two PI controllers

By rearranging, we have

$$u_2 = \frac{A_2}{\gamma_2 k_2}\left(\frac{v_2}{K_c} + \frac{a_2}{A_2}\sqrt{2gx_2} - \frac{a_4}{A_4}\sqrt{2gx_4}\right) \tag{25}$$

Therefore, the transfer function from v_2 to y_2 is given by

$$\frac{y_2}{v_2} = \frac{1}{s} \tag{26}$$

This method is applicable as long as all of the state variables remain stable. It can be shown that all variables will remain stable if, $1 < \gamma_1 + \gamma_2 < 2$ i.e., the system is in minimum phase mode. The schematic diagram of the feedback linearization unit with two proportional controllers is shown in Fig. 6.

In Fig. 6, output is compared with reference input, and then, error goes to PI controller, and futher output of PI controller goes to FL unit.
The model with feedback linearization scheme has been tested for various cases and is as follows where reference $h_1 = 11$ cm and $h_2 = 4$ cm for all cases.

Case I: When all tanks are empty
For $h_1 = 0$ cm, $h_2 = 0$ cm $h_3 = 0$ cm, $h_4 = 0$ cm

Simulation result of QTP with feedback linearization is shown in Fig. 7. It is attaining desired reference height in very less time with almost zero overshoot.

Case II: When all tanks are half filled
For $h_1 = 10$ cm, $h_2 = 10$ cm $h_3 = 10$ cm, $h_4 = 10$ cm Simulation result shown in Fig. 8 with less settling time.

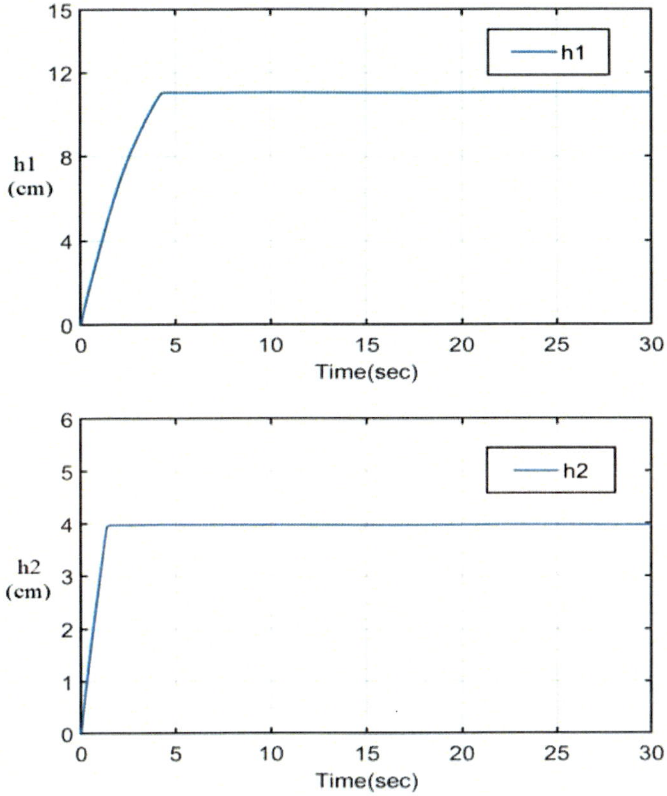

Fig. 7 Variation of level with respect to time of QTP with FL when all tanks are empty in FL

5 Results and Discussion

To validate the mathematical model, values for simulation are considered from the experimental setup. All required values from experimetal setup are shown in Table 1.

The comparison between PI controller and feedback linearization scheme for quadruple tank process is done in this section. Open loop response, closed loop response with PI controller and feedback linearization scheme with PI controller are studied.

Simulation results for case I and case II of PI controller and feedback linearization scheme with PI are given in Table 2. Comparison carried out further.

On comparing results of PI controller and feedback linearization scheme with PI shown in Table 2, PI controller required larger controller gains as compared to feedback linearization scheme for both the cases. In addition, quadruple tank

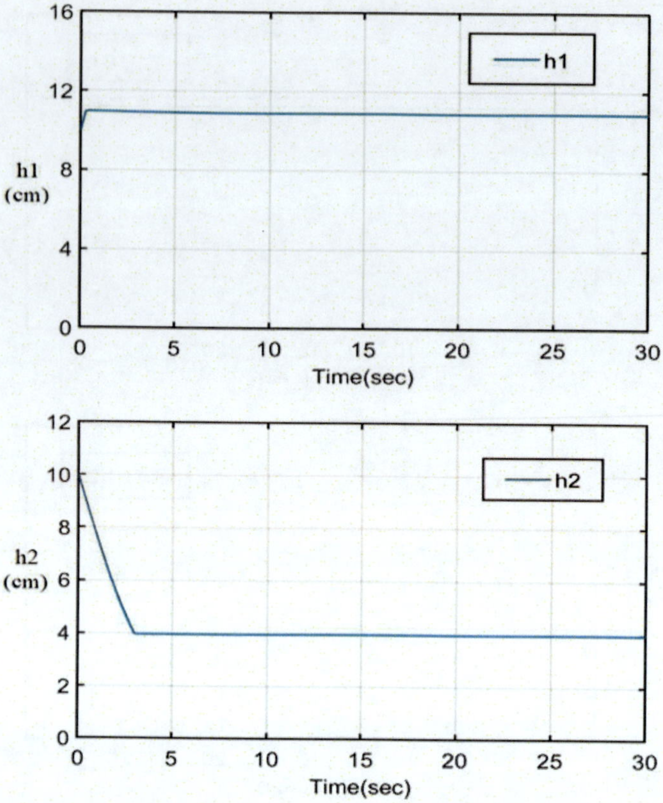

Fig. 8 Variation of level with respect to time of QTP with FL when all tanks are half filled in FL

Table 1 Parameters used in simulation

Parameters	Values under consideration	Unit
k_1,k_2	3.33	$\frac{(cm)^3}{s}$
γ_1,γ_2	0.7,0.6	–
Kc	0.5	$\frac{volt}{cm}$
A_1,A_2,A_3,A_4	71	$(cm)^2$
a_1,a_2,a_3,a_4	1.5	$(cm)^2$
g	981	$\frac{cm}{s^2}$

process with feedback linearization shows better results with very less peak overshoot and settle in very less time than only with PI controller. This proved that feedback controller linearized and decouple the system at the same time. This results Feedback Linearization is beneficial.

Table 2 Controller gains and simulation result comparison

Controller	Cases	K_p		K_i		Overshoot		Settling time	
		h_1	h_2	h_1	h_2	h_1	h_2	h_1 (s)	h_2 (s)
PI controller	I	60.5	63.7	7.297	15.82	22.058	7.488	16.698	8.59
	II					0.9567	150.2	35.55	34.202
Feedback linearization with PI controller	I	9.64	15.297	0.04987	0.0865	0.097	0.227	4.126	1.351
	II					0.684	151.8	4.342	2.908

6 Conclusion

In this paper, design of feedback linearization scheme for the liquid level control of the lower two tanks of the system has been presented. The proposed method is compared with PI controller, and simulation results proved that feedback linearization scheme shows better performance as compared to PI controller.

The advantage of the proposed scheme over the other controller is the capability to linearize and decouple the system. The successful application of the proposed scheme to the quadruple tank process shows the applicability of the controller in multivariable applications. Further, results can be modified using techniques like backstepping to make system more stable.

References

1. Johansson, K.H.: The quadruple-tank process: a multivariable laboratory process with an adjustable zero. IEEE Trans. Control Syst. Technol. **8**(3), 456–465 (2000)
2. Numsomran, A., Tipsuwanporn, V., Tirasesth, K.: Modeling of the modified quadruple-tank process. In: 2008 SICE Annual Conference, pp. 818–823. IEEE (2008)
3. Jayaprakash, J., Kumar, M.H.: State variable analysis of four tank system. In: 2014 International Conference on Green Computing Communication and Electrical Engineering (ICGCCEE), pp. 1–8. IEEE (2014)
4. Pourmohammad, S., Yazdizadeh, A.: Decoupling control of the quadruple-tank system. Citeseer
5. Sivakumaran, N., Kirubakaran, V., Radhakrishnan, T.: Neural model predictive controller for multivariable process. In: 2006 IEEE International Conference on Industrial Technology, pp. 3072–3077. IEEE (2006)
6. Abidin, N.Z., Sahlan, S., Wahab, N.A.: Optimization tuning of pi controller of quadruple tank process. In: 2013 Australian Control Conference, pp. 331–335. IEEE (2013)
7. Raj, R.A., Deepa, S.: Modeling and implementation of various controllers used for Quadruple-Tank. In: 2016 International Conference on Circuit, Power and Computing Technologies (ICCPCT), pp. 1–5. IEEE (2016)
8. Effendi, A.R., Rameli, M., Iskandar, E., Baihaqi, M.: Linearization and decoupling controller for quadruple tank. In: 2017 International Seminar on Intelligent Technology and Its Applications (ISITIA), pp. 233–237. IEEE (2017)

Analysis of Motorcycle Rider's Posture Using sEMG Signals

Aiswarya Varghese, K. B. Akshaya, S. Akshay Prakash, S. Sreehari, Divya Sasidharan, and G. Venugopal

1 Introduction

Motorcycling is an activity of recreation in most parts of the world. Police, fast-food delivery, sellers e.t.c fulfil most of their professional needs using motorcycle. In India, most of the middle class families use motorcycle as the mode of transport. But for motorcycle riders, possibility of incidence fatigue is really high compared to drivers of other transport modes. When performing a task repetitively, fatigue development in the involved muscle will occur [1]. Long time postural fatigue will lead to incidents of fatalities with motorcycle riders. Comfortable riding postures for different types of motorcycles will be different. Besides postural fatigue, drunken driving and lack of driving skills will also lead to motor cycle accidents [2].

Unsuitable riding posture is one of the major factors for rider's muscular fatigue. The long term postural distress and the resulting muscle fatigue might lead to difficulty in navigation of the motorcycle and altered riding behaviour along with diversion of attention. Muscle fatigue, drunken driving, and lack of skills are the factors that become the potential cause for motorcycle accidents.

Development of muscle fatigue is due to a number of reasons. Body postures maintained long time by drivers may cause not only disturbances of physiological functions but also the fatigue of musculoskeletal system. The factors such as rider's physical characteristics (anthropometrics, age, and gender), riding experience, and duration are important determinants of adopted posture during riding. While riding; the body performs two major functions, they are: maintaining the body posture, absorbing road shocks and generate the required forces to operate and control the motorcycle. The existing methods to identify the fatigue are limited to subjective experience. The measurement of the vehicle parameters such as steering torque, shock absorption level, acceleration level etc. are used to identify the fatigue.

A. Varghese · K. B. Akshaya · S. A. Prakash (✉) · S. Sreehari · D. Sasidharan · G. Venugopal
Department of Instrumentation and Control Engineering, NSS College of Engineering (Affliated to APJ Abdul Kalam Technological University Kerala), Palakkad, India

S. K V and K. Guruprasad Rao (eds.), *Smart Sensors Measurements and Instrumentation*, Lecture Notes in Electrical Engineering 750,
https://doi.org/10.1007/978-981-16-0336-5_39

Heart rate, blood pressure are some of the other potential indicators of fatigue [3]. Magnetic resonance imaging (MRI), muscle biopsy and electromyography (EMG) are commonly used to diagnose these conditions. EMG is a procedure used to assess the health condition of muscles and the nerve cells that control the motor neurons. It is a technique for evaluating and recording the electrical activity produced by the muscles. An EMG detects the tiny amount of electricity generated by muscle cells when they are simulated by the nerves connected to them. This signal can be easily estimated than the signals from nerves and brains [4]. Electrical signals produced by the contraction of skeletal muscles are called surface EMG (sEMG) signals [5]. Specialists working in this area analyse these signals to detect medical non conformities and abnormalities in the spinal cord, nerves and the muscles that are connected to these specific nerves. 'Riding Simulation: The Setup and Procedures' has provides a reference platform based on the Postura Motergo motorcycle simulator for setting up a facility that could be used for studying difficulties related to motorcycle ergonomics, and procedures to record muscle activities of motorcyclists using a sEMG device. The sEMG raw data is processed and analysed. Average RMS and MPF could be extracted and further studied especially on motorcyclists' muscle fatigue [6].

Nerve dysfunction, muscle dysfunction or problems with nerve-to-muscle signal transmission can be revealed by EMG. Electrical signals are transmitted by motor neurons which results in muscle contraction. An EMG uses electrodes which translate these signals into graphs, sounds or numerical values that are then interpreted by a specialist. Surface electromyography, which is a non-invasive technique for measuring muscle activity is used. sEMG electrodes record from a wide area of muscle territory, have a relatively narrow frequency band (range, 20–500 Hz). The muscle groups were chosen based on preliminary studies for implication in riding, intensity of the signal and the accessibility of the muscle for electrode application. The following muscle groups are affected mostly: extensor carpi radialis (ECR), biceps brachii (BB), trapezius medial (TM), sternocleidomastoid (S), Latissimus dorsi medial (LDM), erector spinae (ES) and Flexor carpi Ulnaris (FCU). Significant difference was observed in the mean RMS values of the sixth level wavelet coefficients from the decomposed sEMG signals across the Flexor Carpi Ulnaris, Latissimus Dorsi medial and the Trapezius Medial muscles bilaterally during the course of motorcycle riding [3]. As per the above results, Latissimus Dorsi and Flexor Carpi Ulnaris muscles are chosen for this study.

2 Methodology

2.1 Protocol

The Surface EMG signals are taken from twenty healthy male subjects from the age group 20–23 years with an average weight of 75 kg and height of 175 cm.The subjects were instructed not to engage in any physical activity for 24 h prior to the

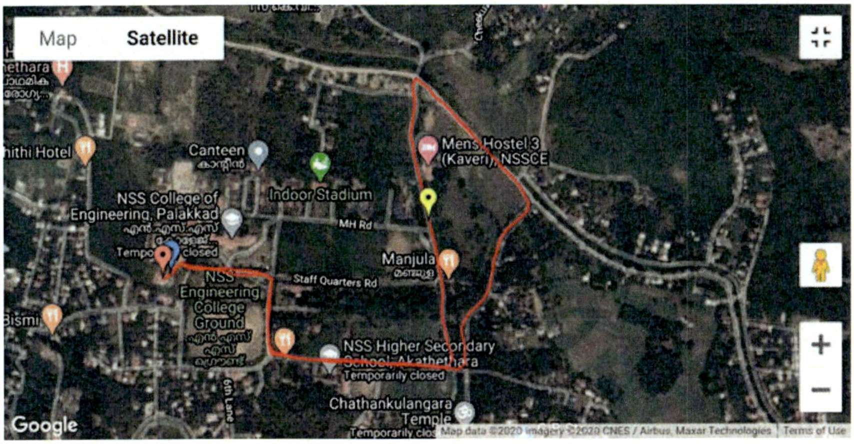

Fig. 1 Satelliteview of the track used for the study

experiment and proper consent is procured. The selected subjects did not have any prior medical history for muscle disorders. A pilot study is conducted prior to the experiment to finalize the rounds for fatigue development. A well-defined path of 3 km is selected and the information regarding the average speed, elevation and time taken to complete the path are recorded using the "speed view GPS pro" application as shown in Fig. 1.

The selected track has slopes, curves, gravel and straight road at regular intervals of the journey. They are asked to complete 3 rounds in the specified track of 3 km while maintaining an average speed of 25 km/hr in the scooter provided. A common load of 80 kg is provided in the back seat. The muscles Latissimus Dorsi and Flexor Carpi Ulnaris are identified in each subject according to the SENIAM standards. Inter electrode distance is 3 cm and a common reference electrode is placed at the bony portion of left elbow. The block diagram is shown in Fig. 2. Electrodes were connected to the Vernier EKG sensor which is powered by 5 V DC supply in an IC trainer kit.

The EMG signals are taken before beginning the experiment which is denoted as Stage 1 (S1). The subjects are instructed to stop after each round and sEMG signals are taken for 4 s, this is denoted as stage 2, 3, 4 (S2, S3, S4) respectively and the values are recorded.

The samples signals obtained from each muscles are given in Figs. 3, 4 and 5 represents the circuit set up for the experiment.

2.2 Instrumentation and Signal Processing

Two signals are acquired from two muscles named Latissimus Dorsi and Flexor Carpi Ulnaris. The sampling rate is 5000 Hz. The signals are filtered online using

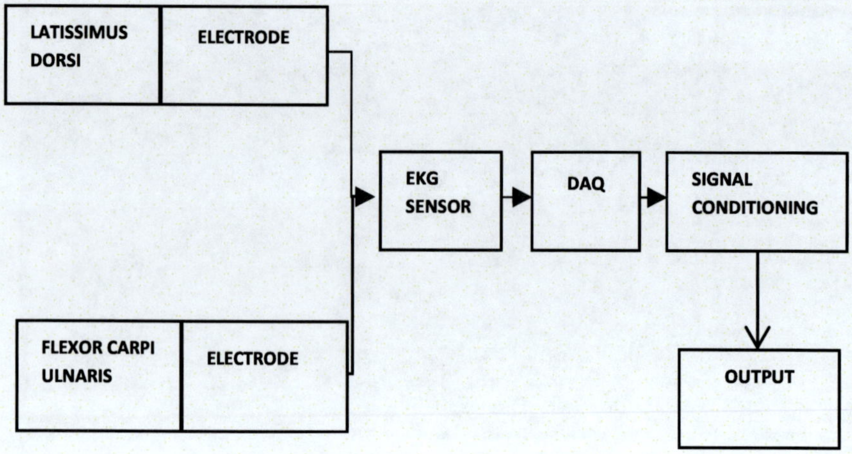

Fig. 2 Block diagram of the experimental setup

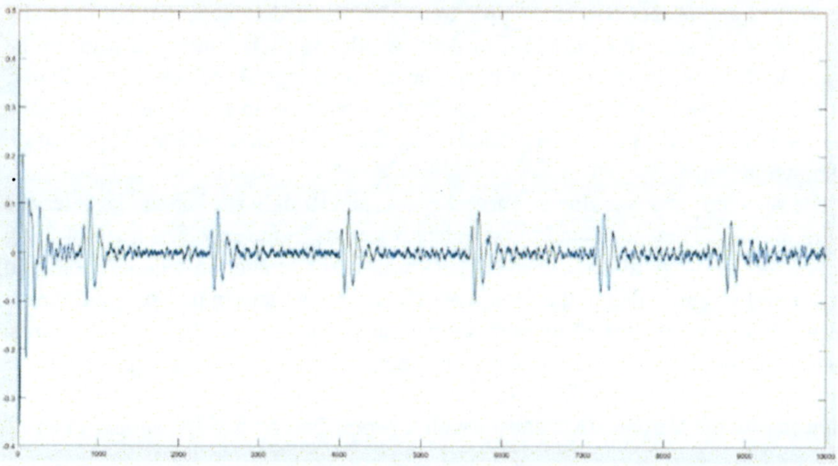

Fig. 3 Signal obtained from Latissimus Dorsi

Butterworth band pass filter with a pass band of 20–400 Hz and Notch filter of 50 Hz. The features in time and frequency domain taken for this analysis are Root mean square (RMS), Mean frequency (MNF) and Entropy. These are the most commonly used simple features for classifying fatigue and non fatigue muscle condition.At the onset of fatigue the commonly observed trends during analysis is that the RMS value of each muscle increases and the mean frequency decreases. The features are extracted using MATLAB.

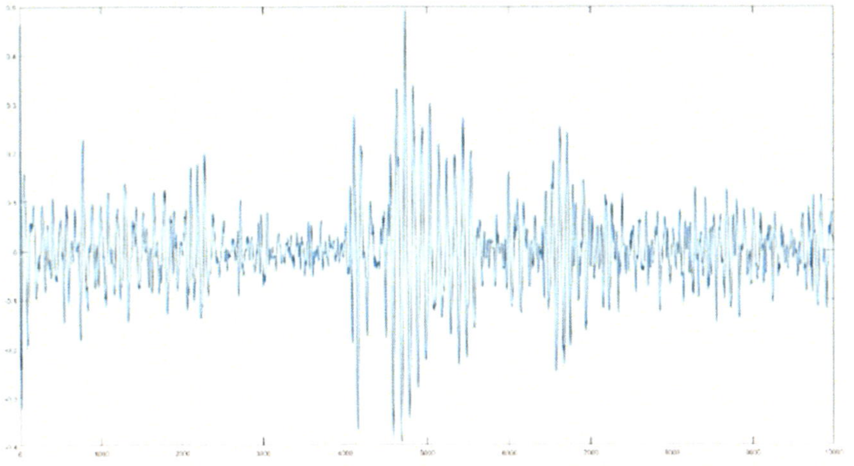

Fig. 4 Signal obtained from Flexor Carpi Ulnaris

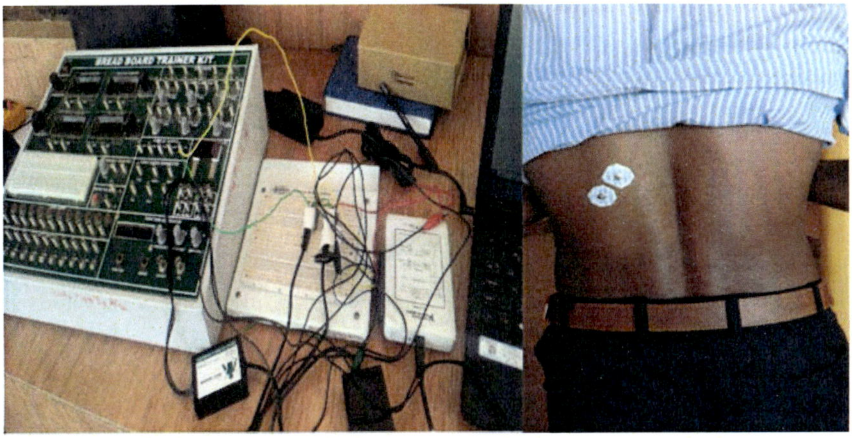

Fig. 5 Circuit setup

2.3 *Feature Extracton*

The features Root mean square (RMS), Mean frequency (MNF), and Entropy are taken.

Root Mean Square (RMS): It represents the square of the root of the average power of the EMG signal for a given period of time (viz. 4 s). It is known a time domain variable. It can be calculated by the formula:

$$\text{RMS} = \sqrt{\frac{1}{N}\sum_{1}^{N}|x_i|^2} \quad (1)$$

where x_i are the values of each sample and N is the number of samples.

Mean frequency (MNF): It is that frequency where the product of the frequency value and the amplitude of the spectrum is equal to the average of all such products throughout the complete spectrum of the signal.

$$\text{MNF} = \sum_{j=1}^{M} P_{ij} f_{ij} \Bigg/ \sum_{j=1}^{M} P_{ij}, \text{ for } i = 1,\ 2\ldots \quad (2)$$

where, P_{ij} is the EMG power spectrum at the frequency bin j for the ith segment, f_{ij} is the frequency value of EMG power spectrum at the frequency bin j for the ith segment, and M is the length of frequency window [7].

Entropy: It is a measure of randomness of the acquired signal. SampEn has been applied in physiological time series analysis including diagnosis of cardiovascular diseases and detection and classification of motor behavior. In order to compute the SampEn, the scalar time series $\{x1, x2, \ldots, xi, \ldots, xn\}$ is first embedded in a delayed m-dimensional space, where vectors are constructed as:

$$x(p) = [x(p+k)]m - 1k = 0,\ p = 1, 2, \ldots, n - m + 1 \quad (3)$$

The probability Bm(r) that two sequences match for m points is computed by counting the average number of vector pairs, for which the distance is lower than the tolerance r. Similarly, $Am(r)$ is defined for an embedding dimension of $m + 1$. The SampEn is then calculated as:

$$SampEn(x,\ m,\ r) = -\ln\big(Am(r)/\,Bm(r)\big) \quad (4)$$

3 Results and Discussions

The signals obtained from the experiment on the muscles Latissimus dorsi and Flexor carpi Ulnaris are used to extract features like RMS value, mean frequency and entropy of the signal. Using the values of the above features at different stages, graphs were generated in order to understand the implications of these features during the analysis at various stages of the experiment. The plots and tables are obtained after the study on the above mentioned muscles.

Generally, muscle fatigue is recognized by analysing the variation in certain features like RMS value, mean frequency and entropy.

The RMS value of sEMG signals acquired from the muscles shows an increasing trend. This is due to the increase in amplitude of the sEMG signal since more motor units are required in order to maintain the same physiological condition. The increasing RMS value of the signals implies the onset of fatigue in the muscles under study. Figures 6 and 7 shows the RMS value characteristics for Lattisimus dorsi and

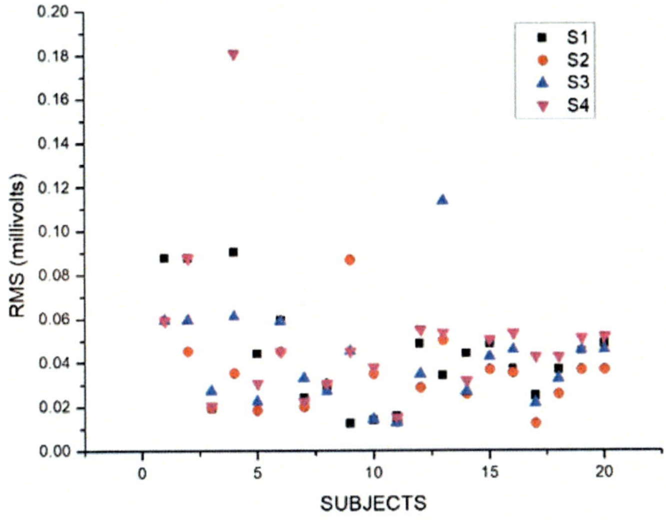

Fig. 6 Variation in RMS of Latissimus Dorsi

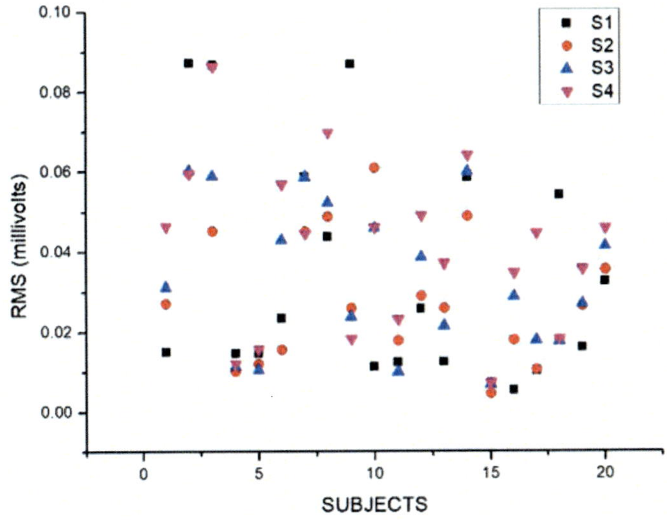

Fig. 7 Variation in RMS of Flexor Carpi Ulnaris

Flexor carpi Ulnaris respectively.

The frequency domain analysis conducted on the signals shows a decreasing trend in mean frequency in kHz. This trend can be due to a number of reasons like modulation of motor unit firing rate, grouping and slowing of conduction velocity or synchronization of signal. Figures 8 and 9 shows the mean frequency characteristics

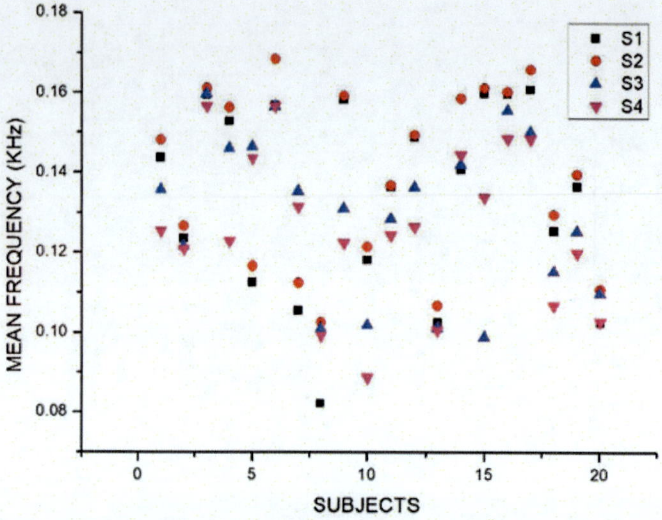

Fig. 8 Variation of mean frequency of Latissimus Dorsi

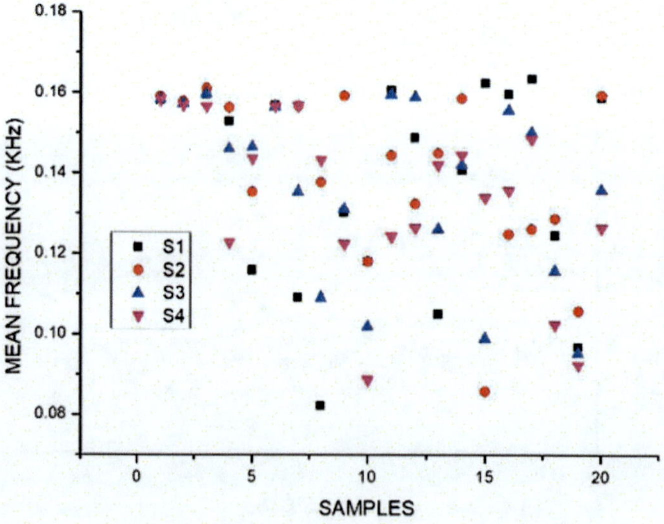

Fig. 9 Variation of mean frequency of Flexor Carpi Ulnaris

for Lattisimus dorsi and Flexor carpi Ulnaris respectively.

Entropy is considered as a measure of order, complexity or randomness of a dynamic system. The entropy of the sEMG signals is found to be decreasing for Latissimus Dorsi and increasing for Flexor Carpi Ulnaris. This is because of the synchronization in muscle firing and reduction in the randomness of the signal. Figures 10 and 11 shows the entropy characteristics for the muscles under study.

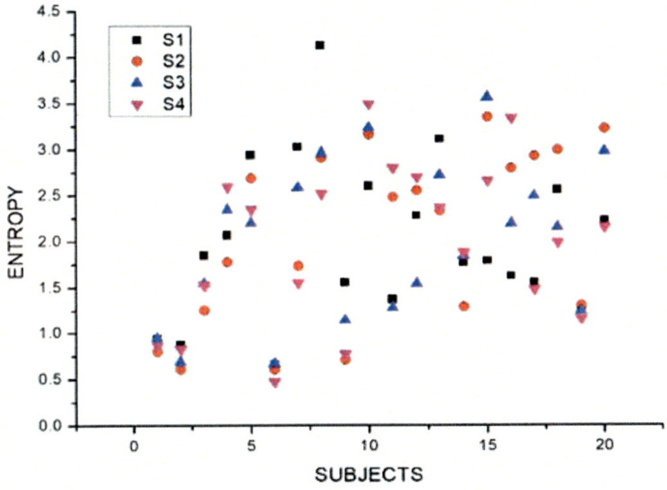

Fig. 10 Variation of entropy in Latissimus Dorsi

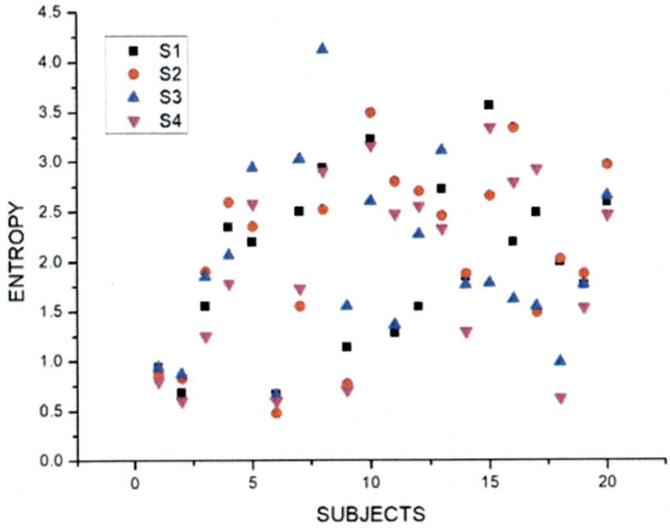

Fig. 11 Variation of entropy in Flexor Carpi Ulnaris

Table 1 Features of Flexor Carpi Ulnaris

Feature	Stage 1	Stage 2	Stage 3	Stage 4
RMS (mV)	0.04696	0.03788	0.04606	0.04875
Mean frequency (kHz)	0.14202	0.14204	0.14079	0.13942
Entropy	2.00246	1.98561	1.9882	2.03348

Table 2 Features of Latissimus Dorsi

Feature	Stage 1	Stage 2	Stage 3	Stage 4
RMS (mV)	0.04226	0.03737	0.04092	0.05017
Mean frequency (kHz)	0.12953	0.12839	0.12967	0.12242
Entropy	2.49971	2.27719	2.43136	2.34881

Table 3 Percentage changes in parameters

Feature	LD	FCU
RMS (mV)	18.7	3.8
Mean frequency (kHz)	5.8	1.8
Entropy	6.03	1.5

In addition to the above method, a statistical data is developed from signals taken in all the 4 stages, S1, S2, S3, and S4 of each subject.

Tables 1 and 2 show the values of various parameters at different stages for the muscles Flexor carpi ulnaris and Latissimus Dorsi.

From Table 3, it is observed that there is an increase of 18.7% and 3.8% RMS value, 5.8% and 1.8% decrease in mean frequency and 6.04% decrease and 1.4% increase in entropy for the muscles Latissimus dorsi and Flexor Carpi Ulnaris respectively.

If the ride continues, there is a further increase in the RMS value and a decrease in the mean frequency which is an indication of the fatigue and can lead to permanent damage of the muscles.

4 Conclusions

Riders tend to experience health issues due to prolonged riding in everyday life. The riders experience muscle fatigue after each round. The present study illustrates the muscle fatigue developed in latissimus dorsi and flexor carpi ulnaris. From the study, it is found that the muscle fatigue experienced in former is greater than the latter. An increase of 18.7% in RMS, 5.8% decrease in mean frequency and a 6.04% decrease in entropy is found in LD muscle and an increase of 3.8% in RMS, decrease of 1.8% in mean frequency and increase of 1.4% in entropy is found in FCU muscle. The findings could contribute for the research study that is associated with analysis of motorcycle riders' posture using surface EMG signals.

References

1. Fuller, J.R., Fung, J., Côté, J.N.: Posture movement responses to stance perturbations and upper limb fatigue during a repetitive pointing task. Hum. Mov. Sci. **32**(4), 618–632 (2013)
2. Arunachalam, M., Mondal, C., Singh, G., Karmakar, S.: Motorcycle riding posture: a review. Measurement (2018)
3. Balasubramanian, V. Jagannath.: Detecting motorcycle rider local physical fatigue and discomfort using surface electromyography and seat interface pressure. Transp. Res. Part F: Traffic Psychol. Behav. **22**, 150–158 (2014)
4. Khamis, N.K., Basrah, N., Deros, B., Nuawi, M.: Electromyography as an objective measurement for evaluating physical fatigue: an overview. JurnalTeknologi, 77 (2015). https://doi.org/10.11113/jt.v77.6888
5. Venugopal, G., Navaneethakrishna, M., Ramakrishnan, S.: Extraction and analysis of multiple time window features associated with muscle fatigue conditions using sEMG signals. Expert Syst. Appl. **41**(6), 2652–2659 (2014)
6. Riding simulation: the setup and procedures. J. Mech. Eng. **4**(5), 197–208 (2017)
7. Thongpanja, S., Phinyomark, A., Phukpattaranont, P., Limsakul, C.: Mean and median frequency of EMG signal to determine muscle force based on time dependent power spectrum. Elektron. Ir Elektrotechnika, **19**(3), (2013) ISSN 1392-1215
8. Mackenzie, Colin: The action of muscles: including muscle rest and muscle re-education, p. 2015. Paul B. Hoeber, England (1918)

Glove Mouse: A Gesture Controlled Wireless Computer Mouse

Vivek Kumar Singh, Abhijeet Sarkar, Vidit Vijay Chaturvedi, Md. Sayeed Ansari, and Tushar S. Narsimpur

1 Introduction

Research in the field of Human Computer Interaction (HCI) seeks to improve the usability of computer interfaces. A mouse is one of the most popular computer interfaces. Its technology has evolved with more responsive, wireless and power efficient enhancements. But it has not seen significant innovation in the fundamental way a user interacts with it. A conventional mouse requires a flat and smooth surface to operate, which cannot satisfy many new usage scenarios. The surface requirement becomes a demanding restriction for the user. For example, a speaker wants to move around and click on different objects in his slide, or a smart TV user wants to control his TV from any part of a room [1]. To achieve this, the computer mouse should work without a surface and implement accurate motion tracking.

Carpal Tunnel Syndrome (CTS) is a musculoskeletal occupational disease caused due to the prolonged usage of desk mouse [2]. It happens because of strain on the median nerve in the wrist (Fig. 1). It causes numbness, tingling, or weakness in the user's hand. To reduce CTS, various ergonomic devices such as gel mouse pads and vertical mouse are developed. However, studies have concluded that none of these ergonomic devices can reduce the strain [3].

The Glove Mouse is an innovative solution to all these issues. It gives the users a next-generation experience, as they can operate their mouse pointer in mid-air and without the need of any surfaces or wires. Its mid-air operation also reduces the strain in the wrist area. Besides, it is a better option for 3D sketching and gaming, and for a person having motion disability [4, 5].

V. K. Singh (✉) · A. Sarkar · V. V. Chaturvedi · Md. S. Ansari · T. S. Narsimpur
Ramaiah Institute of Technology, Bengaluru, India

T. S. Narsimpur
e-mail: vks3111996@gmail.com

S. K V and K. Guruprasad Rao (eds.), *Smart Sensors Measurements and Instrumentation*, Lecture Notes in Electrical Engineering 750,
https://doi.org/10.1007/978-981-16-0336-5_40

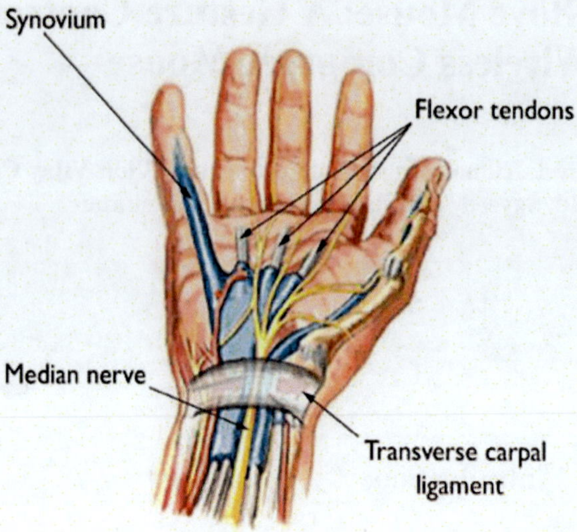

Fig. 1 Carpal Tunnel Syndrome is caused by pressure on the median nerve. *Source* OrthoInfo, Carpal Tunnel Syndrome

2 Working Principle

The Glove Mouse is based on the technology of MEMS accelerometer. MEMS technology has led to the miniaturization of accelerometers, which makes them suitable for the design of wearable devices that require motion tracking [6]. Figure 2 shows the block diagram of the proposed system. It divides the system into two principal parts: a glove unit and a base unit [6–8]. On the glove unit, MMA7361 sensor reads the acceleration along the three axes [9] and sends it to ATmega16 micro-controller. The state of metallic contacts and switches is also fed to the micro-controller. ATmega16 processes all the data and forwards it to HC-05 Bluetooth transmitter that sends the data wirelessly to the receiver. The receiver forwards the data to Arduino Pro Micro

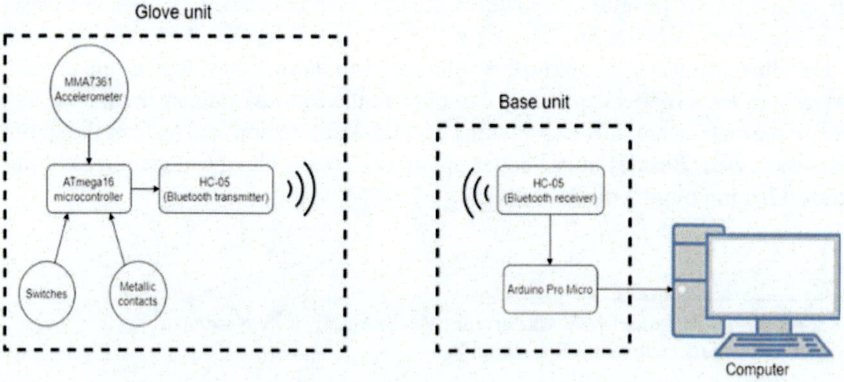

Fig. 2 Block diagram of complete system

that converts it into HID format and sends it to the computer through serial port [10]. Mouse pointer moves according to the data received by the computer.

The whole circuitry of glove unit is mounted on a glove and it has negligible weight. The next section presents a brief explanation of all the hardware and software components of the system.

3 Methodology

3.1 *Hardware*

The glove unit constitutes of accelerometer, micro-controller, transmitter, metallic contacts, switches, and power supply. The base unit constitutes of receiver and micro-controller. The base unit is powered by the serial port of the computer. These components are briefly described below.

MEMS Accelerometer: In the present work, a 3-Axis Accelerometer module by DFRobot is used to detect user's hand motions (Fig. 3). The module carries a Freescale MMA7361 micro-capacitive acceleration sensor (MEMS) [9]. This sensor can measure static (earth gravity) or dynamic acceleration in all three axes. The output of the sensor is analog voltage for each of the three axes. The output voltage is proportional to acceleration along the axes [11].

Glove unit micro-controller: An ATmega16 micro-controller by Atmel is used. It is a low-power CMOS 8-Bit micro-controller based on the AVR enhanced RISC architecture. It has 32 programmable I/o lines, 32 × 8 general purpose working registers,

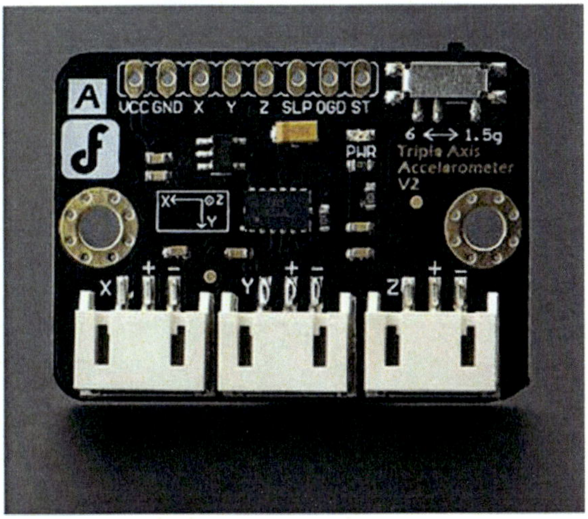

Fig. 3 3-Axis Accelerometer module (*Source* DFRobot, MMA7361)

10-bit ADC, and 16 Kbytes of In-System Self-programmable Flash program memory [12]. This micro-controller is a cost-effective solution in the present requirement.

Transmitter: Two HC-05 Bluetooth modules (make Itead) are used as transmitter and receiver. HC-05 is designed for transparent wireless serial communication [13]. Serial communication sends data one bit at a time over a communication channel or computer bus. It's easy-to-use Bluetooth Serial Port Protocol [13] makes it simple to interface with ATmega16 micro-controller.

Metallic contacts: We used metallic contacts for obtaining mouse clicks and for scrolling [14]. Six metallic contacts were installed on different fingers of a glove as shown in Fig. 4. Metallic contacts on thumb, index, middle, and ring finger function as ground, left mouse button (LMB), middle mouse button (MMB), and right mouse button (RMB) respectively [14]. Metallic contact on little finger is the Scroll Enable contact which is used exclusively for scrolling. On index finger, we have another contact called the Move Enable contact which enables movement of pointer on the screen. Any metallic contact is triggered only when it is touched to the ground contact on the thumb.

Switches: The present design also uses four DIP switches to increase the functionality of the device. These switches are sensitivity, x-inversion, y-inversion, and Rapid fire [8]. The sensitivity switch increases the speed of mouse pointer. This makes the device flexible enough for presentations on big screens and for day-to-day operations on the computer screen. The x-inversion and y-inversion switches invert axes of mouse

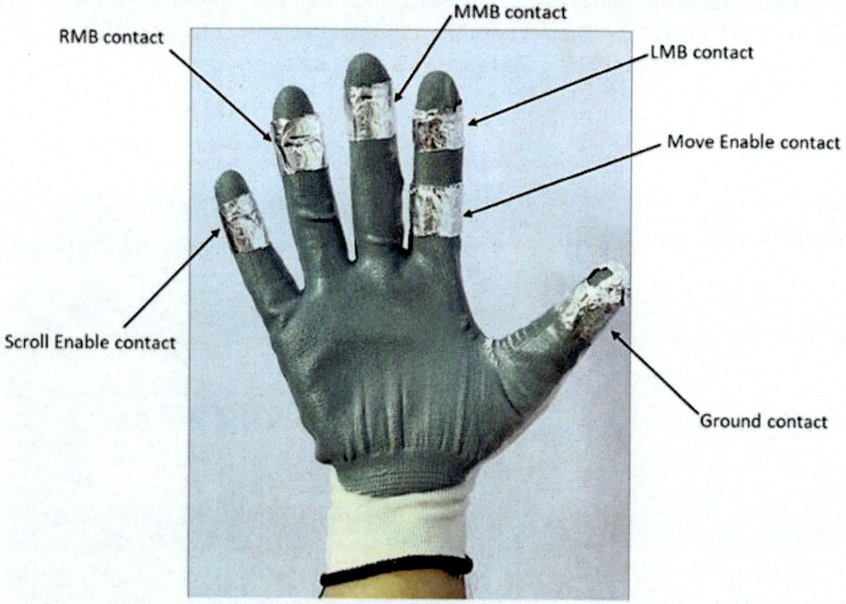

Fig. 4 Metallic contacts on glove (front side)

movement that gives freedom to users with different directional preferences. Rapid Fire switch is used to enter Rapid Fire mode. Many computer video games involve a lot of rapid mouse clicking. With the Rapid Fire switch ON, it triggers the LMB at regular intervals, thus eliminating the need for rapid clicking of LMB [8].

Power supply: The power supply unit provides power to all components within their working range. The components MMA7361, Atmega16 and HC-05 Blue-tooth module, work in the voltage range of 2.2–3.6 V, 4.5–5.5 V and 3.6–6 V respectively. A single 9 V battery is used and its output voltage is stepped down using two voltage regulators. The first voltage regulator, L7805CV, steps down 12–5 V [15]. The second voltage regulator, AMS1117, steps down 5–3.3 V [16].

Receiver: As mentioned earlier, a second HC-05 Bluetooth module acts as the receiver. It receives the data and forwards it to the base unit micro-controller.

Base unit micro-controller: We used Arduino Pro Micro (Fig. 5) for converting the data into Human Interface Device (HID) format. The Pro Micro is an Arduino-compatible micro-controller board based on the ATmega32U4 micro-controller. ATmega32U4 complies with Universal Serial Bus Specification Rev 2.0 [17]. Pro Micro is used to simulate a mouse, keyboard, or other HID-class USB device [10]. We have used it to simulate a mouse. It can operate on Windows, Mac and Linux operating systems as all of them meet the USB specifications [6].

Fig. 5 Arduino Pro Micro (*Source* Sparkfun Pro Micro—5 V/16 MHz)

3.2 Software

In the present work, we use two major software: Atmel Studio 7 for developing glove unit code and Arduino IDE for developing base unit code.

Atmel Studio 7: The code for glove unit was written as GCC C Executable Project in Atmel Studio 7. Its integrated development platform provides a seamless and easy-to-use environment to write, build and debug codes written in C/C ++ or assembly language [18]. The glove unit code contains an Interrupt Service Routine (ISR) which was written to read the accelerometer data and transmit it every 4 ms. This time interval is barely noticeable in actual usage.

Arduino IDE: We wrote the code for base unit as GCC C Executable Project in Arduino IDE. It is a platform to write and upload programs to Arduino compatible boards (here Arduino Pro Micro) [19]. The serial data received by Arduino Pro Micro was displayed on the terminal screen using Serial monitor. The Serial plotter was also used to plot the data received. This gave an advantage of tracking and troubleshooting the device operation.

3.3 Mouse Pointer Movement Calculation

MMA7361 sensor gives acceleration along the three axes due to the user's hand movements. These values need to be scaled for appropriate movement of mouse pointer on the computer screen. Equations (1) and (2) were used to achieve this task [8].

$$\text{if}(|\text{sample}_{\text{ADC}}| < \text{threshold})\text{sample}_{\text{ADC}} = 0 \tag{1}$$

$$\text{cursor move} = \left(\frac{\text{sample}_{\text{ADC}}}{2^{\text{sensitivity}}}\right) * \left(\frac{\text{sample}_{\text{ADC}} - \text{sample}_{\text{ADC}} + 8}{8}\right) * (\text{invert flag}) \tag{2}$$

Here, $\text{sample}_{\text{ADC}}$ is the present set (x, y and z axes) of accelerometer readings with offset eliminated and $\text{sample}_{\text{ADC}_{\text{prev}}}$ is the previous set of accelerometer readings with offset eliminated. We used the parameters threshold, sensitivity and invert flag to account for the drift, sensitivity and axis inversion in the movement of mouse pointer respectively. The parameter cursor move stores the scaled value for the movement of mouse pointer.

Equation (1) was used to eliminate drift in the movement of mouse pointer. The equation dictates that if the magnitude of the present set of accelerometer readings is less than the threshold value (determined experimentally), the present set of readings is set to zero. It in turn sets the cursor move to zero and there is no movement of mouse pointer on the screen. Hence, drift in mouse pointer due to small vibrations of hand, generated during triggering of metallic contacts, was eliminated.

Equation (2) was used to scale the accelerometer readings to mouse pointer movement. The first term on the right side of the equation reduces sample_{ADC} by a factor of $2^{\text{sensitivity}}$ where the value of sensitivity depends on the present state of sensitivity switch (controlled by the user). The second term calculates the relative displacement of hand by subtracting the present set and previous set of accelerometer readings and then scaling it down. The last term, invert flag is used to change the sign of cursor move depending on the present state of x and y axis inversion switches (controlled by user). We multiply these three terms to get the final value of mouse pointer movement. It is stored in cursor move and transmitted to the base station by HC-05 Bluetooth module.

3.4 Device Testing

The glove unit was fabricated for right hand. It was worn on the right hand, its circuit board was strapped on right arm using a belt-buckle mechanism, and the base unit was connected to serial port of the computer as shown in Fig. 6. The Glove Mouse was tested with data transmission set at 57,600 baud rate with one stop bit and no parity. The stepwise testing procedure is described below.

(i) The glove unit was turned ON using the power switch and the Bluetooth modules got paired.
(ii) At this stage, movement of hand did not result in any pointer movement since the Move Enable contact was not triggered. However, clicking and scrolling operations could be performed by triggering their respective contacts.
(iii) To move the mouse pointer, the user laid out his hand forward with palm down and triggered the Move Enable contact. Subsequently mouse pointer on the computer screen moved according to the user's hand movement:

- Hand was bent upwards—pointer moved up

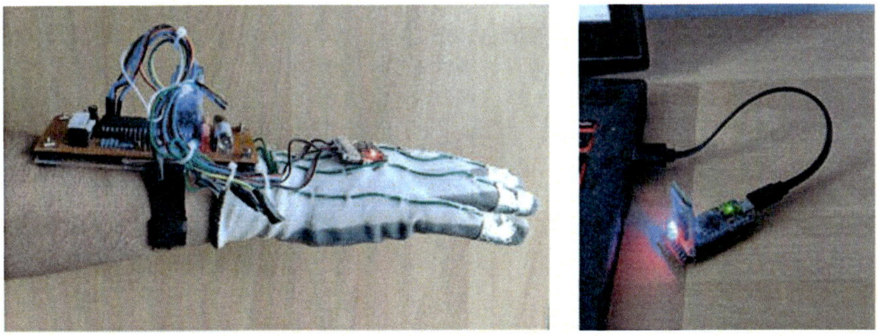

Fig. 6 The Glove unit worn on the right hand (left) and the Base unit connected to the computer (right)

- Hand was bent downwards—pointer moved down
- Hand was rolled clockwise—pointer moved right
- Hand was rolled anticlockwise—pointer moved left

(iv) To move the mouse pointer diagonally, the user performed a two-step hand motion: either the hand was moved vertically (up/down) and then horizontally (left/right), or vice versa.

(v) To perform exclusive scrolling, the user triggered the Scroll Enable contact. It overrides the movement of mouse pointer and hence there is no motion of mouse pointer. Alternatively, the MMB was also used for scrolling. In either case, scrolling was performed by only upwards and downwards motion of the hand.

(vi) By default, the sensitivity was set at normal but could be increased through the Sensitivity switch.

(vii) The x-inversion and y-inversion switches inverted the axes as per directional preferences.

(viii) The Rapid Fire switch performed the repeated clicking of LMB. It is a useful feature for playing video games.

(ix) The power switch was pressed again to turn OFF the device, and the base unit was disconnected from the serial port. This completed the testing of Glove Mouse.

4 Results and Discussion

Figure 7 shows a snapshot of the data received by the base unit on Serial monitor. The output on the screen was divided into four columns 'X', 'Y', 'W' and 'C' which represent the data received for x-axis pointer movement, y-axis pointer movement, scroll movement and mouse click, respectively.

The value 255 received alternately is the synchronization value that ensures the correct mapping of data to their respective columns (or variables). As we can see, columns 'X' and 'Y' are showing some values that indicates the pointer is moving on the computer screen. Column 'W' shows the value zero that means scrolling is disabled. Column 'C' also shows zero, because no clicks have been performed.

Figure 8 shows the snapshot of data received by the base unit when scrolling was enabled and mouse pointer movement was disabled. Here, columns 'X' and 'Y' are showing the value zero, since no mouse pointer movement is happening. Column 'W' is showing values 1 and −1 for upward and downward scrolling, respectively. The change in the set of values from box 1 to box 2 shows that scrolling changed direction from downwards to upwards. Column 'C' again shows zero, because no clicks have been performed. The data in tabular form made it is easy to observe the response of the device to different hand movements and operations.

The snapshot of the data plotted on Serial Plotter is presented in Fig. 9. Here x-axis and y-axis represent the data point and data value respectively. The blue line graph

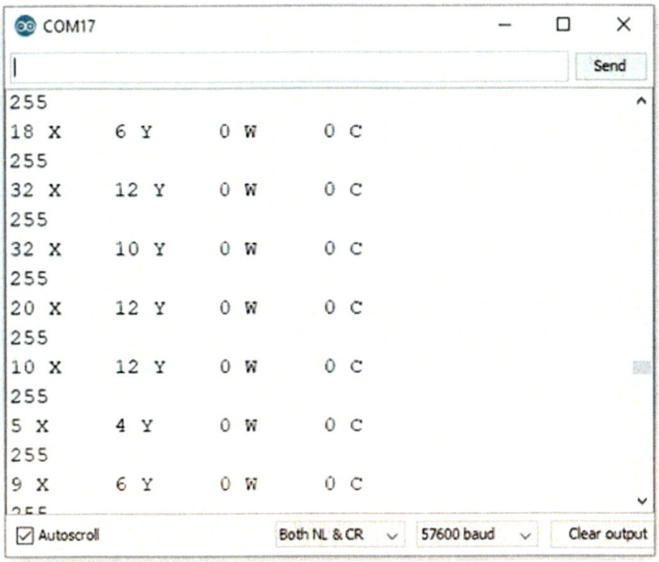

Fig. 7 Snapshot of Serial Monitor with movement of mouse pointer enabled

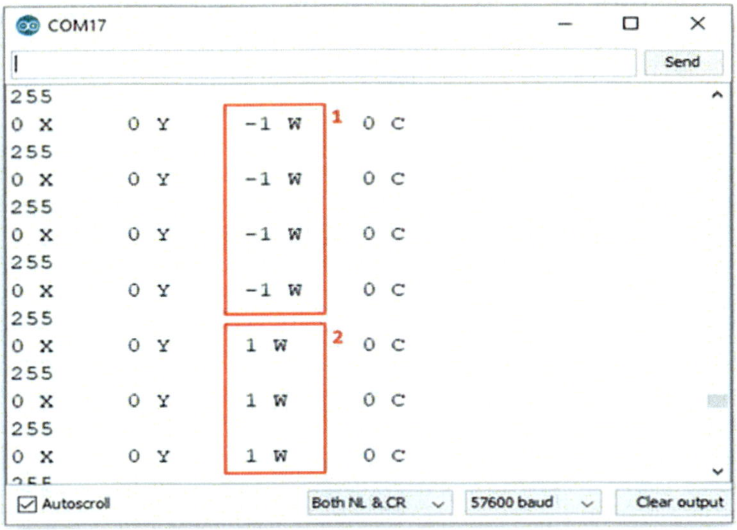

Fig. 8 Snapshot of Serial Monitor with scrolling enabled

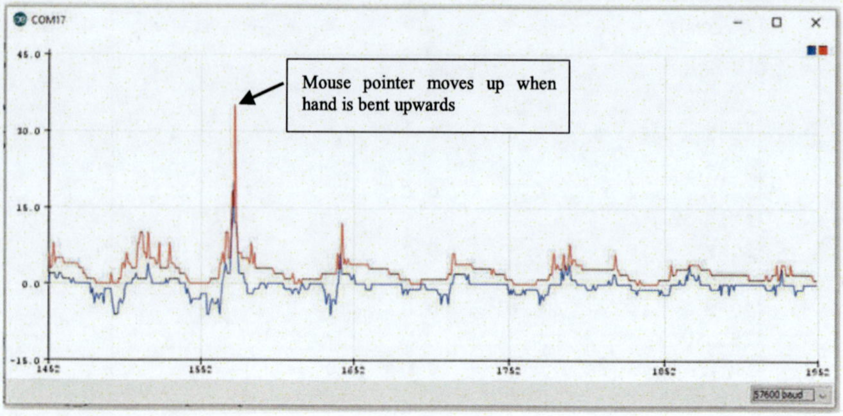

Fig. 9 Snapshot of Serial Plotter

shows the left-right pointer movement, and the red line graph shows the up-down pointer movement.

The upward movement of the mouse pointer on the computer screen resulted in a peak in the red line graph around 1577. This movement of pointer was generated by bending the hand upwards. We observed similar peaks for movement of mouse pointer in other directions. As seen from the graph, at all other points, the data value of the both the line graphs are close to zero that shows the mouse pointer was almost stationary. This graphical representation helped us to understand the pointer movement under different settings such as inversion of x-axis axis.

The equations used for calculating mouse pointer movement (in Sect. 3.3) resulted in an accurate mapping of the mouse pointer to the hand movements of the user. We performed several tests to optimize different parameters like $threshold$ and $sensitivity$ to achieve best device operation. There is a slight delay, of the order of few milliseconds, during data transmission. However, it is barely noticeable in actual usage. The accuracy of the device is good as seen by the results generated. Accelerometer placed on the glove measures tilt very well. The metallic contacts and switches eliminate debouncing, and are very accurate. But we lose some accuracy due to dropped packets between the transmission of data from glove unit to base unit. This can be eliminated in future modifications of the device. The compactness of the device can be increased using ASIC design technologies.

There is a small learning curve to controlling the mouse pointer using Glove Mouse, and it comes intuitively with use. The lightweight design of glove unit allows the user to move, type or work while wearing it. The cost of Glove Mouse is kept in check with the standard wireless mouse available in the market.

5 Conclusion

In this paper, we design a wireless wearable computer mouse named Glove Mouse which works on hand gestures of the user without the need of a surface. Hand gestures are sensed by an accelerometer sensor and its output goes to a micro-controller for processing. It is then transmitted wirelessly to the computer via a pair of Bluetooth modules.

The x-inversion, y-inversion and sensitivity settings extends usability of the device to many users. The Rapid Fire mode is an exclusive feature used for playing video games. Its platform independent design makes it flexible for use across common platforms like Windows, Mac and Linux. With little improvement in aesthetics, Glove Mouse will become a lucrative alternative to replace the conventional mouse.

Acknowledgements We would like to express our gratitude to our project guide Sri. Tushar S. Narsimpur who guided us throughout the project by providing all the necessary ideas, information and knowledge for developing a functioning model. Our special thanks to Dr. Pradipkumar Dixit, HOD, EEE Department for his constant guidance and encouragement in helping us complete our project successfully. Last but not the least, our thanks and appreciation to our parents and friends for their constant encouragement and inspiration.

References

1. Yun, S., Chen, Y.C., Qiu, L.: Turning a mobile device into a mouse in the air. In: Proceedings of the 13th annual international conference on mobile systems, applications, and services, pp. 15–29. Association for Computing Machinery, New York (2015)
2. Giersiepen, K., Spallek, M.: Carpal tunnel syndrome as an occupational disease. In: Deutsches arzteblatt international, pp. 238–242 (2011)
3. Schmid, A.B., Kubler, P.A., Johnston, V., Coppieters, M.W.: A vertical mouse and ergonomic mouse pads alter wrist position but do not reduce carpal tunnel pressure in patients with carpal tunnel syndrome. In: Applied ergonomics, pp. 151–156. Elsevier (2015)
4. Navale, S.P., Tatar, P.S., Teli, P.R., Salunke, D.B.: Electronic mouse for dis-abled people. In: Int. J. Comput. Eng. Res. Trends, pp. 251–254 (2016)
5. Tonape, S., Mane, A., Patil, D., Umate, S., Khan, P.G.M.: Hand action recognition using accelerometer. Int. J. Innov. Res. Comput. Commun. Eng., 3701–3707 (2015)
6. Barot, T., Karhadkar, S.: Development of platform independent hand gesture controlled wearable mouse: a comparative study using accelerometer and gyroscope. In: 3rd international conference for convergence in technology, pp. 1–5. IEEE, Pune (2018)
7. Tarannum, S., Munna, M.S.: Gesture based wireless pc control with gyroscope and accelerometer. In: 1st international conference on advanced information and communication technology. ResearchGate (2016)
8. Glove Mouse: A wireless, wearable computer pointing device. https://people.ece.cornell.edu/land/courses/ece4760/FinalProjects/f2012/as986_hl525/as986_hl525/index.htm
9. ± 1.5 g, $\pm$ 6 g Three axis low-g micromachined accelerometer. https://www.nxp.com/docs/en/data-sheet/MMA7361LC.pdf
10. Turn your ProMicro into a USB Keyboard/Mouse. https://www.sparkfun.com/tutorials/337
11. Ahmed, S., Zubair, M.A., Shaik, I.B.: Accelerometer based wireless air mouse using Arduino micro-controller board. In: Global conference on communication technologies, pp. 348–350. IEEE, Thuckalay (2015)

12. ATmega16—Microchip technology. http://ww1.microchip.com/downloads/en/devicedoc/doc2466.pdf
13. HC-05—Bluetooth to Serial Port Module. https://components101.com/sites/default/files/component_datasheet/HC-05%20Datasheet.pdf
14. Berlia, R., Santosh, P.: Mouse brace: a convenient computer mouse using accelerometer, flex sensors and microcontroller. In: International conference on contemporary computing and informatics, pp. 558–561. IEEE, Mysore (2014)
15. Positive voltage regulators. https://cdn-shop.adafruit.com/product-files/2164/L7805CV.pdf
16. Advanced Monolithic Systems. http://www.advancedmonolithic.com/pdf/ds1117.pdf
17. ATmega16U4/ATmega32U4 DATASHEET. http://ww1.microchip.com/downloads/en/devicedoc/atmel-7766-8-bit-avr-atmega16u4-32u4_datasheet.pdf
18. Atmel Studio 7—Microchip technology. https://www.microchip.com/mplab/avr-support/atmel-studio-7
19. Arduino IDE. https://en.wikipedia.org/wiki/Arduino_IDE

Design and Simulation of a Sliding Mode-Based Control Design for a Quadrotor UAV

Suresh Kumar Suraj, Vadakkekkara Itty George, I. Thirunavukkarasu, and Tinu Valsa Paul

1 Introduction

The latest technological advancements in quadrotor systems has led to design of autonomous guidance systems and navigation systems. Quad rotors are basically designed to operate under high mobility, rapid manoeuvring and greater agility. These are envisioned mainly for surveillance, mapping, search and rescue operation and aviation research. A quad-rotor is categorized as a multi-rotor helicopter that uses four rotors to manoeuvre and propel itself. It can be stated that quad-rotors have been a prevalent research topic in past decades due to their special inflight abilities like monitoring traffic information, reconnaissance, area mapping and more. The key requirement for the design of quad-rotor are safety, reliability and structure simplicity. The analysis of quad-rotor mainly involves owing to non-linear components. The position of these rotors is controlled by a remote-control system. A major challenge in quad-rotor operation is that it demands strong coupling between its translational and rotational dynamic subsystems while tracking and calculation. Some parameters associated with quadrotor dynamic model such as inertial forces and moments, and aerodynamic coefficients cannot be obtained precisely. The benefits of Sliding Mode Control are used to design high order non-linear system, which in turn reduces the insensitiveness to disturbance. The main contribution of this paper is to design a nonlinear controller using Sliding Mode Controller with the requisite disturbance and at the same time achieving a desired orientation along the respective coordinate axes [1].

S. K. Suraj · V. I. George (✉) · I. Thirunavukkarasu · T. Valsa Paul
Instrumentation & Control Engineering, Manipal Institute of Technology, Manipal, Karnataka, India
e-mail: vi.george@manipal.edu

I. Thirunavukkarasu
e-mail: it.arasu@manipal.edu

S. K V and K. Guruprasad Rao (eds.), *Smart Sensors Measurements and Instrumentation*, Lecture Notes in Electrical Engineering 750,
https://doi.org/10.1007/978-981-16-0336-5_41

1.1 Quad-Rotor Configuration

The quad-rotor is a four motors multicopter mounted on four arms each with 90 degrees apart. The four propellers are configured in cross shape as shown in Fig. 1; the propellers (1, 3) and (2, 4) are in two pairs that rotate in the opposite direction, resulting in zero resultant torque at the quadrotor center of mass at the equilibrium position (hovering). Lift and maneuvers can be attained by variation in the rotor-speeds. Varying the rotor velocities for propellers (2, 4) produces roll rotation while varying the rotor velocities for propellers (1, 3) produces pitch rotation and also varying the rotor velocities of propellers (1, 3) and (2, 4) respectively produces yaw motion.

2 Quad-Rotor Dynamic Modeling

Initially a performance index of the carefully chosen mathematical model is defined. This mathematical model is further used for analysis and design of the controller. This system is later developed by the analysis of the equation of motion of quad-rotor taking into account the following assumptions [2]:

- The inertial properties of the quadrotor vary in time.
- The quad-rotor airframe is rigid and symmetric, and also the propellers are considered to be thin, flat spinning discs.

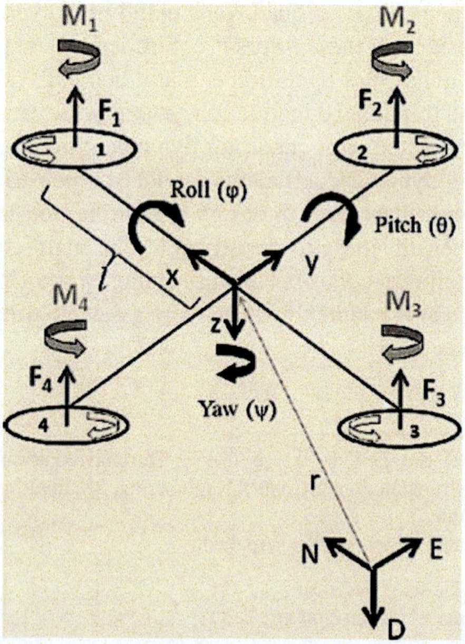

Fig. 1 Quadrotor reference frames

- Pitch and roll angles are small.
- The axial aerodynamic forces acting are proportional to the propeller's speed squared.

Here, change of rotor angular velocity is considered as input and the vehicle position in the x, y and z direction and the angle of pitch (θ), roll (Ψ), and yaw (Φ) act as output of the system. The kinematic and dynamic models of the quadrotor system are derived with the help of above assumptions, based Newtonian and Eulerian formulations.

The initial step in deriving the model of the quadrotor system is to establish the reference coordinate frames for the measurements. The reference frames used are the earth fixed reference frame with north, east and down as axes and the body frame with the quadrotor axes [3]. The distance from the earth fixed frame to the body frame represents the absolute distance of the quadrotor center of mass to the origin of the NED frame, denoted by: $r = \begin{bmatrix} x \; y \; z \end{bmatrix}^{\mathrm{T}}$

$$R = \begin{bmatrix} \cos\theta\cos\psi & \sin\phi\sin\theta\cos\psi & \cos\phi\sin\theta\cos\psi + \sin\phi\sin\psi \\ \cos\theta\sin\psi & \sin\phi\sin\theta\sin\psi + \cos\phi\sin\psi & \cos\phi\sin\theta\sin\psi + \sin\theta\cos\psi \\ -\sin\theta & \sin\phi\cos\theta & \cos\phi\cos\theta \end{bmatrix} \quad (1)$$

The direction cosine matrix R from Eq. (1) is required to derive the quadrotor mathematical model and it is significant because some of the system states are obtained in the body frame (like non-gravitational forces and moments) while some of them are obtained in the inertial frame (like forces due to gravity). Euler angle rates in the inertial frame is given by: $\dot{\eta} = \begin{bmatrix} \dot{\phi} \; \dot{\theta} \; \dot{\psi} \end{bmatrix}^{\mathrm{T}}$ and the corresponding angular body rates is given by: $\omega = \begin{bmatrix} p \; q \; r \end{bmatrix}^{\mathrm{T}}$

The quadrotor dynamic model, divided into two, constitutes: translational subsystem represented by the x, y, and z positions, and rotational subsystem represented by the roll, pitch, and yaw angles [4]. The equations of rotational motion are obtained using the Newtonian and Eulerian formulation in the body frame as given below in Eq. (2):

$$J\dot{\omega} + \omega \times J\omega + M_G = M_B \quad (2)$$

where J is the diagonal inertia matrix of the quadrotor, M_{G} is the gyroscopic moments due to rotors' inertia J_{r}, M_{B} is the moments acting in the body reference frame, and $J\dot{\omega}$ and $\omega \times J\omega$ represent the time derivative of angular momentum in the body frame. M_{G} can be written as $\begin{bmatrix} 0 \; 0 \; J_r\Omega_r \end{bmatrix}^{\mathrm{T}}$ where $\Omega_r = -\Omega_1 + \Omega_2 - \Omega_3 + \Omega_4$ is the rotors' relative speed calculated based on the configuration of the quadrotor, i.e., in this case, the rotors 2 and 4 rotate is clockwise and is considered positive, and the rotation of rotors 1 and 3 is counter clockwise and is considered negative; hence there rotational speeds assume the corresponding signs [5].

State space representation of the quadrotor mathematical model can reduce the complexity and easier to manipulate. The state vector, which defines the position and orientation in space and its angular and linear velocities and mapping the degrees of freedom, gives the following matrix (3) and (4):

$$X = \left[x_1\ x_2\ x_3\ x_4\ x_5\ x_6\ x_7\ x_8\ x_9\ x_{10}\ x_{11}\ x_{12}\right]^{\mathrm{T}} \tag{3}$$

$$X = \left[\phi\ \dot{\phi}\ \theta\ \dot{\theta}\ \psi\ \dot{\psi}\ x\ \dot{x}\ y\ \dot{y}\ z\ \dot{z}\right]^{\mathrm{T}} \tag{4}$$

The input vector for the system, defined by the input forces U_1, U_2, U_3, U_4, is given by Eq. (5)

$$\begin{bmatrix} U_1 \\ U_2 \\ U_3 \\ U_4 \end{bmatrix} = \begin{bmatrix} K_{\mathrm{f}} & K_{\mathrm{f}} & K_{\mathrm{f}} & K_{\mathrm{f}} \\ 0 & -K_{\mathrm{f}} & 0 & K_{\mathrm{f}} \\ K_{\mathrm{f}} & 0 & -K_{\mathrm{f}} & 0 \\ -\mathrm{K_M} & \mathrm{K_M} & -\mathrm{K_M} & \mathrm{K_M} \end{bmatrix} \begin{bmatrix} \Omega_1^2 \\ \Omega_2^2 \\ \Omega_3^2 \\ \Omega_4^2 \end{bmatrix} \tag{5}$$

where K_{f} and K_{m} are thrust constant and moment contant, respectively. The rotational and translational subsystems can be written as given in Eqs. (6, 7, 8, 9, 10, 11):

$$\ddot{\phi} = \frac{1}{I_{zz}}U_4 + \frac{J_r}{I_{xx}}\dot{\theta}\Omega_r + \frac{I_{yy}}{I_{xx}}\dot{\psi}\dot{\theta} - \frac{I_{zz}}{I_{xx}}\dot{\theta}\Omega_r \tag{6}$$

$$\ddot{\theta} = \frac{l}{I_{yy}}U_3 - \frac{J_r}{I_{yy}}\dot{\phi}\Omega_r + \frac{I_{yy}}{I_{yy}}\dot{\phi}\dot{\psi} - \frac{I_{zz}}{I_{yy}}\dot{\psi}\dot{\phi} \tag{7}$$

$$\ddot{\psi} = \frac{l}{I_{zz}}U_4 + \frac{I_{xx}}{I_{zz}}\dot{\theta}\dot{\phi} - \frac{I_{yy}}{I_{zz}}\dot{\phi}\dot{\theta} \tag{8}$$

$$\ddot{x} = -\frac{U_1}{m}(\cos\phi\sin\psi\cos\theta + \sin\phi\sin\psi) \tag{9}$$

$$\ddot{y} = -\frac{U_1}{m}(\cos\phi\sin\psi\sin\theta - \cos\psi\sin\phi) \tag{10}$$

$$\ddot{z} = g - \frac{U_1}{m}(\cos\phi\cos\theta) \tag{11}$$

From the above equations, the translational subsystem can be considered under-actuated as it has the translational and rotational state variables as deciding factors [6, 7]. Using the above equations of translational and rotational motion, we can obtain the final state space model of the system, as given below in Eq. (12):

$$f(X, U) = \begin{bmatrix} x_2 \\ x_4x_6a_1 - x_4\Omega_r a_2 + b_1U_2 \\ x_4 \\ x_2x_6a_3 + x_2\Omega_r a_4 + b_2U_3 \\ x_6 \\ x_2x_4a_5 + b_3U_4 \\ x_8 \\ \frac{U_1}{m}(\sin x_1 \cos x_5 - \cos x_1 \sin x_3 \sin x_5) \\ x_{10} \\ -\frac{U_1}{m}(\sin x_1 \sin x_5 - \cos x_1 \sin x_3 \cos x_5) \\ x_{12} \\ g - \frac{U_1}{m}(\cos x_1 \cos x_3) \end{bmatrix} \quad (12)$$

3 Controller Design

3.1 Sliding Mode Controller

The sliding mode control is categorized under Variable Structure Control. It uses the control law to manipulate the system inputs to force the system trajectories on to the sliding hypersurface with the help of a switching mechanism. Preliminary design of the controller involves design of the sliding surface and the design of the control law. The sliding surface is a subset of the system state space, which can be represented as a single line or as convergence or divergence of several surfaces. The sliding surface is designed such that it addresses all the mission constraints and requirements. It designed such that the system adapts in the desired way [8].

The attitude control for the system is achieved by separately controlling each rotational degree of freedom, i.e., roll, pitch and yaw [8]. The roll control for the quadrotor is attained by tracking the system performance and comparing it with the reference sliding surface, which gives the error, $e = \phi_d - \phi$. This error is used to derive the Lyapunov function, used to determine an exponential reaching law, $\dot{s} = -k_1\mathrm{sgn}(s) - k_2 s$. The obtained reaching law is used to derive the control law for input vector U_2, which is given by Eq. (13),

$$U_2 = \frac{1}{b_1}\left[k_1\mathrm{sgn}(s) + k_2 s + c_1\left(\dot{\phi}_d - \dot{\phi}\right) + \ddot{\phi}_d - a_2\dot{\theta}\Omega_r - a_1\dot{\theta}\dot{\psi}\right] \quad (13)$$

The pitch and yaw controls are obtained similar to that of the roll control as given below in Eqs. (14) and (15),

$$U_3 = \frac{1}{b_2}\left[k_1\mathrm{sgn}(s) + k_2 s + c_1\left(\dot{\theta}_d - \dot{\theta}\right) + \ddot{\theta}_d - a_4\dot{\phi}\Omega_r - a_3\dot{\phi}\dot{\psi}\right] \quad (14)$$

$$U_4 = \frac{1}{b_3}\left[k_1 \text{sgn}(s) + k_2 s + c_1\left(\dot{\psi}_d - \dot{\psi}\right) + \ddot{\psi}_d - a_5\dot{\phi}\dot{\theta}\right] \quad (15)$$

The pitch and yaw are controlled by the control inputs U_3 and U_4 respectively. A similar pattern is followed for the sliding mode control design for the altitude control is given by Eq. (16),

$$U_1 = \frac{m}{\cos\phi\cos\theta}\left[k_1 \text{sgn}(s) + k_2 s + c_1(\dot{z} - \dot{z}_d) + g - \ddot{z}_d\right] \quad (16)$$

3.2 *Proportional Integral Differential Controller*

The linear control performance of the sliding mode controller is compared with that of a modified PID control algorithm tuned with a genetic algorithm to optimize the controller. The PID controller as such is very simple to design and modify as per the system requirements. The PID controller is designed such that it produces the quadrotor control inputs. The main objective of the tuning algorithm is to determine the control gains for an optimal settling time for the quadrotor control inputs. The PID control algorithm is used in the same way as in the case of the SMC control algorithm. The general form of the control law is as given below in Eq. (17),

$$U_j = k_P\left(x_i - x_{i,d}\right) + k_d\left(\dot{x}_i - \dot{x}_{i,d}\right) + k_I \int \left(x_i - x_{i,d}\right)dt \quad (17)$$

where k_P is the proportional gain, k_d is the derivative gain, k_i is the integral gain, x_i is the state vector to be manipulated (altitude, roll, pitch, or yaw), $x_{i,d}$ is the desired value of the state vector.

4 Simulation and Results

4.1 *Proportional Integral Differential Controller*

The GA generated the control design parameters for the altitude PID control algorithm for desired altitudes. The PID controller was reduced to a PD controller by maintaining the integral gain at zero, due to zero steady state error. The differential and proportional gains generated by GA were $k_d = 4.9$ and $k_P = 1.9$ respectively. The GA generated gains produced a settling time of 0.9 s and an overshoot of 1.7%. Similarly, the attitude control gains were also generated by the GA. Fig. 2 shows the responses of the system according to the design parameters generated by the GA. The pitch and roll response of the system are identical, and the GA produce the

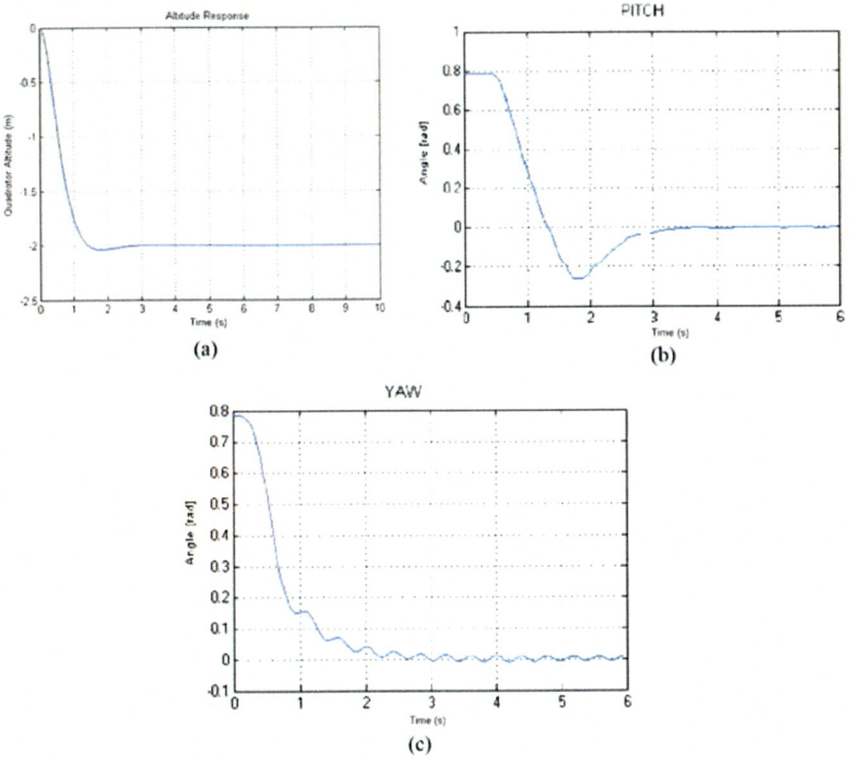

Fig. 2 System response for the PID controller: **a** altitude response, **b** pitch response, **c** yaw response

same design parameters in both cases. The PID controller parameters for the system were tuned by a GA for altitude control and orientation control as mentioned below (Fig. 3).

Table 1 shows the performances characteristics for the PID controller.

4.2 Sliding Mode Controller

As in the case of PID controller, the design parameters for the SMCs is generated by the GA to obtain the lowest settling time. The design parameters produced by the GA for the system is as shown in Table 2 and it also shows the settling time and overshoot of the system. The responses of the system for the generated control parameters is shown graphically in Fig. 4. As in the case of the PID controllers, the pitch and roll response of the system are identical and the GA produce the same design parameters in both cases.

Fig. 4 clearly displays the SMC controller chattering effects which is common in SMCs due to their switching mechanism. The signum function in the control law

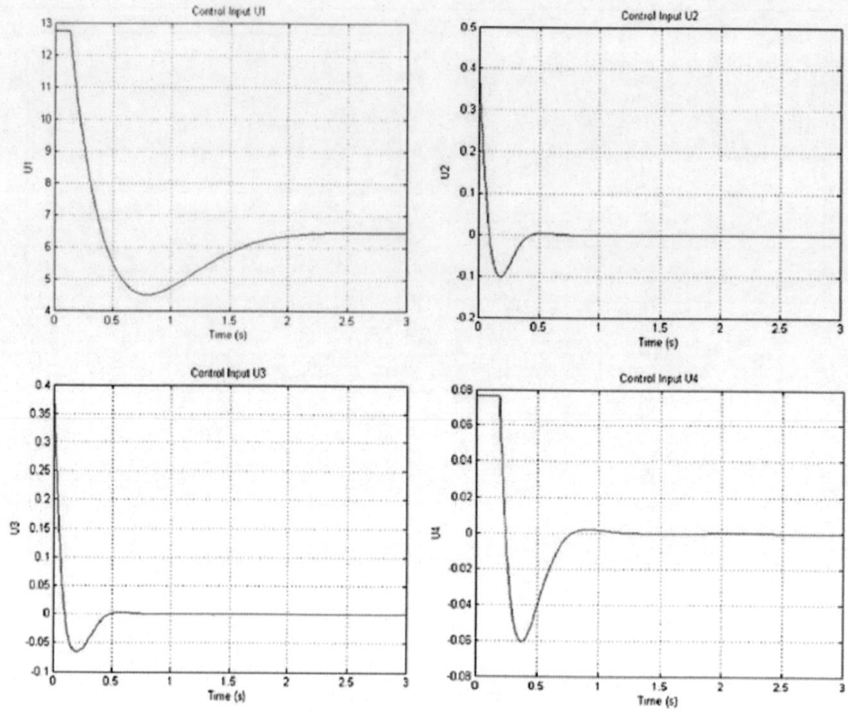

Fig. 3 Control output for PID controller

Table 1 PID controller parameters and performannce characteristics

	Desired value	k_P	k_d	Settling time (s)	Overshoot (%)
Altitude (z)	2 m	4.9	1.9	0.9	1.7
Attitude (ϕ and θ)	1°	4.1	0.3	0.42	2.1
Heading (ψ)	1°	3.9	0.5	0.37	1.82

Table 2 Sliding mode controller parameters and performannce characteristics with chattering effects

	Desired value	c	k_1	k_2	Settling time (s)	Overshoot (%)
Altitude (z)	2 m	6.98	5.2	1.3	1.57	1.4
Attitude (ϕ and θ)	5°	4.68	4.5	0.5	0.8	1.9
Heading (ψ)	5°	5.24	3.9	0.7	0.74	1.7

result in a discontinuous controller which causes distinct instantaneous variations in the system output due to sampling delays. When the system output trajectory is positives ($s > 0$), the control law forces the trajectory to reach $s = 0$. This trajectory

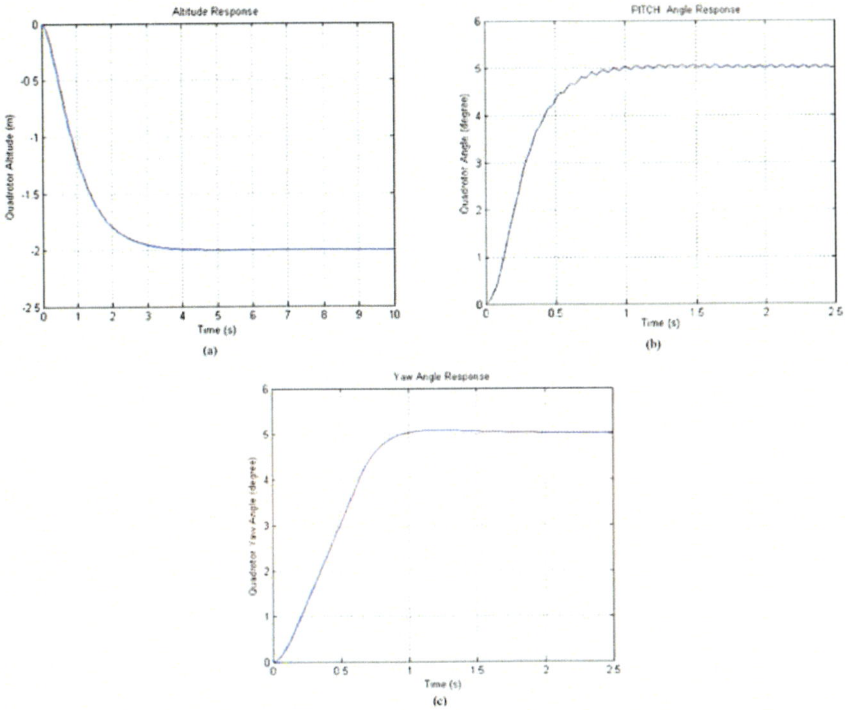

Fig. 4 System response for SMC controller with chattering effects: **a** altitude response, **b** pitch response, **c** yaw response

continues in the negative direction ($s < 0$) which in turn resulting the process called chattering. The main disadvantage of chattering is that it results in unstable system dynamics. This could as a result in decrease in the efficiency of the system and subsequent increase in power consumption.

The system can be tuned with Eq. 18 by the GA to accommodate for the chattering effects. In order to perform the chattering reduction, a mathematical equation is used to quantify chattering effect which is based on the differential area under desired and actual response curves.

$$\text{chattering} = \max_{T_s \to T_{\text{end}}} \dot{s} \tag{18}$$

Implementing the above equation into the GA, the system appears to produce negligible amount of chatter in the system response. Table 3 and Figs. 5, 6 shows the resultant response outputs of the SMC under different conditions.

The comparison between the SMC and PID controller is done by comparison between the response graphs shown in Figs. 2, 3, 4, 5, and 6. Both the controllers for the quadrotor system produced comparable dynamic performances indicated by the settling time and overshoot in case of hover stability of the quadrotor. The initial

Table 3 SMC controller response with chatter reduction

	Desired value	c	k_1	k_2	Settling time (s)
Altitude (z)	2 m	3.73	0.0022	2.58	1.793
Attitude (ϕ and θ)	5°	1.49	0.0073	0.0386	1.877
Heading (ψ)	5°	5.85	0.0027	0.6445	1.946

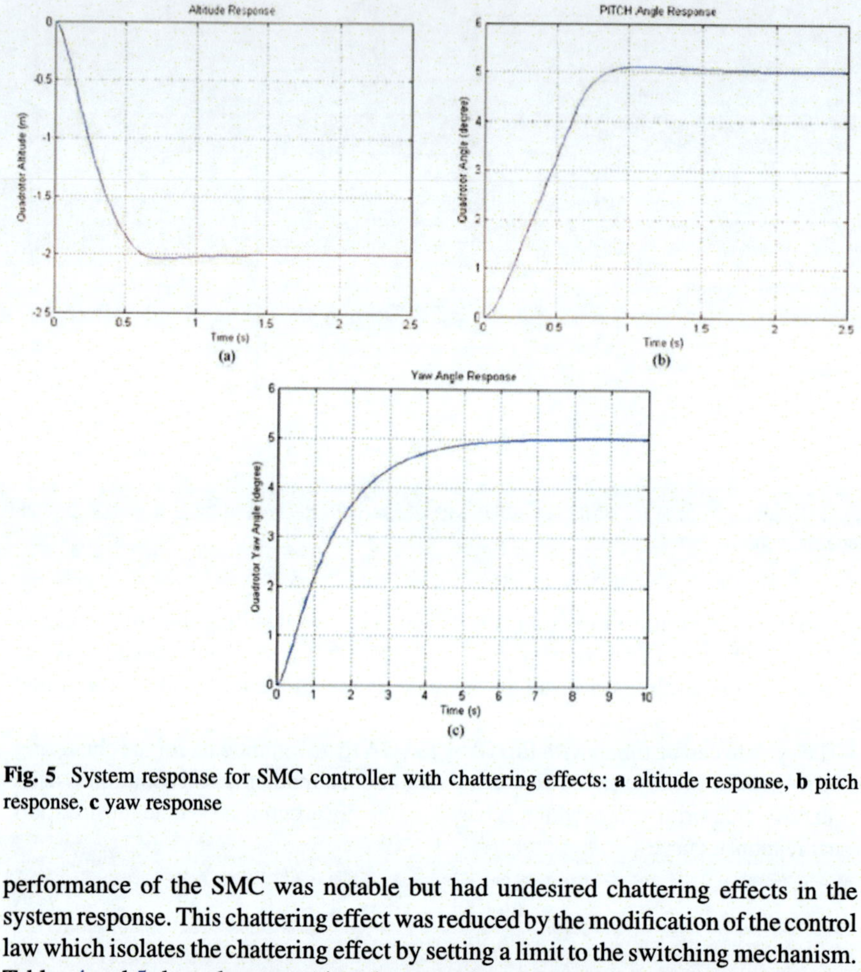

Fig. 5 System response for SMC controller with chattering effects: **a** altitude response, **b** pitch response, **c** yaw response

performance of the SMC was notable but had undesired chattering effects in the system response. This chattering effect was reduced by the modification of the control law which isolates the chattering effect by setting a limit to the switching mechanism. Tables 4 and 5 show the comparison between the performances of the SMC and PID controller based their settling time and overshoot for linear operation.

When the system is put through the nonlinear region of operation, the PID controller cannot control the output as the PID is a linear controller. The GA optimized PID controller also failed to control the system as it is just a modified version of the above mentioned PID controller. The SMC was able to produce stability in the system with acceptable dynamic performance. The main advantage of PID controller

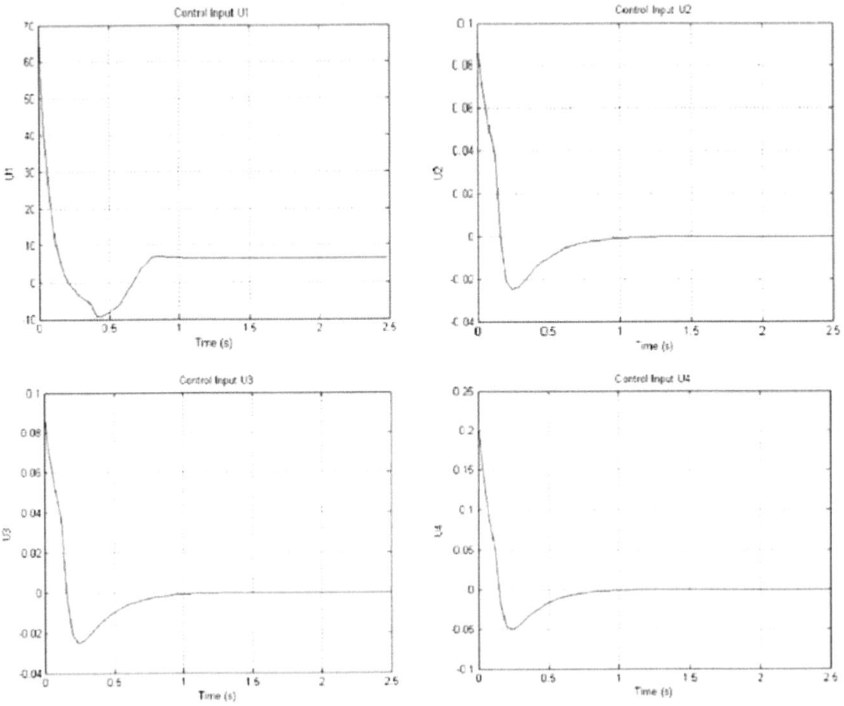

Fig. 6 Control output sliding mode controller

Table 4 Performance comparison in linear operation: settling time

Controller	Altitude (z) 2 m	Attitude (ϕ and θ)	Heading (ψ)
PID	1.24 s	0.24 s	0.52 s
SMC	0.57 s	0.8 s	0.74 s

Table 5 Performance comparison in linear operation: overshoot

Controller	Altitude (z) 2 m	Attitude (ϕ and θ)	Heading (ψ)
PID	1.4%	2%	1.9%
SMC	2%	1.9%	1.7%

is that the control law is not associated with the system parameters, instead it is derived from the state error and its derivative resulting in less computational intensity and also easier implementation. The PID controller is also has lower vulnerability to uncertainties in the system, making it less adaptive. The GA tuned PID performance characteristics are similar to classical PID controller. However, the tuned PID controller exhibits better performances than the classical PID controller. A major disadvantage of the use of GA is that the switching time from one set of control gains to another is very high. This would result in large computational times when the switching is done between infinitesimally small variations in the control gains, which in turn would result in large overshoots and instabilities.

5 Conclusion

The quadrotor mathematical model was developed in detail which included the associated aerodynamic effects and rotor dynamics. A linear Proportional-Integral-Derivative (PID) control algorithm and a nonlinear Sliding Mode Control (SMC) algorithm were then developed. Further, a comprehensive simulation was developed on the MATLAB/Simulink platform based on the thus obtained quadrotor mathematical model. The simulation parameters were varied to evaluate the mentioned controllers and establish the differences in their dynamic performances in different types of mission conditions.

The design parameters for the controllers were tuned and optimized using the Genetic Algorithm (GA) technique which focused on the dynamic response of the system in terms of its settling time and overshoot. Both controllers produced comparable performance characteristics in linear hovering operation of the quadrotor in the range of 2°–20° of attitude and heading. The PID controller optimized by the GA gave better performance than the basic PID controller in terms of overshoot and settling time. The performance of the SMC algorithm was further optimized using the chattering reduction technique and the GA.

6 Future Work

Future work on the topic may include testing the controllers in hardware and real-world environment and load conditions. Further, the sensors assumed to be used in the simulation were perfect, which is an ideal scenario. Hence, sensor modelling, and noise signal ratios of the sensors are required to be taken into consideration. Also, GA tuned PID controller could be replaced with a modified PID controller which use the error and its derivatives, instead of the desired system output, for further comparison of performance and stability. This could provide a much precise performance comparison between the two controllers. Another important addition to performance of the system under uncertainties would be making the system adaptive to disturbances.

Acknowledgements This project has been funded by Indian Space Research Organisation/Department of Space, Government of India, Respond Program, under the project code: ISRO/RES/3/822/19-20, dated on August 8th, 2019.

References

1. L'Afflitto, A., Anderson, R.B., Mohammadi, K.: An introduction to nonlinear robust control for unmanned quadrotor aircraft: how to design control algorithms for quadrotors using sliding mode control and adaptive control techniques [Focus on education]. IEEE Control Syst. Mag. **38**, 102–121 (2018)
2. Zulu, A., John, S.: A review of control algorithms for autonomous quadrotors. Open J. Appl. Sci. **4**, 547–556 (2014)
3. Liu, P., Islam, S., El Saddik, A.: Adaptive sliding mode control of unmanned four rotor flying vehicle. Int. J. Robot. Autom. **30**, 140–148 (2015)
4. Sahul, M.P.V., Chander, V.N., Kurian, T.: A novel method on disturbance rejection pid controller for quadcopter based on optimization algorithm. IFAC Proc. Vol. **47**, 192–199 (2014)
5. Mofid, O., Mobayen, S.: Adaptive sliding mode control for finite time stability of quad-rotor UAVs with parametric uncertainties. ISA Trans. **72**, 1–14 (2018)
6. Islam, S., Faraz, M., Ashour, R.K., Cai, G., Dias, J., Seneviratne, L.: Adaptive sliding mode control design for quadrotor unmanned aerial vehicle. In: 2015 international conference on unmanned aircraft systems (ICUAS) 38, 34–39 (March 2015)
7. Bouabdallah, S., Siegwart, R.: Backstepping and sliding-mode techniques applied to an indoor micro quadrotor. In: Proceedings of the 2005 IEEE international conference on robotics and automation 2247–2252 (2005)
8. Bandyopadhyay, B., Deepak, F., Kim, K.S.: Sliding mode control using novel sliding surfaces. Lect. Notes Control. Inf. Sci. 392, XVI–144 (2009)
9. Chen, S., Mulgrew, B., Grant, P.M.: A clustering technique for digital communications channel equalization using radial basis function networks. IEEE Trans. Neural Netw. **4**, 570–578 (1993)
10. Duncombe, J.U.: Infrared navigation—part I: an assessment of feasibility. IEEE Trans. Electron Devices, ED-11, 34–39 (1959)
11. Lin, C.Y., Wu, M., Bloom, J.A., Cox, I.J., Miller, M.: Rotation, scale, and translation resilient public watermarking for images. IEEE Trans. Image Process. **10**, 767–782 (2001)